**EMIS DATA REVIEWS SERIES 14**

Series Editor: Professor B.L. Weiss



Properties of

# Metal Silicides

**ELECTRONIC MATERIALS INFORMATION SERVICE**
**Other books in the EMIS Data Reviews Series from INSPEC:**

| | |
|---|---|
| No. 6 | **Properties of Indium Phosphide** (1991) |
| No. 7 | **Properties of Aluminium Gallium Arsenide** (1993) |
| No. 8 | **Properties of Lattice-Matched and Strained Indium Gallium Arsenide** (1993) |
| No. 10 | **Properties of Narrow Gap Cadmium-based Compounds** (1994) |
| No. 13 | **Properties of Silicon Carbide** (1995) |
| No. 14 | **Properties of Metal Silicides** (1995) |
| No. 15 | **Properties of III-V Quantum Wells and Superlattices** (1996) |
| No. 17 | **Properties of Wide Bandgap II-VI Semiconductors** (1997) |
| No. 18 | **Properties of Porous Silicon** (1997) |
| No. 19 | **Properties of Amorphous Silicon and its Alloys** (1998) |
| No. 20 | **Properties of Crystalline Silicon** (1999) |
| No. 21 | **Properties, Processing and Applications of Indium Phosphide** (1999) |
| No. 22 | **Properties, Processing and Applications of Glass and Rare-Earth-Doped Glasses for Optical Fibres** (1998) |
| No. 23 | **Properties, Processing and Applications of Gallium Nitride and Related Semi-Conductors** (1999) |
| No. 24 | **Properties of Silicon Germanium and SiGe:Carbon** (2000) |
| No. 25 | **Physical Properties of Liquid Crystals: Nematics** (2000) |
| No. 26 | **Properties, Growth and Applications of Diamond** (2000) |
| No. 28 | **Properties of Lithium Niobate** (2002) |
| No. 29 | **Properties of Amorphous Carbon** (2002) |

Details of books in the EMIS Data Reviews Series are available at www.theiet.org/publications/booksandconferences

Properties of

# Metal Silicides

Edited by Karen Maex and Marc Van Rossum
IMEC, Leuven, Belgium

The Institution of Engineering and Technology

Published by The Institution of Engineering and Technology, London, United Kingdom



First published 1995 (0 85296 859 0)
Paperback edition 2007



The Institution of Engineering and Technology
Michael Faraday House
Six Hills Way, Stevenage
Herts, SG1 2AY, United Kingdom

www.theiet.org



**British Library Cataloguing in Publication Data**
A catalogue record for this product is available from the British Library

**ISBN 978-0-86341-776-4**

Printed in the UK by Lightning Source UK Ltd, Milton Keynes

# Contents

# Foreword

Modern electronics technology has changed the world in a short time span. A good share of the credit should go to microelectronics. From the invention of the first transistor we have come a long way towards integrating billions of components on a single chip.

Continuous advancements in technology have resulted in integrated circuits with smaller device dimensions and larger area and complexity. The trends show that process and device technology evolution is proceeding towards lateral dimensions below 0.1 μm. The number of components per chip increased at a rate of 100% per year during the sixties and 59% during the seventies and eighties. Will these trends continue in the future? Experts have examined various limits to the evolution of microelectronics: fundamental limits imposed by physics as well as practical limits imposed by the device structure, circuit and system configuration. It is believed that we are far from encountering the fundamental limits of physics and for the foreseeable future the scaling of the device dimensions will continue and IC chips with increasing complexity will be built in the future.

Overall circuit performance in the past has depended primarily on device properties. To enhance the circuit and system speed the primary effort had been on improving the device speed. During the last decade that has changed. The parasitic resistance, capacitance and inductance associated with interconnections and contacts are now beginning to influence circuit performance and will be one of the primary factors in the evolution of deep submicron ULSI technology. It is now a foregone conclusion that for submicron feature size the impact of parasitics seriously hurts circuit and system performance. RC time delay, IR voltage drop, power consumption and crosstalk noise due to these parasitics are becoming increasingly significant. Thus even with very fast devices the overall performance of a large circuit could be seriously affected by the limitations of interconnections and contacts. During the last few years a great deal of work has been done on new and innovative materials, device structures and fabrication technology to overcome the problems of conventional ULSI circuits.

In order for a certain conductor to be used to form multilayer interconnections, several requirements must be met which are imposed by fabrication technology and required circuit performance. The main requirements are good conductivity, reliability and manufacturability. In general, several requirements are imposed on interconnection materials by fabrication technology. In a multilayer interconnection structure, the layers incorporated early in the process sequence might be subjected to several fabrication steps, to which layers incorporated later might not be subjected. The most rigorous set of requirements are: low resistivity; ease of formation of thin films of the material; ability to withstand the chemicals and high temperatures required in the fabrication process; good adhesion to other layers; ability to be thermally oxidised; stability of electrical contacts to other layers; ability to contact shallow junctions; good MOS properties; resistance to electromigration; and ability to be defined into fine patterns.

The materials which have been used or proposed for forming interconnections can be broadly classified into four categories: low temperature metals; heavily doped polysilicon; high temperature refractory metals; and metal silicides. Since the main requirement for interconnections specified by circuit performance is good conductivity aluminium and its alloys have long been used for this application. However, as device dimensions are scaled down, its reliability has become a major issue. In general, several other requirements are imposed by fabrication technology, device compatibility, reliability, etc. Refractory metals and silicides are playing increasingly important roles in meeting these requirements. It must be pointed out that these materials will not replace aluminium but complement it in a multilayer structure.

During the last few years, use of refractory metal silicides has been heavily investigated for many applications, such as ohmic contacts, MOS gate electrodes, and silicidation of diffusions, and the results have been very exciting. Silicides of platinum, palladium, tungsten, molybdenum, titanium, cobalt, nickel, tantalum and other metals have reasonably good compatibility with IC fabrication technology for one or more applications cited above. They have fairly high conductivity and resistance to electromigration, they can make low resistance and reliable contacts to shallow p-n junctions, they can withstand the chemicals normally encountered during the fabrication process, thermal oxidation of most of the silicides can be conducted in oxygen and steam to produce a passivating layer of $SiO_2$, and fine features can be formed in these materials.

However, to incorporate the silicides in a microelectronics structure many problems must be solved for proper functioning of the device. Important issues are that the silicides involve non-homogeneous materials, grains and grain boundaries, and often they contain secondary and metastable phases. Equally important is the presence of adjoining layers and the interfaces between them, and the role they play in the different processes and mechanisms. As the semiconductor structures get smaller, other layers and interfaces can dominate the various phenomena in interconnect processing. Reactions and interactions with other layers in these structures are common. In addition, the interfaces themselves can act as separate phases and greatly affect the film properties and processing.

Low resistivity silicide layers are obtained by depositing the silicide directly, or by depositing the metal on silicon and reacting the materials to form the silicide. In either case, a relatively high temperature anneal is needed - in the second case to react the metal and silicide, and in both cases to allow crystallisation/grain growth to occur. In all of these cases a detailed knowledge of the formation techniques, properties of as-formed films and changes during the subsequent processing is absolutely necessary. The electrical properties of silicide interconnects and contacts are determined by factors such as phase formation (including reactions, recrystallisation, grain growth, transformation, and agglomeration), and dopant redistribution and segregation. Properties which are important during processing of the layers may depend on the topography, morphology, and composition of the as-deposited films. Properties such as dopant diffusion, phase transformation, physical stress characteristics, and many others may depend heavily on the grain size, structure, doping, composition, and texture of each layer.

This book brings leading experts in this field together to share the results of their research and experience. It will be a valued reference source to other researchers, manufacturing engineers and academicians working on further development and use of silicides.

Krishna C. Saraswat
Department of Electrical Engineering
Stanford University, Stanford, CA 94305, USA
September 1995

# Introduction

Silicides have been for many decades a fruitful subject of metallurgical studies. The large variety of compounds and their complex phase transitions were amply sufficient reasons to motivate a continuing effort in fundamental investigations, resulting in a steady stream of publications and by now an impressive body of metallurgical literature. Moreover, the growing interest for microelectronic applications over the last twenty years has widened the scope of silicide work and has opened new research avenues, aiming at a better understanding of the behaviour of silicides when combined with other device materials. Today, several silicides belong to the category of strategic materials for the electronics industry, having been introduced in the production of virtually all advanced silicon integrated circuits (see the Foreword by Professor K. Saraswat for more examples). Applications in so-called "niche markets" such as sensors, detectors and other electronic devices are also developing at a fast pace, their success being mainly based on the excellent compatibility of silicides with the basic silicon material. Nevertheless, the silicidation process remains very complex and reflects how closely materials properties are influencing technological developments.

The unusual width and diversity of our subject creates special problems in reviewing the data. It has been impossible to provide an exhaustive coverage of all relevant materials aspects for the many silicide compounds which have been listed in the literature. Moreover, a special difficulty with silicides is the problem of phase identification, since many of the compounds mentioned in research papers have not been unambiguously identified. Although the main categories of silicides (near-noble metal, refractory metal and rare earth metal) have been treated in some detail, our overall approach has been synthetic rather than analytical. As a result, this book is organised along the main categories of materials properties, as they can be found in most solid-state textbooks: structural and thermal properties (ch.1), growth and phase transformations (ch.2), thermochemical properties and phase diagrams (ch.3), electronic properties (ch.4), electrical transport (ch.5), optical properties (ch.6), diffusion (ch.7) and hyperfine interactions (ch.8). All chapters try to introduce a sense of perspective to the wealth of available data by separating the essential from the accessory.

We hope that this book will become a useful working tool for the many researchers who today contribute to the development of the silicide field. It is our pleasure to thank all experts who were involved in this enterprise by writing review articles. We also acknowledge the refereeing support of numerous researchers and in particular would like to thank the following:

Dr. S. Mantl, Forschungszentrum Jülich, Germany
Dr. S. Ogawa, Matsushita Electric, Japan
Dr. A. Vantomme, Physics Department, Leuven University, Belgium
Dr. K. Larsen, University of Catania, Italy

Finally, a special word of thanks is due to John Sears (Managing Editor of the EMIS Datareviews Series) for his support and encouragement during the whole editing process.

Karen Maex
Marc Van Rossum
IMEC
September 1995

# Contributing Authors


| | | |
|---|---|---|
| M. Affronte | Universita Degli Studi di Modena, Dipartimento di Fisica, Via G. Campi, 213/A, 41100 Modena, Italy | 5.1 |
| T. Barge | SOITEC SA, Site Technologique Astec, 15 Avenue des Martyrs, 38054 Grenoble, France | 7.2 |
| L.A. Clevenger | IBM Thomas J. Watson Laboratory, PO Box 218, Yorktown Heights, NY 10598, USA | 2.1, 2.2 |
| J. Derrien | CRMC2-CNRS, Case 913, Campus de Luminy, 13288 Marseille Cedex 09, France | 4.1, 4.2, 4.4 |
| P. Gas | CNRS, Laboratoire de Metallurgie, Faculte des Sciences et Techniques de Saint Jerome, Avenue Escadrille Normandie-Niemen, 13397 Marseille, France | 7.1-7.3 |
| U. Gottlieb | ENSPG, Laboratoire des Materiaux et du Genie Physique, BP46, Domain Universitaire, 38402 St. Martin d'Heres Cedex, France *and* CNRS, Centre de Recherches sur les Tres Basses Temperatures, BP166, 38042 Grenoble Cedex 9, France | 5.1, 5.2 |
| G. Guizetti | Dipartimento di Fisica 'A. Volta', Universita di Pavia, via Basi 6, I-27100 Pavia, Italy | 6.1-6.8 |
| F.M. d'Heurle | IBM Thomas J. Watson Laboratory, PO Box 218, Yorktown Heights, NY 10598, USA | 7.1-7.3 |
| C.C. Hsu (Xu Zhenjia) | Chinese Academy of Sciences, Institute of Semiconductors, PO Box 912, Beijing 100083, China | 1.4, 2.3-2.6, 4.3, 4.5, 5.3 6.9 |
| O. Laborde | CNRS, Centre de Recherches sur les Tres Basses Temperatures, BP166, 38042 Grenoble Cedex 9, France *and at the same address* Laboratoire des Champs Magnetiques Intenses | 5.1, 5.2 |
| G. Langouche | IKS, KU Leuven, Celestijnenlaan 200D, 3001 Leuven, Belgium | 8.1-8.4 |

| | | |
|---|---|---|
| R. Madar | Laboratoire des Materiaux et du Genie Physique, Ecole Nationale Superieure de Physique de Grenoble, BP46, 38402 St. Martin d'Heres Cedex, France | 3.4, 3.5 |
| K. Maex | IMEC, Kapeldreef 75, B-3001 Leuven, Belgium | 1.1 |
| R.W. Mann | IBM Thomas J. Watson Laboratory, PO Box 218, Yorktown Heights, NY 10598, USA | 2.1, 2.2 |
| F. Marabelli | Dipartimento di Fisica 'A. Volta', Universita di Pavia, via Basi 6, I-27100 Pavia, Italy | 6.1-6.8 |
| F. Nava | Universita Degli Studi di Modena, Dipartimento di Fisica, Via G. Campi, 213/A, 41100 Modena, Italy | 5.1, 5.2 |
| M. Ostling | Royal Institute of Technology, Electrum 229, Solid State Electronics, PO Box 1298, S-16440 Kista, Sweden | 1.2, 1.3 |
| A. Reader | Philips Research Laboratories, Building WAM, Prof. Hostlaan 4, 5656 AA Eindhoven, The Netherlands | 1.1 |
| M. Setton | Lam Research Sarl, EUROPOLE, Batiment Eurennepolis, 4, Place R. Schumann, 38000 Grenoble, France | 3.7 |
| R. Sinclair | Stanford University, Department of Materials Science and Engineering, Building 550, Stanford, CA 94305, USA | 3.1-3.3, 3.6, 3.8 |
| O. Thomas | MATOP, associe au CNRS, Faculte des Sciences et Techniques de Saint Jerome, Avenue Escadrille Normandie-Niemen, 13397 Marseille, France | 7.3 |
| M. Van Rossum | IMEC, Kapeldreef 75, B-3001 Leuven, Belgium | 1.1 |
| C. Zaring | Royal Institute of Technology, Electrum 229, Solid State Electronics, PO Box 1298, S-16428 Kista, Sweden | 1.2, 1.3 |

## Abbreviations

The following abbreviations are used in this book.

| | |
|---|---|
| AES | Auger electron spectroscopy |
| ASTM | American Society for Testing and Materials |
| | |
| BIS | Bremsstrahlung isochromat spectroscopy |
| | |
| CSS | complete solid solubility |
| C-V | capacitance-voltage |
| CVD | chemical vapour deposition |
| | |
| DDS | dominant diffusing species |
| DMR | deviation from Mathiessen's rule |
| DOS | density of states |
| DVM | discrete vibrational method |
| | |
| EDC | energy distribution curve |
| EDX | energy dispersive X-rays |
| EFG | electric field gradient |
| EHT | extended Huckel-theory |
| EMP | electron microprobe |
| EPR | electron paramagnetic resonance |
| | |
| FEBA | fast e-beam annealing |
| FIR | far infrared |
| FWHM | full width at half maximum |
| | |
| H | hexagonal |
| Hμ | micro hardness |
| HP | high pressure |
| HRA | Rockwell hardness |
| HRTEM | high resolution transmission electron microscopy |
| HT | high temperature |
| HV | Vickers hardness |
| | |
| IBS | ion beam synthesis |
| IC | integrated circuit |
| IPC | inverse photoemission spectroscopy |
| IR | infrared |
| I-V | current-voltage |
| | |
| JCPDS | Joint Committee on Powder Diffraction Standards |
| | |
| LA | laser annealing |
| LDF | local density formalism |
| LEED | low energy electron diffraction |
| LP | low pressure |

| | |
|---|---|
| LPE | liquid phase epitaxy |
| LT | low temperature |
| | |
| MBE | molecular beam epitaxy |
| MBT | metal base transistor |
| MIGS | metal-induced gap states |
| ML | monolayer |
| MS | Mossbauer spectroscopy |
| | |
| NIR | near infrared |
| NMR | nuclear magnetic resonance |
| NO | nuclear orientation |
| NOS | number of states |
| | |
| O | orthorhombic |
| | |
| PAC | perturbed angular correlations |
| PBT | permeable base transistor |
| PECVD | plasma-enhanced chemical vapour deposition |
| PES | photoemission spectroscopy |
| | |
| R | reflectance |
| RBS | Rutherford back-scattering |
| RDE | reactive deposition epitaxy |
| RE | rare earth |
| RH | resistive heating |
| RHEED | reflection high energy electron diffraction |
| RRR | residual resistance ratio |
| RS | Raman scattering |
| RT | room temperature |
| RTA | rapid thermal annealing |
| RTP | rapid thermal processing |
| | |
| SALICIDE | self-aligned process |
| SB | Schottky barrier |
| SC | single crystal |
| SE | ellipsometric spectroscopy |
| SEM | scanning electron microscopy |
| SIMS | secondary ion mass spectrometry |
| SOI | silicon-on-insulator |
| SPE | solid phase epitaxy |
| SR | synchrotron radiation |
| | |
| T | tetragonal or transmittance |
| TDOS | thermodynamic density of states |
| TE | thermal evaporation |
| TEM | transmission electron microscopy |
| TM | transition metal |
| TT | template technique |

| | |
|---|---|
| UHV | ultra high vacuum |
| ULSI | ultra large scale integration |
| UPS | ultraviolet photoemission spectroscopy |
| UV | ultraviolet |
| | |
| VLSI | very large scale integration |
| VUV | vacuum ultraviolet |
| | |
| WF | work function |
| | |
| $X_{min}$ | minimum channelling yield |
| XAES | X-ray induced Auger electron spectroscopy |
| XARS | X-ray absorption resonance spectroscopy |
| XPD | X-ray photoelectron diffraction |
| XPS | X-ray photoemission spectroscopy |
| XRD | X-ray diffraction |
| XTEM | X-ray cross-section transmission electron microscopy |

CHAPTER 1

# BASIC PHYSICAL PROPERTIES

# 1.1 Crystal structure of TM silicides

K. Maex, M. Van Rossum and A. Reader

May 1995

## A INTRODUCTION

This Datareview contains a table of binary metal silicides with their most important crystal parameters. The silicides have been ordered according to their mass number and silicon content. Since the task of providing an exhaustive list of all known silicide compounds is not achievable in practice (even when limited to the binary compounds), we have based our selection on the following procedure. We started from current JCPDS files which we consulted for all the silicide-forming metals. We then cross-checked the JCPDS data with an extensive list of recent references, which we obtained either from the JCPDS files themselves or from a computer search. This reference list contains both review articles and original research papers. The latter were checked systematically each time we met with questionable phase assignments or ambiguous citations. Although we tried to the best of our ability to eliminate doubtful phases, we cannot guarantee that our screening procedure has been entirely secure. We therefore strongly recommend the reader to consult the original references whenever possible; in addition to structural data, many of those papers also contain useful information on thermodynamic parameters and phase diagrams.

As a general rule, we have not included metastable compounds in our list of crystal parameters, because in many cases their experimental evidence is not sufficiently established. Nevertheless, we have listed general references on metastable phases whenever we estimated that this information would be a worthwhile addition to the present Datareview.

## B DATA ON BINARY SILICIDES

TABLE 1 presents the relevant crystal parameters for the binary metal silicides.

TABLE 1 Crystal parameters for binary metal silicides.

| Compound | | Crystal system | Space group | Prototype | Lattice constants (Å) | | | Density g/cm$^3$ | Ref |
|---|---|---|---|---|---|---|---|---|---|
| | | | | | a | b | c | | |
| Sc-Si: | β-Sc | Cubic | $Im\bar{3}m$ | W | 3.73 | | | 2.985 | [3c] |
| | α-Sc | Hexagonal | $P6_3/mmc$ | Mg | 3.3088 | | 5.2680 | 3.032 | [3b,3e] |
| | Sc (HP) | Tetragonal | | | 3.758(11) | | 4.761(32) | 4.441 | [3d] |
| | $Sc_5Si_3$ | Hexagonal | $P6_3/mcm$ | $Mn_5Si_3$ | 7.861 | | 5.812 | 3.30 | [1,3b] |
| | ScSi | Orthorhombic | Cmcm | BCr | 3.998(2) | 9.882(5) | 3.659(2) | 3.36 | [2,3b] |
| | $ScSi_{2-x}$ | Hexagonal | P6/mmm | $AlB_2$defect | 3.66(4) | | 3.87(7) | 3.40 | [3a,3b,3c] |
| Ti-Si: | α-Ti | Hexagonal | $P6_3/mmc$ | Mg | 2.950 | | 4.686 | 4.504 | [4b] |
| | β-Ti | Cubic | | | 3.330 | | | 4.31 | [4c] |
| | $Ti_5Si_3$ | Hexagonal | $P6_3/mcm$ | $Mn_5Si_3$ | 7.444 | | 5.143 | 4.36 | [4a] |
| | $Ti_5Si_4$ | Tetragonal | $P4_12_12$ | $Zr_5Si_4$ | 6.713 | | 12.190 | 4.25 | [5] |
| | $Ti_5Si_4$ (HT) | Orthorhombic | Pcnm | $Ge_4Sm_5$ | 6.645 | 6.506 | 12.69 | 4.26 | [5] |
| | TiSi | Orthorhombic | Pnma | BFe | 6.544 | 3.638 | 4.997 | 4.24 | [6] |
| | $TiSi_2$ (C49) | Orthorhombic | Cmcm | $Si_2Zr$ | 3.62 | 13.76 | 3.61 | 3.85 | [7] |
| | $TiSi_2$ (C54) | Orthorhombic | Fddd | $Si_2Ti$ | 8.2687(5) | 8.5534(6) | 4.7983(3) | 4.07 | [8] |
| V-Si: | V | Cubic | Im3m | W | 3.0274 | | | 6.097 | [9b,9c] |
| | $V_3Si$ | Cubic | Pm3n | $Cr_3Si$ | 4.725 | | | 5.69 | [9a,9b] |
| | $V_5Si_3$ (HT) | Tetragonal | I4/mcm | $Si_3W_5$ | 9.428 | | 4.756 | 5.33 | [10,9b] |
| | $V_5Si_3$ (C) | Hexagonal | $P6_3/mcm$ | $Si_3Mn_5$ | 7.135 | | 4.842 | 5.27 | [11,9b] |
| | $V_6Si_5$ | Orthorhombic | Immm | $Ge_5Ti_6$ | 4.858 | 7.501 | 15.966 | 5.09 | [12,9b] |
| | $VSi_2$ | Hexagonal | $P6_222$ | $CrSi_2$ | 4.57230(16) | | 6.3730(4) | 4.62 | [13,9b] |
| Cr-Si: | Cr | Cubic | Im3m | W | 2.8847 | | | 7.2 | [14c,14g] |
| | Cr | Cubic | Pm3 | $Cr_3Si$ | 4.5880 | | | 7.152 | [14h] |
| | $Cr_3Si$ | Cubic | Pm3n | $Cr_3Si$ | 4.5580(2) | | | 6.46 | [14a,14b,14c] |
| | α-$Cr_5Si_3$ | Tetragonal | I4/mcm | $W_5Si_3$ | 9.170 | | 4.636 | 5.90 | [14c,14d,14e] |
| | CrSi | Cubic | $P2_13$ | FeSi | 4.620 | | | 5.44 | [14c,14f] |
| | $CrSi_2$ | Hexagonal | $P6_222$ | $CrSi_2$ | 4.4281(2) | | 6.3691(4) | 4.98 | [14c,15] |

Table 1 continued

| Compound | | Crystal system | Space group | Prototype | Lattice constants (Å) | | | Density g/cm$^3$ | Ref |
|---|---|---|---|---|---|---|---|---|---|
| | | | | | a | b | c | | |
| Mn-Si: | δ-Mn | Cubic | $Im\bar{3}m$ | W | 8.9139 | | | | [16b,16c] |
| | γ-Mn | Cubic | $Fm\bar{3}m$ | Cu | 3.862 | | | | [16b,16c] |
| | β-Mn | Cubic | $P4_132$ | β-Mn | 6.315 | | | | [16b,16c] |
| | α-Mn | Cubic | $I\bar{4}3m$ | α-Mn | 8.9139 | | | | [16b,16c] |
| | R-$Mn_6Si$ | Hexagonal | $R\bar{3}$ | $F_7W_6$ | 10.874 | | 19.177 | 5.445 | [18,16c] |
| | μ-$Mn_9Si_2$ | Orthorhombic | Immm | | 16.992(4) | 28.634(7) | 4.656(1) | 6.795 | [21,16c] |
| | β-$Mn_3Si$ | Cubic | $Fm\bar{3}m$ | $BiF_3$ | 5.722(1) | | | | [16b] |
| | $Mn_5Si_2$ (LT) | Tetragonal | $P4_12_12$ | $Mn_5Si_2$ | 8.9097(5) | | 8.7153(3) | 6.353 | [20,22,16c] |
| | $Mn_5Si_3$ | Hexagonal | $P6_3/mcm$ | $Mn_5Si_3$ | 6.912(5) | | 4.812(3) | 6.024 | [23a,16c,16b] |
| | MnSi | Cubic | $P2_13$ | FeSi | 4.548(2) | | | 5.815 | [23b,16c,16b] |
| | $Mn_{15}Si_{26}$ | Tetragonal | $I\bar{4}2d$ | $Mn_{15}Si_{26}$ | 5.525 | | 65.550 | 5.159 | [24] |
| | $Mn_{27}Si_{47}$ | Tetragonal | $P\bar{4}n2$ | $Mn_{27}Si_{47}$ | 5.530 | | 117.900 | 5.164 | [25] |
| | $Mn_{11}Si_{19}$ | Tetragonal | $P\bar{4}n2$ | | 5.518 | | 4.814 | | [16b,16c] |
| Fe-Si: | α-Fe | Cubic | Im3m | | 2.8664 | | | 7.88 | [27b] |
| | γ-Fe | Cubic | | | 3.6468 | | | 7.65 | [4c] |
| | δ-Fe | Cubic | | | 2.94 | | | 7.297 | [4c] |
| | ε-Fe | Hexagonal | $P6_3/mmc$ | | 2.45(1) | | 3.93(1) | 9.079 | [27c] |
| | $Fe_3Si$ | Cubic | Im3m | W | 2.841(5) | | | 7.08 | [27a] |
| | $Fe_2Si$ | Cubic | Pm3m | ClCs | 2.81 | | | 5.23 | [28] |
| | $Fe_5Si_3$ | Hexagonal | $P6_3/mcm$ | $Mn_5Si_3$ | 6.759(5) | | 4.720(5) | 6.46 | [29] |
| | FeSi | Cubic | $P2_13$ | | 4.48798(9) | | | 6.17 | [13] |
| | α-$FeSi_2$ | Tetragonal | P4/mmm | | 2.69392(2) | | 5.1361(3) | 4.99 | [30] |
| | β-$FeSi_2$ | Orthorhombic | C*** | $FeSi_2$ | 9.879 | 7.799 | 7.839 | 4.93 | [31] |
| Co-Si: | α-Co | Cubic | $Fm\bar{3}m$ | Cu | 3.5446 | | | 8.789 | [32b,32d] |
| | ε-Co | Hexagonal | $P6_3/mmc$ | Mg | 2.5071 | | 4.0695 | 8.871 | [32b,32c] |
| | α-$Co_3Si$ | Tetragonal | | | 8.420 | 3.738 | 5.810 | | [32b] |
| | α-$Co_2Si$ | Orthorhombic | Pnma | $Co_2Si$ | 4.918 | 3.737 | 7.109 | 7.42 | [32a,32b] |
| | CoSi | Cubic | $P2_13$ | FeSi | 4.447 | | | 6.65 | [33,32b] |
| | $CoSi_2$ | Cubic | Fm3m | $CaF_2$ | 5.3640(2) | | | 4.95 | [13,32b] |
| | metastable phases | | | | | | | | [32b] |

Table 1 continued

| Compound | | Crystal system | Space group | Prototype | Lattice constants (Å) | | | Density g/cm$^3$ | Ref |
|---|---|---|---|---|---|---|---|---|---|
| | | | | | a | b | c | | |
| Ni-Si: | Ni | Cubic | Fm3m | Cu | 3.5241 | | | 8.91 | [34b,34c] |
| | $\beta_1$-$Ni_3Si$ | Cubic | Pm3m | $AuCu_3$ | 3.506 | | | 7.87 | [34a,34b] |
| | $\beta_2$-$Ni_3Si$ | Monoclinic | | $GePt_3$? | 6.79(2) | 6.25(4) | 5.07(8) | | [35,34b] |
| | $\beta_3$-$Ni_3Si$ | Monoclinic | P | $GePt_3$? | 7.04(7) | 6.26(4) | 5.08(1) | 8.05 | [35,34b] |
| | $\gamma$-$Ni_{31}Si_{12}$ | Hexagonal | $P3_21$ | $Ni_{31}Si_{12}$ | 6.667 | | 12.277 | 7.58 | [36,37a,37b] |
| | $\delta$-$Ni_2Si$ (LT) | Orthorhombic | | $Co_2Si$ | 7.060 | 4.990 | 3.720 | | [38] |
| | $\theta$-$Ni_2Si$ (HT) | Hexagonal | | | 3.797 | | 4.893 | 7.91 | [38] |
| | $Ni_3Si_2$ | Orthorhombic | Cmc21 | $Ni_3Si_2$ | 12.229 | 10.805 | 6.924 | 6.75 | [39] |
| | NiSi | Orthorhombic | Pnma | MnP | 5.233 | 3.258 | 5.659 | | [40,34b] |
| | $\zeta$-$NiSi_2$ | Cubic | Fm3m | $CaF_2$ | 5.406 | | | 4.859 | [40b,40c,34b] |
| Cu-Si: | Cu | Cubic | Fm3m | Cu | 3.6149 | | | 8.94 | [41b,42] |
| | $\kappa$ ($Cu_7Si$) | Hexagonal | $P6_3/mmc$ | Mg | 2.5622 | | 4.1823 | | [41b] |
| | $\beta$ (85.1 at % Cu) | Cubic | I3m3 | W | 2.854 | | | | [41b] |
| | $\gamma$ ($Cu_5Si$) | Cubic | $P4_132$ | $\beta$-Mn | 6.222 | | | 7.94 | [41a,41b] |
| | $\delta$ ($Cu_{0.83}Si_{0.17}$) | Tetragonal | | | 8.815 | | 7.903 | | [43a] |
| | $\varepsilon$ ($Cu_{15}Si_4$) | Cubic | | | 9.615 | | | 9.55 | [43a] |
| | $\eta$ ($Cu_3Si$) | Rhombohedral | $R\bar{3}m$ | | 2.47 | | | | [43b] |
| | $\eta'$ ($Cu_3Si$) | Rhombohedral | $R\bar{3}$ | | 4.72 | | | | [43b] |
| | $\eta''$ ($Cu_3Si$) | Orthorhombic | | | 76.76 | 7.00 | 21.94 | | [43b] |
| | $\eta''$ (Cu.Si) | Tetragonal | | | 7.267 | | 7.829 | | [43a] |
| | $Cu_3Si$ | Hexagonal | | | 4.040 | | 2.440 | | [43b] |
| | metastable phases | | | | | | | | [41b] |
| Y-Si: | $\beta$-Y | Cubic | $Im\bar{3}m$ | W | 4.07 | | | | [44c] |
| | $\alpha$-Y | Hexagonal | $P6_3/mmc$ | Mg | 3.6482 | | 5.7318 | 4.474 | [44c,44d] |
| | $Y_5Si_3$ | Hexagonal | $P6_3/mmc$ | $Mn_5Si_3$ | 8.429(3) | | 6.352(4) | 4.46 | [44a,44b,44c] |
| | $Y_5Si_4$ | Orthorhombic | Pnma | $Sm_5Ge_4$ | 7.311(8) | 14.157(8) | 7.627(8) | | [44b,44c] |
| | YSi | Orthorhombic | Cmcm | BCr | 4.269(3) | 10.555(3) | 3.837(3) | 4.54 | [45,44b,44c] |
| | $Y_3Si_5$ | Orthorhombic | | $GdSi_2$ | 4.060(3) | 3.967(3) | 13.39(3) | 4.50 | [46,44b,44c] |
| | $Y_3Si_5$ | Hexagonal | P6/mmm | $AlB_2$ | 3.849(3) | | 4.147(3) | 4.55 | [46,44b,44c] |

Table 1 continued

| Compound | | Crystal system | Space group | Prototype | Lattice constants (Å) | | | Density g/cm$^3$ | Ref |
|---|---|---|---|---|---|---|---|---|---|
| | | | | | a | b | c | | |
| Zr-Si: | α-Zr | Hexagonal | $P6_3/mmc$ | Mg | 3.232 | | 5.147 | 6.506 | [47b] |
| | β-Zr | Cubic | Im3m | W | 3.545 | | | 6.798 | [47c] |
| | ω-Zr | Hexagonal | P6/mmm | $AlB_2$ | 5.039 | | 3.136 | 6.59 | [47d] |
| | $Zr_2Si$ | Tetragonal | I4/mcm | $Al_2Cu$ | 6.609 | | 5.298 | 6.04 | [47a] |
| | $Zr_5Si_3$ | Hexagonal | $P6_3/mcm$ | $Mn_5Si_3$ | 7.885 | | 5.558 | 6.00 | [48] |
| | $Zr_3Si_2$ | Tetragonal | P4/mbm | $Si_2U_3$ | 7.082 | | 3.714 | 5.88 | [52] |
| | ZrSi | Orthorhombic | Pnma | BFe | 6.981 | 3.785 | 5.301 | 5.66 | [49] |
| | (Si,Zr) | Hexagonal | | | 7.010 | | 12.770 | | [50] |
| | $ZrSi_2$ | Orthorhombic | Cmcm | $Si_2Zr$ | 3.6958(3) | 14.751(9) | 3.6654(3) | 4.90 | [51] |
| Nb-Si: | Nb | Cubic | $Im\bar{3}m$ | W | 3.3002 | | | 8.534 | [53b,53c] |
| | $Nb_3Si$ | Tetragonal | $P4_2/n$ | $PTi_3$ | 10.210 | | 5.190 | 7.53 | [53a,53b] |
| | $Nb_3Si$ | Tetragonal | | | 6.98(2) | | 12.72(2) | | [54] |
| | $Nb_3Si$ | Tetragonal | $P4_2/n$ | $Ni_3P$ | 10.39(2) | | 5.17(1) | 7.30 | [54] |
| | α-$Nb_5Si_3$ | Tetragonal | I4/mcm | $B_3Cr_5$ | 6.570 | | 11.887 | 7.11 | [57,53b] |
| | β-$Nb_5Si_3$ | Tetragonal | I4/mcm | $Si_3W_5$ | 10.028 | | 5.070 | 7.15 | [57,53b] |
| | $NbSi_2$ | Hexagonal | $P6_422$ | $CrSi_2$ | 4.797 | | 6.592 | 5.65 | [58,53b] |
| | metastable phases | | | | | | | | [53b] |
| Mo-Si: | Mo | Cubic | $Im\bar{3}m$ | W | 3.1470 | | | 10.221 | [59b,59c] |
| | $Mo_3Si$ | Cubic | $Pm\bar{3}n$ | $Cr_3Si$ | 4.890 | | | 8.97 | [59a,59b] |
| | $Mo_5Si_3$ | Tetragonal | I4/mcm | $Si_3W_5$ | 9.6483(6) | | 4.9135(5) | 8.19 | [60,61,59b] |
| | α-$MoSi_2$ (HT) | Tetragonal | I4/mmm | $MoSi_2$ | 3.200 | | 7.850 | 6.28 | [63,59b] |
| | β-$MoSi_2$ | Hexagonal | $C6_22$ | $CrSi_2$ | 4.596 | | 6.550 | 6.32 | [64,59b] |
| Ru-Si: | Ru | Hexagonal | $P6_3/mmc$ | Mg | 2.7038 | | 4.2816 | 12.36 | [65c,65d] |
| | $Ru_2Si$ | Orthorhombic | Pnma | $CoSi_2$ | 5.279 | 4.005 | 7.418 | 9.75 | [65a] |
| | $Ru_5Si_3$ | Orthorhombic | | $Rh_5Ge_3$ | 5.246(2) | 9.815(2) | 4.023(1) | | [65b] |
| | $Ru_4Si_3$ | Orthorhombic | Pnma | | 5.1936 | 4.0216 | 17.1343 | | [67b,65b] |
| | RuSi | Cubic | $P2_13$ | FeSi | 4.703 | | | 8.25 | [67a,65c] |
| | $Ru_2Si_3$ | Tetragonal | | | 11.073 | 8.954 | | | [65c] |
| | $Ru_2Si_3$ | Orthorhombic | Pnca | $Ru_2Si_3$ | 5.530(1) | 11.060(2) | 8.952(2) | 6.95 | [68,69,69b,65c] |

Table 1 continued

| Compound | Crystal system | Space group | Prototype | Lattice constants (Å) | | | Density g/cm$^3$ | Ref |
|---|---|---|---|---|---|---|---|---|
| | | | | a | b | c | | |
| Rh-Si: Rh | Cubic | $Fm\bar{3}m$ | Cu | 3.8032 | | | 12.43 | [70b,71] |
| $Rh_2Si$ | Orthorhombic | Pnma | $\alpha$-$Co_2Si$ | 5.408 | 3.930 | 7.383 | 9.90 | [70a,73,70b] |
| $Rh_5Si_3$ | Orthorhombic | Pbam | $Rh_5Ge_3$ | 5.317 | 10.124 | 3.902 | 9.47 | [70a,70b] |
| $Rh_{20}Si_{13}$ | Hexagonal | $P6_3/m$ | $Rh_{20}Si_{13}$ | 11.851 | | 3.623 | | [70a,72,70b] |
| $\alpha$-RhSi | Orthorhombic | Pnma | MnP | 5.531 | 3.063 | 6.362 | 6.36 | [73,70b] |
| $\beta$-RhSi | Cubic | $P2_13$ | FeSi | 4.685 | | | | [72,70b] |
| $Rh_4Si_5$ | Monoclinic | $P2_1/m$ | $Rh_4Si_5$ | 12.305 | 3.495 | 5.915 | 7.32 | [73,70b] |
| $Rh_3Si_4$ | Orthorhombic | Pnma | $Rh_3Si_4$ | 18.819 | 3.609 | 5.801 | 7.10 | [73,70b] |
| metastable phases | | | | | | | | [70b] |
| Pd-Si: Pd | Cubic | $Fm\bar{3}m$ | Cu | 3.8901 | | | 12.01 | [74b,74c] |
| $Pd_{100-x}Si_x$ | Cubic | | | 3.081 | | | | [74a] |
| $Pd_5Si$ | Monoclinic | $P2_1$ | $Pd_5P$ | 8.465 | 7.485 | 5.555 | 10.76 | [75,74b] |
| $Pd_9Si_2$ | Orthorhombic | Pnma | | 9.414 | 7.419 | 9.055 | 10.65 | [75,74b] |
| $Pd_4Si$ | Cubic | P | | 4.000 | | | 11.77 | [76] |
| $Pd_3Si$ | Orthorhombic | Pnma | $CFe_3$ | 5.735 | 7.555 | 5.260 | 10.12 | [75,74b] |
| $Pd_2Si$ | Hexagonal | $P\bar{6}2m$ | $Fe_2P$ | 6.493 | | 3.472 | 9.59 | [77,74b] |
| $Pd_2Si$ | Hexagonal | $P\bar{6}2m$ | | 13.055 | | 27.490 | 9.46 | [78] |
| PdSi | Orthorhombic | Pnma | MnP | 5.599 | 3.381 | 6.133 | 7.69 | [79,74b] |
| $Pd_4Si_{20}$ | Tetragonal | | | 3.143 | | 2.898 | | [80] |
| Hf-Si: $\alpha$-Hf | Hexagonal | $P6_3/mmc$ | Mg | 3.1946 | | 5.5051 | 13.245 | [81b,81c] |
| $\beta$-Hf | Cubic | $Im\bar{3}m$ | W | 3.610 | | | | [81b] |
| $Hf_2Si$ | Tetragonal | I4/mcm | $Al_2Cu$ | 6.553 | | 5.186 | 11.49 | [81a,81b] |
| $Hf_5Si_3$ | Hexagonal | $P6_3/mcm$ | $Mn_5Si_3$ | 7.890 | | 5.558 | 10.83 | [82] |
| $Hf_3Si_2$ | Tetragonal | P4/mbm | $Si_2U_3$ | 7.000 | | 3.671 | 10.92 | [83,81b] |
| $Hf_5Si_4$ | Hexagonal | $P4_12_12$ | $Zr_5Si_4$ | 7.039(7) | | 12.83(1) | | [81b] |
| HfSi | Orthorhombic | Pnma | FeB | 6.855 | 3.753 | 5.191 | 10.27 | [82,81b] |
| $HfSi_2$ | Orthorhombic | Cmcm | $ZrSi_2$ | 3.6798(2) | 14.5562(7) | 3.6491(2) | 7.97 | [13,81b] |

Table 1 continued

| Compound | | Crystal system | Space group | Prototype | Lattice constants (Å) | | | Density g/cm$^3$ | Ref |
|---|---|---|---|---|---|---|---|---|---|
| | | | | | a | b | c | | |
| Ta-Si: | Ta | Cubic | $Im\bar{3}m$ | W | 3.3030 | | | 16.634 | [85b,85c] |
| | β-Ta | Tetragonal | $P4_2/mnm$ | CrFe | 10.1940 | | 5.313 | 16.326 | [85d] |
| | $Ta_3Si$ | Tetragonal | P42/n | $PTi_3$ | 10.193 | | 5.175 | 14.11 | [85a,85b] |
| | $Ta_2Si$ | Tetragonal | I4/mcm | $Al_2Cu$ | 6.157 | | 5.039 | 13.56 | [84,85b] |
| | α-$Ta_5Si_3$ | Tetragonal | I4/mcm | $B_3Cr_5$ | 6.516 | | 11.873 | 13.03 | [87,85b] |
| | β-$Ta_5Si_3$ | Tetragonal | I4/mcm | $Si_3W_5$ | 9.88 | | 5.06 | 13.30 | [87,85b] |
| | $TaSi_2$ | Hexagonal | $P6_222$ | $CrSi_2$ | 4.78351(7) | | 6.56980(16) | 9.07 | [88,85b] |
| | metastable phases | | | | | | | | [85b] |
| W-Si: | W | Cubic | Im3m | W | 3.165 | | | 19.262 | [89b,89c] |
| | β-W | Cubic | | | 5.048 | | | 18.97 | [4c] |
| | $W_5Si_3$ | Tetragonal | I4/mcm | $W_5Si_3$ | 9.601 | | 4.972 | 14.54 | [89a,89b] |
| | $WSi_2$ | Tetragonal | I4/mmm | $MoSi_2$ | 3.211 | | 7.829 | 9.88 | [90a,89b] |
| | $WSi_2$ | Hexagonal | | | 4.614 | | 6.414 | | [90b] |
| Re-Si: | Re | Hexagonal | P63/mmc | Mg | 2.760 | | 4.458 | 21.03 | [91b] |
| | $Re_5Si_3$ | Monoclinic | $P2_1/b$ | $Re_5Si_3$ | 6.451 | 5.394 | 9.601 | 15.18 | [20] |
| | ReSi | Cubic | $P2_13$ | FeSi | 4.775 | | | 13.07 | [91a] |
| | $ReSi_2$ | Orthorhombic | Immm | $MoPt_2$ | 3.128 | 3.144 | 7.677 | 10.66 | [92a,92b] |
| Os-Si: | Os | Hexagonal | $P6_3/mmc$ | Mg | 2.7341 | | 4.3197 | 22.59 | [93b] |
| | OsSi | Cubic | $P2_13$ | FeSi | 4.729 | | | 13.71 | [93a] |
| | OsSi | Cubic | Pm3m | ClCs | 2.960 | | | 13.98 | [94a] |
| | $Os_2Si_3$ | Orthorhombic | Pbcn | $Ru_2Si_3$ | 11.124 | 8.932 | 5.570 | 11.15 | [65c] |
| | $OsSi_{1.8}$ | Monoclinic | | | 8.770 | 3.000 | 7.300 | 8.05 | [94a] |
| | $OsSi_2$ | Orthorhombic | Cmca | $FeSi_2$ | 10.150 | 8.117 | 8.223 | 9.66 | [65c] |

Table 1 continued

| | Compound | Crystal system | Space group | Prototype | Lattice constants (Å) | | | Density g/cm$^3$ | Ref |
|---|---|---|---|---|---|---|---|---|---|
| | | | | | a | b | c | | |
| Ir-Si: | Ir | Cubic | Fm3m | Cu | 3.839 | | | 22.56 | [95b] |
| | $Ir_3Si$ | Tetragonal | I4/mcm | SeTl | 5.222 | | 7.934 | 18.52 | [95a] |
| | $Ir_2Si$ | Orthorhombic | Pnma | $Co_2Si$ | 7.615 | 5.284 | 3.980 | 17..11 | [96] |
| | IrSi | Orthorhombic | Pnma | MnP | 5.5579 | 3.2213 | 6.2673 | 13.07 | [93a] |
| | IrSi (HT) | Hexagonal | $P6_3/mmc$ | $InNi_2$ | 3.963 | | 5.126 | | [94a] |
| | $Ir_4Si_5$ | Monoclinic | $P2_1/m$ | $Rh_4Si_5$ | 12.359 | 3.6181 | 5.8805 | | [94b,94c] |
| | $Ir_3Si_4$ | Orthorhombic | Pnma | $Rh_3Si_4$ | 18.871 | 3.6979 | 5.7717 | | [94b,94c] |
| | $IrSi_{1.75}$ | Monoclinic | | | 5.542 | 14.166 | 12.426 | | [94b,94c] |
| | $IrSi_3$ | Hexagonal | $P6_3/mmc$ | $AsNa_3$ | 4.350 | | 6.630 | 8.45 | [94a] |
| | $IrSi_3$ | Orthorhombic | C | | 7.5634(11) | 4.3469(2) | 6.6238(10) | 6.32 | [97] |
| | $IrSi_x$ (x=3) | Monoclinic | C | | 7.6976(3) | 4.3770(3) | 6.5467(4) | 6.25 | [97] |
| Pt-Si: | Pt | Cubic | $Fm\bar{3}m$ | Cu | 3.9242 | | | 21.46 | [98b,98c] |
| | α-$Pt_3Si$ (LT) | Monoclinic | C2/m | $GePt_3$ | 7.702 | 7.765 | 7.765 | 17.56 | [98a,98b] |
| | β-$Pt_3Si$ | Orthorhombic | Pnma | $CFe_3$ | 5.579 | 5.520 | 7.697 | 17.19 | [99,98b] |
| | α-$Pt_{12}Si_5$ (HT) | Tetragonal | P4/n | $Pt_{12}Si_5$ | 13.404 | | 5.451 | 16.83 | [100,98b] |
| | β-$Pt_{12}Si_5$ | Tetragonal | I4/m | $Ni_{12}P_5$ | 9.607 | | 5.542 | 16.11 | [100,98b] |
| | α-$Pt_2Si$ | Tetragonal | I4/mmm | $H_2Th$ | 3.933 | | 5.910 | 15.20 | [98a,98b] |
| | β-$Pt_2Si$ | Hexagonal | $P\bar{6}2m$ | $Fe_2P$ | 6.54 | | 3.606 | | [98b] |
| | $Pt_6Si_5$ | Monoclinic | $P2_1/m$ | | 15.462 | 3.499 | 6.169 | | [101,98b] |
| | PtSi | Orthorhombic | Pnma | MnP | 5.932 | 5.595 | 3.603 | 12.40 | [102,98b] |
| | metastable phases | | | | | | | | [98b] |

**REFERENCES**


[1] J. Arbuckle, E. Parthe [ *Acta Crystallogr. (Denmark)* vol.15 (1962) p.1205 ]

[2] O. Schob, E. Parthe [ *Acta Crystallogr. (Denmark)* vol.19 (1965) p.214 ]

[3a] E.I. Gladyshevskii, E.I. Emes-Misenko [ *Russ. J. Inorg. Chem. (UK)* vol.4 (1963) p.793 ]

[3b] A.B. Gokhale, G.J. Abbaschian [ *Bull. Alloy Phase Diagrams (USA)* vol.7 (1986) p.333 ]

[3c] H. Okamoto [ *J. Phase Equilibria* vol.13 (1992) p.679 ]

[3d] J. Akella, J. Xu, G. Smith [ *Physica (Netherlands)* vol.139 (1986) p.285 ]

[3e] J. Geiselman [ *J. Less-Common Met. (Switzerland)* vol.4 (1962) p.362 ]

[4a] J. Quakernaat, J.W. Visser [ *High Temp.-High Press. (UK)* vol.6 (1974) p.515 ]

[4b] H.E. Swanson, R.K. Fuyat [ *NBS Circ. 539 (USA)* no.3 (1954) p.4 ]

[4c] J.D.M. Donnay, H.M. Ondik [ *Crystal data-determinative tables* vol.2 (NBS, USA, 1973) ]

[5] J. Nickl, H. Sprenger [ *Z. Metallkd. (Germany)* vol.60 (1969) p.136 ]

[6] C. Brukl et al [ *Monatsh. Chem. (Austria)* vol.92 (1961) p.781 ]

[7] P.G. Cotter, J.A. Kohn [ *J. Am. Ceram. Soc. (USA)* vol.39 (1956) p.11 ]

[8] [ *NBS Monograph 25 (USA)* no.21 (1984) p.126 ]

[9a] [ *JCPDS-ICDD* no.19 (1990) p.1405 ]

[9b] J.F. Smith [ *Bull. Alloy Phase Diagrams (USA)* vol.6 (1985) p.266 ]

[9c] [ *NBS Monograph 25 (USA)* no.9 (1971) p.58 ]

[10] E. Parthe et al [ *Monatsh. Chem. (Austria)* vol.86 (1955) p.385 ]

[11] H. Schachner et al [ *Monatsh. Chem. (Austria)* vol.85 (1954) p.245 ]

[12] J. Hallais [ *Ann. Chim. (France)* vol.6 (1971) p.321 ]

[13] W. Wong-Ng, H. McMurdie, B. Paretzkin, C. Hubbard, A. Dragoo [ NBS Grant-in-Aid Report (1987) ]

[14a] [ *NBS Circ. 539 (USA)* no.6 (1956) p.29 ]

[14b] M. Jurisch, G. Behr [ *Acta Phys. Acad. Sci. Hung. (Hungary)* vol.47 (1979) p.201 ]

[14c] A.B. Gokhale, G.J. Abbaschian [ *Bull. Alloy Phase Diagrams (USA)* vol.8 (1987) p.474 ]

[14d] E. Parthe, H. Nowotny, H. Schmid [ *Monatsch. Chem. (Austria)* vol.86 (1955) p.385 ]

[14e] C.H. Dauben, D.H. Templeton, C.E. Myers [ *J. Phys. Chem. (USA)* vol.60 (1956) p.443 ]

[14f] B. Boren [ *Ark. Kem., Min. Geol. A (Sweden)* vol.11 (1933) p.2 ]

[14g] H.E. Swanson et al [ *NBS Circ. 539 (USA)* no.5 (1955) p.20 ]

[14h] K. Kimoto, I. Nishida [ *J. Phys. Soc. Jpn. (Japan)* vol.22 (1967) p.744 ]

[15] [ *NBS Monograph 25 (USA)* no.21 (1984) p.64 ]

[16a] P.F. Weiser, W.D. Forgeng [ *Trans. AIME (USA)* vol.230 (1964) p.1675 ]

[16b] A.B. Gokhale, R. Abbaschian [ *Bull. Alloy Phase Diagrams (USA)* vol.11 (1990) p.168 ]

[16c] H. Okamoto [ *J. Phase Equilibria* vol.12 (1991) p.505 ]

[17] A.M. Bardos, P.A. Beck [ *Trans. AIME (USA)* vol.233 (1965) p.1446 ]

[18] Yu.B. Kuz'ma, E.I. Gladyshevskii [ *Russ. J. Inorg. Chem. (UK)* vol.9 (1964) p.373 ]

[19] E.I. Gladyshevskii et al [ *Inorg. Mater. (USA)* vol.1 (1965) p.996 ]

[20] L. Calvert [ JCPDS Grant-in-Aid Report, NRC of Canada, Ottawa (1978) ]

[21] K.P. Gupta [ *Trans. AIME (USA)* vol.230 (1964) p.253 ]

[22] Smith, Fausey [ JCPDS Grant-in-Aid Report, Pennsylvania State University, University Park (1977) ]

[23a] K. Amark, B. Boren, A. Westgren [ *Sven. Kem. Tidskr. (Sweden)* vol.48 (1936) p.273 ]

[23b] J.E. Jorgensen, S.E. Rasmussen [ *Powder Diffraction (USA)* vol.6 (1991) p.194 ]

[24] G. Flieher et al [ *Monatsh. Chem. (Austria)* vol.98 (1967) p.2173 ]

[25] G. Zwilling, H. Nowotny [ *Monatsh. Chem. (Austria)* vol.102 (1971) p.672 ]

[26] S. Ogawa, Y. Matsuzaki [ *Sci. Rep. Res. Inst. Tohoku Univ. A (Japan)* vol.3 (1951) p.50 ]

[27a] K. Keil et al [ *Am. Mineral. (USA)* vol.67 (1982) p.126 ]

[27b] H.E. Swanson et al [ *NBS Circ. 539 (USA)* no.4 (1955) p.3 ]

[27c] T. Takahashi, W.A. Bassett [ *Science (USA)* vol.145 (1964) p.483 ]

[28] K. Khalaff, K. Schubert [ *J. Less-Common Met. (Switzerland)* vol.35 (1974) p.341 ]

[29] Z. Yu [ *Acta Petro. Mineral. Anal. (China)* vol.3 (1984) p.23 ]
[30] [ *NBS Monograph 25 (USA)* no.21 (1984) p.73 ]
[31] R. Buksch [ *Z. Nat.forsch. A (Germany)* vol.22 (1967) p.2124 ]
[32a] B. Boren et al [ *Z. Phys. Chem., Leipz. (Germany)* vol.29 (1935) p.231 ]
[32b] K. Ishida, T. Nishizawa, M.E. Schlesinger [ *J. Phase Equilibria* vol.12 (1991) p.587 ]
[32c] Hofer, Peebles [ *J. Am. Chem. Soc. (USA)* vol.69 (1974) p.897 ]
[32d] [ *NBS Monograph 25 (USA)* no.25 (1965) p.10 ]
[33] [ *JCPDS-ICDD* no.8 (1990) p.362 ]
[34a] N.F. Lashko [ *Dokl. Akad. Nauk SSSR (Russia)* vol.81 (1951) p.605 ]
[34b] P. Nash, A. Nash [ *Bull. Alloy Phase Diagrams (USA)* vol.8 (1987) p.6 ]
[34c] H.E. Swanson, E. Tatge [ *NBS Circ. 539 (USA)* no.1 (1953) p.13 ]
[35] R.P. Ram, S. Bhan [ *Z. Metallkd. (Germany)* vol.66 (1975) p.521 ]
[36] G.S. Saini, L.D. Calvert, J.B. Taylor [ *Can. J. Chem. (Canada)* vol.42 (1964) p.1511 ]
[37a] D. Smith et al [ JCPDS Grant-in-Aid Report, Pennsylvania State University, University Park (1973) ]
[37b] S.H. Chen, C.B. Carter, Z. Elgat, L.R. Zheng, J.W. Mayer [ *J. Appl. Phys. (USA)* vol.62 (1987) p.1189 ]
[38] K. Toman [ *Acta Crystallogr. (Denmark)* vol.5 (1952) p.329 ]
[39] G. Pilstrom [ *Acta Chem. Scand. (Denmark)* vol.15 (1961) p.893 ]
[40] F. d'Heurle, C.S. Petersson, J.E.E. Baglin, S.J. La Placa, C.Y. Wong [ *J. Appl. Phys. (USA)* vol.55 (1984) p.4208 ]
[41a] S. Arrhenius, A. Westgren [ *Z. Phys. Chem. (Germany)* vol.14 (1931) p.66 ]
[41b] R.W. Olesinski, G.J. Abbaschian [ *Bull. Alloy Phase Diagrams (USA)* vol.7 (1986) p.170 ]
[42] H.E. Swanson, E. Tatge [ *NBS Circ. 539 (USA)* no.1 (1953) p.15 ]
[43a] K.P. Mukherjee, J. Bandyopadhyaya, K.P. Gupta [ *Trans. AIME (USA)* vol.245 (1969) p.2335 ]
[43b] J.K. Solberg [ *Acta Crystallogr. A (Denmark)* vol.34 (1978) p.684 ]
[44a] I. Mayer [ JCPDS Grant-in-Aid Report, Hebrew University of Jerusalem, Israel (1972) ]
[44b] T.W. Button, I.J. McColm, J.M. Ward [ *J. Less-Common Met. (Switzerland)* vol.159 (1990) p.205 ]
[44c] H. Okamoto [ *J. Phase Equilibria* vol.12 (1991) p.397 ]
[44d] [ *NBS Monograph 25 (USA)* no.18 (1981) p.77 ]
[45] E. Parthe [ *Acta Crystallogr. (Denmark)* vol.12 (1959) p.559 ]
[46] S. La Placa [ JCPDS Grant-in-Aid Report, Polytechnic Institute of Brooklyn, New York (1959) ]
[47a] [ *JCPDS* no.25 (1990) p.757 ]
[47b] H.E. Swanson, R.K. Fuyat [ *NBS Circ. 539 (USA)* no.2 (1953) p.11 ]
[47c] I. Ferguson [ UK Atomic Energy Association TRG Report no.2438 (1976) p.1 ]
[47d] B. Olinger, J. Jamieson [ *High Temp.-High Press. (UK)* vol.5 (1973) p.123 ]
[48] H. Schachner, H. Nowotny, Machenschalk [ *Monatsh. Chem. (Austria)* vol.84 (1953) p.677 ]
[49] H. Schachner et al [ *Monatsh. Chem. (Austria)* vol.85 (1954) p.1140 ]
[50] R.E. Riley, W. Rostoker et al [ US Atomic Energy Commission (1952) ]
[51] [ *NBS Monograph 25 (USA)* no.17 (1980) p.86 ]
[52] O. Schob et al [ *Monatsh. Chem. (Austria)* vol.92 (1961) p.1218 ]
[53a] D.K. Deardorff, R.E. Siemens, P.A. Romans, R.A. McCune [ *J. Less-Common Met. (Switzerland)* vol.18 (1969) p.11 ]
[53b] M.E. Schlesinger, H. Okamoto, A.B. Gokhale, R. Abbaschian [ *J. Phase Equilibria* vol.14 (1993) p.502 ]
[53c] [ *NBS Monograph 25 (USA)* no.19 (1984) p.67 ]
[54] R. Waterstrat, F. Haenssler, J. Muller, S.D. Dahlgren, J.O. Willis [ *J. Appl. Phys. (USA)* vol.49 (1978) p.1143 ]
[55] H. Iwasaki [ *Sci. Rep. Res. Inst. Tohoku Univ. A (Japan)* vol.31 (1983) p.1 ]

[56] H. Schachner, E. Cerwenka, H. Nowotny [ *Monatsh. Chem. (Austria)* vol.85 (1954) p.245 ]
[57] [ *NBS Monograph 25 (USA)* no.15 (1978) p.44 ]
[58] [ *NBS Circ. 539 (USA)* no.8 (1959) p.39 ]
[59a] D.H. Templeton, C.H.Dauben [ *Acta Crystallogr. (Denmark)* vol.3 (1950) p.262 ]
[59b] A.B. Gokhale, G.J. Abbaschian [ *J. Phase Equilibria* vol.12 (1991) p.493 ]
[59c] H.E. Swanson, E. Tatge [ *NBS Circ. 539 (USA)* no.1 (1953) p.20 ]
[60] S. Amberg [ *Monatsh. Chem. (Austria)* vol.91 (1960) p.412 ]
[61] [ *NBS Monograph 25 (USA)* no.19 (1982) p.59 ]
[62] L. Brewer, A.W. Searcy, D.H. Templeton [ *J. Am. Ceram. Soc. (USA)* vol.33 (1950) p.291 ]
[63] [ *JCPDS-ICDD* no.6 (1990) p.681 ]
[64] [ *JCPDS-ICDD* no.17 (1990) p.917 ]
[65a] B. Aronsson, J. Aselius [ *Acta Chem. Scand. (Denmark)* vol.15 (1961) p.1571 ]
[65b] I. Engstrom [ *Acta Chem. Scand. (Denmark)* vol.24 (1970) p.1466 ]
[65c] C.S. Petersson, J.E.E. Baglin, J.J. Dempsey, F.M. d'Heurle, S.J. La Placa [ *J. Appl. Phys. (USA)* vol.53 (1982) p.4866 ]
[65d] [ *NBS Circ. 539 (USA)* no.4 (1955) p.5 ]; [ *Am. Mineral. (USA)* vol.61 (1961) p.177 ]
[66] W.L. Korst, L.N. Finnie, A.W. Searcy [ *J. Phys. Chem. (USA)* vol.61 (1957) p.1541 ]
[67a] L.N. Finnie [ *J. Less-Common Met. (Switzerland)* vol.4 (1962) p.24 ]
[67b] I. Engstrom, T. Johnsson [ *Arkiv for Kemi (Sweden)* vol.30 (1968) p.141 ]
[68] Y.S. Chang, J.J. Chu [ *Mater. Lett. (Netherlands)* vol.5 (1987) p.67 ]
[69] [ *JCPDS-ICDD* no.32 (1990) p.978 ]
[70a] I. Engstrom [ *Acta Chem. Scand. (Denmark)* vol.17 (1963) p.775 ]
[70b] M.E. Schlesinger [ *J. Phase Equilibria* vol.13 (1992) p.54 ]
[71] H.E. Swanson, R.K. Fuyat [ *NBS Circ. 539 (USA)* no.III (1953) ]
[72] L. Schellenberg, J. Jorda, J. Muller [ *J. Less-Common Met. (Switzerland)* vol.109 (1985) p.261 ]
[73] C.S. Petersson et al [ *J. Appl. Phys. (USA)* vol.51 (1980) p.373 ]
[74a] [ *JCPDS-ICDD* no.40 (1990) p.1335 ]
[74b] H.C. Baxi, T.B. Massalski [ *J. Phase Equilibria* vol.12 (1991) p.349 ]
[74c] H.E. Swanson, E. Tatge [ *NBS Circ. 539 (USA)* no.1 (1953) p.21 ]
[75] J.A. Wysocki, P.E. Duwez [ *Metall. Trans. A (USA)* vol.12 (1981) p.1455 ]
[76] T. Masumoto, R. Maddin [ *Acta Metall. (USA)* vol.19 (1971) p.725 ]
[77] K. Anderko, K. Schubert [ *Z. Metallkd. (Germany)* vol.44 (1953) p.307 ]
[78] Nylund [ *Acta Chem. Scand. (Denmark)* vol.20 (1966) p.2381 ]
[79] H. Pfisterer, K. Schubert [ *Z. Metallkd. (Germany)* vol.41 (1950) p.358 ]
[80] H. Iwasaki, T. Masumoto [ *J. Mater. Sci. (UK)* vol.13 (1978) p.2171 ]
[81a] [ *JCPDS-ICDD* no.25 (1990) p.371 ]
[81b] A.B. Gokhale, G.J. Abbaschian [ *Bull. Alloy Phase Diagrams (USA)* vol.10 (1989) p.390 ]
[81c] H. McMurdie et al [ *Powder Diffraction* vol.2 (1987) p.46 ]
[82] H. Nowotny, E. Laube, R. Kieffer, F. Benesovsky [ *Monatsh. Chem. (Austria)* vol.89 (1958) p.701 ]
[83] O. Schob, H. Nowotny, F. Benesovsky [ *Monatsh. Chem. (Austria)* vol.92 (1961) p.1218 ]
[84] H. Nowotny et al [ *Monatsh. Chem. (Austria)* vol.84 (1953) p.1 ]
[85a] W. Rossteutscher, K. Schubert [ *Z. Metallkd. (Germany)* vol.56 (1965) p.813 ]
[85b] M.E. Schlesinger [ *J. Phase Equilibria* vol.15 (1994) p.90 ]
[85c] H.E. Swanson, E. Tatge [ *NBS Circ. 539 (USA)* no.1 (1953) p.29 ]
[85d] P.G.N. Moseley, C.J. Seabrook [ *Acta Crystallogr. B (Denmark)* vol.29 (1973) p.1170 ]
[86] L. Brewer, A.W. Searcy, D.H. Templeton, C.H. Dauben [ *J. Am. Ceram. Soc. (USA)* vol.33 (1950) p.291 ]
[87] E. Parthe, H. Nowotny, H. Schmid [ *Monatsh. Chem. (Austria)* vol.86 (1955) p.859 ]
[88] W. Wong-Ng, H. McMurdie, B. Paretzkin, C. Hubbard, A. Dragoo [ NBS JCPDS Grant-in-Aid Report (1986) ]

[89a] [ *JCPDS-ICDD* no.16 (1990) p.261 ]

[89b] S.V. Nagender Naidu, A.M. Shiramamurthy, P. Rama Rao [ *J. Alloy Phase Diagrams* vol.5 (1989) p.149 ]

[89c] H.E. Swanson, E. Tatge [ *NBS Circ. 539 (USA)* no.1 (1953) p.28 ]

[90a] A. Amendola [ JCPDS Grant-in-Aid Report, Polytechnic Institute of Brooklyn, Brooklyn, New York (1959) ]

[90b] F.M. d'Heurle, C.S. Petersson, M.Y. Tsai [ *J. Appl. Phys. (USA)* vol.51 (1980) p.5976 ]

[91a] McNees Jr., Seracy [ *J. Am. Chem. Soc. (USA)* vol.77 (1955) p.5290 ]

[91b] H.E. Swanson, R.K. Fuyat [ *NBS Circ. 539 (USA)* no.II (1953) p.13 ]

[92a] [ *JCPDS-ICDD* no.35 (1990) p.1384 ]

[92b] T. Siegrist, F. Hulliger, G. Travaglini [ *J. Less-Common Met. (Switzerland)* vol.92 (1983) p.119 ]

[93a] W.L. Korst et al [ *J. Phys. Chem. (USA)* vol.61 (1957) p.1541 ]

[93b] [ *NBS Circ. 539 (USA)* no.4 (1955) p.8 ]

[94a] L.N. Finnie [ *J. Less-Common Met. (Switzerland)* vol.4 (1962) p.24 ]

[94b] I. Engstrom, F. Zackrisson [ *Acta Chem. Scand. (Denmark)* vol.24 (1970) p.2109 ]

[94c] S. Petersson et al [ *J. Appl. Phys. (USA)* vol.50 (1979) p.3357 ]

[95a] L. Calvert [ ICPDS Grant-in-Aid Report, NRC of Canada, Ottawa (1980) ]

[95b] [ *NBS Circ. 539 (USA)* no.4 (1955) p.9 ]

[96] D. Smith, Yu [ JCPDS Grant-in-Aid Report, Pennsylvania State University, University Park (1975) ]

[97] I. Engstrom, ?. Zdansky [ *Acta Chem. Scand. A (Denmark)* vol.36 (1982) p.857 ]

[98a] R. Gohle, K. Schubert [ *Z. Metallkd. (Germany)* vol.55 (1964) p.503 ]

[98b] L.E. Tanner, H. Okamoto [ *J. Phase Equilibria* vol.12 (1991) p.571 ]

[98c] H.E. Swanson, E. Tatge [ *NBS Circ. 539 (USA)* no.1 (1953) p.31 ]

[99] R.P. Ram, S. Bhan [ *Z. Metallkd. (Germany)* vol.69 (1978) p.524 ]

[100] W. Gold, K. Schubert [ *Z. Kristallogr., Kristallgeom., Krystalphys., Kristalchem. (Germany)* vol.128 (1969) p.406 ]

[101] G. Majni, M. Costato, F. Panini, G. Celotti [ *J. Phys. Chem. Solids (UK)* vol.46 (1985) p.631 ]

[102] H. Pfisterer, K. Schubert [ *Z. Metallkd. (Germany)* vol.41 (1950) p.358 ]

# 1.2 Mechanical properties of TM silicides

M. Östling and C. Zaring

May 1995

## A INTRODUCTION

The mechanical properties of the transition metal silicides are very important in that they determine how well a certain silicide will work in a specific application. The existing data on the mechanical properties are, however, mostly limited to rather old values based on investigations on bulk material. In a few cases data were found from thin film investigations. One should use the data cautiously since bulk properties can differ substantially from thin film properties.

The data are displayed in SI-units which means that some values have been converted from the original reference since they were given in non-standardised units. In this Datareview data on the following mechanical properties are tabulated: volumetric changes during silicide formation, modulus of elasticity, tensile and compressive strength, and hardness in terms of Vickers-, Rockwell- and micro-hardness.

## B VOLUMETRIC CHANGES DURING SILICIDE FORMATION

During the formation of transition metal silicides by solid phase reactions there is always a net volumetric change. FIGURE 1 illustrates two cases. When 100 nm of cobalt or titanium reacts with silicon to form the different silicide phases the final phases, $CoSi_2$ and $TiSi_2$, will result in a total volume shrinkage. The cobalt disilicide surface is recessed 12 nm below the original silicon surface while, on the contrary, the final titanium disilicide surface is raised

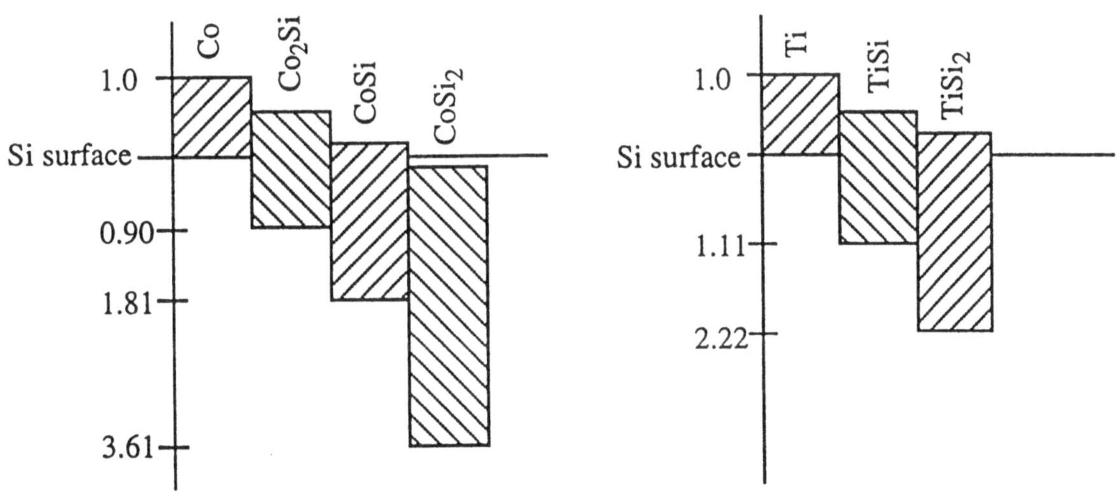


FIGURE 1 The volumetric changes during silicide formation as seen from the original silicon surface are illustrated with the two different cases of cobalt and titanium silicides.

10 nm above the original silicon surface. TABLE 1 shows the metal thickness ($t_M$) normalised to 1, the consumed silicon thickness ($t_{Si}$) in units of $t_M$, and the resulting silicide thickness ($t_{sil}$) in units of $t_M$. The total volumetric change always results in a shrinkage ($t_{sil}/(t_M + t_{Si}) < 1$). The formed silicides are often polycrystalline and the volume change is assumed to be accommodated completely by a change in thickness. In TABLE 1 the values are calculated from the atomic density of the respective elements in the phase formed.

## C MODULUS OF ELASTICITY, TENSILE STRENGTH AND COMPRESSIVE STRENGTH OF TRANSITION METAL SILICIDES

Deformation is here defined as strain, $\varepsilon$, and can be either elastic or plastic. In the elastic region the stress, $\sigma$, is given by Hooke's law, $\sigma = E\varepsilon$, where the proportionality constant E is called the modulus of elasticity, or Young's modulus, and is a measure of the resistance to elastic deformation. Values of the modulus of elasticity for the transition metal silicides are given in TABLE 2. Most of the data are for bulk materials. In the case of a thin film on a substrate of any other type of material the stress behaviour can differ markedly from that given in TABLES 2-9.

A thin film on a substrate can either experience a tensile stress or a compressive stress. If the thin film wants to expand but is attached to the substrate it will experience a compressive stress. On the other hand, if the thin film wants to contract it will experience a tensile stress. A compressive stress is usually considered negative and can cause curling of the film while tensile stress is considered positive and can cause cracks in the film. When a silicide is formed by a solid state reaction there is always an intrinsic stress developed because of the net volume change. These changes are given in TABLE 1.

Upon heating, transition metal silicides will experience a compressive stress as they all have a thermal expansion coefficient larger than that of silicon ($2.6 \times 10^{-6}\ K^{-1}$).

The method of silicide formation (e.g. deposition by physical vapour deposition or chemical vapour deposition), temperature, pressure, deposition rate, ion energy and substrate bias (sputtering) are all factors influencing the microstructure of the silicide and thus also the mechanical properties. The subsequent heat treatment to form the final silicide phase may also influence the microstructure.

As a general guide deposition temperatures lower than 1/3 of the melting point cause intrinsic stress to develop while at higher temperatures, because of the higher atomic mobility, intrinsic stress would not normally occur.

The total stress, $\sigma_{tot}$, in a material can be divided in two parts, the intrinsic stress, $\sigma_i$, and the thermal stress, $\sigma_{th}$.

Total stress: $\sigma_{tot} = \sigma_i + \sigma_{th}$

Thermal stress: $\sigma_{th} = (\alpha_{sil} - \alpha_{Si}) \dfrac{E}{1 - \nu} \Delta T$

where $\alpha_{Si}$ and $\alpha_{sil}$ are the thermal expansion coefficients of silicon and silicide, respectively. E is Young's modulus and ν is Poisson's ratio. The biaxial elastic modulus E/(1-ν) was also found for $WSi_2$ and calculated for $CoSi_2$.

The bend strength value can be used for applications in which the principal stressing mode is bending. For most of these materials the tensile and compressive moduli are different. Since the bending modulus is a combination of the tensile and compressive moduli, it is often different from each of them. The bend strength is here defined as the nominal stress that results in a specific permanent strain upon unloading.

## D HARDNESS OF TRANSITION METAL SILICIDES

A common feature for most of the transition metal silicides is that they show large hardness values. There are several methods of determining hardness and in TABLES 7-9 three different measures are given, namely Vickers hardness, Rockwell hardness and microhardness. Most of the data are given for hot pressed and sintered polycrystalline compounds and in no cases were data on thin film structures found. The microhardness measurements utilised an indenter in the form of a square pyramid with an apical angle of 136°. The load during testing is given when available. Only a few recent investigations exist.

## E CONCLUSION

A variety of mechanical properties for the transition metal silicides have been compiled. The majority of data are given for sintered or compressed bulk silicides with varying impurity content. Thin film applications clearly call for a lot more basic research to generate reliable mechanical properties of the transition metal silicides. We believe these data will emerge at a slow rate controlled by specific application requirements.

In summary, the compiled data should be useful for comparison among the different silicides, but the absolute values should be treated with some caution.

## ACKNOWLEDGEMENT

The authors would like to thank professor Simon Wong at the Center for Integrated Systems, Stanford University, CA, for his great hospitality during the writing of this Datareview. Financial support from the Fulbright Commission, the Royal Institute of Technology, Marcus Wallenbergs Stiftelse and Wenner-Grens Centers Stiftelse are acknowledged.

TABLE 1 Volumetric changes during silicide formation. The number density of formula units is taken from [1]. The number density of formula units for silicon is 5 x $10^{22}$ $cm^{-3}$. The metal thickness, $t_M$, is normalised to 1 and the silicon and silicide thickness, $t_{Si}$ and $t_{sil}$, are in units of $t_M$.

| Metal thickness, $t_M$, normalised to 1. Number density of formula units ($10^{22}$ $cm^{-3}$) | | Consumed silicon thickness, ($t_{Si}$), in units of metal thickness $t_M$ | Thickness of the resulting silicide, ($t_{sil}$), in units of $t_M$, for the respective silicide phase. Number density of formula units ($10^{22}$ $cm^{-3}$) | | | Ratio of thicknesses of silicide formed to that of the metal, the silicon and the total thickness of the metal and the silicon | | |
|---|---|---|---|---|---|---|---|---|
| | | | | | | $\frac{t_{sil}}{t_M}$ | $\frac{t_{sil}}{t_{Si}}$ | $\frac{t_{sil}}{t_M + t_{Si}}$ |
| α-Co | 9.03 | 0.90 | $Co_2Si$ | 1.47 | 3.062 | 1.47 | 1.63 | 0.77 |
| β-Co | 8.85 | 1.81 | CoSi | 1.98 | 4.562 | 1.98 | 1.09 | 0.70 |
| | | 1.80 | | 1.98 | 4.548 | 1.98 | 1.10 | 0.71 |
| | | 3.61 | $CoSi_2$ | 3.49 | 2.590 | 3.49 | 0.97 | 0.76 |
| Cr | 8.328 | 0.55 | $Cr_3Si$ | 1.32 | 2.10 | 1.32 | 2.40 | 0.85 |
| | | 0.64 | $Cr_5Si_3$ | 1.62 | 1.03 | 1.62 | 2.53 | 0.99 |
| | | 1.67 | CrSi | 2.04 | 4.090 | 2.04 | 1.22 | 0.76 |
| | | 3.33 | $CrSi_2$ | 3.00 | 2.777 | 3.00 | 0.90 | 0.69 |
| Cu | 3.724 | 0.74x | κ-$CuSi_x$ (≈12% Si) | 0.88 | 4.213 | 0.88 | 1.19/x | 0.51/x |
| | | 0.15 | γ-$Cu_5Si$ | 0.54 | 1.383 | 0.54 | 3.60 | 0.47 |
| | | 0.20 | ε-$Cu_{15}Si_4$ | 0.57 | 0.4364 | 0.57 | 2.85 | 0.47 |
| | | 0.25 | $Cu_3Si$ | 0.65 | 1.907 | 0.65 | 2.60 | 0.52 |
| α-Fe | 8.479 | 0.57 | $Fe_3Si$ | 1.28 | 2.211 | 1.28 | 2.24 | 0.81 |
| γ-Fe | 8.246 | 1.02 | $Fe_5Si_3$ | 1.58 | 1.073 | 1.58 | 1.55 | 0.78 |
| δ-Fe | 7.869 | 1.69 | FeSi | 1.91 | 4.428 | 1.91 | 1.13 | 0.71 |
| ε-Fe | 9.453 | 3.64 | α-$FeSi_2$ | 3.36 | 2.706 | 3.36 | 0.99 | 0.76 |
| | | 3.39 | β-$FeSi_2$ | 3.20 | 2.649 | 3.20 | 0.94 | 0.73 |
| α-Hf | 4.479 | 0.45 | $Hf_2Si$ | 1.22 | 1.828 | 1.22 | 2.71 | 0.84 |

Table 1 continued

| Metal thickness, $t_M$, normalised to 1. Number density of formula units ($10^{22}$ $cm^{-3}$) | | Consumed silicon thickness, ($t_{Si}$), in units of metal thickness $t_M$ | Thickness of the resulting silicide, ($t_{sil}$), in units of $t_M$, for the respective silicide phase. Number density of formula units ($10^{22}$ $cm^{-3}$) | | | Ratio of thicknesses of silicide formed to that of the metal, the silicon and the total thickness of the metal and the silicon | | |
|---|---|---|---|---|---|---|---|---|
| | | | | | | $\frac{t_{sil}}{t_M}$ | $\frac{t_{sil}}{t_{Si}}$ | $\frac{t_{sil}}{t_M + t_{Si}}$ |
| β-Hf | 4.665 | 0.54 | $Hf_5Si_3$ | 1.34 | 0.6674 | 1.34 | 2.48 | 0.87 |
| | | 0.60 | $Hf_3Si_2$ | 1.34 | 1.112 | 1.34 | 2.23 | 0.84 |
| | | 0.71 | $Hf_5Si_4$ | 1.42 | 0.629 | 1.42 | 2.00 | 0.83 |
| | | 0.90 | HfSi | 1.50 | 2.995 | 1.50 | 1.67 | 0.79 |
| | | 1.79 | $HfSi_2$ | 2.30 | 1.945 | 2.30 | 1.28 | 0.82 |
| Ir | 7.096 | 0.47 | $Ir_3Si$ | 1.28 | 1.844 | 1.28 | 2.72 | 0.87 |
| | | 0.71 | $Ir_2Si$ | 1.42 | 2.492 | 1.42 | 2.00 | 0.83 |
| | | 1.42 | IrSi | 1.99 | 3.572 | 1.99 | 1.40 | 0.82 |
| | | 4.26 | $IrSi_3$ | 3.84 | 1.847 | 3.84 | 0.90 | 0.73 |
| α-La | 2.672 | 0.35 | $La_3Si_2$ | 1.24 | 0.718 | 1.24 | 3.54 | 0.92 |
| β-La | 2.675 | 0.53 | LaSi | 1.38 | 1.942 | 1.38 | 2.60 | 0.90 |
| γ-La | 2.593 | 0.53 | | 1.36 | 1.959 | 1.36 | 2.57 | 0.89 |
| La(HP) | 2.894 | 1.07 | $LaSi_2$ | 1.68 | 1.587 | 1.68 | 1.57 | 0.81 |
| α-Mn | 8.193 | 0.54 | $Mn_3Si$ | 1.28 | 2.135 | 1.28 | 2.37 | 0.83 |
| β-Mn | 7.991 | 0.66 | $Mn_5Si_2$ | 1.42 | 1.156 | 1.42 | 2.15 | 0.85 |
| γ-Mn | 7.914 tetr | 0.98 | $Mn_5Si_3$ | 1.64 | 1.00 | 1.64 | 1.67 | 0.83 |
| | 6.942 cub | 1.64 | MnSi | 1.94 | 4.226 | 1.94 | 1.18 | 0.73 |
| δ-Mn | 6.841 | 2.83 | $Mn_{11}Si_{19}$ | 2.73 | 0.2729 | 2.73 | 0.96 | 0.71 |

Table 1 continued

| Metal thickness, $t_M$, normalised to 1. Number density of formula units ($10^{22}$ $cm^{-3}$) | | Consumed silicon thickness, ($t_{Si}$), in units of metal thickness $t_M$ | Thickness of the resulting silicide, ($t_{sil}$), in units of $t_M$, for the respective silicide phase. Number density of formula units ($10^{22}$ $cm^{-3}$) | | | Ratio of thicknesses of silicide formed to that of the metal, the silicon and the total thickness of the metal and the silicon | | |
|---|---|---|---|---|---|---|---|---|
| | | | | | | $\frac{t_{sil}}{t_M}$ | $\frac{t_{sil}}{t_{Si}}$ | $\frac{t_{sil}}{t_M + t_{Si}}$ |
| Mo | 6.415 | 0.43 | $Mo_3Si$ | 1.25 | 1.710 | 1.25 | 2.91 | 0.87 |
| | | 0.77 | $Mo_5Si_3$ | 1.45 | 0.8821 | 1.45 | 1.88 | 0.82 |
| | | 2.57 | $MoSi_2$ | 2.60 | 2.470 | 2.60 | 1.01 | 0.73 |
| | | 2.56 | | 2.57 | 2.491 | 2.57 | 1.00 | 0.72 |
| | | 2.57 | | 2.25 | 2.852 or 2.584 | 2.25 | 0.87 | 0.63 |
| Nb | 5.53 | 0.66 | $\alpha$-$Nb_5Si_3$ | 1.41 | 0.7837 | 1.41 | 2.14 | 0.85 |
| | | 0.66 | $\beta$-$Nb_5Si_3$ | 1.41 | 0.7857 | 1.41 | 2.14 | 0.85 |
| | | 2.21 | $NbSi_2$ | 2.39 | 2.311 | 2.39 | 1.08 | 0.74 |
| | | 2.21 | | 2.42 | 2.283 | 2.42 | 1.09 | 0.75 |
| Ni | 9.14 | 0.61 | $\beta_1$-$Ni_3Si$ | 1.31 | 2.324 | 1.31 | 2.15 | 0.81 |
| | | 0.36 | | 0.79 | 2.312 | 0.79 | 2.19 | 0.58 |
| | | 0.91 | $\delta$-$Ni_2Si$ | 1.49 | 3.065 | 1.49 | 1.64 | 0.78 |
| | | 0.91 | $\theta$-$Ni_2Si$ | 1.40 | 3.262 | 1.40 | 1.54 | 0.73 |
| | | 1.21 | $Ni_3Si_2$ | 1.74 | 1.747 | 1.74 | 1.44 | 0.79 |
| | | 1.83 | NiSi | 2.01 | 4.552 | 2.01 | 1.10 | 0.71 |
| | | 3.66 | $NiSi_2$ | 3.59 | 2.547 | 3.59 | 0.98 | 0.77 |
| Os | 7.165 | 1.43 | OsSi | 1.89 | 3.782 | 1.89 | 1.32 | 0.78 |
| | | 2.15 | $Os_2Si_3$ | 5.00 | 0.7169 | 5.00 | 2.32 | 1.59 |
| | | 2.87 | $OsSi_2$ | 3.07 | 2.337 | 3.07 | 1.07 | 0.79 |
| Pd | 6.813 | 0.30 | $Pd_9Si_2$ | 1.19 | 0.6342 | 1.19 | 3.97 | 0.91 |
| | | 0.45 | $Pd_3Si$ | 1.29 | 1.755 | 1.29 | 2.87 | 0.89 |
| | | 0.68 | $Pd_2Si$ | 1.42 | 2.397 | 1.42 | 2.09 | 0.84 |
| | | 0.68 | | 1.44 | 2.366 | 1.44 | 2.12 | 0.86 |
| | | 1.36 | PdSi | 1.98 | 3.445 | 1.98 | 1.45 | 0.84 |

Table 1 continued

| Metal thickness, $t_M$, normalised to 1. Number density of formula units ($10^{22}$ $cm^{-3}$) | | Consumed silicon thickness, ($t_{Si}$), in units of metal thickness $t_M$ | Thickness of the resulting silicide, ($t_{sil}$), in units of $t_M$, for the respective silicide phase. Number density of formula units ($10^{22}$ $cm^{-3}$) | | | Ratio of thicknesses of silicide formed to that of the metal, the silicon and the total thickness of the metal and the silicon | | |
|---|---|---|---|---|---|---|---|---|
| | | | | | | $\frac{t_{sil}}{t_M}$ | $\frac{t_{sil}}{t_{Si}}$ | $\frac{t_{sil}}{t_M + t_{Si}}$ |
| Pt | 6.622 | 0.44 | α-$Pt_3Si$ | 1.28 | 1.724 | 1.28 | 2.91 | 0.89 |
| | | 0.44 | β-$Pt_3Si$ | 1.29 | 1.707 | 1.29 | 2.93 | 0.89 |
| | | 0.55 | $Pt_{12}Si_5$ | 1.35 | 0.4084 | 1.35 | 2.45 | 0.87 |
| | | 0.66 | α-$Pt_2Si$ | 1.51 | 2.188 | 1.51 | 2.29 | 0.91 |
| | | 0.66 | β-$Pt_2Si$ | 1.41 | 2.343 | 1.41 | 2.14 | 0.85 |
| | | 1.10 | $Pt_6Si_5$ | 1.84 | 0.6005 | 1.84 | 1.67 | 0.88 |
| | | 1.32 | PtSi | 1.98 | 3.342 | 1.98 | 1.50 | 0.85 |
| Re | 6.805 | 0.82 | $Re_5Si_3$ | 1.49 | 0.9135 | 1.49 | 1.82 | 0.82 |
| | | 1.36 | ReSi | 1.85 | 3.682 | 1.85 | 1.36 | 0.78 |
| | | 2.72 | $ReSi_2$ | 2.56 | 2.659 | 2.56 | 0.94 | 0.69 |
| Rh | 7.226 | 0.72 | $Rh_2Si$ | 1.42 | 2.549 | 1.42 | 1.97 | 0.82 |
| | | 0.87 | $Rh_5Si_3$ | 1.52 | 0.9531 | 1.52 | 1.75 | 0.81 |
| | | 1.44 | RhSi | 1.95 | 3.711 | 1.95 | 1.35 | 0.80 |
| | | 1.44 | | 1.87 | 3.862 | 1.87 | 1.30 | 0.77 |
| | | 1.44 | | 1.85 | 3.908 | 1.85 | 1.28 | 0.76 |
| | | 1.80 | $Rh_4Si_5$ | 2.28 | 0.792 | 2.28 | 1.27 | 0.81 |
| | | 1.93 | $Rh_3Si_4$ | 2.38 | 1.012 | 2.38 | 1.23 | 0.81 |
| Ru | 7.370 | 0.73 | $Ru_2Si$ | 1.44 | 2.55 | 0.73 | 1.97 | 0.83 |
| | | 1.47 | α-RuSi | 1.81 | 4.062 | 1.47 | 1.23 | 0.73 |
| | | 1.48 | β-RuSi | 1.92 | 3.845 | 1.48 | 1.30 | 0.77 |
| | | 2.21 | $Ru_2Si_3$ | 2.53 | 1.457 | 2.21 | 1.14 | 0.79 |
| α-Sc | 3.999 | 0.48 | $Sc_5Si_3$ | 1.24 | 0.647 | 1.24 | 2.58 | 0.84 |

Table 1 continued

| Metal thickness, $t_M$, normalised to 1. Number density of formula units ($10^{22}$ cm$^{-3}$) | | Consumed silicon thickness, ($t_{Si}$), in units of metal thickness $t_M$ | Thickness of the resulting silicide, ($t_{sil}$), in units of $t_M$, for the respective silicide phase. Number density of formula units ($10^{22}$ cm$^{-3}$) | | | Ratio of thicknesses of silicide formed to that of the metal, the silicon and the total thickness of the metal and the silicon | | |
|---|---|---|---|---|---|---|---|---|
| | | | | | | $\frac{t_{sil}}{t_M}$ | $\frac{t_{sil}}{t_{Si}}$ | $\frac{t_{sil}}{t_M + t_{Si}}$ |
| β-Sc | 4.287 | 0.80<br>1.33 | ScSi<br>$ScSi_{2-x}$ | 1.44<br>1.79 | 2.774<br>0.7423 | 1.44<br>1.79 | 1.80<br>1.34 | 0.80<br>0.77 |
| Ta | 5.534 | 0.37<br>0.55<br>0.66<br>0.66<br>2.21 | $Ta_3Si$<br>$Ta_2Si$<br>α-$Ta_5Si_3$<br>β-$Ta_5Si_3$<br>$TaSi_2$ | 1.24<br>0.55<br>1.35<br>1.39<br>2.40 | 1.49<br>5.039<br>0.8186<br>0.7935<br>2.304 | 1.24<br>0.55<br>1.35<br>1.39<br>2.40 | 3.35<br>1.00<br>2.04<br>2.11<br>1.08 | 0.90<br>0.35<br>0.81<br>0.84<br>0.75 |
| α-Ti | 5.566 | 0.67 | $Ti_5Si_3$ | 1.37 | 0.814 | 1.37 | 2.04 | 0.82 |
| β-Ti | 5.419 | 1.11 | TiSi | 1.62 | 3.44 | 1.62 | 1.46 | 0.77 |
| Ti(HP) | | 2.22<br>2.22 | C49 $TiSi_2$<br>C54 $TiSi_2$ | 2.33<br>2.44 | 2.388<br>2.278 | 2.33<br>2.44 | 1.05<br>1.10 | 0.72<br>0.76 |
| V | 7.118 | 0.47<br>0.85<br>0.86<br>2.85 | $V_3Si$<br>$V_5Si_3$ tetra<br>$V_5Si_3$ hexa<br>$VSi_2$ | 1.25<br>1.50<br>1.65<br>2.74 | 1.899<br>0.946<br>0.8647<br>2.602 | 1.25<br>1.50<br>1.65<br>2.74 | 2.66<br>1.76<br>1.92<br>0.96 | 0.85<br>0.81<br>0.89<br>0.71 |
| α-W | 6.305 | 0.76 | $W_5Si_3$ | 1.46 | 0.8650 | 1.46 | 1.92 | 0.83 |
| β-W | 6.214 | 0.76<br>2.52<br>2.52 | $WSi_2$ | 1.37<br>2.58<br>2.48 | 0.9211<br>2.446<br>2.537 | 1.37<br>2.58<br>2.48 | 1.80<br>1.02<br>0.98 | 0.78<br>0.73<br>0.70 |

Table 1 continued

| Metal thickness, $t_M$, normalised to 1. Number density of formula units ($10^{22}$ $cm^{-3}$) | | Consumed silicon thickness, ($t_{Si}$), in units of metal thickness $t_M$ | Thickness of the resulting silicide, ($t_{sil}$), in units of $t_M$, for the respective silicide phase. Number density of formula units ($10^{22}$ $cm^{-3}$) | | | Ratio of thicknesses of silicide formed to that of the metal, the silicon and the total thickness of the metal and the silicon | | |
|---|---|---|---|---|---|---|---|---|
| | | | | | | $\frac{t_{sil}}{t_M}$ | $\frac{t_{sil}}{t_{Si}}$ | $\frac{t_{sil}}{t_M + t_{Si}}$ |
| Y | 3.031 | 0.36 | $Y_5Si_3$ | 1.18 | 0.5143 | 1.18 | 3.28 | 0.87 |
| | | 0.61 | YSi | 1.30 | 2.336 | 1.30 | 2.13 | 0.81 |
| | | 1.01 | $\beta$-$Y_3Si_5$ | 0.74 | 1.370 | 0.74 | 0.73 | 0.37 |
| $\alpha$-Zr | 4.334 | 0.43 | $Zr_2Si$ | 1.16 | 1.874 | 1.16 | 2.70 | 0.81 |
| $\beta$-Zr | 4.219 | 0.52 | $Zr_5Si_3$ | 0.13 | 6.680 | 0.13 | 0.25 | 0.08 |
| Zr(HP) | 4.395 | 0.51 | | 0.14 | 6.044 | 0.14 | 0.27 | 0.09 |
| | | 0.58 | $Zr_3Si_2$ | 1.35 | 1.073 | 1.35 | 2.33 | 0.85 |
| | | 0.87 | ZrSi | 1.52 | 2.852 | 1.52 | 1.75 | 0.81 |
| | | 0.87 | | 1.46 | 2.969 | 1.46 | 1.68 | 0.78 |
| | | 1.73 | $ZrSi_2$ | 2.17 | 1.993 | 2.17 | 1.25 | 0.79 |

TABLE 2 Modulus of elasticity (Young's modulus), E, of transition metal silicides.

| Silicide phase | Modulus of elasticity, E (GPa) | Temperature (K) | Ref |
|---|---|---|---|
| $Co_2Si$ | 150 | - | [10] |
| $CoSi_2$ | 160<br>E/(1-ν) = 145 | - | [10]<br>[6] the biaxial elastic modulus E/(1-ν) was calculated from a tensile stress diagram |
| $Cr_2Si$ | 240.41 | - | [10] |
| CrSi | 269.78 | - | [10] |
| $Fe_3Si$ | 203.07 | - | [10] |
| $Fe_5Si_3$ | 295.44 | - | [10] |
| FeSi | 310 | - | [10] |
| $FeSi_2$ | 205 | - | [10] |
| $Mo_3Si$ | 30<br>304.11 | 293<br>- | [2]<br>[10] |
| $Mo_3Si_2$ | 155.98 | - | [10] |
| $MoSi_2$ | 430<br>440.47 | 293<br>- | [2]<br>[10] |
| $NbSi_2$ | 255.1 | - | [10] |
| $Nb_{3.34}Si$ | 65.5<br>65.15<br>64.8<br>64.3<br>63.6<br>62.9<br>62.3<br>61.6 | 80<br>120<br>160<br>200<br>240<br>280<br>320<br>360 | [10]<br>[10]<br>[10]<br>[10]<br>[10]<br>[10]<br>[10]<br>[10] |
| $Ni_3Si$ | 139.30 | - | [10] |
| $Ni_2Si$ | 160.88 | - | [10] |
| $Ni_5Si_3$ | 176.58 | - | [10] |
| $Ni_3Si_2$ | 166.77 | - | [10] |
| $Ti_5Si_3$ | 156<br>140.5 | 293<br>1273 | [3]<br>[3] |
| TiSi | 214.84 | - | [10] |
| $TiSi_2$ | 264<br>255.6<br>200<br>355.06 | 293<br>293<br>1273<br>- | [2]<br>[4] hot pressed with grain size 20-30 μm<br><br>[10] |
| $WSi_2$ | 530.72 | - | [10] |
| $WSi_{2.4-2.5}$<br>Sputter deposited films<br>50 nm - 250 nm | 137<br>270 - 300<br>306 - 310<br>312 - 319 | As deposited<br>1023 K annealed<br>1123 K annealed<br>1223 K annealed | [11]<br>[11]<br>[11]<br>[11] |
| $WSi_2$<br>(co-sputter deposited) | 290 biaxial elastic modulus (E/(1-ν)) | | [12] measured on micro-machined diaphragm |

Table 2 continued

| Silicide phase | Modulus of elasticity, E (GPa) | Temperature (K) | Ref |
|---|---|---|---|
| $ZrSi_2$ | 268<br>234.8<br>~200<br>259.97 | 293<br>293<br>1273<br>- | [2]<br>[4] hot pressed with grain size<br>20-30 μm<br>[10] |

TABLE 3 Poisson's ratio, ν, for transition metal silicides.

| Silicide phase | Poisson's ratio, ν (GPa) | Comment | Ref |
|---|---|---|---|
| $TiSi_2$ | 0.22 ± 0.07<br><br>0.24 | Average in temperature range 293 - 1193 K | [7] assumed bulk value of Young's modulus E = 259 GPa<br>[4] |
| $WSi_2$ | 0.3 | | [11] |
| $ZrSi_2$ | 0.22 | | [4] |

TABLE 4 Tensile strength, $\sigma_{tens}$, of transition metal silicides.

| Silicide phase | $\sigma_{tens}$ (MPa) | Temperature (K) | Ref |
|---|---|---|---|
| $MoSi_2$ | 280 | 1253 | [2] |
| | 294 | 1473 | [2] |
| | 275 | - | [8] |
| | 370 | 1023 | [10] |
| | 400 | 1273 | [10] |
| | 420 | 1573 | [10] |
| | 100 | 1623 | [10] |
| $Ni_2Si$ | 6.0 | 293 | [2,10] |
| | 112 | 873 | [2,10] |
| | 142 | 923 | [2,10] |
| | 59 | 1023 | [2] |
| NiSi | 6.0 | 293 | [2,10] |
| | 8.0 | 773 | [2,10] |
| | 20.0 | 823 | [2,10] |
| | 11.0 | 923 | [2,10] |
| | 5.3 | 1023 | [2,10] |
| $TiSi_2$ | 150 | 293 | [2,10] |

TABLE 5 Compressive strength, $\sigma_{comp}$, of transition metal silicides.

| Silicide phase | $\sigma_{comp}$ (MPa) | Temperature (K) | Ref |
|---|---|---|---|
| CoSi | 38 | 293 | [2,10] |
| | 63 | 773 | [2,10] |
| | 340 | 1023 | [2,10] |
| $CoSi_2$ | 100 | 293 | [2,10] |
| | 152 | 773 | [2,10] |
| | 600 | 1023 | [2,10] |
| $MoSi_2$ | 1130 | 293 | [2] |
| | 405 | 1273 | [2] |
| | 350 | 1473 | [2] |
| | 390 | 1673 | [2] |
| | 45 | 1873 | [2] |
| | 1378 | - | [8] |
| | 1500 | 293 | [10] |
| | 400 | 1273 | [10] |
| | 390 | 1473 | [10] |
| | 420 | 1673 | [10] |
| | 350 | 1723 | [10] |
| | 80 | 1773 | [10] |
| | 90 | 1873 | [10] |
| $Ni_2Si$ | 316 | 293 | [2,10] |
| | 579 | 773 | [2,10] |
| | 760 | 873 | [2,10] |
| NiSi | 158 | 293 | [2,10] |
| | 467 | 773 | [2,10] |
| | 507 | 873 | [2,10] |
| | 625 | 1023 | [2,10] |
| $TiSi_2$ | 117.9 | 293 | [2] |
| | 39.7 | 1273 | [2] |
| | 10.5 | 1373 | [2] |
| | 5.5 | 1473 | [2] |
| $WSi_2$ | 126.9 | 293 | [2] |
| | 59.5 | 1273 | [2] |

TABLE 6 Bend strength, $\sigma_{bend}$, of some transition metal silicides.

| Silicide phase | $\sigma_{bend}$ (MPa) | Temperature (K) | Ref |
|---|---|---|---|
| $LaSi_2$ | 272 | 293 | [10] |
| $TiSi_2$ | 210 | 293 | [10] |
| $MoSi_2$ | 210 | 293 | [10] |
| | 320 | 1023 | [10] |
| | 360 | 1173 | [10] |
| | 430 | 1473 | [10] |
| | 480 | 1523 | [10] |

TABLE 7 Mechanical hardness: Vickers hardness, HV, for transition metal silicides.

| Silicide phase | HV (GPa) | Temperature (K) | Ref |
|---|---|---|---|
| $Cr_3Si$ | 9.0 - 9.8 | 293 | [2,10,13] |
| $Cr_3Si_2$ | 10.5 - 12.0 | 293 | [2,10,13] |
| CrSi | 9.5 - 10.5 | 293 | [2,10,13] |
| $CrSi_2$ | 8.8 - 11.0 | 293 | [2,10,13] |
| $Mo_3Si$ | 13.2 - 15.5 | 293 | [2,10] |
| $Mo_5Si_3$ | 12.0 - 13.2 | 293 | [2,10] |
| $MoSi_2$ | 13.2 - 15.5 | 293 | [2,10] |
| $Nb_4Si$ | 4.7 - 5.5 | - | [2] |
| $Nb_5Si_3$ | 4.0 - 6.0 | - | [2] |
| $NbSi_2$ | 6.6 - 7.0<br>~10 | -<br>room temperature | [2]<br>[9] microindentation load < 0.6 N |
| $Ta_{4.5}Si$ | 10.0 - 12.0 | - | [2] load p = 400 N |
| $Ta_2Si$ | 12.0 - 15.0 | - | [2] for all |
| $Ta_5Si_3$ | 12.0 - 15.0 | - | [2] |
| $TaSi_2$ | 10.0 - 12.0 | 293 | [2,10] |
| $Ti_5Si_3$ | 9.68 | - | [3] |
| $TiSi_2$ | 8.70<br>~10.5 | -<br>room temperature | [4]<br>[9] microindentation load < 0.6 N |
| $W_3Si_2$ | 7.7 | | [2] load = 100 g |
| $Y_5Si_3$ | 4.45 - 4.79 | room temperature | [14] |
| $Y_5Si_4$ | 3.43 - 3.75 | room temperature | [14] |
| YSi | 4.18 - 4.60 | room temperature | [14] |
| $Y_3Si_5$ | 6.76 - 7.32 | room temperature | |
| $Zr_2Si$ | 11.8 - 12.8 | | [2] |
| $Zr_5Si_3$ | 12.8 - 13.9 | | [2] |
| ZrSi | 10.2 - 11.8 | | [2] |
| $ZrSi_2$ | 8.3 - 9.8<br>9.42 | | [2]<br>[4] |

TABLE 8 Mechanical hardness: Rockwell hardness, HRA, for transition metal silicides.

| Silicide phase | HRA (GPa) | Ref |
|---|---|---|
| $Cr_3Si$ | 0.85 | [2] |
| CrSi | 0.82 | [2] |
| $LaSi_2$ | 0.31 | [2] |
| $Mo_5Si_3$ | 0.74 | [2] |
| $MoSi_2$ | 0.74 | [2] |
| $TiSi_2$ | 0.81 | [2] |
| $V_3Si$ | 0.78 | [2] |
| $V_5Si_3$ | 0.79 | [2] |
| $VSi_2$ | 0.32 | [2] |

TABLE 9 Mechanical hardness: microhardness, Hμ, for some transition metal silicides.

| Silicide phase | Hμ (GPa) | Load (N) | Temperature (K) and comments | Ref |
|---|---|---|---|---|
| CoSi | 10.0 | 0.2 | 293 | [2,10] |
| | 3.0 | 0.2 | 773 | [2,10] |
| | 1.15 | 0.2 | 1273 | [2,10] |
| $CoSi_2$ | 5.52 | 0.2 | 293 | [2,10] |
| | 3.22 | 0.2 | 773 | [2,10] |
| | 0.77 | 0.2 | 1273 | [2,10] |
| | 8.89 | 0.5 | 293 | [10] |
| $Cr_3Si$ | 10.05 | 0.5 | 293 | [2] |
| $Cr_3Si_2$ | 12.8 | 0.5 | 293 | [2] |
| CrSi | 10.05 | 0.5 | 293 | [2] |
| $CrSi_2$ | 11.3 | 0.5 | 293 | [2,10] |
| | 798 | 0.5 | Annealed | [2] |
| | 704 | 0.5 | Cast | [2] |
| | 996 - 1150 | 0.5 | 293 | [10] |
| $FeSi_2$ | 10.74 | 0.5 | 293 | [2] |
| β-$FeSi_2$ | 12.60 ± 0.5 | 0.5 | 293 | [10] |
| $HfSi_2$ | 9.30 | 0.5 | | [2] |
| | 9.12 ± 0.21 | 0.5 | 293 | [10] |
| $La_5Si_3$ | 5.95 ± 0.05 | 0.5 - 0.8 | 293 | [2,10] |
| $La_3Si$ | 5.25 ± 0.13 | 0.5 - 0.8 | 293 | [2,10] |
| LaSi | 6.32 ± 0.18 | 0.5 - 0.8 | 293 | [2,10] |
| $LaSi_2$ | 4.76 ± 0.14 | 0.5 - 0.8 | 293 | [2,10] |
| $Mo_3Si$ | 13.1 | 1.0 | 293 | [2,10] |
| $Mo_5Si_3$ | 11.7 | 1.0 | 293 | [2,10] |
| $MoSi_2$ | 12.0 | 0.5 | 293 | [2,10] |
| | 12.0 - 13.5 | 1.0 | 293 | [10] |
| | 7.07 | 0.5 | Cast | [2] |
| | 7.35 | 0.5 | Annealed | [2] |
| $Nb_4Si$ | 6.9 - 8.2 | 1.0 | 293 | [2] |
| $Nb_5Si_3$ | 7.0 | 1.0 | 293 | [2] |
| $NbSi_2$ | 10.5 | 0.5 | | [2] |
| | 10.82 | 0.5 | 293 | [10] |
| $Ni_3Si$ | 4.00 | 0.2 | 293 | [2,10] |
| $Ni_2Si$ | 4.40 | 0.2 | 293 | [2,10] |
| | 3.20 | 0.2 | 773 | [2,10] |
| | 1.20 | 0.2 | 1273 | [2,10] |
| NiSi | 4.00 | 0.2 | 293 | [2,10] |
| | 2.56 | 0.2 | 773 | [2,10] |
| ζ-$NiSi_2$ | 15.6 ± 0.3 | 0.5 | 293 | [2,10] |
| $NiSi_2$ | 10.19 | 0.5 | 293 | [2,10] |
| $OsSi_2$ | 19.5 | 0.5 | 293 | [5,10] |

Table 9 continued

| Silicide phase | Hμ (GPa) | Load (N) | Temperature (K) and comments | Ref |
|---|---|---|---|---|
| $ReSi_2$ | 15.0 | 0.5 | | [2] |
| $Sc_5Si_3$ | 9.73 ± 0.13 | 0.2 | 293 | [2,10] |
| ScSi | 10.05 ± 0.6 | 0.5 - 1.0 | 293 | [2,10] |
| $ScSi_{1.66}$ | 8.97 ± 0.4 | 0.5 - 1.0 | 293 | [2,10] |
| $TaSi_2$ | 14.07 | 0.5 | 293 | [2,10] |
| $Ti_5Si_2$ | 12.83 ± 0.67 | - | 293 | [10] |
| $Ti_5Si_3$ | 9.86 | 1.0 | 293 | [2,10] |
| TiSi | 10.39<br>18.0 ± 1.0 | 1.0<br>- | 293<br>293 | [2,10]<br>[10] |
| $TiSi_2$ | 7.76<br>8.92<br>8.82 | -<br>0.5<br>- | 293<br>293<br>293 | [10]<br>[2,10]<br>[10] |
| $V_3Si$ | 14.3 - 15.6 | 0.5 | 293 | [2] |
| $V_5Si_3$ | 13.5 - 15.1 | 0.5 | 293 | [2] |
| $VSi_2$ | 8.9 - 9.6 | 0.5 | 293 | [2] |
| $W_3Si_2$ | 7.7 | 0.5 | 293 | [2,10] |
| $WSi_2$ | 10.74<br>12.5<br>13.0 - 14.0 | 0.5<br>0.5<br>1.0 | <br>293<br>293 | [2]<br>[10]<br>[10] |
| YSi | 11.06 ± 0.2 | 0.3 - 0.4 | 293 | [10] |
| $Y_3Si_5$ | 8.45 ± 0.12 | 0.3 - 0.4 | 293 | [10] |
| $YSi_2$ | 8.04 ± 0.09 | 0.3 - 0.4 | 293 | [10] |
| $Zr_2Si$ | 11.8 - 12.8 | 0.5 | 293 | [2] |
| $Zr_5Si_3$ | 12.8 - 13.9 | 0.5 | 293 | [2] |
| ZrSi | 10.2 - 11.8 | 0.5 | 293 | [2] |
| $ZrSi_2$ | 10.63<br>11.19 ± 0.51 | 0.5<br>0.5 | 293<br>293 | [2]<br>[10] |

## REFERENCES


[1] M.-A. Nicolet, S.S. Lau [ in *VLSI Electronics: Microstructure Science* vol.6 Eds N.G. Einspruch, G.B. Larrabee (Academic Press Inc., 1983) ch.6 p.329-464 ]

[2] G.V. Samsonov, I.M. Vinitskii [ *Handbook of Refractory Compounds* (IFI/Plenum Data Company, 1980) ]

[3] G. Frommeyer, R. Rosenkranz, C. Lüdecke [ *Z. Met.kd. (Germany)* vol.81 no.5 (1990) p.307-13 ]

[4] R. Rosenkrantz, G. Frommeyer [ *Z. Met.kd. (Germany)* vol.83 no.9 (1992) p.685-9 ]

[5] I. Engstrom [ *Structural Chemistry of Platinum Metal Silicides* (Acta. Univ. Uppsala, 1970) p.54 ]

[6] A.H. van Ommen, C.W.T. Bulle-Lieuwma, C. Langereis [ *J. Appl. Phys. (USA)* vol.64 (1988) p.2706 ]

[7] P.J.J. Wessels, J.F. Jongste, G.C.A.M. Janssen, A.L. Mulder, S. Radelaar, O.B. Loopstra [ *J. Appl. Phys. (USA)* vol.63 no.10 (1988) p.4979-82 ]

[8] T.A. Kircher, E.L. Courtright [ *Mater. Sci. Eng. A (Switzerland)* vol.155 (1992) p.67-74 ]

[9] J. Guille, L. Matini [ *J. Mater. Sci. Lett. (UK)* vol.7 (1988) p.952-4 ]

[10] T.Y. Kosolapova [ *Handbook of High Temperature Compounds: Properties, Production, Applications* (Hemisphere Publishing Corporation, 1990) ]

[11] G.M. Crean, A. Golanski, J.C. Oberlin, A. Perio [ *Appl. Phys. Lett. (USA)* vol.51 no.20 (1987) p.1591 ]

[12] M.-L. Ger, R.B. Brown [ *Mater. Res. Soc. Symp. Proc. (USA)* vol.276 (1992) p.283-94 ]

[13] S. Okada, K. Kudou, M. Miyamoto, Y. Hikichi [ *Nippon Kagaku Kaishi (Japan)* vol.12 (1991) p.1612-7 ]

[14] T.W. Button, I.J.McColm, J.M. Ward [ *J. Less-Common Met. (Switzerland)* vol.159 (1990) p.205-22 ]

# 1.3 Thermal properties of TM silicides

M. Östling and C. Zaring

May 1995

## A INTRODUCTION

Information about the thermal properties of the transition metal silicides is important since they reveal the materials behaviour during subsequent processing at elevated temperature. In this Datareview data are presented on the melting points of the respective silicides, their thermal linear expansion coefficient, their thermal conductivity and finally their specific heat.

## B MELTING POINTS OF TRANSITION METAL SILICIDES

TABLE 1 gives the melting points of the transition metal silicides. The listing is in alphabetical order and the respective silicide system is listed in order of increasing Si content. The nature of fusion is also given except for a few cases where there is controversy or no information.

TABLE 1 Melting point or transformation point for transition metal silicides (p = peritectic, c = congruent, e = eutectic).

| Silicide phase | Melting point (°C) | Nature of fusion | Ref |
|---|---|---|---|
| $Co_3Si$ | 1212 | p | [1] |
| | 1214 | p | [4] |
| $\alpha$-$Co_2Si$ | 1320 | p | [1,4] |
| $\beta$-$Co_2Si$ | 1332 | c | [1,2] |
| | 1334 | c | [4] |
| CoSi | 1460 | c | [1,2,4] |
| $CoSi_2$ | 1326 | c | [1,2,4] |
| $Cr_5Si$ | 1720 | c | [1] |
| $Cr_3Si$ | 1770 | c | [1] |
| | $1770 \pm 10$ | c | [4] |
| | $1750 \pm 20$ | - | [2] |
| $Cr_5Si_3$ | $1680 \pm 20$ | c | [4] |
| | $1650 \pm 30$ | - | [2] |
| CrSi | 1475 | p | [1] |
| | $1450 \pm 30$ | - | [2] |
| | $1413 \pm 5$ | p | [4] |
| $CrSi_2$ | $1460 \pm 20$ | - | [2] |
| | 1475 | c | [1] |
| | $1490 \pm 20$ | c | [4] |
| $Fe_2Si$ | 1215 | c | [1] |
| FeSi | 1410 | c | [1] |
| $\alpha$-$FeSi_2$ | 1220 | c | [1] |
| $\alpha$-$FeSi_2$ to $\beta$-$FeSi_2$ | 937 | polymorphic | [4] |

Table 1 continued

| Silicide phase | Melting point (°C) | Nature of fusion | Ref |
|---|---|---|---|
| $Hf_2Si$ | $2430 \pm 30$ | p | [1,2] |
| | $2083 \pm 12$ | p | [4] |
| $Hf_5Si_3$ | $2600 \pm 50$ | p | [1,2] |
| $Hf_3Si_2$ | >2600 | c | [1] |
| | $2650 \pm 50$ | - | [2] |
| $Hf_5Si_4$ | $2480 \pm 20$ | c | [4] |
| HfSi | $2320 \pm 15$ | p | [4] |
| | $2200 \pm 20$ | c | [6,2] |
| $HfSi_2$ | $2142 \pm 15$ | p | [4] |
| | $1900 \pm 10$ | p | [1,2] |
| | $1543 \pm 8$ | p | [4] |
| $Ir_3Si_2$ | 500 - 750 | phase diagram | [10] |
| $\alpha$-$IrSi_3$ to $\beta$-$IrSi_3$ | 900 - 1000 | not available | |
| $Mn_9Si_2$ | 1060 | p | [1] |
| $Mn_3Si$ | 1075 | p | [1,5] |
| $Mn_5Si_3$ | 1285 | c | [1] |
| | 1283.2 | c | [5] |
| | $1300 \pm 20$ | - | [2] |
| MnSi | 1275 | c | [1] |
| | $1270 \pm 10$ | - | [2] |
| $Mn_xSi_{2-x}$ | 1269.6 | c | [5] |
| $Mn_{11}Si_{19}$ | 1152 | p | [1] |
| | 1145 | c | [1] |
| | 1150.2 | p | [5] |
| $Mo_3Si$ | 2025 | p | [1,4] |
| | $2030 \pm 20$ | - | [2] |
| $Mo_5Si_2$ | 2180 | c | [1,4] |
| $Mo_5Si_3$ | $2180 \pm 20$ | - | [2] |
| $MoSi_2$ | $2020 \pm 20$ | c | [1,2,4] |
| $\alpha$-$MoSi_2$ to $\beta$-$MoSi_2$ | 1900 | polymorphic | [4] |
| $Nb_4Si$ | 1950 | p | [1] |
| $Nb_3Si$ | 1980 | p | [4] |
| $Nb_5Si_3$ | 2480 | c | [1] |
| | $2450 \pm 50$ | - | [2] |
| $\alpha$-$Nb_5Si_3$ | 1940 | p | [4] |
| $\beta$-$Nb_5Si_3$ | 2520 | c | [4] |
| $NbSi_2$ | 1930 | c | [1] |
| | 1940 | c | [4] |
| $Ni_4Si$ | 1035 | peritectoid | [4] |
| $Ni_3Si$ | 1170 | p | [4] |
| $Ni_{31}Si_{12}$ | 1242 | c | [4] |
| $Ni_5Si_2$ | 1282 | c | [1] |
| $\theta$-$Ni_2Si$ | 1306 | c | [4] |
| $\delta$-$Ni_2Si$ | 1255 | p | [4] |
| $\beta$-$Ni_2Si$ | 1318 | c | [1] |
| NiSi | 992 | c | [1,4] |
| $\beta$-$NiSi_2$ | 993 | p | [1,4] |
| $\alpha$-$NiSi_2$ to $\beta$-$NiSi_2$ | 981 | polymorphic | [4] |

Table 1 continued

| Silicide phase | Melting point (°C) | Nature of fusion | Ref |
|---|---|---|---|
| OsSi | $1730 \pm 10$ | p | [4] |
| $Os_2Si_3$ | $1840 \pm 20$ | c | [4] |
| $OsSi_2$ | $1640 \pm 20$ | p | [4] |
| $Pd_5Si$ | 818 | c | [5] |
| $Pd_9Si_2$ | 823 | p | [5] |
| $Pd_3Si$ | 960 | p | [1] |
| | 1045 | p | [5] |
| $Pd_2Si$ | 1330 | c | [1] |
| | 1331 | c | [5] |
| PdSi | 1100 | c | [1] |
| | 972 | c | [5] |
| $Pt_3Si$ | ~870 | p | [1] |
| α-$Pt_3Si$ to β-$Pt_3Si$ | $360 \pm 20$ | polymorphic | [4] |
| β-$Pt_{12}Si_5$ | $986 \pm 4$ | p | [4] |
| α-$Pt_{12}Si_5$ to β-$Pt_{12}Si_5$ | 280 | polymorphic | [4] |
| $Pt_7Si_3$ | 986 | p | [1] |
| $Pt_2Si$ | 1100 | c | [1,4] |
| α-$Pt_2Si$ to β-$Pt_2Si$ | $695 \pm 5$ | polymorphic | [4] |
| $Pt_6Si_5$ | 975 | peritectoid | [4] |
| PtSi | 1229 | c | [1,4] |
| $Re_2Si$ | $1810 \pm 10$ | c | [4] |
| $Re_5Si_3$ | 1960 | c | [1] |
| ReSi | 1880 | p | [1] |
| | $1820 \pm 10$ | p | [4] |
| $ReSi_{1.8}$ | $1940 \pm 10$ | c | [4] |
| $ReSi_2$ | 1980 | c | [1] |
| $Rh_2Si$ | 1622 | melting | [5] |
| | $1650 \pm 10$ | c | [4] |
| $Rh_5Si_3$ | 1470 | p | [4,5] |
| $Rh_3Si_2$ | $820 \pm 20$ | p | [4] |
| β-RhSi | 1452 | melting | [5] |
| | $1450 \pm 10$ | c | [4] |
| $Rh_4Si_5$ | $1030 \pm 10$ | p | [4] |
| $Rh_3Si_4$ | $1040 \pm 10$ | p | [4] |
| $Ru_4Si_3$ | 1695 | e | [4] |
| β-RuSi | ~1730 | c | [4] |
| β-$Ru_2Si_3$ | ~1710 | c | [4] |
| $Sc_5Si_3$ | 2060 | c | [5] |
| ScSi | 1785 | p? | [5] |
| β-$ScSi_{2-x}$ | 1260 | p | [5] |

Table 1 continued

| Silicide phase | Melting point (°C) | Nature of fusion | Ref |
|---|---|---|---|
| $Ta_9Si_2$ | 2510 ± 100 | - | [2] |
| $Ta_5Si$ | 2500 | c | [1] |
| $Ta_{4.5}Si$ | 2510 | c | [1] |
| $Ta_3Si$ | 2340 | p | [4] |
| $Ta_2Si$ | 2440 | p | [4] |
| α-$Ta_5Si_3$ to β-$Ta_5Si_3$ | 2160 | polymorphic | [4] |
| β-$Ta_5Si_3$ | 2500 ± 100 | - | [2] |
| | 2550 | c | [2] |
| $Ta_2Si_3$ | 2450 | p | [1] |
| | 2460 ± 100 | - | [2] |
| $TaSi_2$ | 2200 ± 100 | c | [1,2] |
| | 2040 | c | [4] |
| $Ti_3Si$ | 1170 | peritectoid | [4] |
| $Ti_5Si_3$ | 2130 ± 10 | c | [1,2,4] |
| $Ti_5Si_4$ | 1920 ± 20 | p | [1,2,4] |
| TiSi | 1570 ± 10 | p | [1,4] |
| $TiSi_2$ | 1500 ± 10 | c | [1,2,4] |
| $V_3Si$ | 1935 ± 10 | c | [1,2] |
| | 1925 | c | [4] |
| $V_5Si_3$ | 2010 | c | [1,4] |
| | 2125 ± 25 | - | [2] |
| $V_5Si_4$ | 1670 | p | [1] |
| $V_6Si_5$ | 1670 | p | [4] |
| $VSi_2$ | 1680 ± 30 | c | [1,2] |
| | 1677 | c | [4] |
| $W_5Si_3$ | 2370 | p | [1] |
| | 2350 ± 30 | - | [2] |
| | 2320 | c | [4] |
| $WSi_2$ | 2160 ± 30 | c | [1,2,4] |
| $Y_5Si_3$ | 1855 ± 20 | - | [2] |
| | 1875 | c | [5] |
| YSi | controversy | | [5] |
| | 1840 | | [9] |
| | 1835 ± 20 | c or p | [2,9] |
| $Y_3Si_4$ | 1845 ± 20 | - | [2] |
| $Y_3Si_5$ | 1635 | p | [5] |
| | 1640 ± 20 | - | [2] |
| $YSi_2$ | 1520 | | [9] |

Table 1 continued

| Silicide phase | Melting point (°C) | Nature of fusion | Ref |
|---|---|---|---|
| $Zr_4Si$ | 1630 ± 15 | p | [1,2] |
| $Zr_2Si$ | 2110 ± 25 | p | [1,2] |
| | 1925 | p | [4] |
| $Zr_3Si_2$ | 2215 | p | [4] |
| $Zr_5Si_3$ | 2210 ± 25 | p | [1,2] |
| $Zr_4Si_3$ | 2225 | p | [1] |
| $\alpha$-$Zr_5Si_4$ to $\beta$-$Zr_5Si_4$ | 1860 | polymorphic | [4] |
| | 2250 | c | [4] |
| $\beta$-$Zr_5Si_4$ | 2250 | c | [1] |
| $Zr_6Si_5$ | 1860 | polymorphic | [4] |
| | 2250 ± 25 | - | [2] |
| $\alpha$-ZrSi to $\beta$-ZrSi | 2095 | p | [1] |
| ZrSi | 1520 | p | [1] |
| | 2095 ± 25 | - | [2] |
| $ZrSi_2$ | 1520 ± 15 | - | [2] |
| | 1620 | p | [4] |

## C THERMAL LINEAR EXPANSION COEFFICIENT OF TRANSITION METAL SILICIDES

The thermal linear expansion coefficient is listed in alphabetical order and for increasing silicon content for the respective silicide systems in TABLE 2. In several cases a selection of temperatures was made. Whenever found, the expansion coefficient is given for different crystal orientations and in the form $\alpha(T) = a + bT$. The data in [3,8] were obtained using an X-ray diffraction camera with high temperature vacuum heating capability and a Guinier-Hägg geometry.

TABLE 2 Thermal linear expansion coefficient $\alpha$ of transition metal silicides.

| Silicide | $\alpha$ ($10^{-6}$ $K^{-1}$) | Temperature range (°C) | Refs and comments |
|---|---|---|---|
| $Co_3Si$ | 13.4 | 20 - 520 | [1] |
| | 16.6 | 520 - 970 | [1] |
| CoSi | 11.1 ± 1.0 | 20 - 800 | [1] |
| $CoSi_2$ | 10.14 | - | [1] |
| | 9.469 ± 0.007379T | 25 - 927 | [3] |
| $Cr_3Si$ | 10.5 | 20 - 1070 | [1] |
| $Cr_5Si_3$ | 6.0 | 10 - 170 | [1] |
| | 10.6 | 170 - 720 | [1] |
| $Cr_5Si_3$ | 14.2 | 720 - 1070 | [1] |
| CrSi | 11.3 | 20 - 770 | [1] |
| $CrSi_2$ | 8.177 + 0.009903T | 27 - 1127 | [8] in a direction |
| | 8.986 + 7.617 x $10^{-4}$T | 27 - 1127 | [8] in c direction |
| | 25.30 + 0.02055T | 27 - 1127 | [8] volumetric |

Table 2 continued

| Silicide | α ($10^{-6}$ $K^{-1}$) | Temperature range (°C) | Refs and comments |
|---|---|---|---|
| $Fe_3Si$ | 14.4 | 20 - 870 | [1] |
| | 11.0 | 870 - 1070 | [1] |
| $Fe_5Si_3$ | 13.0 + 0.008T | 20 - 600 | [1] |
| FeSi | 15.7 + 0.0025T | 20 - 1000 | [1] |
| $FeSi_2$ | 7.7 + 0.006T | 20 - 900 | [1] |
| β-$FeSi_2$ | 16.697 + 0.001442T | 25 - 950 | [3] in a direction |
| | 0.026 + 0.001673T | 25 - 950 | [3] in b direction |
| | 0.415 + 0.001872T | 25 - 950 | [3] in c direction |
| $FeSi_{2.3}$ | 8.3 + 0.0067T | 20 - 500 | [1] |
| $HfSi_2$ | 10.08 + 0.004920T | 27 - 1127 | [8] in a direction |
| | 8.380 - 5.376 x $10^{-4}$T | 27 - 1127 | [8] in b direction |
| | 4.892 + 0.007143T | 27 - 1127 | [8] in c direction |
| | 23.90 + 0.01108T | 27 - 1127 | [8] volumetric |
| MnSi | 16.3 ± 1.0 | 20 - 800 | [1] |
| $Mn_{15}Si_{26}$ | 5.289 + 0.007900T | 27 - 1127 | [8] in a direction |
| | 9.024 + 0.005006T | 27 - 1127 | [8] in c direction |
| | 19.05 + 0.02131T | 27 - 1127 | [8] volumetric |
| $Mo_3Si$ | 3.4 | 20 - 170 | [1] |
| | 6.5 | 170 - 1070 | [1] |
| $Mo_5Si_3$ | 4.3 | 20 - 270 | [1] |
| | 6.7 | 270 - 1070 | [1] |
| $MoSi_2$ | 8.25 | 20 - 1070 | [1] |
| | 5.617 + 0.003544T | 27 - 1127 | [8] in a direction |
| | 4.115 + 0.005565T | 27 - 1127 | [8] in c direction |
| | 15.30 + 0.01268T | 27 - 1127 | [8] volumetric |
| $Nb_5Si_3$ | 7.3 | 20 - 650 | [1] |
| | 4.6 | 20 - 650 | [1] |
| $NbSi_2$ | 8.4 | 20 - 370 | [1] |
| | 11.7 | 370 - 1070 | [1] |
| | 9.078 + 5.738 x $10^{-4}$T | 27 - 1127 | [8] in a direction |
| | 8.653 + 0.002060T | 27 - 1127 | [8] in c direction |
| | 24.22 + 0.006663T | 27 - 1127 | [8] volumetric |
| $Ni_3Si$ | 9.0 | 20 - 370 | [1] |
| | 11.5 | 370 - 770 | [1] |
| | 14.85 | 770 - 1070 | [1] |
| $Ni_2Si$ | 16.5 | 20 - 870 | [1] |
| | 19.0 | 870 - 1070 | [1] |
| $Ni_{1.04}Si_{1.93}$ | 12.06 | - | [1] |
| $NiSi_2$ | 3.932 + 0.001082T | 25 - 1077 | [3] |
| $OsSi_2$ | 6.483 + 0.003537T | 25 - 1279 | [3] in a direction |
| | 5.871 + 0.003538T | 25 - 1279 | [3] in b direction |
| | 5.709 + 0.003401T | 25 - 1279 | [3] in c direction |
| ReSi | 2.7 | 20 - 220 | [1] |
| | 4.9 | 220 - 1070 | [1] |
| $ReSi_2$ | 6.6 | 20 - 1070 | [1] |
| | 4.239 + 0.004396T | 27 - 1127 | [8] in a direction |
| | 5.604 + 0.003853T | 27 - 1127 | [8] in b direction |
| | 7.530 + 0.001208T | 27 - 1127 | [8] in c direction |
| | 17.39 + 0.009420T | 27 - 1127 | [8] volumetric |

Table 2 continued

| Silicide | α ($10^{-6}$ K$^{-1}$) | Temperature range (°C) | Refs and comments |
|---|---|---|---|
| $Ru_2Si_3$ | 7.384 + 0.005504T | 25 - 777 | [3] in a direction |
| | 4.806 + 0.006310T | 25 - 777 | [3] in b direction |
| | 6.923 + 0.008924T | 25 - 777 | [3] in c direction |
| $Sc_5Si_3$ | 9.8 ± 0.2 | 20 - 450 | [1] |
| | 10.7 ± 0.21 | 450 - 900 | [1] |
| ScSi | 9.4 ± 0.19 | 20 - 580 | [1] |
| | 9.5 ± 0.19 | 580 - 900 | [1] |
| $ScSi_{1.66}$ | 8.6 ± 0.17 | 20 - 600 | [1] |
| | 9.0 ± 0.18 | 600 - 900 | [1] |
| $Ta_5Si_3$ | 5.5 | 20 - 1000 | [1] |
| | 8.0 | 20 - 1000 | [1] |
| | 6.3 | 20 - 1000 | [1] |
| | 6.6 | 20 - 1000 | [1] |
| $TaSi_2$ | 8.9 | 20 - 1000 | [1] |
| | 8.8 | 20 - 1000 | [1] |
| | 9.5 | 20 - 320 | [1] |
| | 10.7 | 320 - 1070 | [1] |
| | 6.783 + 0.004116T | 27 - 1127 | [8] in a direction |
| | 6.127 + 0.005672T | 27 - 1127 | [8] in c direction |
| | 19.67 + 0.01390T | 27 - 1127 | [8] volumetric |
| $Ti_5Si_3$ | 11.0 | 170 - 1070 | [1] |
| | 9.1 | 298 - 473 | [12] grain size 20-50 μm |
| TiSi | 8.8 | 20 - 370 | [1] |
| | 10.4 | 370 - 1070 | [1] |
| $TiSi_2$ | 12.5 | 200 - 1200 | [1] |
| | 6.875 + 0.007262T | 27 - 1127 | [8] in a direction |
| | 6.925 + 0.006344T | 27 - 1127 | [8] in b direction |
| | 5.409 + 0.008196T | 27 - 1127 | [8] in c direction |
| | 18.89 + 0.02212T | 27 - 1127 | [8] volumetric |
| $V_3Si$ | 8.0 | 20 - 620 | [1] |
| | 12.0 | 620 - 820 | [1] |
| | 14.1 | 820 - 1070 | [1] |
| $V_5Si_3$ | 9.5 | 20 - 770 | [1] |
| | 11.1 | 770 - 1070 | [1] |
| $VSi_2$ | 11.2 | 20 - 770 | [1] |
| | 14.65 | 770 - 1070 | [1] |
| | 8.016 + 0.004598T | 27 - 1127 | [8] in a direction |
| | 7.485 + 0.004865 | 27 - 1127 | [8] in c direction |
| | 23.44 + 0.01410T | 27 - 1127 | [8] volumetric |
| $WSi_2$ | 6.25 | 20 - 420 | [1] |
| | 7.90 | 420 - 1070 | [1] |
| | 6.512 + 0.002513T | 27 - 1127 | [8] in a direction |
| | 8.800 + 0.002404T | 27 - 1127 | [8] in c direction |
| | 22.00 + 0.007254T | 27 - 1127 | [8] volumetric |
| $WSi_2$ (co-sputtered thin film on Si and GaAs) | 6.0 (average) | 298 - 598 | [11] |

Table 2 continued

| Silicide | α ($10^{-6}$ $K^{-1}$) | Temperature range (°C) | Refs and comments |
|---|---|---|---|
| $Zr_2Si$ | ~7 | 200 - 1000 | [1] |
| $Zr_5Si_3$ | ~7.2 | 200 - 1000 | [1] |
| $Zr_3Si_2$ | ~9 | 200 - 1000 | [1] |
| $Zr_5Si_4$ | ~10 | 200 - 1000 | [1] |
| ZrSi | ~8 | 200 - 1000 | [1] |
| $ZrSi_2$ | ~8.3 | 200 - 1000 | [1] |
| | 11.08 + 0.003449T | 27 - 1127 | [8] in a direction |
| | 8.634 - 8.807 x $10^{-5}$T | 27 - 1127 | [8] in b direction |
| | 6.022 + 0.005558T | 27 - 1127 | [8] in c direction |
| | 26.65 + 0.007882T | 27 - 1127 | [8] volumetric |

## D THERMAL CONDUCTIVITY OF TRANSITION METAL SILICIDE FILMS

TABLE 3 shows the thermal conductivity of the transition metal silicides and the listing is in alphabetical order. Each silicide is listed in order of increasing Si content. The samples are polycrystalline except for those cases where a crystal orientation is given.

TABLE 3 Thermal conductivity, λ, of the transition metal silicides.

| Silicide phase | Thermal conductivity λ (W $cm^{-1}$ $K^{-1}$) | Temperature (°C) | Ref |
|---|---|---|---|
| CoSi | 0.405 | 65 | [2] in direction <210> |
| | 0.143 | 300 | [2] in direction <210> |
| | 0.124 | 700 | [2] in direction <210> |
| | 0.194 | 1600 | [2] in direction <210> |
| $CoSi_2$ | 0.33 | 60 | [2] in direction <111> |
| | 0.38 | 360 | [2] in direction <111> |
| | 0.719 | 80 | [2] |
| | 0.510 | 350 | [2] |
| CrSi | 0.152 | 80 | [2] |
| | 0.13 | 150 | [2] |
| | 0.133 | 300 | [2] |
| | 0.184 | 1000 | [2] |
| | 0.231 | 1700 | [2] |
| $CrSi_2$ | 0.054 | 300 | [2] |
| | 0.042 | 1000 | [2] |
| | 0.063 | 1200 | [2] |
| FeSi | 0.248 | 65 | [2] in direction <210> |
| | 0.098 | 300 | [2] in direction <210> |
| | 0.109 | 900 | [2] in direction <210> |
| | 0.171 | 1600 | [2] in direction <210> |
| $FeSi_2$ | 0.658 | 70 | [2] |
| | 0.122 | 300 | [2] |
| | 0.037 | 1000 | |
| | 0.080 | 800 - 900 | [6] |
| $FeSi_{2.43}$ | 0.16 | 70 | [2] calculated |
| | 0.097 | 300 | |

Table 3 continued

| Silicide phase | Thermal conductivity λ ($W cm^{-1} K^{-1}$) | Temperature (°C) | Ref |
|---|---|---|---|
| MnSi | 0.085 | 80 | [2] |
| | 0.10 | 300 | [2] |
| | 0.123 | 800 | [2] |
| | 0.152 | 1400 | [2] |
| $MnSi_{1.72}$ | 0.037 | 300 | [2] in direction <100> |
| | 0.018 | 300 | [2] in direction <001> |
| $MnSi_{1.73}$ | 0.075 | 70 | [2] |
| | 0.040 | 300 | [2] |
| | 0.042 | 700 | [2] |
| | 0.060 | 1100 | [2] |
| $Mo_3Si$ | 0.397 | 300 | [2] |
| $Mo_5Si_3$ | 0.217 | 300 | [2] |
| $MoSi_2$ | 0.485 | 300 | [2] |
| TiSi | 0.125 | 300 | [2] |
| | 0.154 | 900 | [2] |
| | 0.167 | 1500 | [2] |
| $TiSi_2$ | 0.405 | 300 | [2] |
| | 0.334 | 900 | [2] |
| | 0.293 | 1500 | [2] |
| $VSi_2$ | 0.502 | 70 | [2] |
| | 0.251 | 300 | [2] |
| | 0.274 | 1100 | [2] |
| | 0.251 | 1600 | [2] |

## E SPECIFIC HEAT OF TRANSITION METAL SILICIDES

TABLE 4 lists the specific heat $c_p$ for the transition metal silicides. The data are presented in alphabetical order and each silicide is listed in order of increasing Si content. The data also represent a selection of temperatures. The most application oriented interval was selected. In a few cases the specific heat is given in the form $c_p = a + bT + cT^2$ for a certain temperature range. The specific heat is given for a mole of respective phase of the silicides.

TABLE 4 Specific heat $c_p$ of transition metal silicides.

| Silicide phase | $c_p$ ($J mol^{-1} K^{-1}$) | Temperature (K) | Ref |
|---|---|---|---|
| $Co_2Si$ | 70.67 | 298.15 | [2] |
| | 81.48 | 500 | |
| | 88.37 | 700 | |
| | 97.31 | 1000 | |
| | 102.99 | 1200 | |
| CoSi | 44.335 | 298.15 | [2] |
| | 50.80 | 500 | |
| | 54.30 | 700 | |
| | 58.50 | 1000 | |
| | 61.00 | 1200 | |

Table 4 continued

| Silicide phase | $c_p$ (J mol$^{-1}$ K$^{-1}$) | Temperature (K) | Ref |
|---|---|---|---|
| $CoSi_2$ | 65.47<br>75.41<br>80.30<br>85.83<br>89.15 | 298.15<br>500<br>700<br>1000<br>1200 | [2] |
| $Cr_3Si$ | 90.16<br>106.68<br>119.18<br>130.70<br>137.42 | 298.15<br>500<br>700<br>1000<br>1200 | [2] |
| $Cr_5Si_3$ | 184.50<br>213.01<br>227.88<br>245.33<br>255.97 | 298.15<br>500<br>700<br>1000<br>1200 | [2] |
| CrSi | 44.88<br>51.36<br>55.40<br>58.78<br>61.03 | 298.15<br>500<br>700<br>1000<br>1200 | [2] |
| $CrSi_2$ | 63.58<br>73.76<br>79.78<br>87.34<br>92.98 | 298.15<br>500<br>700<br>1000<br>1200 | [2] |
| $Fe_3Si$ | 98.30<br>$71.35 + 87.38 \times 10^{-3}T + 148.93 \times 10^{2}T^{2}$ | 293<br>273 - 800 | [1]<br>[1] |
| FeSi | 45.16<br>52.10<br>55.24<br>59.88<br>62.95 | 298.15<br>500<br>700<br>1000<br>1200 | [2] |
| $FeSi_2$ | 66.15<br>$60.96 + 0.017T$ | 293<br>273 - 1238 | [1] |
| $HfSi_2$ | 66.53<br>74.56<br>78.92<br>84.15<br>87.36 | 298.15<br>500<br>700<br>1000<br>1200 | [2] |
| $Mn_3Si$ | 99.94<br>$100.99 + 0.052T - 1.47 \times 10^{4}T^{-2}$ | 293<br>298 - 950 | [1] |
| $Mn_5Si_3$ | 195.49<br>220.61<br>235.28<br>253.56<br>264.99 | 298.15<br>500<br>700<br>1000<br>1200 | [2] |

Table 4 continued

| Silicide phase | $c_p$ (J mol$^{-1}$ K$^{-1}$) | Temperature (K) | Ref |
|---|---|---|---|
| MnSi | 45.93<br>53.15<br>56.96<br>61.45<br>64.20 | 298.15<br>500<br>700<br>1000<br>1200 | [2] |
| $Mn_{0.37}Si_{0.63}$ | 21.73<br>$26.67 + 1.71 \times 10^{-3}T - 4.84 \times 10^5T^{-2}$ | 293<br>298 - 1425 | [1] |
| $Mo_3Si$ | 86.59<br>102.41<br>109.74<br>117.69<br>122.37 | 298.15<br>500<br>700<br>1000<br>1200 | [2] |
| $Mo_5Si_3$ | 205.86<br>221.47<br>229.46<br>238.73<br>244.36 | 298.15<br>500<br>700<br>1000<br>1200 | [2] |
| $MoSi_2$ | 63.79<br>72.29<br>76.84<br>82.27<br>85.60 | 298.15<br>500<br>700<br>1000<br>1200 | [2] |
| $Nb_4Si$ | 117.15<br>121.98<br>131.60<br>147.77<br>158.92 | 298.15<br>500<br>700<br>1000<br>1200 | [2] |
| $Nb_5Si_3$ | 181.40<br>198.53<br>207.64<br>218.45<br>225.07 | 298.15<br>500<br>700<br>1000<br>1200 | [2] |
| $NbSi_2$ | 64.60<br>69.73<br>73.35<br>78.25<br>81.41 | 298.15<br>500<br>700<br>1000<br>1200 | [2] |
| $Ni_2Si$ | 72.57<br>76.21<br>81.14<br>89.02<br>94.38 | 298.15<br>500<br>700<br>1000<br>1200 | [2] |
| NiSi | 45.63<br>$48.74 + 6.11 \times 10^{-3} - 6.53 \times 10^5T^{-2}$ | 293<br>298 - 1265 | [1] |
| $Ni_{0.35}Si_{0.65}$ | 144.13<br>$25.04 + 3.68 \times 10^{-3}T - 3.61 \times 10^5T^{-2}$ | 293<br>298 - 1200 | [1] |

Table 4 continued

| Silicide phase | $c_p$ (J mol$^{-1}$ K$^{-1}$) | Temperature (K) | Ref |
|---|---|---|---|
| $Re_5Si_3$ | 188.45 | 298.15 | [2] |
| | 207.76 | 500 | |
| | 219.55 | 700 | |
| | 234.57 | 1000 | |
| | 244.04 | 1200 | |
| $ReSi_2$ | 65.27 | 298.15 | [2] |
| | 67.90 | 500 | |
| | 71.34 | 700 | |
| | 76.82 | 1000 | |
| | 80.53 | 1200 | |
| $Ta_5Si_3$ | 181.34 | 298.15 | [2] |
| | 195.70 | 500 | |
| | 205.27 | 700 | |
| | 217.93 | 1000 | |
| | 226.03 | 1200 | |
| $TaSi_2$ | 65.34 | 298.15 | [2] |
| | 73.48 | 500 | |
| | 76.80 | 700 | |
| | 80.05 | 1000 | |
| | 81.87 | 1200 | |
| $Ti_5Si_3$ | 126.21 | 298.15 | [2] |
| | 211.32 | 500 | |
| | 237.82 | 700 | |
| | 256.55 | 1000 | |
| | 264.73 | 1200 | |
| TiSi | 22.99 | 298.15 | [2] |
| | 49.78 | 500 | |
| | 57.02 | 700 | |
| | 60.86 | 1000 | |
| | 61.99 | 1200 | |
| $TiSi_2$ | 51.47 | 298.15 | [2] |
| | 72.30 | 500 | |
| | 82.99 | 700 | |
| | 95.42 | 1000 | |
| | 102.96 | 1200 | |
| $V_3Si$ | 91.40 | 298.15 | [2] |
| | 100.13 | 500 | |
| | 105.14 | 700 | |
| | 111.35 | 1000 | |
| | 115.22 | 1200 | |
| $V_5Si_3$ | 171.12 | 298.15 | [2] |
| | 216.57 | 500 | |
| | 228.98 | 700 | |
| | 235.75 | 1000 | |
| | 237.84 | 1200 | |

Table 4 continued

| Silicide phase | $c_p$ (J mol$^{-1}$ K$^{-1}$) | Temperature (K) | Ref |
|---|---|---|---|
| $VSi_2$ | 63.45 | 298.15 | [2] |
| | 71.26 | 500 | |
| | 75.55 | 700 | |
| | 80.74 | 1000 | |
| | 83.94 | 1200 | |
| $W_5Si_3$ | 181.38 | 298.15 | [2] |
| | 195.73 | 500 | |
| | 205.31 | 700 | |
| | 217.99 | 1000 | |
| | 226.10 | 1200 | |
| $WSi_2$ | 64.27 | 298.15 | [2] |
| | 70.91 | 500 | |
| | 74.31 | 700 | |
| | 78.26 | 1000 | |
| | 80.65 | 1200 | |
| $Y_5Si_3$ | 187.1 | 293.8 | [14] |
| $Zr_2Si$ | 75.37 | - | [1] |
| $Zr_5Si_3$ | 188.42 | - | [1,14] |
| ZrSi | 48.15 | - | [1] |
| $ZrSi_2$ | 64.71 | 298.15 | [2] |
| | 62.88 | 500 | |
| | 68.62 | 700 | |
| | 79.97 | 1000 | |
| | 88.10 | 1200 | |

## F ELECTRONIC HEAT CAPACITY

TABLE 5 displays data obtained from low-temperature measurements and shows the electronic contribution to the specific heat $c_p$. The Debye temperature $\Theta_D$ is also given.

TABLE 5 The electronic contribution to the specific heat and the Debye temperature for transition metal silicides.

| Silicide phase | Specific heat $c_p$ (mJ mol$^{-1}$ K$^{-2}$) | Debye temperature $\Theta_D$ (K) | Ref |
|---|---|---|---|
| $CoSi_2$ | $3.70 \pm 0.20$ | $520 \pm 10$ | [7] |
| $MoSi_2$ | $0.57 \pm 0.01$ | $350 \pm 7$ | [7] |
| $NbSi_2$ | $3.45 \pm 0.02$ | $608 \pm 3$ | [7] |
| $Pd_2Si$ | $1.66 \pm 0.02$ | $337 \pm 2$ | [7] |
| $TaSi_2$ | $3.22 \pm 0.03$ | $465 \pm 5$ | [7] |
| $VSi_2$ | $6.05 \pm 0.05$ | $660 \pm 10$ | [7] |
| $Y_5Si_3$ | 22.2 | 348 | [14] |
| $Zr_5Si_3$ | 23.4 | 434 | [14] |

## G CONCLUSION

The compiled data above are based on the presently available reference literature. Various data exist and their validity can sometimes be questionable since limited information only can be found on the sample purity, formation conditions and measurement techniques. We have tried to select data useful from an application oriented approach and from respected sources. The majority of data are given for bulk samples of mostly polycrystalline structure. Using these data in thin film structures should be possible but with some caution taken in the particular application.

## REFERENCES

[1] G.V. Samsonov, I.M. Vinitskii [ *Handbook of Refractory Compounds* (IFI/Plenum Data Company, New York, 1980) ]

[2] T.Y. Kosolapova [ *Handbook of High Temperature Compounds: Properties, Production, Applications* (Hemisphere Publishing Corporation, 1990) ]

[3] M. Andersson-Söderberg [ Master's Thesis, Uppsala University, Sweden (1986) ]

[4] T.B. Massalski [ *Binary Alloy Phase Diagrams 2nd edition* vol.1-3 (ASM, 1990) ]

[5] H. Okamoto [ *Binary Alloy Phase Diagrams Updating Service* (ASM Int., 1992) ]

[6] C.A. Dimitriadis, J.H. Werner [ *J. Appl. Phys. (USA)* vol.68 no.1 (1990) p.93-6 and reference 22 therein ]; G. Waldecker, H. Meinold, U. Birkholz [ *Phys. Status Solidi A (Germany)* vol.15 (1973) p.143 ]

[7] F. Nava et al [ *Mater. Sci. Rep. (Netherlands)* vol.9 (1993) p.141-200 ]

[8] I. Engström, B. Lönnberg [ *J. Appl. Phys. (USA)* vol.63 (1988) p.4476-84 ]

[9] M.-A. Nicolet, S.S. Lau [ *VLSI Electronics: Microstructure Science* vol.6, Eds N.G. Einspruch, G.B. Larrabee (Academic Press Inc., 1983) ch.6 p.329-464 ]

[10] I. Engström, E. Zdansky [ *Acta Chem. Scand. A (Denmark)* vol.36 no.3 (1982) p.857-8 ]

[11] M.-L. Ger, R.B. Brown [ *Mater. Res. Soc. Symp. Proc. (USA)* vol.276 (1992) p.283-94 ]

[12] G. Frommeyer, R. Rosenkranz, C. Lüdecke [ *Z. Met.kd. (Germany)* vol.81 no.5 (1990) p.307-13 ]

[13] A.B. Gokhale, G.J. Abbaschian [ *J. Phase Equilibria* vol.12 (1991) p.493-8 ]

[14] V.N. Safonov, P.V. Gel'd, N.I. Sychev, G.I. Kalishevich [ *Russ. J. Phys. Chem. (UK)* vol.59 (1985) p.1273-5 ]

# 1.4 Basic physical properties of RE silicides

Xu Zhenjia (C.C. Hsu)

June 1994

## A INTRODUCTION

The properties given in this Datareview have been compiled from data in the open literature. Most of the results reported have been measured on bulk silicides and not on thin silicide films of the order of or smaller than 1000 Å as in electronics applications. This fact should be borne in mind when using the data for thin films of RE silicides. Because of the very limited studies on the RE silicides, the available information for the properties of RE silicides is far from complete. In spite of this limitation, it is hoped that these properties will prove useful since, to the author's knowledge, no other compilation of the various properties of RE silicides has been published collectively before.

## B CRYSTALLOGRAPHIC PROPERTIES

### B1 Crystal Structure

The $RESi_2$ compounds are the largest family among the RE silicides. They crystallise in three different types: tetragonal lattice of the $ThSi_2$ type (I4/amd space group), hexagonal lattice of the $AlB_2$ type (P6/mmm space group) and orthorhombic lattice of the $GdSi_2$ type (Imma space group). The orthorhombic lattice of the $GdSi_2$ structure should correspond exactly to $RESi_2$. The major open question concerns the composition of these phases, the available data being not sufficient to clarify this point. When the structure adopted is of the $AlB_2$ type, it has been recognised by several authors that the phases have random defects in the Si sublattice.

The $ThSi_2$ type structure, which appears mainly for light RE silicides from La to Eu, is shown in FIGURE 1.

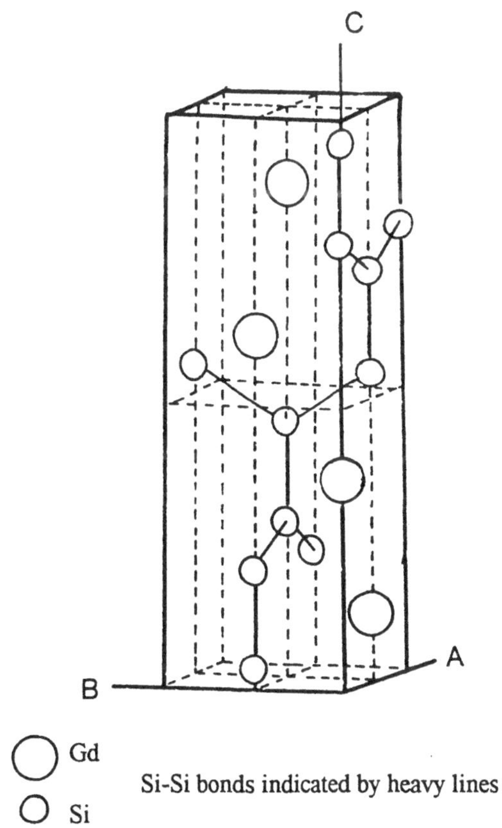


FIGURE 1 Tetragonal structure of $RESi_2$ ($ThSi_2$ type) (after [4]).

The atomic positions of the 4 RE atoms and 8 silicon atoms are [4],

| | | | | |
|---|---|---|---|---|
| 4 RE in | $0\,0\,0$ | $0\,\frac{1}{2}\,\frac{1}{4}$ | $\frac{1}{2}\,\frac{1}{2}\,\frac{1}{2}$ | $\frac{1}{2}\,0\,\frac{3}{4}$ |
| 8 Si in | $0\,0\,Z$ | $0\,0\,Z$ | $0\,\frac{1}{2}\left(\frac{1}{4}+Z\right)$ | $0\,\frac{1}{2}\left(\frac{1}{4}-Z\right)$ |
| | $\frac{1}{2}\,\frac{1}{2}\left(\frac{1}{2}+Z\right)$ | $\frac{1}{2}\,\frac{1}{2}\left(\frac{1}{2}-Z\right)$ | $\frac{1}{2}\,0\left(\frac{3}{4}+Z\right)$ | $\frac{1}{2}\,0\left(\frac{3}{4}-Z\right)$ |

The $AlB_2$ type has been shown to be a representative of a transposition structure of the $ThSi_2$ type and appears only for heavy RE silicides from Gd to Lu (the yttrium subgroup), as shown in FIGURE 2 [5].

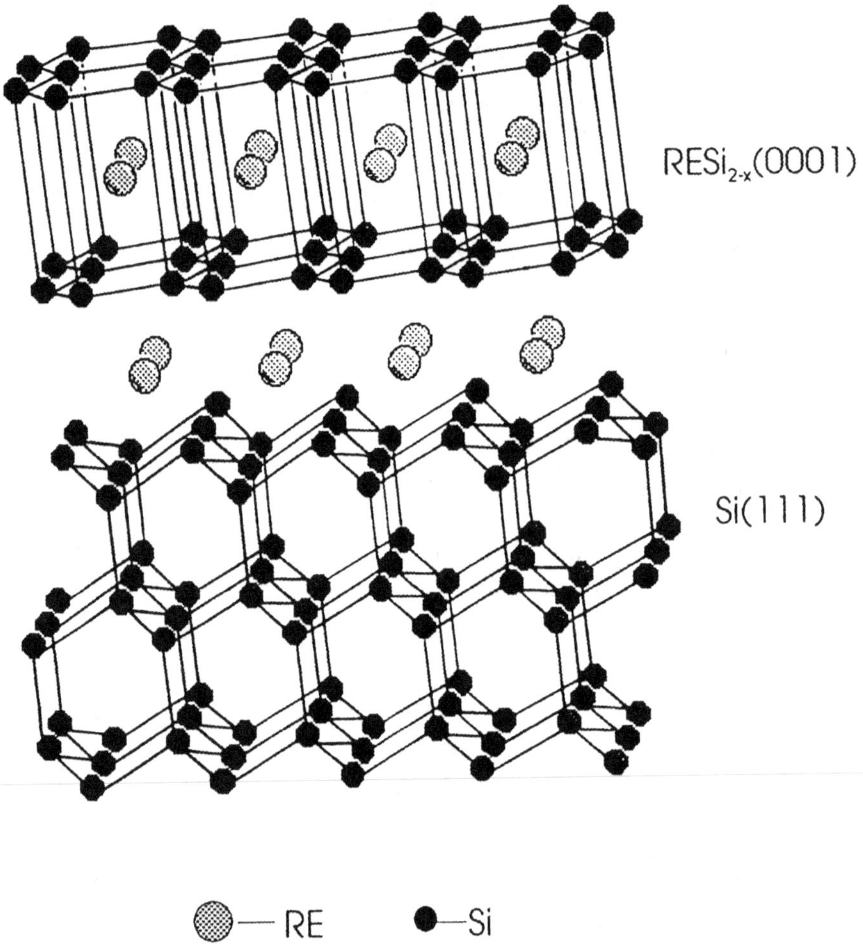


FIGURE 2 $AlB_2$ type structure of $RESi_{2-x}$.

$RESi_2$ compounds of the $AlB_2$ type are composed of alternating layers of silicon and RE atoms, but leaving a number of vacant sites. In fact, the stoichiometry for perfect $AlB_2$ RE silicides should be $RESi_2$, but the actual compositions vary between 1.7 and 1.85 atoms per RE atom. In most RE silicides, no fixed lattice sites are vacant. This hexagonal lattice is closely related (in both the lattice type and the lattice parameter) to the silicon lattice. This means that during deposition of RE silicide thin films onto a silicon substrate it is possible to obtain a sharp, stable and defect-free RE disilicide/Si interface.

The $GdSi_2$ type can be considered as an orthorhombic deformation of the tetragonal $ThSi_2$ type or distorted $AlB_2$ type. Silicon-rich Yb silicides display the distorted $AlB_2$ type structure for compositions like $YbSi_{1.8}$ whilst for slightly lower Si concentrations the crystal structure is of the hexagonal $Th_3Pd_5$ type. $LaSi_2$ has a homogeneity range from $LaSi_2$ accompanied by a change of structure from the $ThSi_2$ to the $GdSi_2$ type.

The light lanthanide silicides $Ln_5Si_3$, where Ln = La, Ce, Pr or Nd, are described as tetragonal with the $Cr_5B_3$ structure [8]. In contrast, lanthanide silicides from samarium to lutetium with this composition crystallise in the hexagonal $Mn_5Si_3$-type structure. The light lanthanides have not been definitely established, probably because the equilibrium is difficult to achieve and a strong composition dependence exists [8].

## B2 Crystallographic Data for RE Silicides

TABLES 1 to 6 show the lattice parameters, structure type and space group of the various RE silicides.

TABLE 1 $RX_2$ silicides with tetragonal $ThSi_2$ type structure [1].

| Silicide | a (Å) | c (Å) |
|---|---|---|
| $LaSi_2$ | 4.295 | 13.75 |
| $CeSi_2$ | 4.192 | 13.86 |
| $PrSi_2$ | 4.22 | 13.71 |
| $NdSi_2$ | 4.111 | 13.56 |
| $SmSi_2$ | 4.08 | 13.51 |
| $EuSi_2$ | 4.29 | 13.66 |
| $GdSi_2$ | 4.10 | 13.61 |
| $DySi_2$ | 4.03 | 13.38 |

TABLE 2 $RX_2$ silicides with orthorhombic $GdSi_2$ type structure [1].

| Silicide | a (Å) | b (Å) | c (Å) |
|---|---|---|---|
| $LaSi_2$ | 4.271 | 4.182 | 14.035 |
| $CeSi_2$ | 4.19 | 4.13 | 13.92 |
| $PrSi_2$ | 4.20 | 4.16 | 13.76 |
| $NdSi_2$ | 4.155 | 4.125 | 13.67 |
| $SmSi_2$ | 4.105 | 4.035 | 13.46 |
| $GdSi_2$ | 4.09 | 4.01 | 13.44 |
| $TbSi_2$ | 4.057 | 3.97 | 13.375 |
| $DySi_2$ | 4.04 | 3.945 | 13.34 |
| $HoSi_2$ | 4.03 | 3.944 | 13.30 |

TABLE 3 $RX_2$ silicides with the hexagonal $AlB_2$ structure.

| Silicide | a (Å) | c (Å) | Ref |
|---|---|---|---|
| $GdSi_{2-x}$ | 3.877 | 4.172 | [5] |
| | 3.877 | 4.172 | [9,10] |
| $TbSi_{2-x}$ | 3.847 | 4.146 | [5] |
| | 3.847 | 4.146 | [9,10] |
| $TbSi_2$ | 3.84 | 4.14 | [2] |
| $DySi_{2-x}$ | 3.831 | 4.121 | [5] |
| | 3.831 | 4.121 | [9,10] |
| $DySi_2$ | 3.83 | 4.12 | [2] |
| $HoSi_{2-x}$ | 3.816 | 4.107 | [5] |
| | 3.816 | 4.107 | [9,10] |
| $HoSi_2$ | 3.80 | 4.10 | [2] |
| | 3.81 | 4.10 | [6] |
| $ErSi_{2-x}$ | 3.799 | 4.090 | [5] |
| | 3.798 | 4.088 | [9,18] |
| $ErSi_2$ | 3.782 | 4.090 | [5] |
| $TmSi_{2-x}$ | 3.773 | 4.070 | [5] |
| | 3.768 | 4.070 | [9,10] |
| $TmSi_2$ | 3.768 | 4.071 | [6] |
| $YbSi_{2-x}$ | 3.771 | 4.098 | [5] |
| | 3.784 | 4.098 | [9,10] |
| $YbSi_2$ | 3.77 | 4.10 | [6] |
| $LuSi_{2-x}$ | 3.745 | 4.050 | [5] |
| | 3.747 | 4.046 | [9,10] |
| $LuSi_2$ | 3.745 | 4.050 | [6] |

TABLE 4 RX silicides.

| Silicide | Type | Space group | a (Å) | b (Å) | c (Å) | Ref |
|---|---|---|---|---|---|---|
| LaSi | Orthor FeB | Pnma | 8.404 | 4.010 | 6.059 | [6] 19-661 |
| CeSi | Orthor FeB | Pnma | 8.302 | 3.962 | 5.964 | [6] 18-320 |
| PrSi | Orthor FeB | Pnma | 8.243 | 3.941 | 5.918 | [6] 31-1146 |
| NdSi | Orthor FeB | Pnma | 8.158 | 3.918 | 5.887 | [6] 31-883 |
| SmSi | Orthor FeB | Pnma | 8.13 | 3.90 | 5.83 | [6] 18-320 |
| EuSi | Orthor CrB | Cmcm | 4.694 | 11.14 | 3.981 | [6] 23-248 |
| GdSi | Orthor FeB | Pnma | 8.00 | 3.85 | 5.73 | [6] 18-320 |
| TbSi | Orthor FeB | Pnma | 7.97 | 3.82 | 5.69 | [6] 18-320 |
| DySi | Orthor FeB | Pnma | 7.87 | 3.80 | 5.65 | [6] 18-320 |
| HoSi(LT) | Orthor CrB | Cmcm | 4.228 | 10.429 | 3.801 | [6] 19-561 |
| HoSi(HT) | Orthor FeB | Pnma | 7.808 | 3.801 | 5.633 | [6] 19-562 |
| ErSi | Orthor FeB | Cmcm | 4.19 | 10.40 | 3.79 | [6] 18-511 |
| TmSi | Orthor CrB | Cmcm | 4.18 | 10.35 | 3.77 | [6] 18-511 |
| YbSi | Orthor CrB | Cmcm | 4.178 | 10.31 | 3.768 | [6] 18-511 |
| LuSi | Orthor CrB | Cmcm | 4.15 | 10.24 | 3.75 | [6] 18-511 |

TABLE 5 $RE_5Si_3$ silicides.

| Silicide | Type | Space group | a (Å) | c (Å) | Ref |
|---|---|---|---|---|---|
| $La_5Si_3$ | | $P4_12_12$ | 8.110 | 15.41 | [8] |
| | Tetra. $Cr_5B_3$ | | 7.953 | 14.04 | [6] 20-555 |
| $Ce_5Si_3$ (I) | | $P4_12_12$ | 7.971 | 14.97 | [8] |
| | | | 7.89 | 13.77 | [6] 19-297 |
| (II) | | 4/mmm (Lau Group) | 11.01 | 4.03 | [8] |
| $PrSi_3$ | Tetra. $Cr_5B_3$ | $P4_12_12$ | 7.81 | 13.75 | [6] 19-1018 |
| $Nd_5Si_3$ (I) | Tetra. (1450°C) | | 7.651 | 4.541 | [8] |
| (II) | Hexa. (950°C) | | 15.045 | 6.623 | [8] |
| (III) | Tetra. (850°C) | | 7.832 | 14.683 | [8] |
| (IV) | Orth. (800°C) | | a (7.621); b (15.31); c (7.930) | | [8] |
| $Sm_5Si_3$ | Hexa. $Mn_5Si_3$ | $P6_3/mcm$ | 8.56 | 6.45 | [1] |
| $Gd_5Si_3$ | Hexa. $Mn_5Si_3$ | $P6_3/mcm$ | 8.510 | 6.390 | [1] |
| | | | 8.507 | 6.402 | [6] 24-1288 |
| $Tb_5Si_3$ | Hexa. $Mn_5Si_3$ | $P6_3/mcm$ | 8.42 | 6.29 | [1] |
| | | | 8.426 | 6.330 | [6] 26-1496 |
| $Dy_5Si_3$ | Hexa. $Mn_5Si_3$ | $P6_3/mcm$ | 8.37 | 6.26 | [1] |
| | | | 8.392 | 6.284 | [6] 24-1367 |
| $Ho_5Si_3$ | Hexa. $Mn_5Si_3$ | $P6_3/mcm$ | 8.34 | 6.22 | [1] |
| | | | 8.341 | 6.251 | [6] 26-748 |
| $Er_5Si_3$ | Hexa. $Mn_5Si_3$ | $P6_3/mcm$ | 8.293 | 6.207 | [1] |
| | | | 8.291 | 6.220 | [6] 26-613 |
| $Tm_5Si_3$ | Hexa. $Mn_5Si_3$ | $P6_3/mcm$ | 8.25 | 6.18 | [1] |
| | | | 8.240 | 6.163 | [6] 26-1497 |
| $Yb_5Si_3$ | Hexa. $Mn_5Si_3$ | $P6_3/mcm$ | 8.23 | 6.19 | [1] |
| | | | 8.265 | 6.276 | [6] 24-1025 |
| $Lu_5Si_3$ | Hexa. $Mn_5Si_3$ | $P6_3/mcm$ | 8.21 | 6.14 | [1] |
| | | | 8.241 | 6.130 | [6] 26-868 |

Tetra. = Tetragonal. Orth. = Orthorhombic. Hexa. = Hexagonal.

TABLE 6 $RE_3Si_2$ silicides [2].

| Silicide | a (Å) | c (Å) |
|---|---|---|
| $La_3Si_2$ | 7.87 | 4.50 |
| $Ce_3Si_2$ | 7.79 | 4.36 |
| $Pr_3Si_2$ | 7.75 | 4.38 |

# C THERMOMECHANICAL PROPERTIES

## C1 Melting Points of RE Silicides

TABLE 7 Melting points of RE silicides [2].

| Silicide | Melting point (°C) |
|---|---|
| YSi | 1870 |
| $YSi_2$ | 1520 |
| $Y_5Si_3$ | 1895 |
| $Y_3Si_5$ | 1635 |
| $LaSi_2$ | 1520 |
| CeSi | 1530 |
| $CeSi_2$ | 1420 |
| $NdSi_2$ | 1525 |
| $EuSi_2$ | 1500 |
| $GdSi_2$ | 1540 |
| $DySi_2$ | 1550 |

## C2 Microhardness of RE Silicides

TABLE 8 Microhardness of RE silicides [2].

| Silicide | Microhardness (kg/mm$^2$) |
|---|---|
| ScSi | 1005 |
| $Sc_5Si_3$ | 973 |
| $Sc_3Si_5$ | 897 |
| YSi | 1106 |
| $YSi_2$ | 804 |
| $Y_3Si_5$ | 845 |
| LaSi | 632 |
| $LaSi_2$ | 476 |
| $La_3Si_2$ | 525 |
| $La_5Si_3$ | 592 |
| CeSi | 628 |
| $CeSi_2$ | 520 |
| $Ce_3Si_2$ | 460 |
| $Ce_5Si_3$ | 597 |
| $PrSi_2$ | 875 |
| $NdSi_2$ | 698 |

## C3 Density of RE Silicides

TABLE 9 Density of RE silicides.

| Silicide | Density ($g/cm^3$) | | Ref |
|---|---|---|---|
| | $D_x$ | $D_m$ | |
| $LaSi_2$(T) | 5.14 | 5.05 | [3] |
| $LaSi_2$(O) 800°C | 5.166 | | [3] |
| LaSi(O) | 5.386 | | [3] |
| $CeSi_2$(T) | 5.45 | | [2] |
| CeSi | 5.67 | | [2] |
| $PrSi_2$(O) | 5.38 | | [2] |
| $PrSi_2$(T) | 5.64 | | [2] |
| $NdSi_2$(O) | 5.62 | | [2] |
| $NdSi_2$(T) | 5.84 | | [2] |
| $SmSi_2$(O) | 6.13 | | [2] |
| $SmSi_2$(T) | 6.26 | | [2] |
| $EuSi_2$(T) | 5.50 | | [2] |
| $GdSi_2$(O) | 6.43 | | [2] |
| $GdSi_2$(H) | 6.19 | | [2] |
| $DySi_2$(O) | 6.18 | | [2] |
| $DySi_2$(H) | 6.68 | | [2] |
| $ErSi_2$ | 7.31 | | [2] |
| $TmSi_2$ | 7.48 | | [2] |
| $YbSi_2$ | 7.54 | | [2] |
| $LuSi_2$ | 7.81 | | [2] |

T - tetragonal; O - orthorhombic; H - hexagonal; $D_x$ - density calculated from the X-ray data; $D_m$ - density as measured.

## C4 Heat of Formation of RE Silicides

TABLE 10 Heat of formation of RE silicides [2].

| Silicide | $H^0_{298}$ (kcal/mole) |
|---|---|
| YSi | -32.2 |
| LaSi | -30.0 |
| $LaSi_2$ | -44.4 |
| $CeSi_2$ | -50.0 |

When considering the thin film formation of RE silicides, some thermodynamical parameters, such as heat of formation, are of great importance. Unfortunately, to our knowledge there exist only very limited thermodynamical data for RE silicides. Fujimori et al [11] estimated the heat of formation $H_f$ for bulk RE disilicides. The authors assumed that $H_f$ of the divalent RE bulk disilicide per mole metal atoms ($H_f^{II}$) was independent of the atomic number and equal to the heat of formation of Ca silicide:

$$H_f^{II} = H_f(CaSi_2) = -362 \text{ kcal/mole}.$$

Based on the estimation from two methods, $H_f$ of trivalent RE bulk disilicides per mole metal atoms ($H_f^{III}$) is also estimated to be independent of the atomic number and is equal to -462 kcal/mole.

## C5 Coefficient of Thermal Linear Expansion of RE Silicide

TABLE 11 Coefficient of thermal linear expansion of RE silicide [3].

| Silicide | Temp (K) | $\Delta L/L_o$ (%) | $\alpha$ ($10^{-6}$ $K^{-1}$) |
|---|---|---|---|
| $LaSi_2$ | 293 | 0.000 | |
| | 673 | 0.290 | |
| | 1073 | 0.673 | |
| | 1338 | 0.972 | |
| | 293 | 0.000 | |
| | 843 | 0.4218 | |
| | 293 - 843 | | 7.8 |
| | 843 - 1343 | | 11.0 |

## REFERENCES


[1] K.A. Gschneider, L.-R. Eyring (Eds) [ *Handbook on the Physics and Chemistry of Rare Earths* (North-Holland Publishing Company, 1978) vol.1 *Metals*; vol.2 *Alloy and Intermetallics* (1979); and references therein ]

[2] Department of Metals, Zhongshan University [ *Physical and Chemical Constants of Rare Earths* (Industry of Metallurgy Press, Beijing, 1978); and references therein ]

[3] M.-A. Nicolet, S.S. Lau [ *Formation and Characterization of Transition Metal Silicides* Eds N.G. Einspruch, G.B. Larrabee *VLSI Electronics-Microstructure Science* ch.6 and references therein ]

[4] V.G. Brauer, H. Hagg [ *Z. Anorg. Allg. Chem. (Germany)* vol.267 (1952) p.198-212 ]; A. Perri, I. Binder, B. Bos [ *J. Phys. Chem. (USA)* vol.63 (1959) p.616-9 ]

[5] V.M. Koleshko, V.F. Belitoky, A.A. Khodin [ *Thin Solid Films (Switzerland)* vol.141 (1986) p.277-85 ]

[6] [ ASTM. Joint Committee on Powder Diffraction Standards, 1988 ]

[7] E.I. Gladyshevskii, P.I. Krypyakevich [ *Izv. Akad. Nauk SSSR Neorg. Mater. (Russia)* vol.1 (1965) p.702 ]

[8] G.Y.M. Al-Shahery, I.J. McColm [ *J. Less-Common Met. (Switzerland)* vol.98 (1984) p.L5-8 ]

[9] J.A. Knapp, S.T. Picraux [ *Appl. Phys. Lett. (USA)* vol.48 (1986) p.466-8 ]

[10] A. Iandelli, A. Palenzena, G.L. Olcese [ *J. Less-Common Met. (Switzerland)* vol.64 (1986) p.213-20 ]

[11] A. Fujimori, M. Grioni, J.H. Weaver [ *Phys. Rev. B (USA)* vol.33 (1986) p.726-35 ]

CHAPTER 2

# GROWTH AND PHASE TRANSFORMATIONS

# 2.1 Thermodynamic parameters of TM silicides

R.W. Mann and L.A. Clevenger

June 1995

## A INTRODUCTION

Thermodynamics deals primarily with how energy and matter interact and the prediction of what reactions are energetically possible. To predict what reactions are energetically favourable, one must have the necessary thermodynamic parameters. As many of the practical applications where silicides are used involve thin films, where surface energies and thin film stress can play a significant role, these effects must also be taken into account. It is important to keep in mind that available thermodynamic parameters are typically given for bulk conditions at equilibrium rather than for thin film interactions.

## B HEAT OF FORMATION AND MELTING POINT DATA FOR TM SILICIDES

In the following compilation heats of formation at 25°C are converted to kcal $mole^{-1}$. The heat of formation data are taken from a range of sources [1,2,15-20]. The melting point data are taken from [21]. Silicides are listed only when sufficient data exist.

TABLE 1 Heats of formation (25°C) and melting points of transition metal silicides.

| Transition metal | Silicide phase | $-\Delta H_f$ (kcal $mole^{-1}$) | $T_M$ (°C) |
|---|---|---|---|
| Ti | $Ti_3Si$ | - | 1170 |
| | $Ti_5Si_3$ | 138.4[a], 138.5[h] | 2130 |
| | $Ti_5Si_4$ | - | 1920 |
| | TiSi | 31.0[a,h] | 1570 |
| | $TiSi_2$ | 32.1[a,h] | 1500 |
| Zr | $Zr_3Si$ | - | 1650 |
| | $Zr_2Si$ | 50.1[a,c], 81.0[h] | 1925 |
| | $Zr_5Si_3$ | 146.7[h] | 2180 |
| | $Zr_3Si_2$ | - | 2215 |
| | $Zr_5Si_4$ | - | 2250* |
| | ZrSi | 37[a], 35.5[c], 35.3[h] | 2210* |
| | $ZrSi_2$ | 38.1[a], 55.5[c] | 1620 |
| Hf | $Hf_2Si$ | 45[g] | 2083 |
| | $Hf_3Si_2$ | - | 2480 |
| | $Hf_5Si_3$ | 133.5[g] | NP |
| | $Hf_5Si_4$ | - | 2320 |
| | HfSi | 34[g] | 2142 |
| | $HfSi_3$ | 54[g] | 1543 |

* for high temperature phase. Reference code: [a] = [15]; [b] = [2]; [c] = [16]; [d] = [17]; [e] = [18]; [f] = [19]; [g] = [1]; [h] = [20]; NP = not present in the equilibrium phase diagram.

Table 1 continued

| Transition metal | Silicide phase | $-\Delta H_f$ (kcal $mole^{-1}$) | $T_M$ ($^oC$) |
|---|---|---|---|
| V | $V_3Si$ | 27.2[a], 27.9[g], 36[h] | 1925 |
| | $V_2Si$ | 73.8[a] | NP |
| | $V_5Si_3$ | 96[a,g], 110.5[h] | 2010 |
| | $V_6Si_5$ | - | 1670 |
| | $VSi_2$ | 73.8[a], 75[g], 30[h] | 1677 |
| Nb | $Nb_3Si$ | - | 1980 |
| | $Nb_5Si_3$ | 63.2[a], 87.2[b] | 2520* |
| | $NbSi_2$ | 21.9[a], 32.1[b] | 1940 |
| Ta | $Ta_3Si$ | 36.9[g] | 2340 |
| | $Ta_2Si$ | 30.3[a], 30[g,h] | 2440 |
| | $Ta_5Si_3$ | 72[a], 76[b], 80[h] | 2550* |
| | $TaSi_2$ | 24[a], 27.9[b], 26.1,28.5[g,h] | 2040 |
| Cr | $Cr_3Si$ | 33[a], 30[b], 33[g], 22[h] | 1770 |
| | $Cr_5Si_3$ | 78.4[a], 64[b], 78[g], 50.5[h] | 1680* |
| | CrSi | 19[a], 15[b], 12.7[h] | 1413 |
| | $CrSi_2$ | 28.8,23.1[b] | 1490 |
| Mo | $Mo_3Si$ | 23.2[a], 22.4[b] | 2025 |
| | $Mo_5Si_3$ | 68[a,b], 74.1[h] | 2180 |
| | $MoSi_2$ | 27.9[a], 31.5[b], 26.1[b], 31.5[c,h] | 2020* |
| W | $W_5Si_3$ | 22.5[a], 21.9[b], 22.0[c], 22.2[h] | 2320 |
| | $WSi_2$ | 36.9[g] | 2160 |
| Mn | $Mn_3Si$ | 25.6[a], 19.0[h] | 1070* |
| | $Mn_5Si_3$ | 48[a,h] | 1300 |
| | MnSi | 26.6[a], 14.5[h] | 1276 |
| | $MnSi_2$ | 7.8[a] | 1155 |
| Re | $Re_2Si$ | - | 1810 |
| | $Re_5Si_3$ | 38.4[a] | NP |
| | ReSi | 10.2[a], 30[c] | 1820 |
| | $ReSi_2$ | 16.5[a] | 1940 |
| Fe | $Fe_3Si$ | 22.4[a] | NP |
| | $Fe_2Si$ | - | 1212 |
| | $Fe_5Si_3$ | 36.8[a] | 1060 |
| | FeSi | 19.2[a], 17.6[b], 18.4[h] | 1410 |
| | $FeSi_2$ | 16.8[a], 18.6[b] | 1220* |
| Ru | $RuSi_2$ | 16[f] | 1550 |
| | $Ru_5Si_3$ | - | 1550 |
| | $Ru_4Si_3$ | - | 1695 |
| | RuSi | 16[a], 15.9[f] | 1750* |
| | $Ru_2Si_3$ | 32[h] | 1171* |
| Os | OsSi | 15.6[a] | 1730 |
| | $Os_2Si_3$ | 49.5[h] | 1840 |
| | $OsSi_2$ | 24.6[h] | 1640 |
| Co | $Co_3Si$ | - | 1210 |
| | $Co_2Si$ | 27.6[a,b] | 1334* |
| | CoSi | 24[a] | 1460 |
| | $CoSi_2$ | 24.6[a] | 1326 |

* for high temperature phase. Reference code: [a] = [15]; [b] = [2]; [c] = [16]; [d] = [17]; [e] = [18]; [f] = [19]; [g] = [1]; [h] = [20]; NP = not present in the equilibrium phase diagram.

Table 1 continued

| Transition metal | Silicide phase | $-\Delta H_f$ (kcal mole$^{-1}$) | $T_M$ ($^oC$) |
|---|---|---|---|
| Rh | $Rh_5Si_3$ | 55[f] | 1470 |
| | $Rh_2Si$ | 18.4[f] | 1650 |
| | RhSi | 16.2[b] | 1450* |
| Ir | $Ir_3Si$ | 36.7[d] | - |
| | $Ir_3Si_4$ | 52.2[d] | - |
| | $Ir_4Si_5$ | 68.4[d] | - |
| | $Ir_3Si_5$ | 55.2[d] | - |
| | $IrSi_2$ | 18.3[d] | - |
| | $IrSi_3$ | 18.4[d] | - |
| Ni | $Ni_3Si$ | 35.6 | 1170* |
| | $Ni_2Si$ | 31.5[a,b,c], 34.1[h] | 1255 |
| | $Ni_5Si_2$ | 72.1[a] | 1230 |
| | $Ni_3Si_2$ | 53.5[a,c], 55.5[h] | 845* |
| | NiSi | 20.4[a], 20.6[b], 20.5[c], 21.4[h] | 992 |
| | $NiSi_2$ | 20.7[a], 22.5[h] | 993* |
| Pd | $Pd_3Si$ | 20.7[f] | 1070 |
| | $Pd_2Si$ | 20.8[f], 20.7[c] | 1394 |
| | PdSi | 12.9[f], 13.8[b] | 901 |
| Pt | $Pt_2Si$ | 20.7[b,e] | 1100* |
| | PtSi | 15.7[e], 15.8[b] | 1229 |

* for high temperature phase. Reference code: [a] = [15]; [b] = [2]; [c] = [16]; [d] = [17]; [e] = [18]; [f] = [19]; [g] = [1]; [h] = [20]; NP = not present in the equilibrium phase diagram.

## C HEAT OF FORMATION RELATIONSHIP TO THE SCHOTTKY BARRIER HEIGHT

The relationship between the Schottky barrier height, $\Phi_B$, and the heat of formation, $\Delta H_f$, was first shown by Andrews and Philips [18]. The heat of formation, when normalised per metal atom per formula unit, was found to give a remarkably good straight line fit where

$$\Phi_B = 0.83 - 0.18 \times \Delta H_f$$

The barrier height $\Phi_B$ is of considerable importance in the formation of contacts between the silicide and the silicon. For relatively high (>1 x $10^{19}$ atoms cm$^{-3}$) doping levels in the silicon, the specific contact resistance between the silicide and the silicon can be expressed as:

$$R_c = C_1 \exp\left(\frac{\Phi_{Bn}}{\sqrt{N_D}}\right)$$

where $\Phi_{Bn}$ is the barrier height on n-type silicon, $C_1$ is a constant and $N_D$ is the substrate doping concentration [22].

## D HEAT OF FORMATION RELATIONSHIP TO THE Si/TM RATIO

The heat of formation (when normalised per metal atom) is observed to increase as the Si/TM ratio increases [1]. With the exception of Ir, which forms a stable $IrSi_3$ phase, a Si/TM ratio of two is the most silicon-rich phase which forms. As the Si/TM ratio increases above one the change in $-\Delta H_f$ with Si/TM becomes smaller, although $VSi_2$ and $HfSi_2$ appear to be exceptions to this rule and do exhibit a significantly more negative $-\Delta H_f$ as the silicon ratio goes from one to two. As the Si/TM ratio increases above one the change in $-\Delta H_f$ with Si/TM increases only slightly. This is true, for example, in the case of Ti, Zr, Co, Ni and Fe silicides.

## E THERMODYNAMIC ASPECTS OF TM SILICIDE THIN FILM THERMAL STABILITY

The thermal stability of the TM silicides has received attention particularly for silicides used in thin film applications [4,23-32]. Thermal instability or morphological degradation can occur as the TM silicide/silicon system attempts to migrate, via thermally activated diffusion, to its lowest energy configuration. Depending upon the surface energies, film thickness and the grain size, this minimum energy or equilibrium configuration can result in either a continuous or discontinuous silicide film on the substrate material. For the case of $TiSi_2$ on a silicon substrate, given sufficient thermal energy, silicon diffuses through the silicide and epitaxially precipitates out at points of highest interface curvature, i.e. at the silicide grain boundaries. The surface or interfacial energy, the grain size and film thickness are the dominant factors in determining the thermal stability of the film. With extended annealing at sufficient temperatures, the $TiSi_2$ grains tend to spheriodise and the $TiSi_2$ film becomes discontinuous.

Van den hove [4] first proposed using the grain boundary grooving model, as originally developed by Mullins [33,34], to describe the thermal instability in $TiSi_2$ thin films. The grooving model provides a means of relating the groove depth, d, to both time and temperature, i.e.

$$d = \tan\beta \left( D_{Si} \nu_s \Omega^2 C_0 \frac{t}{kT} \right)^{\frac{1}{n}}$$

where $\beta$ = groove angle, $D_{Si}$ = diffusivity of Si in $TiSi_2$, $\nu_s$ = silicide/silicon interface energy, $\Omega$ = atomic volume of silicon, $C_0$ = equilibrium concentration of silicon in $TiSi_2$, n = 3 for bulk diffusion or 4 for interface diffusion, t = time, T = temperature and k = Boltzmann's constant. In the case of silicide thin films, the silicide becomes discontinuous when the groove depth becomes equivalent to the film thickness. It is difficult, however, to locate accurate values for the surface energies in question.

## F CONCLUSION

We have compiled the TM silicide heats of formation and melting points where data were available. Certain thermodynamic parameters were discussed in relationship to practical engineering problems. The relationship of the heat of formation to the Schottky barrier height and the role of surface energy in thermal stability in thin films was discussed.

## REFERENCES


[1] S.P. Murarka [ *Silicides for VLSI Applications* (Academic Press Inc., New York, 1983) ]

[2] J.M. Poate, K.N. Tu, J.W. Mayer [ *Thin Films-Interdiffusion and Reactions* (John Wiley and Sons, New York, 1978) ]

[3] E.P. George, M. Yamaguchi, K.S. Kumar, C.T. Liu [ *Annu. Rev. Mater. Sci. (USA)* vol.24 (1994) p.409 ]

[4] L Van den hove [ Ph.D. Thesis, Katholieke Universiteit Leuven, Leuven, Belgium (1988) ]

[5] M.E. Alperin, T.C. Holloway, R.A. Haken, C.D. Gosmeyer, R.V. Karnaugh, W.D. Parmantie [ *IEEE Trans. Electron Devices (USA)* vol.32 (1985) p.141 ]

[6] D. Chen et al [ *IEDM Tech. Dig. (USA)* vol.118 (1985) p.118 ]

[7] C. Koburger, M. Ishaq, H.J. Geipel [ *J. Electrochem. Soc. (USA)* vol.129 (1982) p.1307 ]

[8] A.K. Vasudevan, J.J. Petrovic [ *Mater. Sci. Eng. A (Switzerland)* vol.155 (1992) p.1 ]

[9] P. Villars, L.D. Calvert [ *Pearson's Handbook of Crystallographic Data for Intermetallic Phases* (American Society for Materials, Metals Park, Ohio, 1985) ]

[10] A.K. Green, E. Bauer [ *J. Appl. Phys. (USA)* vol.47 (1976) p.1984 ]

[11] J.W. Mayer, S.S. Lau [ *Electronic Materials Science: For Integrated Circuits in Si and GaAs* (Macmillan Publishing Co., New York, 1990) ]

[12] R. Beyers, R. Sinclair [ *J. Appl. Phys. (USA)* vol.57 (1985) p.5240 ]

[13] H.J.W. van Houtum, I.J.M.M. Raaijmakers [ *Mater. Res. Soc. Symp. Proc. (USA)* vol.54 (1986) p.37-42 ]

[14] R.J. Nemanich, H. Jeon, C.A. Sukow, J.W. Honeycutt, G.A. Rozgonyi [ *Mater. Res. Soc. Symp. Proc. (USA)* vol.260 (1992) p.195 ]

[15] R. Pretorius, J.M. Harris, M.A. Nicolet [ *Solid-State Electron. (UK)* vol.21 (1978) p.667-75 ]

[16] C.J. Smithells [ *Metals Reference Book* 4th edition (Plenum Press, New York, 1967) ]

[17] F.M. d'Heurle [ *J. Mater. Res. (USA)* vol.3 (1988) p.167 ]

[18] J.M. Andrews, J.C. Phillips [ *Phys. Rev. Lett. (USA)* vol.35 (1975) p.56 ]

[19] A.W. Searcy, L.N. Finnie [ *J. Am. Ceram. Soc. (USA)* vol.45 (1962) p.268 ]

[20] O. Kubaschewski, C.B. Alcock [ *Metallurgical Thermochemistry* 5th edition (Pergamon Press, Oxford, 1979) ]

[21] T.B. Massalski, L.H. Bennett, J.L. Murray, H. Baker (Eds) [ *Binary Alloy Phase Diagrams* (ASM International, Metals Park, Ohio, 1990) ]

[22] S.M. Sze [ *Physics of Semiconductor Devices* 2nd edition (John Wiley and Sons, New York, 1981) p.304 ]

[23] R.W. Mann, C.A. Racine, R.S. Bass [ *Mater. Res. Soc. Symp. Proc. (USA)* vol.224 (1991) p.115 ]

[24] K. Maex et al [ *J. Mater. Res. (USA)* vol.4 (1989) p.1209 ]

[25] C.A. Sukow, R.J. Nemanich [ *Mater. Res. Soc. Symp. Proc. (USA)* vol.260 (1992) p.251 ]

[26] R.K. Shukla, J.S. Multani [ *Thermal Stability of Titanium Silicide Thin Films* V-MIC Conf. Proc. (1987) p.470 ]

[27] T.P. Nolan, R. Sinclair, R. Beyers [ *J. Appl. Phys. (USA)* vol.71 (1992) p.720 ]

[28] C.Y. Wong, L.K. Wang, P.A. McFarland, C.Y. Ting [ *J. Appl. Phys. (USA)* vol.60 (1986) p.243 ]

[29] P. Revesz, L.R. Sheng, L.S. Hung, J.W. Mayer [ *Appl. Phys. Lett. (USA)* vol.48 (1986) p.1591 ]

[30] H. Norstrom, K. Maex, P. Van den Abeele [ *Thin Solid Films (Switzerland)* vol.198 (1991) p.53 ]

[31] R.W. Mann, C.A. Racine [ *Ext. Abstr. Electrochem. Soc. (USA)* Abstract no.272 (1992) ]

[32] J.B. Lasky, J.S. Nakos, O.B. Cain, P.J. Geiss [ *IEEE Trans. Electron Devices (USA)* vol.38 (1991) p.262 ]

[33] W.W. Mullins [ *J. Appl. Phys. (USA)* vol.28 (1957) p.333 ]

[34] W.W. Mullins [ *Trans. Metall. Soc. AIME (USA)* vol.218 (1960) p.354 ]

[35] R.W. Cahn (Ed.) [ *Physical Metallurgy* ch.7 (John Wiley and Sons, New York, 1965) ]

[36] C. Vahlas, P.Y. Chevalier, E. Blanquet [ *CALPHAD (UK)* vol.13 (1989) p.273 ]

[37] A.L. Vazquez de Parga, J. de la Figuera, C. Ocal, R. Miranda [ *Ultramicroscopy (Netherlands)* vol.42 (1992) p.845 ]

[38] W.J. Boettinger, J.H. Perepezko, P.S. Frankwicz [ *Mater. Sci. Eng. A (Switzerland)* vol.155 (1992) p.33 ]

[39] W.B. Pearson [ *The Crystal Chemistry and Physics of Metals and Alloys* (Wiley-Interscience, New York, 1972) ]

[40] M. Setton, J. Van der Spiegel [ *J. Appl. Phys. (USA)* vol.69 (1991) p.994 ]

[41] M. Setton, J. Van der Spiegel, R. Madar, O. Thomas [ *Mater. Res. Soc. Symp. Proc. (USA)* vol.181 (1990) p.497 ]

[42] M. Setton, J. Van der Spiegel [ *J. Appl. Phys. (USA)* vol.70 (1991) p.193 ]

[43] J.S. Reid, E. Kolawa, M.-A. Nicolet [ *J. Mater. Res. (USA)* vol.7 (1992) p.2424 ]

[44] R. Beyers [ *Mater. Res. Soc. Symp. Proc. (USA)* vol.47 (1985) p.143 ]

[45] H. Jiang, C.S. Petersson [ *Thin Solid Films (Switzerland)* vol.140 (1986) p.115 ]

[46] V. Ya Markiv, E.K. Gladyshevskii, T.I. Fedoruk [ *Russ. Metall. (UK)* vol.3 (1966) p.118 ]

# 2.2 Formation of epitaxial TM silicides

L.A. Clevenger and R.W. Mann

June 1995

## A INTRODUCTION

Epitaxial metal silicides have many uses in the areas of semiconductor devices [1-5], detectors [3] and optical electronics [6]. For these applications, it is important that the thin (<50 nm) defect free epitaxial layers be formed on top of or buried in (100) or (111) Si substrates [7,8]. Forming good quality epitaxial silicide thin films can be very difficult and a variety of deposition approaches have been developed to address this problem. A number of review articles on the formation of epitaxial metal silicides have been published. Several references are von Kanel [8], Chen and Tu [9], Tung [10] and Mantl [11,12]. A list of silicides which can be grown epitaxially on silicon is presented in FIGURE 1 [12,13]. This figure also includes electrical resistivities and bandgaps (both direct and indirect) of the epitaxial silicides.

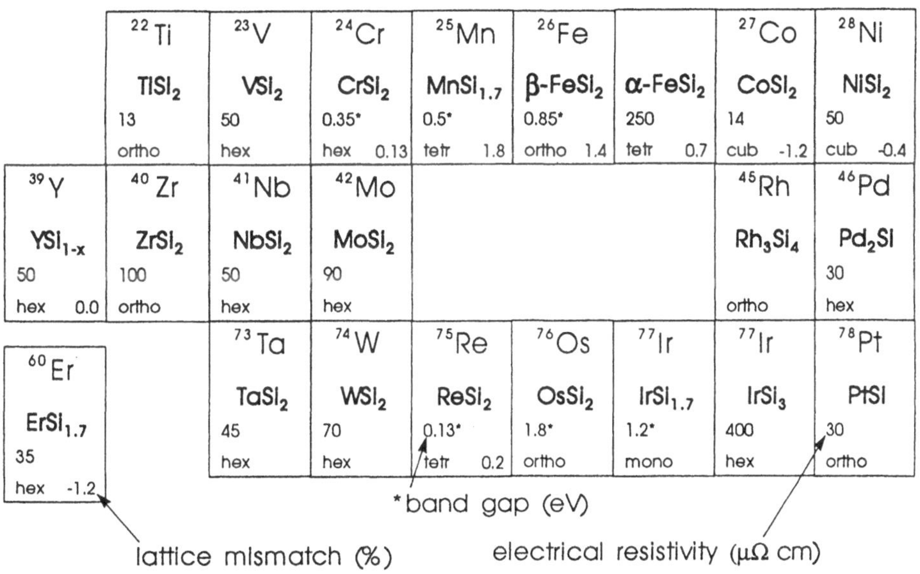


FIGURE 1 Crystal structures, lattice mismatches and electrical resistivities of selected metallic silicides and bandgap energies of semiconducting silicides $CrSi_2$, $MnSi_{1.7}$, β-$FeSi_2$, $ReSi_2$, $OsSi_2$ and $IrSi_{1.7}$. After [12,13].

## B LATTICE MISMATCH AND EPITAXIAL ORIENTATION

A key parameter that determines the ability of a silicide to be grown epitaxially on silicon is the lattice mismatch of the silicide to the single crystal silicon substrate, $\eta$, which is defined as:

$$\eta = (a_{silicide} - a_{si})/a_{si} \tag{1}$$

where $a_{silicide}$ is the lattice parameter of the crystalline plane of the silicide that is lattice matched to the silicon and $a_{Si}$ is the lattice parameter of silicon. For good quality epitaxial films, the lattice mismatch between the silicide and the silicon substrate is typically less than 3% [9,11]. FIGURE 2 demonstrates the lattice mismatch between various transition metal silicides and silicon on (100), (110) and (111) silicon. The silicides that produce high quality epitaxial films, and are the easiest to grow, are the ones with the smallest unit cell areas and smallest lattice mismatches (i.e. $NiSi_2$ and $CoSi_2$) [1,8,14].

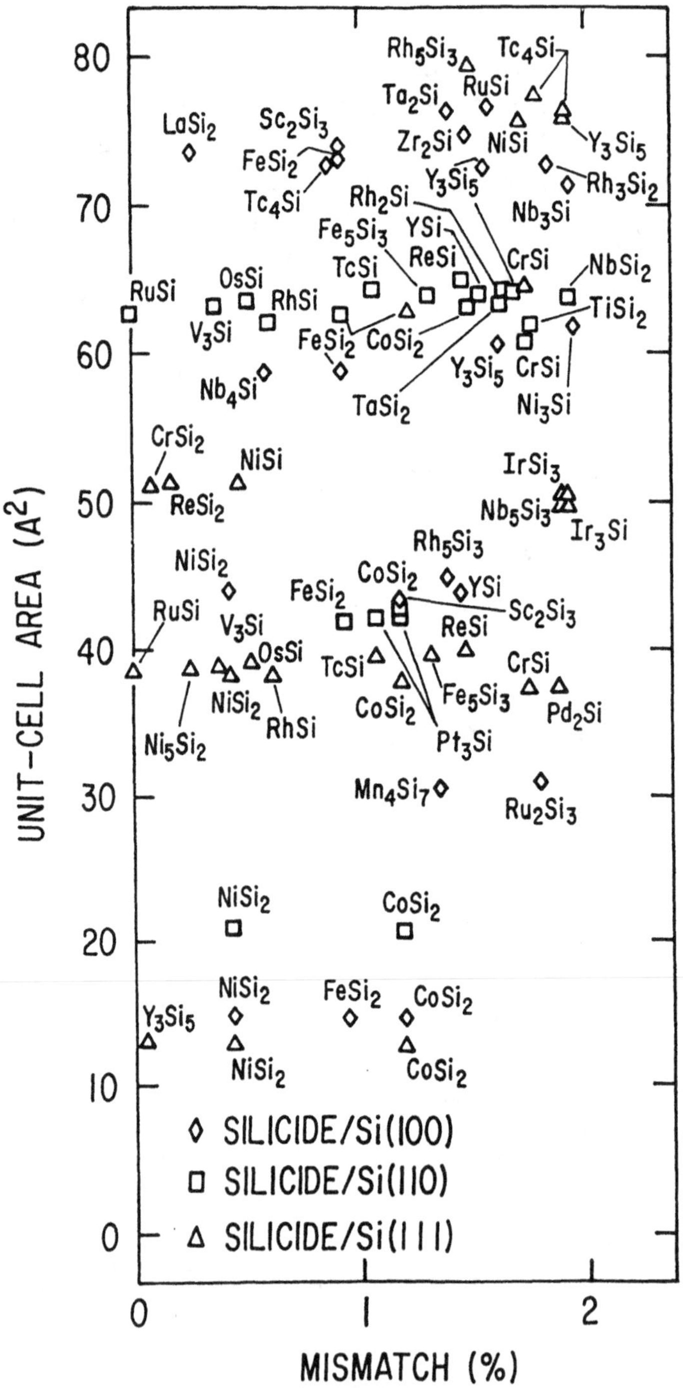


FIGURE 2 Lattice matches of transition-metal silicides on silicon (100), (110) and (111) at room temperature. All the possible mismatches smaller than 2% and unit-cell areas less than 80 Å are considered [11].

Epitaxial silicide films tend to form in a number of different orientations with respect to (111) or (100) silicon substrates. Examples are A- and B-type $NiSi_2$ or $CoSi_2$ ($CaF_2$ structure) grown epitaxially on (111) silicon, where the A-type is a continuation of the Si lattice planes in the silicide and the B-type results from a single stacking fault caused by a 180° rotation of the silicide with respect to the substrate [1]. A more complicated example of the possible epitaxial orientations of orthorhombic β-$FeSi_2$ is detailed in TABLE 1 [12,15-17].

TABLE 1 Possible orientations for the epitaxy of β-$FeSi_2$ on (100) and (111) silicon (compiled from [16,61,33,70]).

| (100) silicon | | (111) silicon | | | |
|---|---|---|---|---|---|
| Matching plane | | Matching planes | | | |
| β-$FeSi_2$(100)$\parallel$Si(100) | | β-$FeSi_2$(101) $\parallel$Si(111) | β-$FeSi_2$(110) $\parallel$Si(111) | β-$FeSi_2$(100) $\parallel$Si(111) | β-$FeSi_2$(010) $\parallel$Si(111) |
| Azimuthal orientations | | Azimuthal orientations | | | |
| A-type | B-type | A-type | B-type | | |
| β-$FeSi_2$[001] $\parallel$Si$\langle 110 \rangle$ | β-$FeSi_2$[010] $\parallel$Si$\langle 100 \rangle$ | β-$FeSi_2$[010] $\parallel$Si[$1\bar{1}0$] | β-$FeSi_2$[001] $\parallel$Si[$1\bar{1}0$] | | |
| β-$FeSi_2$[010] $\parallel$Si$\langle 110 \rangle$ | β-$FeSi_2$[001] $\parallel$Si$\langle 100 \rangle$ | | | | |

## C GROWTH PROCESSES

There are various growth techniques that are used to produce thin (<50 nm) epitaxial silicide layers on top of silicon. Some of the more common are:

(1) Solid Phase Epitaxy (SPE) [18-20]. A thin metal layer (typically a few monolayers) is deposited on an atomically clean substrate at room temperature under UHV (ultra high vacuum) conditions. The substrate is then heated from 400 to 700°C to form the epitaxial silicide phase. This results in a series of reactions terminating with the most silicon-rich phase.

(2) Molecular Beam Epitaxy (MBE) [21,22]. The metal and the silicon are co-deposited under UHV conditions onto the silicon substrate and the film is heated to form the epitaxial silicide.

(3) Template method [16,23]. Typically a combination of SPE and MBE, a two-step procedure where a well-controlled epitaxial template layer is first produced at low temperature by SPE and the thickness is then increased via an MBE deposition step performed at higher temperatures.

(4) Ion Beam Synthesis (IBS) [11]. High dose ion implantation is used to produce a buried silicide underneath the silicon surface which is then heated to between 900 and 1100°C to anneal out implantation damage and form a continuous silicide layer.

(5) Bilayer [22,23]. Metal bilayer thin films are used to form epitaxial silicides (usually $CoSi_2$) on (100) silicon.

In all the deposition processes listed above (except for the bilayer case) a UHV deposition and annealing environment is used. In addition, the silicon wafers are normally subjected to a Shiraki clean [25] and a high temperature in-situ anneal (>800°C) [10,26] with a slight silicon flux to remove any silicon oxide on the surface. Usually an epitaxial buffer Si layer (typically 50 nm) is then grown to eliminate the possibility of carbon contamination of the silicide/silicon interface before the growth of the epitaxial layer [26].

## D GROWTH OF EPITAXIAL SILICIDES

The epitaxial silicides of most interest to the scientific community are $NiSi_2$, $CoSi_2$, $FeSi_2$ and $TiSi_2$: $NiSi_2$ because it has the best lattice match to silicon, $CoSi_2$ and $TiSi_2$ because of their low resistivities (15 to 20 μΩ cm) and widespread use in the microelectronics industry [27] and $FeSi_2$ because of its direct energy gap of 0.87 eV [28-30]. As shown in FIGURE 1, other transition metal and rare earth silicides such as $CrSi_2$ and $YSi_{2-x}$ can be grown epitaxially on crystalline silicon. These material sets have low Schottky barrier heights to n-type silicon and low indirect bandgaps [31-33] but have not been as extensively studied as $NiSi_2$, $CoSi_2$, $FeSi_2$ or $TiSi_2$. The reader is referred to [31-40] for more information.

### D1 Epitaxial $TiSi_2$

$TiSi_2$ comes in two forms, a metastable C49-$TiSi_2$ (base centre orthorhombic) low temperature phase and a higher temperature stable C54-$TiSi_2$ phase [41]. The C49-$TiSi_2$ phase has a high resistivity (60 μΩ cm) and the C54 phase has a low resistivity (15 μΩ cm), so it is desirable to form the high temperature C54 phase for most applications. There are a few reports in the literature of SPE being used to grow epitaxial C49 and C54 $TiSi_2$ on (111) Si [41-49]. For titanium film thicknesses less than 40 nm and annealing temperatures of 800°C or less, only epitaxial C49-$TiSi_2$ has been achieved [42,46] while for film thicknesses greater than 40 nm and annealing temperatures (after Ti deposition) of >750°C, the formation of epitaxial C54-$TiSi_2$ on (111) silicon has been reported [47]. This thickness effect has been reported in the literature [41] and is attributed to the high surface to volume ratio of the films coupled with the low surface energy of the C49 phase and the low driving force for the C49 to C54 transformation [50,51]. In both the C49 and C54 cases, several azimuthal epitaxial orientations have been reported as being present in the same epitaxial film, leading to an increased defect density ($[3\bar{1}1]$C49||[112]Si and (130)C49||$(11\bar{1})$Si and (111)C45||(111)Si and (311)C54||(111)Si and (022)C54||(111)Si) [42,47,48]. Epitaxial C49 $TiSi_2$ has a tendency to form in isolated islands rather than uniform thin layers [42].

### D2 Epitaxial $NiSi_2$

In early work on epitaxial nickel silicide formation, it was found that thick (few hundred nm) $NiSi_2$ forms from the reaction

$$Ni \overset{\sim 250^{\circ}C}{\rightarrow} Ni_2Si \overset{350^{\circ}C}{\rightarrow} NiSi \overset{800^{\circ}C}{\rightarrow} NiSi_2$$

with the formation of $NiSi_2$ being nucleation-limited due to a small driving force for its formation [50]. Nucleation-limited formation of $NiSi_2$ leads to the preference for forming a mixture of A- and B-type and even polycrystalline $NiSi_2$ layers at high temperatures [8,52]. The first report of drastically altered behaviour was given by Tung et al [10-16,49] who showed, by using SPE, that the orientation of epitaxial $NiSi_2$ layers (A-type or B-type) formed depends on the deposition temperature, deposition rate, amount of nickel deposited, and the annealing temperature [8,27,54]. For example, when 0.1 to 0.7 nm is deposited on atomically clean Si(111) at room temperature and then heated to 500°C, B-type oriented $NiSi_2$ forms. When initial Ni thicknesses were 1.6 to 2.0 nm, A-type $NiSi_2$ formed and with intermediate coverages or thicker Ni layers, both orientations form [53,55-60]. This is attributed to a different phase formation sequence with very thin nickel layers [8].

For MBE grown films, Tung and Schrey [59] have shown that the deposition of Ni and Si at room temperature followed by annealing always forms B-type $NiSi_2$. The quality of the B-type layer is improved by first depositing a small amount of Ni (two to three monolayers) before the deposition of the Ni and Si. There has been a limited amount of work on using IBS to form epitaxial $NiSi_2$ in (100) or (111) silicon [11,60]. This is because $NiSi_2$ has a eutectic reaction at 950°C which prevents any thermal treatments at temperatures higher than 950°C to anneal out ion implantation damage. Recently there have been reports [61] of the formation of a double buried layer of B-type $NiSi_2$ using IBS by lowering the implant temperature from 350 to 150°C. The reason for this behaviour is not yet understood. In general, a detailed understanding of the kinetics of epitaxial $NiSi_2$ formation by different techniques is lacking. However, Hesse et al [62] have shown that the growth rate of A-type $NiSi_2$ is about ten times faster than that of B-type $NiSi_2$ due to differences in interplanar step heights.

**D3 Epitaxial $CoSi_2$ Formation**

As with $NiSi_2$, the reaction of Co and Si to form $CoSi_2$ goes through the following stages [63]:

$$Co \overset{\sim 250^\circ C}{\rightarrow} Co_2Si \overset{350^\circ C}{\rightarrow} CoSi \overset{650^\circ C}{\rightarrow} CoSi_2$$

with the major differences being that $CoSi_2$ forms at about 600°C (compared to $NiSi_2$ forming at 800°C) and that its formation is usually not nucleation-controlled [50]. The use of SPE, MBE and the template techniques has been applied in various ways to the formation of epitaxial $CoSi_2$ on (111) single crystal silicon. When the SPE technique is used to grow epitaxial $CoSi_2$ on (111) silicon, a mixture of A- and B-type $CoSi_2$ is often observed [64-66]. In addition, a high density of pinholes in the epitaxial layer ($>10^9$ $cm^{-2}$) forms due to the need to stabilise silicon-rich $CoSi_2$ surfaces [10-32]. The co-deposition of Co and Si at room temperature by either evaporation or MBE techniques has been shown to allow the formation of high quality (>10 nm) B-type epitaxial layers with a minimum of pinholes [54,67]. For this approach, a thin $CoSi_2$ layer (<3 nm) is deposited at room temperature, annealed at 300 to 500°C, and then used as a template for further growth of the epitaxial films by the co-deposition of Co and Si [67,68]. If further growth is done in multiple steps (i.e. the co-deposition of a few nm of Co and Si, annealing, then repetition of the process), good quality epitaxial films of over 100 nm in thickness can be grown [66]. However, in any annealed epitaxial $CoSi_2$ film grown on (111) Si more than 3 to 5 nm thick, there is evidence of the formation of misfit dislocations to relieve part of the misfit strain due to lattice mismatch [8,69]. The epitaxial growth of $CoSi_2$ on (100) Si is far more difficult than that for (111) silicon substrates due to the higher interfacial energy of $CoSi_2$(100)/Si(100) [70]. In many cases, (110) misoriented $CoSi_2$ grains are contained within the epitaxial $CoSi_2$ layers [71]. It is much easier to grow epitaxial $CoSi_2$ films on (100) silicon substrates using the bilayer technique discussed below.

The use of IBS to form buried epitaxial $CoSi_2$ layers in either (100) or (111) silicon was first reported by White et al [72,73]. With this technique the Co is first implanted into a Si substrate which is heated at $\approx$350°C to prevent amorphisation of the silicon. The sample is then heated to temperatures over 1000°C to form a continuous epitaxial layer and to anneal out crystalline defects [11]. A key parameter is the implant dose. The peak concentration of the implant should be above about half the stoichiometry of $CoSi_2$ [11]. Experimentally, the critical implant dose has been shown to range from about 0.5 to 4.0 x $10^{17}$ $cm^{-2}$ [74]. The thickness of the buried layer formed depends linearly on the ion dose and for $CoSi_2$ it can be expressed as $d = (\phi/2.59) \times 10^{22}$ cm, where $\phi$ is the implant dose [11]. The amounts of A- and B-type $CoSi_2$ formed by the IBS depend on the thickness of the buried layer and on the implant conditions [11,75]. For example, using the implant energies and doses in the following ranges, 30 - 200 keV and 0.47 - 1.6 x $10^{17}$ $cm^{-2}$ (310°C), and anneals at 600°C for 1 hr and 1000°C for 1/2 hr, Vanderstraeten et al [75] reported that layers thicker than 35 nm are A-type and layers thinner than this are mostly B-type. Hsieh et al [76] have reported similar results.

One of the easiest ways to form epitaxial $CoSi_2$ on (100) Si is to use the 'bilayer' approach first demonstrated by Wei et al and Dass et al [23,24]. With this method, a thin Ti layer (usually 1 to 5 nm thick) is deposited onto a (100) silicon substrate before the Co thin film (usually 7.5 to 100 nm thick). The sample is then rapidly thermally annealed (RTA) at temperatures close to 900°C and the Co diffuses through the Ti and forms epitaxial $CoSi_2$ on (100) Si. Other similar approaches that have been used to attain the same result are a Co/Ta bilayer structure [77], a $Co_xW_{1-x}$ (x = 0.6 to 0.75) thin film [78], and Co/Ti multilayer thin films [79]. While the detailed mechanisms which make this technique work are still under debate [80], it is thought that the presence of the Ti underlayer during thermal annealing delays the cobalt-silicon reaction to high temperatures where the formation of the epitaxial phase is preferred [60,81]. An activation energy of 3.5 eV for the formation of $CoSi_2$ in a bilayer approach has been reported [71]. This is higher than the value of ~2.0 eV for a standard Co-Si structure [63].

### D4 Epitaxial $FeSi_2$ Formation

The orthorhombic $\beta$-$FeSi_2$ phase (a = 0.986 nm, b = 0.779 nm and c = 0.788 nm [15]) is thermodynamically stable at room temperature and can be grown epitaxially in or on either (100) or (111) Si [19,20,82,83]. Unfortunately, a number of different possible epitaxial orientations [16,17], a complicated phase formation sequence which depends on film thickness [22,84,85], and a reversible phase transition to the tetragonal $\alpha$-$FeSi_2$ phase (a = b = 0.269 nm, c = 0.509 nm [61]) at 970°C all have contributed to make the formation of large epitaxial areas (>10 μm) of $\beta$-$FeSi_2$ on either (111) or (100) silicon very difficult [16,85,86].

Some of the reported phases seen so far in the epitaxial growth of β-$FeSi_2$ are cubic FeSi, β-$FeSi_2$, α-$FeSi_2$, metastable γ-$FeSi_2$ (fcc, a = 0.539 nm) and metastable s-$FeSi_2$ (bcc, a = 0.27 nm) [15]. The metastable γ-$FeSi_2$ and s-$FeSi_2$ phases are not present in the equilibrium phase diagram. The epitaxial orientations that are possible for the growth of β-$FeSi_2$ on (100) and (111) silicon were given in TABLE 1. The lattice mismatch with silicon for β-$FeSi_2$ varies from about 1.2 to 5.5% depending on the type of silicon substrate and its epitaxial relationship to β-$FeSi_2$ [15,16]. An example of the effect of film thickness on the phase formation sequence during epitaxial growth of β-$FeSi_2$ is presented in [22]. In this work, a varying number of monolayers of iron were deposited onto clean (111) silicon at evaluated temperatures ranging from 300 to 550°C and in-situ RHEED analysis was used to detect the phases present which were: e-FeSi, s-$FeSi_2$ or β-$FeSi_2$ depending on the deposited Fe thickness and annealing temperature.

There are numerous reports in the literature on the epitaxial formation of β-$FeSi_2$ by SPE [17,86,88], MBE [15,21], template [27,14] and IBS techniques [2,82]. For SPE techniques, A-type $FeSi_2$ or a mixture of A- and B-type β-$FeSi_2$ is commonly grown [89] and formation temperatures for β-$FeSi_2$ have been reported from 200 to 500°C, with decreasing $FeSi_2$ thickness from 20 to 1.4 nm [27]. The maximum thickness for high quality SPE grown β-$FeSi_2$ layers without any pinhole formation is about 30 nm. Radermacher et al [90] have measured the activation energy of formation of SPE grown FeSi, β-$FeSi_2$ and IBS formed $FeSi_2$ to be 1.4, 2.6 and 3.6 eV. For MBE or template films deposited at 500 to 600°C, there are several reports of α-$FeSi_2$ formation even though at this temperature it is not thermodynamically stable [15,21]. In addition, Derrien et al [16] have shown that heating rate, in addition to initial Fe or co-deposited $FeSi_2$ film thickness, plays an important role in determining the phase sequence for forming epitaxial β-$FeSi_2$ on both (111) and (100) silicon. IBS formation of β-$FeSi_2$ is performed in a similar manner to IBS of $CoSi_2$ described earlier. A typical process example is $4 \times 10^{17}$ $cm^{-2}$ Fe (300 keV) at 350°C into (111) silicon, an 850°C anneal to form a stable β-$FeSi_2$ layer, a 1050°C anneal to transform the β-$FeSi_2$ into α-$FeSi_2$ and anneal out any defects and then a long anneal (ten hr) at 800°C to convert the α-$FeSi_2$ into epitaxial β-$FeSi_2$ [91]. Lin et al [92] have reported that the coarsening in the plane of the IBS formed β-$FeSi_2$ layer has an activation energy consistent with Oswald ripening (3.5 eV) and that the thickening of the layer has an activation energy consistent with interface-controlled growth (2.2 eV). In addition, the grain size of β-$FeSi_2$ formed by IBS in (100) silicon has been determined to be about an order of magnitude less than formed in (111) silicon [69]. α-$FeSi_2$ and γ-$FeSi_2$ can be produced by IBS with low temperature annealing if the initial implant is done at room temperature [12,93].

## E CONCLUSION

We have discussed the formation of thin epitaxial layers of $NiSi_2$, $CoSi_2$, $TiSi_2$ and $FeSi_2$ on both (111) and (100) silicon substrates using SPE, MBE, template and bilayer techniques. We have also detailed which silicides have been reported to form epitaxial thin films on crystalline silicon substrates. There has been a large amount of effort to create process recipes for producing these layers and significant progress has been made in developing a fundamental understanding of the formation of epitaxial silicide thin films on silicon.

**REFERENCES**


[1] A. Vantomme, M.-A. Nicolet, G. Bai, D. Fraser [ *Appl. Phys. Lett. (USA)* vol.62 (1993) p.243 ]

[2] J. Desimoni, H. Bernas, M. Behar, X.W. Lin, J. Washburn, Z. Liliental-Weber [ *Appl. Phys. Lett. (USA)* vol.62 (1993) p.18 ]

[3] J.C. Hensel, A.F.J. Levi, R.T. Tung, J.M. Gibson [ *Appl. Phys. Lett. (USA)* vol.47 (1985) p.151 ]

[4] R.W. Fathauer, T. George, W.T. Pike [ *Mater. Res. Soc. Symp. Proc. (USA)* vol.263 (1992) p.149 ]

[5] Y.-F. Hsieh, R. Hull, A.E. White, K.T. Short [ *Appl. Phys. Lett. (USA)* vol.58 (1991) p.122 ]

[6] D.R. Peale, R. Haight, H. Ott [ *Appl. Phys. Lett. (USA)* vol.62 (1993) p.1402 ]

[7] A.H. Reader, J.P.W.B. Duchateau, J.E. Crombeen [ *Semicond. Sci. Technol. (UK)* vol.8 (1993) p.1204 ]

[8] H. von Känel [ *Mater. Sci. Rep. (Netherlands)* vol.8 (1991) p.193 ]

[9] L.J. Chen, K.N. Tu [ *Mater. Sci. Rep. (Netherlands)* vol.6 (1991) p.53 ]

[10] R.T. Tung [ *Mater. Chem. Phys. (Switzerland)* vol.32 (1992) p.107 ]

[11] S. Mantl [ *Mater. Sci. Rep. (Netherlands)* vol.8 (1992) p.1 ]

[12] S. Mantl [ *Nucl. Instrum. Methods Phys. Res. B (Netherlands)* vol.84 (1994) p.127 ]

[13] S.P. Murarka [ *Silicides for VLSI Applications* (Academic Press, New York, 1983) ]

[14] N. Onda, H. Henz, E. Müller, K.A. Mäder, H. von Känel [ *Appl. Surf. Sci. (Netherlands)* vol.56 (1992) p.421 ]

[15] I. Berbezier, J. Chevrier, J. Derrien [ *Surf. Sci. (Netherlands)* vol.315 (1994) p.27 ]

[16] J. Derrien, J. Chevrier, V. Le Thanh, J.E. Mahan [ *Appl. Surf. Sci. (Netherlands)* vol.56 (1993) p.382 ]

[17] W. Raunau, H. Niehus, G. Comsa [ *Surf. Sci. (Netherlands)* vol.284 (1993) p.L375 ]

[18] A.L. Vazquez de Parga, J. de la Figuero, C. Ocal, R. Miranda [ *Ultramicroscopy (Netherlands)* vol.42 (1992) p.845 ]

[19] K. Konuma, J. Vrijmoeth, P.M. Zagwijn, J.W.M. Frenkne, E. Vlieg, J.F. van der Veen [ *J. Appl. Phys. (USA)* vol.73 (1993) p.1104 ]

[20] P. Stacker, J. Gay, J. Natolli [ *Physica B* vol.198 (1994) p.240 ]

[21] L. Vinh, J. Chevrier, H. Derrien [ *Phys. Rev. B (USA)* vol.46 (1992) p.15946 ]

[22] R.T. Tung, J.L. Batstone [ *Appl. Phys. Lett. (USA)* vol.52 (1988) p.1611 ]

[23] C.-S. Wei, D.B. Fraser, M.L.A. Dass, M. Frost, T. Brat [ *V-MIC* (1990) p.233 ]

[24] M.L.A. Dass, D.B. Fraser, C. Wei [ *Appl. Phys. Lett. (USA)* vol.58 (1991) p.1308 ]

[25] A. Ishizaka, Y. Shiraki [ *J. Electrochem. Soc. (USA)* vol.133 (1986) p.666-71 ]

[26] H. von Känel, R. Stalder, H. Sirringhas, N. Onda, J. Henz [ *Appl. Surf. Sci. (Netherlands)* vol.53 (1991) p.196 ]

[27] L.A. Clevenger, R.W. Mann [ *Mater. Res. Soc. Symp. Proc. (USA)* vol.320 (1994) p.14 ]

[28] J. Alvarez, J.J. Hinorejas, E.G. Michel, G.R. Castro, R. Miranda [ *Phys. Rev. B (USA)* vol.45 (1992) p.14042 ]

[29] M.C. Bost, J.E. Mahan [ *J. Appl. Phys. (USA)* vol.64 (1988) p.2034 ]

[30] H.C. Cheng, T.R. Yew, L.J. Cheng [ *J. Appl. Phys. (USA)* vol.57 (1985) p.5246 ]

[31] R.W. Fathauer, C.W. Nieh, Q.F. Xiao, S. Haashimoto [ *Mater. Res. Soc. Symp. Proc. (USA)* vol.198 (1990) p.583 ]

[32] N.W. Christensen [ *Phys. Rev. B (USA)* vol.42 (1990) p.7148 ]

[33] T.L. Lee, W.D. See, J.H. Lin, C.H. Luo, L.J. Chen [ *Mater. Res. Soc. Symp. Proc. (USA)* vol.312 (1992) p.125 ]

[34] F.H. Koatz, J. Van der Spiegel, W.R. Graham [ *Mater. Res. Soc. Symp. Proc. (USA)* vol.198 (1990) p.601 ]

[35] N. Guerfi, T.A. Nguyen, T.J.Y. Veuillen, D.B. Lollman [ *Appl. Surf. Sci. (Netherlands)* vol.56 (1992) p.501 ]

[36] R.G. Lang, M.C. Bost, J.E. Mahan [ *Thin Solid Films (Switzerland)* vol.162 (1988) p.29 ]
[37] F.Y. Shain, H.C. Cheng, L.J. Chen [ *J. Appl. Phys. (USA)* vol.59 (1988) p.2784 ]
[38] L. Haderbache, P. Wetzel, C. Pirri, I.C. Peruchelti, D. Bolmenot, G. Gewinrer [ *Surf. Sci. (Netherlands)* vol.209 (1989) p.L139 ]
[39] J.E. Mahan et al [ *Appl. Phys. Lett. (USA)* vol.56 (1990) p.2439 ]
[40] G. Bai, M.-A. Nicolet, J.E. Mahan, K. Geib [ *Phys. Rev. B (USA)* vol.41 (1990) p.8603 ]
[41] R. Beyers, R. Sinclair [ *J. Appl. Phys. (USA)* vol.57 (1985) p.5240 ]
[42] H. Jean, J.W. Honeycutt, A. Sukow, T.P. Humphreys, R.J. Nemanich, G.A. Rozgonyi [ *Mater. Res. Soc. Symp. Proc. (USA)* vol.198 (1990) p.595 ]
[43] A. Cantana, P.E. Schmid, M. Heintze, F. Levy [ *J. Appl. Phys. (USA)* vol.67 (1990) p.1820 ]
[44] I.C. Wu, J.J. Chu, L.J. Chen [ *J. Appl. Phys. (USA)* vol.60 (1986) p.3172 ]
[45] M.S. Fung, H.C. Cheng, L.J. Chen [ *Appl. Phys. Lett. (USA)* vol.47 (1985) p.1312 ]
[46] R.J. Nemanich, H. Jean, G.A. Sukow, J.W. Honeycutt, G.A. Rozgonyi [ *Mater. Res. Soc. Symp. Proc. (USA)* vol.260 (1992) p.195 ]
[47] K.H. Kim et al [ *J. Appl. Phys. (USA)* vol.71 (1992) p.3812 ]
[48] T.C. Chou, C.Y. Wang, K.N. Tu [ *J. Appl. Phys. (USA)* vol.62 (1987) p.2275 ]
[49] M.S. Fung, H.C. Cheng, L.J. Chen [ *Appl. Phys. Lett. (USA)* vol.47 (1985) p.1312 ]
[50] F. d'Heurle [ *J. Mater. Res. (USA)* vol.3 (1988) p.167 ]
[51] R.W. Mann, C.A. Racine, R.S. Bass [ *Mater. Res. Soc. Symp. Proc. (USA)* vol.224 (1991) p.115 ]
[52] J.M. Gibson, J.C. Bean, J.M. Poate, R.T. Tung [ *Appl. Phys. Lett. (USA)* vol.41 (1982) p.818 ]
[53] R.T. Tung, J.M. Gibson, J.M. Poate [ *Phys. Rev. Lett. (USA)* vol.50 (1983) p.429 ]
[54] R.T. Tung [ *Mater. Chem. Phys. (Switzerland)* vol.32 (1992) p.107 ]
[55] R.T. Tung, J.M. Gibson, J.M. Poate [ *Appl. Phys. Lett. (USA)* vol.42 (1983) p.888 ]
[56] R.T. Tung, J.M. Gibson [ *Mater. Res. Soc. Symp. Proc. (USA)* vol.67 (1986) p.211 ]
[57] R.T. Tung [ *J. Vac. Sci. Technol. A (USA)* vol.5 (1987) p.1840 ]
[58] D. Hesse, R. Mattheis [ *Phys. Status Solidi B (Germany)* vol.116 (1989) p.67 ]
[59] R.T. Tung, F. Schrey [ *Appl. Phys. Lett. (USA)* vol.55 (1989) p.256 ]
[60] A.E. White, K.T. Short, R.C. Dynes, J.P. Gorno, J.M. Gibson [ *Mater. Res. Soc. Symp. Proc. (USA)* vol.93 (1987) p.93 ]
[61] J. Vanhellemont et al [ *Appl. Phys. Lett. (USA)* vol.62 (1993) p.2795 ]
[62] D. Hesse, P. Werner, R. Matteis, J. Heydenrich [ *Appl. Phys. A, Solids Surf. (Germany)* vol.57 (1993) p.415 ]
[63] H. Miura, E. Ma, C.V. Thompson [ *J. Appl. Phys. (USA)* vol.70 (1991) p.4287 ]
[64] F.A. d'Avitaya, J.A. Chroboczek, C. d'Anterroches, G. Glastre, Y. Campidelli, E. Rosencher [ *J. Cryst. Growth (Netherlands)* vol.81 (1987) p.463 ]
[65] S. Saitoh, H. Ishiwara, S. Furukawa [ *Appl. Phys. Lett. (USA)* vol.37 (1980) p.203 ]
[66] R.T. Tung, J.M. Gibson, J.C. Bean, J.M. Poate, D.C. Jacobson [ *Appl. Phys. Lett. (USA)* vol.40 (1982) p.684 ]
[67] R.T. Tung, F. Schrey [ *Appl. Phys. Lett. (USA)* vol.54 (1989) p.852 ]
[68] S.A. Chambers, S.B. Anderson, H.W. Chen, J.H. Weaver [ *Phys. Rev. B (USA)* vol.34 (1986) p.913 ]
[69] C.W.T. Bulle-Lieuwma, A.H. von Ommen, J. Honstra [ *Mater. Res. Soc. Symp. Proc. (USA)* vol.102 (1988) p.377 ]
[70] S.M. Yalsoue, R.T. Tung, D. Lorretto [ *J. Vac. Sci. Technol. A (USA)* vol.6 (1989) p.1472 ]
[71] A. Vantomme, M.-A. Nicolet, N.D. Theodore [ *J. Appl. Phys. (USA)* vol.75 (1994) p.3882 ]
[72] A.E. White, K.T. Short, R.C. Dynes, J.P. Garno, J.M. Gibson [ *Appl. Phys. Lett. (USA)* vol.95 (1987) p.50 ]
[73] A.E. White, K.T. Short, J.L. Batstone, D.C. Jacobson, J.M. Poate, K.W. West [ *Appl. Phys. Lett. (USA)* vol.50 (1979) p.479 ]

[74] S. Mantl, R. Jebasinski, D. Hartmann [ *Nucl. Instrum. Methods Phys. Res. B (Netherlands)* vol.59 (1991) p.666 ]

[75] H. Vanderstraeten, Y. Bruynseraede, M.F. Wu, A. Vantomme, G. Langouche, J.M. Phillips [ *Appl. Phys. Lett. (USA)* vol.57 (1990) p.135 ]

[76] Y.-F. Hsieh, R. Hull. A.E. White, K.T. Short [ *Appl. Phys. Lett. (USA)* vol.53 (1991) p.122 ]

[77] J.S. Byun, S.B. Kang, H.J. Kim [ *J. Appl. Phys. (USA)* vol.74 (1993) p.3156 ]

[78] A.H. Reader, J.P.W.B. Duchateau, E. Crambeen, E.P. Naburgh, M.A.J. Soners [ *Appl. Surf. Sci. (Netherlands)* vol.53 (1991) p.92 ]

[79] F. Hong, G. Rozgonyi, B.K. Patnaik [ *Appl. Phys. Lett. (USA)* vol.64 (1994) p.2241 ]

[80] S.-L. Zhang, J. Cordenas, F.M. d'Heurle, B.G. Svensoon, C.S. Petersson [ *Appl. Phys. Lett. (USA)* vol.66 (1995) p.58 ]

[81] K. Barmak et al [ *Mater. Res. Soc. Symp. Proc. (USA)* vol.238 (1992) p.575 ]

[82] N. Jedrecy, Y. Zheng, A. Waldhauer, M. Sauvage-Simkin, R. Pinchaux [ *Phys. Rev. B (USA)* vol.48 (1993) p.8801 ]

[83] J. Desimani, M. Behar, H. Bernas, X.W. Lin, Z. Liliental-Weber, J. Washburn [ *Nucl. Instrum. Methods Phys. Res. B (Netherlands)* vol.80 (1993) p.755 ]

[84] N. Onda, J. Henz, E. Müller, K.A. Mäder, H. von Känel [ *Appl. Surf. Sci. (Netherlands)* vol.56 (1992) p.421 ]

[85] T.E. Crumbaker, J.Y. Natoli, I. Berbezier, J. Derrien [ *J. Cryst. Growth (Netherlands)* vol.127 (1993) p.158 ]

[86] N. Motta, A. Sgarlata, G. Gaggiotti, F. Patella, A. Balzorotti, M. DeCrescenzi [ *Surf. Sci. (Netherlands)* vol.284 (1993) p.257 ]

[87] A.L. Vazquez de Parga, J. de la Figuero, C. Ocal, R. Miranda [ *Ultramicroscopy (Netherlands)* vol.42 (1992) p.845 ]

[88] H. Moritz, B. Rösen, S. Popuiá, A. Rizzi, H. Lüth [ *J. Vac. Sci. Technol. B (USA)* vol.10 (1992) p.1704 ]

[89] C.W.T. Bulle-Lieuwma, D.J. Oostra, D.E.W. Vandenhoudt [ *Mater. Res. Soc. Symp. Proc. (USA)* vol.283 (1993) p.885 ]

[90] K. Radermacher, S. Mantl, C. Dieker, H. Luth, C. Freiburg [ *Thin Solid Films (Switzerland)* vol.215 (1992) p.76 ]

[91] H. Reuther, E. Wieser, D. Ponkin, R. Grotzschel, W. Skorupa, G. Querner [ *Nucl. Instrum. Methods Phys. Res. B (Netherlands)* vol.68 (1992) p.241 ]

[92] X.W. Lin, J. Washburn, Z. Liliental-Weber, H. Bernas [ *J. Appl. Phys. (USA)* vol.75 (1994) p.4686 ]

[93] D.J. Oostra, C.W.T. Bulle-Lieuwma, D.E.W. Vandenhoudt, F. Felten, J.C. Jons [ *J. Appl. Phys. (USA)* vol.74 (1993) p.4347 ]

# 2.3 Phase diagrams of RE silicides

Xu Zhenjia (C.C. Hsu)

June 1995

Known phase diagrams (complete or partial) for binary rare earth (RE)-Si systems are very limited. In FIGURES 1 to 5, we reproduce systems for which the binary phase diagrams have been determined. For the REs, the more studied systems are La, Ce, Pr, Sm, Gd, Er and Yb, while for the others (perhaps for economic reasons), only a few systems have been examined so far.

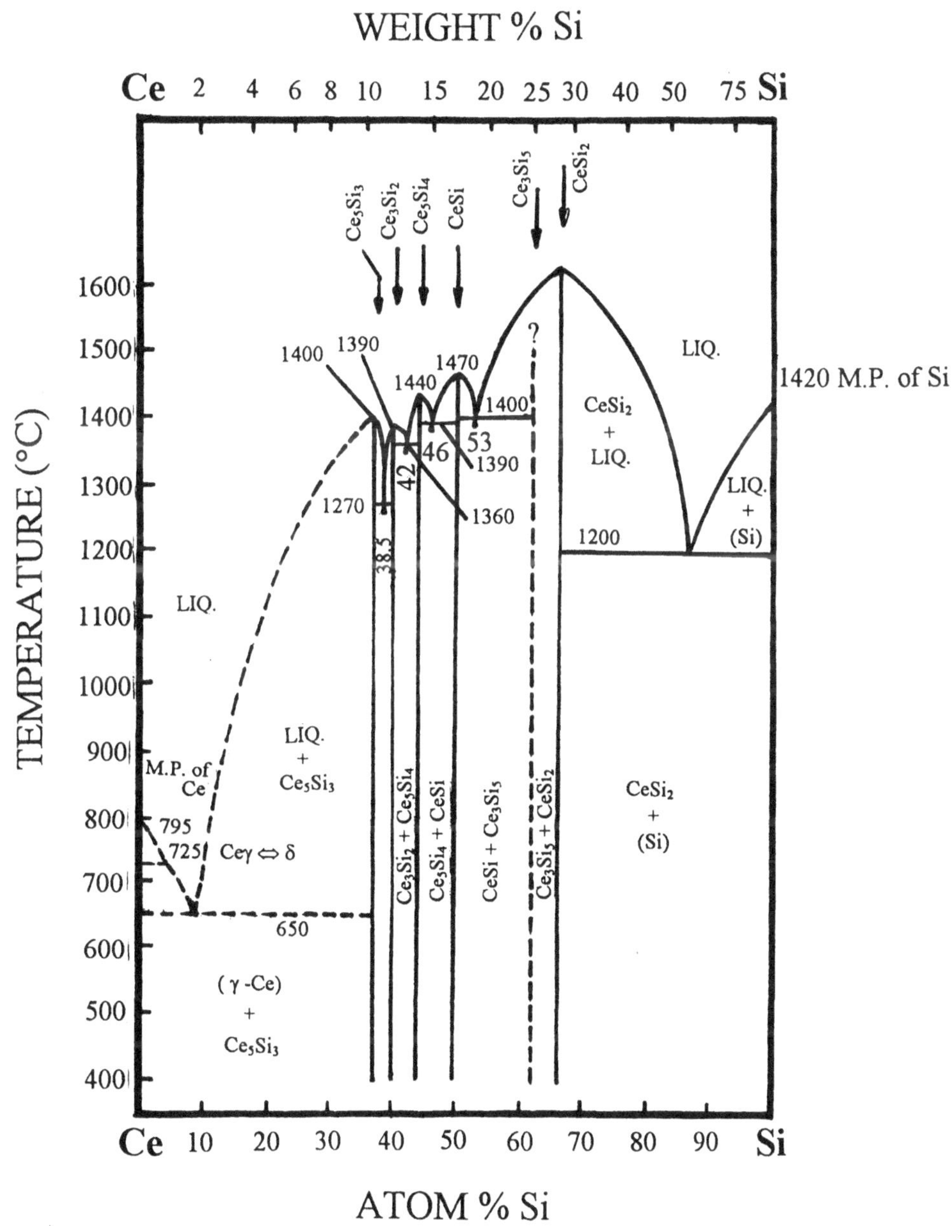


FIGURE 1 Ce-Si phase diagram [1].

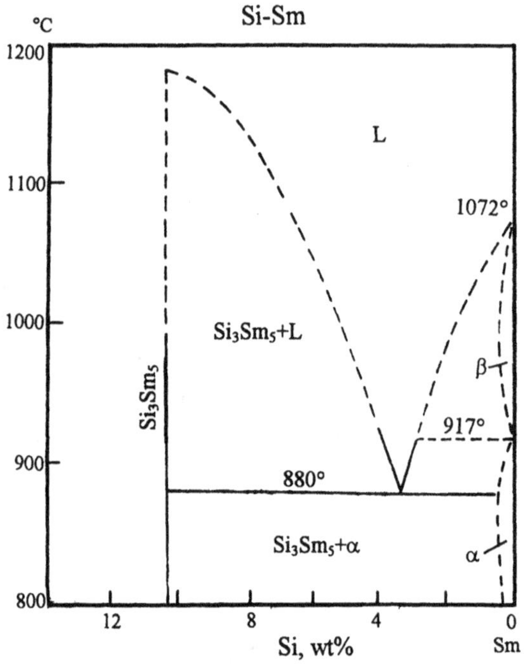


FIGURE 2 Sm-Si phase diagram [1].

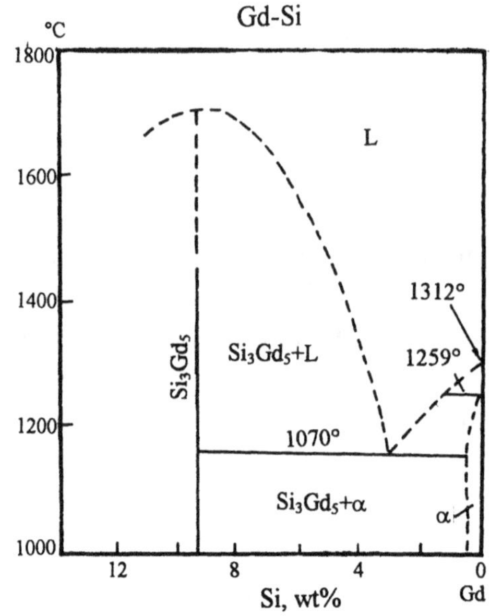


FIGURE 3 Gd-Si phase diagram [2].

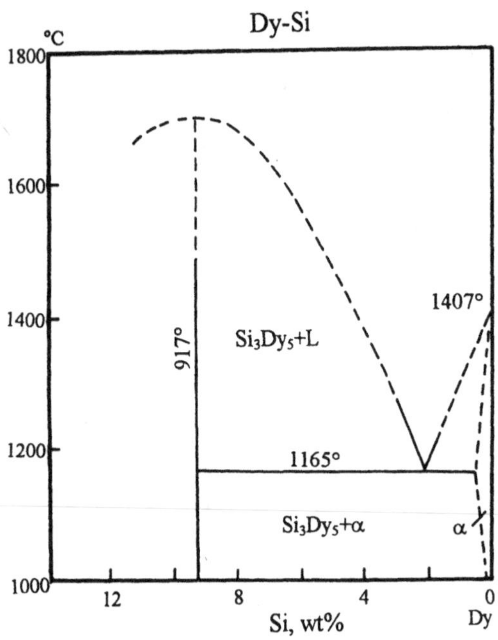


FIGURE 4 Dy-Si phase diagram [2].

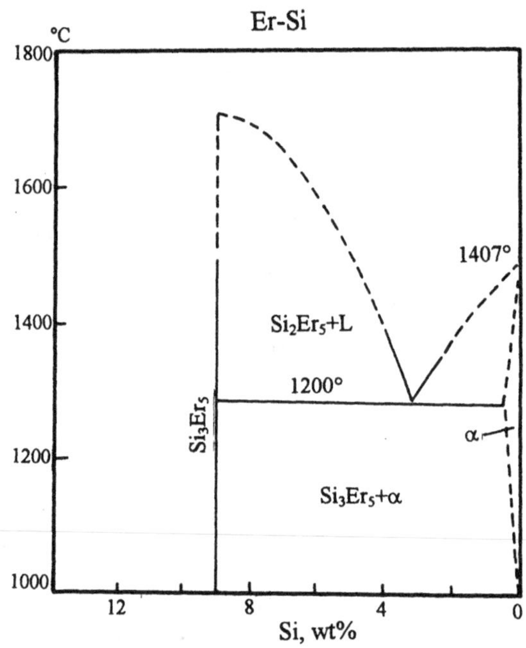


FIGURE 5 Er-Si phase diagram [2].

## REFERENCES

[1] F.A. Shunk [ *Constitution of Binary Alloys* 2nd Suppl. (McGraw-Hill, New York, 1969) ]; W.G. Moffat [ *The Handbook of Binary Phase Diagrams* vol.1 (General Electric Co., 1976, 1977) ]

[2] Yu Jueqi, Yi Wenzhi, Chen Bangdi, Chen Hongjian (Eds) [ *Constitution of Binary Alloys* (Shanghai Science and Technology Press, 1987) ]

# 2.4 Thin film formation of RE silicides

Xu Zhenjia (C.C. Hsu)

June 1995

## A INTRODUCTION

During the last ten years, a number of investigators have considered the thin film formation of RE silicides [1-11]. These studies are of considerable technological and fundamental interest. From a practical point of view, it is quite obvious that the incentive for thin film studies originates from the demands of technology. Prior to any application or measurements of a physical characteristic, it is necessary to have a thin film and it is necessary also to identify the phases that are formed and possibly the respective mechanisms of phase formation. On the academic side, RE silicides are a novel class of materials. Recently, the thin film formation of RE silicides has received much attention for several reasons. First, the reactions of RE silicides show markedly different growth kinetics from those observed in the formation of transition metal silicides. A second reason is that some of the REs can grow epitaxially on Si(111) with a small lattice mismatch of the order of 1%, and form interfaces of high crystalline quality. In this Datareview, we will discuss the formation of RE silicide thin films having a thickness of $10^2$ - $10^4$ Å.

## B THIN FILM CONTACT REACTION AND GROWTH

As RE metals are very chemically reactive, oxidation of the RE is a severe problem in forming RE silicides. Therefore, comparatively little work has been done and the understanding of the formation of RE silicides is far from complete.

Many deposition techniques can be applied to form thin transition metal silicide films, but cannot be applied to form thin RE silicide films. The most common method for forming RE silicide thin films is evaporation with a filament or e-gun under ultrahigh vacuum (UHV) conditions. The RE silicide thin films are usually prepared by a contact reaction with the RE metal deposited on a pre-cleaned Si surface and then heated. As a protective coating, a thin layer of Si deposited on top of the RE layer is used to protect the RE layer from oxidation. This method usually gives rough surfaces, unless heating is very rapid. Recently, the other current methods, such as molecular beam epitaxy (MBE) [7,8], template techniques and reactive MBE [9,10] have also been used.

The standard annealing method for the RE silicide thin film formation is steady-state annealing (resistive heating) but transient treatments (such as light beam rapid thermal annealing (RTA)) have also been investigated. The RE silicide is produced by subjecting the sample (a thin RE film deposited on a single-crystal Si substrate) to a heat treatment that will promote the reaction between the film and the substrate. Because of the severe problem of oxidation, the sample should be annealed under UHV conditions or in a special device containing a flowing inert gas atmosphere.

The most common methods for analysing the RE silicide thin films are X-ray diffraction (XRD), Rutherford backscattering spectrometry (RBS), Auger electron spectroscopy (AES and Scanning AES), high resolution transmission electron microscopy (HRTEM), scanning electron microscopy (SEM), secondary ion mass spectrometry (SIMS), electron microprobe (EMP) and four-point probe. An adequate characterisation of a film almost always requires a combination of several techniques. TABLE 1 summarises the annealing conditions and the

TABLE 1 Summary of annealing results of RE-Si.

| Element | Annealing conditions | RE silicides | Remarks | Ref |
|---|---|---|---|---|
| La | <275°C, 1 hr | $LaSi_2$ | RH, La-Si (100°C) | [2] |
| | 200°C, 1 hr | $LaSi_{2-x}$ | RH, La-Si (RT) | [5] |
| | 600°C, 10 sec | $LaSi_{2-x}$ | IR-RTA, La-Si (RT) | [5] |
| | 8 W | $LaSi_{2-x}$ | LA, CW $Ar^+$, La-Si (140°C) | [5] |
| Ce | 400°C, 1 hr<br>500°C, 1 hr | $CeSi_2$ | RH, Ce-Si (100°C) | [2] |
| | 300°C, 1 hr<br>400°C, 1 hr | Si-Ce rich intermixed layer | RH, Ce-Si (RT), large amount of $O_2$ in the Ce metal film | [5] |
| Gd | 300°C, 1-2 hr<br>325°C, <30 min | $GdSi_2$ | RH, Gd-Si (100°C) | [2] |
| | 400°C, 30 min | No | IR-RTA, Gd-Si (<100°C) | [4] |
| Tb | 350°C, 30 min | $TbSi_{2-x}$ | IR-RTA, Tb-Si (<100°C) | [5] |
| Dy | 325°C, 2.5 hr<br>350°C, 1 hr | $DySi_2$ | RH, Dy-Si (100°C) | [2] |
| | 300°C, 30 min | $Dy\text{-}Si_{2-x}$ | IR-RTA, Dy-Si (<100°C) | [4] |
| | 423 K, 1 hr | $Dy_5Si_3$ - $DySi_{2-x}$ (graded layer) | RH, Dy-Si (RT) | [6] |
| | 623 K, 1 hr | $DySi_{2-x}$ (dominant) | RH | [6] |
| Ho | 350°C, 2 hr<br><4 hr<br>375°C, 45 min | $HoSi_2$ | RH, Ho-Si (100°C) | [2] |
| | 300°C, 30 min | $HoSi_{2-x}$ | IR-RTA, Ho-Si (<100°C) | [4] |
| Er | 305°C, >24 hr<br>375°C, 75 min<br>400°C, 15 min | $ErSi_2$ | RH, Er-Si (100°C) | [2] |
| | 350°C, 30 min | $ErSi_{2-x}$ | IR-RTA, Er-Si (<100°C) | [4] |
| | Two steps:<br>(1) 400°C<br>(2) >800°C (900°C) | $ErSi_{2-x}$ | MBE, Er-Si (RT), RH | [7,8] |
| | Reactive MBE | $ErSi_{2-x}$ | RH (two steps) | [9,10] |
| | ~900°C | $ErSi_2$ | MBE, RH | [11] |
| Tm | 350°C, 30 min | $TmSi_{2-x}$ | IR-RTA, Tm-Si (<100°C) | [4] |
| Yb | 350°C, 30 min | $YbSi_{2-x}$ | IR-RTA, Yb-Si (<100°C) | [4] |
| Lu | 350°C, 30 min | No $LuSi_{2-x}$ | IR-RTA, Lu-Si (<100°C) | [4] |

RH: Resistive heating; RTA: Rapid thermal annealing; LA: Laser annealing; IR: Infrared; RT: Room temperature annealing results of RE-Si couples. It should be noted that these data perhaps are not complete, since these are only the main experimental results we have been able to document from our literature survey. From TABLE 1, the observation of disilicides as the first phase to form is similar to the situation in the refractory metals.

## C MECHANISM OF THIN FILM REACTION

### C1 Diffusing Species

It is of great interest to know the relative motion of atoms during the contact reaction between Si and RE, as this knowledge sheds light on the microscopic processes that control the reactions. The identification of the dominant diffusion species requires the use of inert markers. During the formation of the silicides of some RE metals the RE atoms remain essentially immobile so one could use the interface between the RE and Si as a marker during the formation of the respective compounds [16].

On the other hand, the silicon atoms are very mobile because they behave like interstitials and the silicon sublattice in the $AlB_2$ structure contains 15% vacancies. However, the composition is $RESi_{1.7}$. These two factors allow the growth of the silicides at low temperatures. However, the metal atoms themselves in the silicides do not become mobile unless extremely high temperatures, of the order of 1000°C [28], are reached. A study of the RE (Tb, Er) silicides obtained by thin film reactions on Si indicates that Si atoms constitute the dominant diffusion species during the formation of these compounds [16]. Recently, a marker study of lutetium silicide prepared by vacuum furnace annealing of a Lu thin layer on Si substrate also shows that Si atoms are the dominant species during the formation of $LuSi_{2-x}$ [29].

### C2 Diffusion Mechanism

Reactions between Si and transition metals have been well established. From similar studies of the RE metals, it is generally recognised that the RE silicides form a new class that is distinct from the transition metal silicides [1-3,13,28]. The formation of RE silicides is characterised by the fact that the solid state interactions between RE and single crystal silicon exhibit a critical temperature dependence. Below the critical annealing temperature, interactions are very sluggish, while above the critical anneal temperature, interactions are fast. RE silicides grow in a very narrow temperature range (or in a very narrow time range). For example, Thompson et al [2] found that the reactions between Si and thin films of RE (Gd, Dy, Ho, Er, Y and La) form the disilicides rapidly within a narrow temperature range (325-400°C). The growth does not follow a layer growth mode. Thompson et al [3] also showed that a 400°C anneal will convert a 3000 Å thick erbium film to $ErSi_2$ in 15 min. If the temperature is lowered to 300°C, the silicide does not begin to form until the film has been annealed for several days. Hsu et al [5] observed similar results when studying the effect of thermal treatment on the thin film reaction of La, Ce and Nd on silicon surfaces. They found that a 200°C anneal will convert a 500 Å thick La film to a $LaSi_2$ intermixed layer in 45 min. If the temperature is lower than a critical temperature, the silicide does not begin to form even when the film has been annealed for a long time (~6 hr) [13]. Because of the extremely fast rate at which the RE film is converted to silicide, it is difficult to ascertain the type of growth rate exhibited by the RE silicides. Because the formation temperature range is narrow, it is quite a problem to calculate the growth kinetics with any certainty within this narrow temperature window. Therefore, only the kinetic rate or activation energy of growth has been determined for RE silicide.

In the study of lanthanum silicide formation, Hsu et al [5] found that with the isothermal annealing, RE metals reacted and intermixed with Si at temperatures as low as 200°C for 6 min, but the disilicide phase, confirmed by X-ray diffraction, did not emerge until the annealing temperature was raised to 600°C for 60 min. Thus they suggested that there is an intermediate stage of silicide growth between 200 and 600°C before the complete formation of lanthanum disilicide [5]. A multilayer sample was used to study this intermediate stage, since a phase which forms to a limited extent is more readily apparent when many interfaces are present in a multilayer sample [14]. The authors of [14] showed that at the intermediate stages of formation of $LaSi_{2-x}$, between 150 and 400°C, there is some evidence for small grains in the selected area diffraction patterns, indicating that $LaSi_2$ crystallites were present in some regions. In the study of cerium silicide formation, Hsu et al observed similar results [15].

From the experimental results observed to date, it is generally accepted that the mechanism of RE silicide thin film formation is dominated by nucleation phenomena [1-3,13,28], although much is left to be done on the formation of these RE silicides.

## REFERENCES


[1] J.E. Baglin, F.M. d'Heurle, C.S. Peterson [ *Appl. Phys. Lett. (USA)* vol.36 (1980) p.594-6 ]

[2] R.D. Thompson, B.Y. Tsaur, K.N. Tu [ *Appl. Phys. Lett. (USA)* vol.38 (1981) p.535-7 ]

[3] R.D. Thompson, K.N. Tu [ *Thin Solid Films (Switzerland)* vol.93 (1982) p.265-74 ]

[4] V.M. Koleshko, V.F. Belitsky, A.A. Khodin [ *Thin Solid Films (Switzerland)* vol.141 (1986) p.277-85 ]

[5] C.C. Hsu, Y.X. Wang, J. Hu, J. Ho, J.J. Qian [ *J. Vac. Sci. Technol. A (USA)* vol.7 (1989) p.3016-22 ]

[6] S. Gokhale, N. Ahmed, S. Mahamuni, V.J. Rao, A.S. Nigavekar, S.K. Kulkarni [ *Surf. Sci. (Netherlands)* vol.210 (1989) p.85-98 ]

[7] J.Y. Duboz et al [ *Appl. Surf. Sci. (Netherlands)* vol.38 (1989) p.171-7 ]

[8] F. Arnaud D'Avitaya, P.A. Badoz, Y. Campidelli, J.A. Chroboczek, J.Y. Duboz, A. Perio [ *Thin Solid Films (Switzerland)* vol.184 (1990) p.283-93 ]

[9] M.P. Siegal, F.H. Kaatz, W.R. Graham, J.J. Santiago, J. Van der Spiegel [ *Appl. Surf. Sci. (Netherlands)* vol.38 (1989) p.162-70 ]

[10] F.H. Kaatz, M.P. Siegal, W.R. Graham, J. Van der Spiegel, J.J. Santiago [ *Thin Solid Films (Switzerland)* vol.184 (1990) p.325-33 ]

[11] C.D. Anteroches, P. Perret, F. Arnaud D'Aritaya, J.A. Chroboczek [ *Thin Solid Films (Switzerland)* vol.184 (1990) p.349-56 ]

[12] C.S. Wu, S.S. Lau, T.F. Kuech, B.X. Liu [ *Thin Solid Films (Switzerland)* vol.104 (1983) p.175-82 ]

[13] J. Ho, C.C. Hsu [ *Acta Phys. Sin. (China)* vol.42 (1993) p.1617-62 ]

[14] C.C. Hsu, J. Ho, J.J. Qian, Y.T. Wang, Y.X. Wang [ *Vacuum (UK)* vol.41 (1990) p.1425-7 ]

[15] C.C. Hsu, J. Ho, J.J. Qian, Y.X. Wang [ *J. Vac. Sci. Technol. A (USA)* vol.9 (1991) p.998-1001 ]

[16] J.E.E. Baglin, F.M. d'Heurle, C.S. Peterson [ *J. Appl. Phys. (USA)* vol.52 (1981) p.2841-6 ]

[17] A. Iandelli, A. Palenzona, G. Olcese [ *J. Less-Common Met. (Switzerland)* vol.64 (1979) p.213-20 ]

[18] J.A. Knapp, S.T. Picraux [ *Appl. Phys. Lett. (USA)* vol.48 (1986) p.466-8 ]

[19] G. Molnár, I. Gerőcs, G. Petö, E. Zsoldos, E. Jároli, J. Gyulai [ *J. Appl. Phys. (USA)* vol.64 (1988) p.6746-9 ]

[20] G. Molnár, I. Gerőcs, G. Petö, E. Zsoldos, J. Gyulai [ *Appl. Phys. Lett. (USA)* vol.58 (1991) p.249-50 ]

[21] I. Gerőcs, G. Molnár, E. Jároli, E. Zsoldos, G. Petö, J. Gyulai [ *Appl. Phys. Lett. (USA)* vol.51 (1987) p.2144-5 ]

[22] A. Travlos, P. Aloupogiannis, E. Rokofllou, C. Papastaikoudis, G. Weber, A. Traverse [ *J. Appl. Phys. (USA)* vol.72 (1992) p.948-52 ]

[23] F.H. Kaatz, J. Van der Spiegel, W.R. Graham [ *J. Appl. Phys. (USA)* vol.69 (1991) p.514-6 ]

[24] F. Arnaud D'Avitaya, A. Perio, J.C. Oherlin, Y. Campidelli, J.A. Chroboczek [ *Appl. Phys. Lett. (USA)* vol.54 (1989) p.2198-200 ]

[25] C. Wigren, J.N. Anderson, R. Nyholm, U.O. Karlsson [ *J. Vac. Sci. Technol. A (USA)* vol.9 (1991) p.1942-5 ]

[26] S.S. Lau, C.S. Pai, C.S. Wu, T.F. Kurch, B.X. Lau [ *Appl. Phys. Lett. (USA)* vol.41 (1982) p.77-80 ]

[27] J.J. Qian, Y.T. Wang, J. Ho, C.C. Hsu [ *Appl. Phys. Lett. (USA)* vol.57 (1990) p.43-5 ]

[28] F.M. d'Heurle, P. Gas [ *J. Mater. Res. (USA)* vol.1 (1986) p.205-21 ]

[29] A. Travlos, P. Aloupogiannis, G. Weber, G. Rohaye [ *J. Appl. Phys. (USA)* vol.70 (1991) p.7620-2 ]

# 2.5 Epitaxial growth of RE silicides

Xu Zhenjia (C.C. Hsu)

June 1995

## A THE LATTICE MISMATCH TO THE Si (111) SURFACE

RE silicides have attracted recent attention as model candidates for heteroepitaxial growth on Si(111), since the structure of the RE silicides is almost an ideal match to that of the (111) silicon bulk lattice.

The RE silicides are usually prepared by a contact reaction method with the RE metal deposited on a pre-cleaned silicon surface and then heated. The RE silicides formed by this method usually are found to form silicon-rich silicides of the form $RESi_{2-x}$. This is a hexagonal phase based on the structure $AlB_2$, but with 15 - 20% vacancies on the Si sublattice (thus the ratio is 1.6 - 1.7 instead of 2). Based on TEM results this deficiency has been suggested to result from vacancies ordered in the silicon-silicide sublattice and the composition has been determined by RBS. The arrangement of Si atoms in a plane normal to the silicide c axis is essentially identical to that on the (111) surface of Si, resulting in a nearly ideal epitaxial arrangement. The lattice mismatch to the Si(111) surface is given in TABLE 1. Although a good lattice match to the substrate is not essential for epitaxy, it is obviously helpful.

TABLE 1 Lattice constants and lattice match to Si(111) for $RESi_{2-x}$*.

| | $a_R$ (Å) | $c_R$ (Å) | F (%) | Ref |
|---|---|---|---|---|
| Si | $a_0$ = 3.840 | | | |
| $YSi_{2-x}$ | 3.842 | 4.140 | +0.052 | [4] |
| $GdSi_{2-x}$ | 3.877 | 4.172 | +0.960 | [4,17] |
| $TbSi_{2-x}$ | 3.847 | 4.146 | +0.182 | [4,17] |
| $DySi_{2-x}$ | 3.831 | 4.121 | -0.234 | [4,17] |
| $HoSi_{2-x}$ | 3.816 | 4.107 | -0.630 | [4,17] |
| $ErSi_{2-x}$ | 3.799 | | -1.07 | [4] |
| | 3.798 | 4.088 | -1.09 | [17] |
| $TmSi_{2-x}$ | 3.773 | 4.070 | -1.75 | [4] |
| | 3.768 | 4.070 | -1.88 | [17] |
| $YbSi_{2-x}$ | 3.771 | 4.098 | -1.80 | [4] |
| | 3.784 | 4.098 | -1.46 | [17] |
| $LuSi_{2-x}$ | 3.745 | 4.050 | -2.47 | [4] |
| | 3.747 | 4.046 | -2.42 | [17] |

* The lattice constant of Si, a = b = c = 5.430 Å. The half-diagonal of the silicon lattice, $a_0 = \frac{\sqrt{2}}{2} a$ = 3.840 Å. The lattice parameters of RE silicides are $a_R$ and $c_R$ (see TABLE 1).

$$\text{Mismatch to Si(111) (\%)} = F = \frac{a_R - a_0}{a_0} \%.$$

## B EPITAXIAL GROWTH CONDITIONS

Apart from the ideal match of RE silicides to the Si(111) bulk lattice, some factors which have been identified as affecting the quality of the crystallinity, such as the phase formation and the quality of the RE silicide/Si interface, must be considered.

The surface of RE silicides formed by the reaction of a thin film metal layer with a silicon substrate is typically dominated by deep, penetrating, regularly shaped pits. It was suggested that contamination at the metal/Si or silicide/Si interface is the primary cause of surface pitting [12,18,26]. Lau et al [26] showed that the surface pits may be reduced in density or eliminated entirely through either the use of a Si substrate surface prepared under UHV conditions prior to metal deposition and silicide formation, or by means of ion irradiation techniques. Silicide layers formed by these techniques possess an almost planar morphology. Wu et al [12] suggested that the surface morphology can be improved significantly via reaction with amorphous silicon (a-Si) on top of a crystalline Si (c-Si) substrate. Planar $ErSi_2$ layers (pit free) were made using this approach with a sample structure a-Si/Er/c-Si. The fast reaction between a-Si and Er leads to a planar sample structure $ErSi_2$/c-Si with little or no reaction between erbium and the c-Si substrate. Knapp and Picraux [18] investigated the whole RE group using an electron beam rapid heating system to react overlayers of RE metals on Si(111), forming epitaxial layers of silicides of Y, Gd, Tb, Dy, Ho, Er, Tm, Yb and Lu. The use of fast electron beam heating not only results in a much smoother surface topology but also promotes epitaxial growth on Si(111) in both solid and liquid phase reactions.

Growth technique and epitaxial conditions are both extremely important for improving epitaxial quality in silicides. Continued in-situ work in UHV systems is necessary, since REs are very reactive. Hsu et al [5] studied the effect of thermal treatment on the thin-film reaction of RE on the silicon surface. They found that rapid thermal annealing and laser annealing are superior to conventional furnace annealing in forming an ideal Re silicide/Si interface. Recently, it was reported that continuous and monocrystalline RE silicide films can be prepared with an MBE system by co-deposition of both RE and Si on either heated or room temperature Si substrates followed by solid phase epitaxy [7,8]. This technique gives films with very satisfactory overall properties. A successful template technique for improving epitaxial quality in RE silicides has been reported [9]. By this approach, a very thin layer (10 - 50 Å) of RE metal is deposited onto a substrate held at 300°C, the lowest temperature conducive to RE-silicide formation, then raised slowly to 1000°C to form epitaxial $RESi_{2-x}$. This method of film growth is similar to reactive deposition epitaxy (RDE) where silicide formation occurs immediately upon metal deposition onto a heated substrate. This layer then acts as a template for the growth of thicker films, also by RDE. Therefore, the formation of an ultra-thin silicide template is a precursor to thicker film growth.

Knapp and Picraux [18] reported that Gd did not form an epitaxial silicide layer in a solid phase reaction, but did show some epitaxy when reacted in the liquid phase with the use of fast electron beam heating. However, Molnár et al [19] showed that the gadolinium thin-film silicon substrate system has two phases, the first being hexagonal $GdSi_{\sim 1.7}$ and the second one orthorhombic $GdSi_2$. The ratio of the two phases depends on the temperature of the heat treatment and, at a given temperature and time of annealing, a dependence on the thickness of the evaporated Gd layer was found. At ~100 nm Gd thickness the dominant phase was orthorhombic $GdSi_2$, while at ~250 nm hexagonal $GdSi_{1.7}$ was formed. In the 300 - 1000 nm interval orthorhombic was the main compound again. RBS analysis showed that the phases were mixed within the layer. Gerőcs et al [21] reported that epitaxial orthorhombic $GdSi_2$ can be grown by in-situ vacuum annealing of a 50 nm Gd layer on (100)Si. The lattice mismatch between the orthorhombic $GdSi_2$ and <100> silicon substrate was found to be 4%. Molnár et al [20] reported that epitaxial hexagonal $GdSi_{1.7}$ was also grown by in-situ vacuum annealing of 50 and 250 nm Gd layers on <111>Si.

In the study of the atomic structure at the $LaSi_{2-x}$/Si(100) interface, Qian et al [27] showed by high resolution cross-sectional transmission electron microscopy that the c axis of the $LaSi_{2-x}$ unit cell points in the <332>Si direction, and they proposed a model suggesting that the {100} planes of $LaSi_{2-x}$ grow on the {113}Si surface.

The epitaxial growth of RE silicides on <111>Si is summarised in TABLE 2.

TABLE 2 Epitaxial growth of RE silicides on <111>Si.

| RE/<111>Si | Epitaxial growth | Growth conditions | Remarks | Ref |
|---|---|---|---|---|
| La, Ce, Pr, Nd, Pm, Sm, Eu | No | | | |
| Gd | Yes | LPE, FEBA, Xmin ~93% | | [18] |
| | | LPE, RTA, $10^{-6}$ Pa<br>50 nm (350°C, 1.5 min)<br>250 nm (600 - 500°C, 4 min) | $GdSi_{1.7}$ | [20] |
| | Yes<br>Gd/<100>Si | LPE, RTA<br>50 nm (500°C, 5 min) | $GdSi_2$ (orthorhombic) | [21] |
| Tb | Yes | LPE, FEBA, Xmin ~85% | 33 LEED pattern | [18] |
| | | SPE (template Tb(3A)/Si(5A, 400°C)), 800 - 850°C, Xmin <10% | [1010]$TbSi_{2-x}$‖[112]Si<br>pinholes <0.15 μm | [25] |
| Dy | | LPE, FEBA, Xmin ~57% | | [18] |
| Ho | Yes | SPE, FEBA, Xmin ~81% | | [18] |
| Er | Yes | SPE, FEBA, Xmin ~35% | | [18] |
| | | SPE, RH, 200 - 400 Å,<br>5 x $10^{-10}$ torr, 850°C (30 min)<br>Xmin ~2.2% | 33 LEED pattern<br>pinholes <1 μm<br>35 cm (RT) | [10] |
| | | SPE, MBE ~330 Å.<br>Two steps: Reaction 400°C;<br>Annealing >800°C,<br>co-deposition (Er:Si = 1:2)<br>Si/$ErSi_{2-x}$/Si (Xmin ~70%) | 33 LEED pattern<br>(0001) plane<br>35 cm (RT) | [7] |
| | | SPE, MBE, co-deposition<br>(Er:Si = 1:2) 200 Å,<br>900 - 650°C | [1230]$ErSi_2$‖[002]Si<br>[1010]$ErSi_2$‖[422]Si | [11] |
| | | SPE, RH (template), 200 Å<br>$10^{-10}$ torr, 700°C, Xmin ~2% | 30 cm (RT) $ErSi_{2-x}$ | [9] |
| | | SPE, MBE, co-deposition<br>780°C, 300 - 5000 Å | Deficiency 20% | [24] |

Table 2 continued

| RE/<111>Si | Epitaxial growth | Growth conditions | Remarks | Ref |
|---|---|---|---|---|
| Tm | Yes | SPE, FEBA, Xmin ~60%<br>LPE, FEBA, Xmin ~75% | $TmSi_{1.7}$<br>(0001)‖Si(111) | [18] |
| Yb | Yes | SPE, FEBA, Xmin ~71%<br>LPE, FEBA, Xmin ~25%<br>SPE, RH 40 ML<br>400°C, 1 x $10^{-10}$ mbar | $YbSi_{1.7}$(0001)‖Si(111)<br><br>11 or 33 LEED pattern<br>$YbSi_{2-x}$ | [18]<br>[25] |
| Lu | Yes | LPE, FEBA, Xmin ~20%<br>SPE, FEBA, Xmin ~78% | $LuSi_{1.7}$(0001)‖Si(111) | [18]<br>[22] |

SPE: Solid Phase Epitaxy; MBE: Molecular Beam Epitaxy; LPE: Liquid Phase Epitaxy;
RDE: Reaction Deposition Epitaxy; RH: Resistive Heating; Xmin: Minimum Channelling Yield;
FEBA: Fast e-Beam Annealing (1200 - 1400 K, dwell time ~400 s);
RTA: Rapid Thermal Annealing (IR irradiation).

## REFERENCES

[1] J.E. Baglin, F.M. d'Heurle, C.S. Peterson [ *Appl. Phys. Lett. (USA)* vol.36 (1980) p.594-6 ]

[2] R.D. Thompson, B.Y. Tsaur, K.N. Tu [ *Appl. Phys. Lett. (USA)* vol.38 (1981) p.535-7 ]

[3] R.D. Thompson, K.N. Tu [ *Thin Solid Films (Switzerland)* vol.93 (1982) p.265-74 ]

[4] V.M. Koleshko, V.F. Belitsky, A.A. Khodin [ *Thin Solid Films (Switzerland)* vol.141 (1986) p.277-85 ]

[5] C.C. Hsu, Y.X. Wang, J. Hu, J. Ho, J.J. Qian [ *J. Vac. Sci. Technol. A (USA)* vol.7 (1989) p.3016-22 ]

[6] S. Gokhale, N. Ahmed, S. Mahamuni, V.J. Rao, A.S. Nigavekar, S.K. Kulkarni [ *Surf. Sci. (Netherlands)* vol.210 (1989) p.85-98 ]

[7] J.Y. Duboz et al [ *Appl. Surf. Sci. (Netherlands)* vol.38 (1989) p.171-7 ]

[8] F. Arnaud D'Avitaya, P.A. Badoz, Y. Campidelli, J.A. Chroboczek, J.Y. Duboz, A. Perio [ *Thin Solid Films (Switzerland)* vol.184 (1990) p.283-93 ]

[9] M.P. Siegal, F.H. Kaatz, W.R. Graham, J.J. Santiago, J. Van der Spiegel [ *Appl. Surf. Sci. (Netherlands)* vol.38 (1989) p.162-70 ]

[10] F.H. Kaatz, M.P. Siegal, W.R. Graham, J. Van der Spiegel, J.J. Santiago [ *Thin Solid Films (Switzerland)* vol.184 (1990) p.325-33 ]

[11] C.D. Anteroches, P. Perret, F. Arnaud D'Avitaya, J.A. Chroboczek [ *Thin Solid Films (Switzerland)* vol.184 (1990) p.349-56 ]

[12] C.S. Wu, S.S. Lau, T.F. Kuech, B.X. Liu [ *Thin Solid Films (Switzerland)* vol.104 (1983) p.175-82 ]

[13] J. Ho, C.C. Hsu [ *Acta Phys. Sin. (China)* vol.42 (1993) p.1617-62 ]

[14] C.C. Hsu, J. Ho, J.J. Qian, Y.T. Wang, Y.X. Wang [ *Vacuum (UK)* vol.41 (1990) p.1425-7 ]

[15] C.C. Hsu, J. Ho, J.J. Qian, Y.X. Wang [ *J. Vac. Sci. Technol. A (USA)* vol.9 (1991) p.998-1001 ]

[16] J.E.E. Baglin, F.M. d'Heurle, C.S. Peterson [ *J. Appl. Phys. (USA)* vol.52 (1981) p.2841-6 ]

[17] A. Iandelli, A. Palenzona, G. Olcese [ *J. Less-Common Met. (Switzerland)* vol.64 (1979) p.213-20 ]

[18] J.A. Knapp, S.T. Picraux [ *Appl. Phys. Lett. (USA)* vol.48 (1986) p.466-8 ]

[19] G. Molnár, I. Geröcs, G. Petö, E. Zsoldos, E. Jároli, J. Gyulai [ *J. Appl. Phys. (USA)* vol.64 (1988) p.6746-9 ]

[20] G. Molnár, I. Geröcs, G. Petö, E. Zsoldos, J. Gyulai [ *Appl. Phys. Lett. (USA)* vol.58 (1991) p.249-50 ]

[21] I. Geröcs, G. Molnár, E. Jároli, E. Zsoldos, G. Petö, J. Gyulai [ *Appl. Phys. Lett. (USA)* vol.51 (1987) p.2144-5 ]

[22] A. Travlos, P. Aloupogiannis, E. Rokofllou, C. Papastaikoudis, G. Weber, A. Traverse [ *J. Appl. Phys. (USA)* vol.72 (1992) p.948-52 ]

[23] F.H. Kaatz, J. Van der Spiegel, W.R. Graham [ *J. Appl. Phys. (USA)* vol.69 (1991) p.514-6 ]

[24] F. Arnaud D'Avitaya, A. Perio, J.C. Oherlin, Y. Campidelli, J.A. Chroboczek [ *Appl. Phys. Lett. (USA)* vol.54 (1989) p.2198-200 ]

[25] C. Wigren, J.N. Anderson, R. Nyholm, U.O. Karlsson [ *J. Vac. Sci. Technol. A (USA)* vol.9 (1991) p.1942-5 ]

[26] S.S. Lau, C.S. Pai, C.S. Wu, T.F. Kurch, B.X. Lau [ *Appl. Phys. Lett. (USA)* vol.41 (1982) p.77-80 ]

[27] J.J. Qian, Y.T. Wang, J. Ho, C.C. Hsu [ *Appl. Phys. Lett. (USA)* vol.57 (1990) p.43-5 ]

[28] F.M. d'Heurle, P. Gas [ *J. Mater. Res. (USA)* vol.1 (1986) p.205-21 ]

[29] A. Travlos, P. Aloupogiannis, G. Weber, G. Rohaye [ *J. Appl. Phys. (USA)* vol.70 (1991) p.7620-2 ]

## 2.6 The RE/Si interface

Xu Zhenjia (C.C. Hsu)

June 1995

### A INTRODUCTION

The physical and chemical properties of the interfaces prepared by depositing RE overlayers on different Si surfaces have been the subject of extensive investigations during the last several decades. This is due to the relatively well established techniques for preparing atomically clean interfaces in ultrahigh vacuum, as opposed to the more difficult growth of stoichiometric single phase bulk compounds of convenient size for spectroscopic analysis. Two major problems have been addressed: the structure and the electronic states of the interfaces. The structural properties of the interfaces have been recognised to be a complex function of several parameters such as the surface orientation, the annealing temperature etc. The experimental evidence for either the evolution of the interface reaction of the RE/silicon interface or the electronic properties of the RE/silicon interface is mainly based on the RE adsorbate photoemission signal intensity as a function of the deposited amount of RE atoms and the chemical shifts of the Si2p core line. Support for this interpretation comes from the RE compound Auger transition intensity and, in the case of Si, the LVV transition, which reflects an intermixing process. Several mechanisms have been proposed in order to model the interface reactions, but none of them alone is capable of explaining all the experimental observations. We will attempt a review of the experimental information obtained in recent years on the early stages of formation of RE-silicon systems. In the following we will deal with the evolution of the interface reaction of the RE-silicon couples, the recognition of the electronic properties of the RE/silicon interface products and the growth mechanism of the interface formation.

### B EVOLUTION OF THE INTERFACE REACTION

#### B1 La/Si Interface

Puppin et al [1] have studied the La/Si interface prepared by depositing increasing amounts of La on Si(111)-2x1 at room temperature. After each deposition a set of core level and valence band photoemission spectra was taken using synchrotron radiation monochromatic radiation in the energy range between 30 and 200 eV. Two different conditions can be identified in the Si2p spectra measured. One contribution comes from the unreacted Si substrate; the second one comes from the Si atoms in which the distribution of the valence electrons has been modified by the interaction with La. The evolution of the La/Si interface reaction can be discussed separately for two different coverage regions:

(1) ~0.1 monolayer (ML). (a) The metal core level is strongly shifted with respect to the pure metal case, whereas the Si2p shift is relatively small. (b) The appearance of a new narrow contribution to the density of states (DOS) which is centred around 1.4 eV below the Fermi level, $E_F$.

(2) ~2 ML. (a) The La5p shift is negligible compared to the pure metal whereas the reacted component of the 2p core level is strongly shifted with respect to the unreacted substrate. (b) The difference EDCs consist of a broad emission covering a region 4 eV wide below $E_F$.

(3) Between these two extreme situations (0.1 - 2 ML), the interface parameters evolve in a more or less gradual way. The Si2p/La5p ratio remains approximately constant between 0.5 and 2 ML.

From the above results, it can be concluded that the interface appears to be reacted from 0.1 ML to 2 ML. Further deposition of La results in the growth of pure metal.

### B2 Ce/Si Interface

Grioni et al [2,3] used a variety of techniques, including synchrotron radiation photoemission, angle-resolved Auger and LEED, to study in detail the heterogeneous growth of Ce deposited onto cleaned Si(111)-2x1 and 7x7 surfaces. At any coverage, their results can be fitted with components representing clean Si (Si-1), a reacted phase (Si-2) and the surface segregated Si (Si-3). The attenuation of the total Si2p emission is a measure of the rate at which Si is masked by the overlayer. The interface reaction demonstrated by the experimental results is complex. It can be discussed according to four different coverage regions.

(1) <0.6 ML. Between zero and 0.6 ML, there is a rigid shift of 200 meV due to a variation in band bending, but the unvarying line shape indicates minimal Ce-Si substrate interaction. From the results of subtraction of normalised EDCs for clean Si and 0.4 ML Ce/Si, the well-defined doublet 1.2 and 3.2 eV below $E_F$ reveals emission from Ce-induced states and is indicative of weakly interacting Ce clusters. These Ce clusters interact weakly with the substrate and the Si2p binding energy does not change.

(2) ~(0.6-3 ML). At 0.6 ML coverage, the reacted Si component is first observed. Analysis of the Si2p line shape for 0.6 ML reveals a reacted Si species shift of 0.7 eV to lower binding energy. This component grows and is most pronounced near 3 ML whereas the substrate attenuates rapidly. The featureless valence-band spectra provide no evidence of a metallic Ce layer. However, both the valence bands and the core levels do indicate that a Ce layer forms shortly thereafter.

(3) ~(3-5 ML). Decomposition of the core emission demonstrates that the third Si species (Si-3) appears at nominal coverage of 3 ML and all these Si components are present in about equal amounts in the surface region for ~5 ML. At this coverage, the valence-band results show a metallic Ce overlayer, and the (Si-2) attenuation reveals submergence of the reacted phase. The persistence and nearly constant concentration of (Si-3) suggests surface segregated Si within an escape depth from the top of the Ce film.

(4) ~(8-10 ML). The almost complete loss of (Si-1) and (Si-2) signals by 8 ML indicates that the Ce-Si reaction is confined to a thin well-defined layer. This is consistent with the completion of silicide formation and Ce submergence.

### B3 Sm/Si Interface

There have been several experimental data on the evolution of the Sm/Si interface at room temperature using synchrotron radiation photoemission, AES and LEED [4-9]. The interface reaction between Sm and cleaved Si(111)-2x1 and Si(100) with increasing Sm coverage has been studied. From the reported experimental results, it was concluded that there are two stages of the interface reaction:

(1) At Sm coverage from 0.5 ML up to 1 ML (~2-3 Å), Sm atoms appear only in a divalent state corresponding to a weak chemisorption first stage. The evolution of the Sm/Si interface reaction at room temperature is characterised by a disruption of the Sm-Si bonding, leading to the diffusion of Si.

(2) Intermixing occurs and silicides are formed in the mixed zone in a deposition range 1-5 ML. Increased Sm deposition above (5-6) ML results in a metallic phase which eventually leads to a thin film displaying a photoemission similar to that of clean Sm. In this stage, the trivalent Sm configuration dominates, and shifts in energy of the Si and Sm core level features reveal a strong chemical interaction. The Sm valence change is related to the onset of the reactive interdiffusion of Sm and Si.

Several LEED structures were observed for Sm/Si(100) with increasing temperature and different initial deposition [8]. One characteristic pattern is the (32) structure, which was observed for a deposition of 0.34 ML and annealing at 475°C.

### B4 Eu/Si Interface

The Eu/Si interface formation has been studied with photoemission spectroscopy for Eu monolayer onto Si(111)-7x7 substrates [10] and for Si(111)-2x1 substrates [11]. Recently, Henle et al [12] studied in detail the formation and evolution of the Eu/Si(111)-7x7 interface at room temperature (RT) and at elevated temperatures using synchrotron radiation photoelectron spectroscopy, Auger electron spectroscopy and LEED.

#### B4.1 Room temperature

Based on the experimental results of [12], the evolution of the Eu/Si(111) interface at RT can be summarised and illustrated in TABLE 1. The Si2p core level spectra are normalised to equal peak height and have been separated by computer analysis into bulk (B), surface (S) and reacted components (R1, R2). The first reacted R1 component appears to be chemically inhomogeneous as indicated by the broadened Gaussian width of the Si2p component. The second reacted R2 component is a well-defined silicide type reaction product as suggested by the remarkably narrow Si2p line width.

TABLE 1 The evolution of Eu/Si(111) interface reaction at room temperature [12].

| Eu, (Å)* | R1 component | R2 component | LEED | Remarks |
|---|---|---|---|---|
| 0 | | | 7x7 reconstruction pattern | |
| 2 (~0.5 ML) | | | 7x7 reconstruction pattern is lifted (1x1) pattern is observed | Onset of chemical reaction between Eu and Si |
| 4 - 5 (1 ML) | Increased to a maximum | | | |
| 7 (~1.8 ML) | Decreased | Increased at the expense of the bulk and R1 | (1x1) pattern destroyed | The surface becomes scrambled by an intermixing of the R1 and R2 |
| 20 (~5.3 ML) | | Mainly R2 | | |
| 7 - 25 (~1.8 - 6.6 ML) | | | | Plateau region in the $Si_{LVV}$ versus (Eu). Strong intermixing region. |
| 50 (~13 ML) | | Weak signal | | |

* 1 ML = 3.8 Å.

## B4.2 Thermal evolution of Eu/Si(111) interface reaction

FIGURE 1 represents a phase diagram of the Eu/Si(111)-7x7 interface in a three-dimensional representation [12].

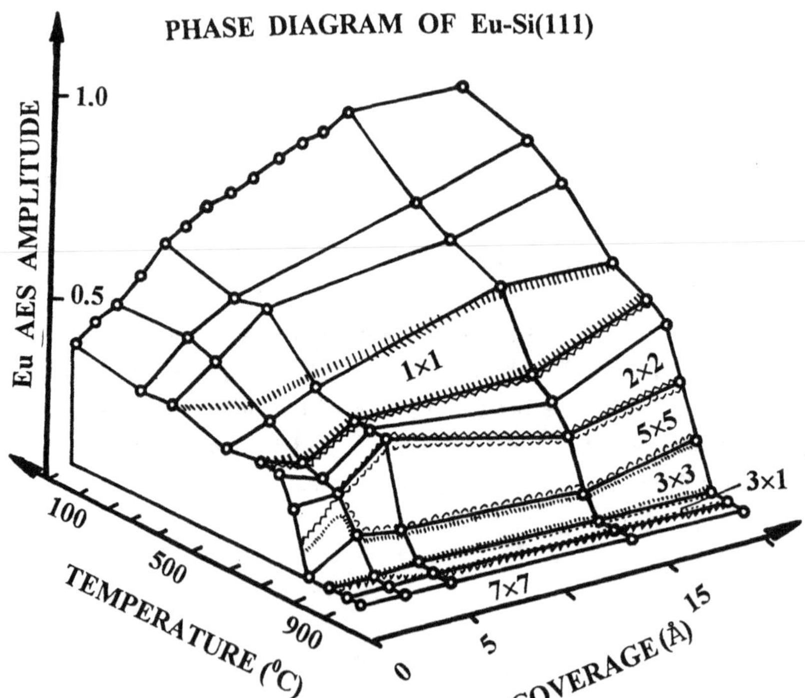


FIGURE 1 Phase diagram of the Eu/Si(111)-7x7 interface (after [12]).

## B5 Gd/Si Interface

Gd is trivalent and has seven 4f electrons as does Eu, its neighbour in the periodic table. The interaction of Gd with Si(111) surfaces [13-16] and with Si(100) surfaces [17] as a function of Gd coverage at RT and at elevated temperatures has been studied using synchrotron radiation photoemission spectroscopy, Auger electron spectroscopy and LEED.

### B5.1 Room temperature

Based on the experimental results of [13-17], the evolution of the Gd/Si interface reaction at RT is summarised in TABLE 2. As expected for a RE metal the reactivity of Gd on Si surfaces is high but, compared to Si(111)-7x7, the intermixing of Gd and Si is more pronounced on the Si(100) surface.

TABLE 2 The evolution of Gd/Si(111) and Gd/Si(100) interface reactions at room temperature [13-17].

| Substrate | Gd (Å) | R component | LEED | Remarks |
|---|---|---|---|---|
| Si(111) 7x7<br>1 ML ~3 Å | 0.8 - 1 | Appearance of R | (7x7) LEED pattern is replaced by a (1x1) pattern at 1 Å | Onset of the reaction |
| | 2 - 2.5 | | (1x1) surface becomes disordered | Disruption of the Si(111) |
| | >2.5 | R dominates | | The growth of the interface after disruption of the Si(111) surface |

| Substrate | Gd (Å) | R1 component | R2 component | LEED | Remarks |
|---|---|---|---|---|---|
| Si(100) 2x1<br>1 ML ~2.2 Å | 0.4 | | | | Surface component disappeared |
| | 1 - 1.5 | Appearance first | | (2x1) pattern turns into a (1x1) pattern | Onset of reaction |
| | 3 - 15 | Intensity maximum at around 10 Å | First observed at 3 Å | Beyond (3-4) Å Gd the surface is disordered | |
| | 5 - 15 | | | | An intermediate region where approximately constant AES signals are obtained |
| | 16 | | Dominates | | |

## B5.2 Thermal evolution of Gd/Si(111) interface reaction

FIGURE 2 represents a phase diagram of the Gd/Si(111)-7x7 interface in a three-dimensional representation [16].

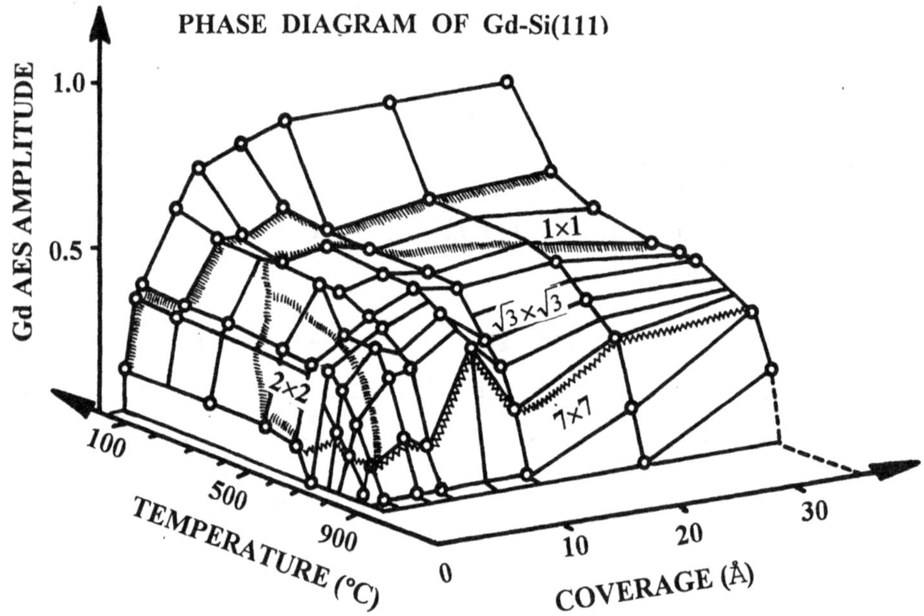


FIGURE 2 Three-dimensional representation of the Gd/Si(111)-7x7 phase diagram (after [16]).

## B6 Tb/Si Interface

Kennou et al [18] studied the interaction of Tb with Si(111)-7x7 at RT for various Tb coverages by photoelectron spectroscopy (UPS, XPS), AES and WF measurements. For the XPS measurements, in addition to the conventional Mg K line (E = 1253.6 eV), a Zr M (E = 151.4 eV) excitation source was also used. The Zr M source results in a high surface sensitivity, since the kinetic energy of the Si2p photoelectron is about 50 eV. The experimental results indicate that Tb forms a two-dimensional layer up to 2 Å as is shown by the shift of Tb 4f and 4d states. Interdiffusion between Si and the Tb overlayer has been observed above 2 Å (0.8 ML) for the RT Tb/Si(111) system. For greater coverages, a reaction of Tb and Si starts evidenced by the reacted Si2p component and the change of Si Auger KLL line shape. The interacted interface extends up to 12 Å of the Tb thickness where metallic Tb begins to build up.

## B7 Dy/Si Interface

Only limited experimental results have been published on the Dy/Si interface. Wang et al [19] studied the Dy/Si(111) interface reaction at RT using LEED, XPS, UPS and AES. Gokhale et al [20] published XPS and XRD investigations of Dy/Si(111) at different temperatures.

Experimental results [19] show that at low Dy coverages (<4 ML) atomic intermixing takes place, causing a chemical shift of the Dy 4f and Dy 3d components. However, no significant chemical shift of Si 2p has been observed at low coverages, suggesting a lack of strong chemical bonds of Dy-Si type. With increasing coverage a (7x7) LEED pattern from a clean Si(111) surface turns into a (1x1) pattern, and finally becomes ambiguous. After annealing for 30 min at 400°C, the Dy silicides formed with an atomic concentration ratio of Dy to Si of ~1:2. The binding energies of Dy 4f ($DySi_2$), Dy 3d ($DySi_2$) and Si 2p ($DySi_2$) are 5.5, 1295.2 and 98.65 eV respectively.

### B8 Er/Si Interface

Lollman et al [21] studied the Er/Si(111)-7x7 interface reaction at RT using LEED, XPS (Zr M source) and XAES. Paki et al [22] investigated the thermal evolution of the Er/Si(111)-7x7 interface reaction in the 300 - 800°C temperature range using LEED, XPS and XPD (X-ray photoelectron diffraction).

### B8.1 Room temperature

From the experimental results, the authors [21] propose a three-step mechanism for the formation at RT of the Er/Si(111) interface:

(1) A chemisorbed Er layer is first formed until (Er) = 0.8 ML.

(2) Intermixing reaction between Er and Si takes place at this critical coverage (Er) and continues to 2.5 ML.

(3) The reacted phase presents a silicide-like (most probably $ErSi_{1.7}$) chemical state. Small amounts of Si atoms are diluted in an Er deposit immediately above the surface silicide layer. The upper part of the deposit is pure Er.

### B8.2 Thermal evolution of the Er/Si(111) interface reaction

FIGURE 3 shows valence band spectra measured on a 3 ML Er deposit as a function of the annealing temperature [22]. All Er coverages are given in terms of the Si(111) atomic density: 1 ML for 7.8 x $10^{14}$ atoms $cm^{-2}$. One ML of Er may react with Si to form about one layer of $ErSi_{1.7}$ (4.09 Å).

The main peak, common to all spectra of FIGURE 3 and located at 2.2 eV for 320°C, is assigned to Er5d-Si3sp derived states. Since in the 380 - 480°C temperature range the UPS spectrum does not change, one might consider that the Er/Si(111) interface is fully reacted and that a well-defined Er silicide is formed. The thin silicide layer is epitaxially grown $ErSi_{1.7}$ attested by the 33 R 30 LEED pattern [22].

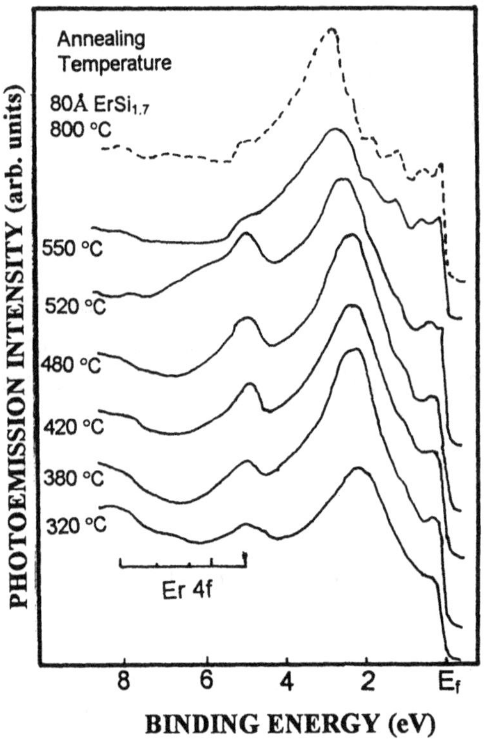


FIGURE 3 Valence band spectra evolution vs. annealing temperature for a 3 ML Er deposit on Si(111) (full lines) and valence band spectrum for an 80 Å thick $ErSi_{1.7}$ layer annealed at 800°C (dashed line) (after [22]).

## B9 Yb/Si Interface

The 4f subshell of Yb is fully occupied with 14 electrons. Therefore, a higher stability of the electronic configuration is expected. The Yb/Si(111) interface reaction has been studied using synchrotron radiation photoemission spectroscopy and Auger electron spectroscopy at RT [23] and at elevated temperatures [24,25].

### B9.1 Room temperature

The experimental results [23] show that the first monolayer is absorbed on the Si(111) surface at RT without apparent mixing or reaction. For 1.5 - 2 monolayers, some reaction is triggered, causing a strong redistribution of the electronic states of the Si atoms involved. The Yb/Si interface, as prepared at RT, is characterised by a narrow mixed-reacted region which involves 2 - 3 layers of material, covered by an almost pure Yb overlayer. The difficulty of intermixing at RT could account for the limited reactivity [23].

### B9.2 Thermal evolution of Yb/Si(111) interface reaction

Braicovich et al [25] reported the detailed synchrotron radiation photoemission results (80 - 200 eV) from Si(111)/Yb interfaces in the coverage range 0.2 - 2 ML before and after heating at 300°C and 450°C. At 1 ML, the interfaces are extremely sensitive to thermal treatments with the formation of Si-rich reaction products, while at submonolayer coverages they are essentially insensitive to the same thermal treatment. This result means that the metal concentration at submonolayer coverages is not enough to make Si displaceable through a dielectric screening effect, or that the quantity of the metal diffused into Si is not enough to make Si displaceable [25]. At higher metal coverages, Si is made available for intermixing and the sensitivity to temperature becomes very strong. The transition between the two regions is rather sharp and in this sense a critical thickness can be defined [25]. The authors also suggested that the concept of critical thickness must be used properly. In fact, it does not imply that nothing happens below this critical coverage since a strong chemical effect with a Si 2p shift is seen from the very beginning of the interface growth [25].

## C INTERFACE FORMATION AND GROWTH MODEL

The RE metals are known to form very reactive interfaces with Si with an interfacial reaction onset taking place at RT and at very low metal coverage. Consequently, these systems may be regarded as prototypes of reactive interfaces, and they are therefore of interest from a fundamental point of view. In general, it is well recognised that a multi-step interface formation process is a characteristic for the RE/Si interface growth process at RT, that is:

(1) An initial chemisorption stage. An overlayer is formed for $_{(C)}$.

(2) Reaction starts. Intermixing occurs for $_{(C)} << _{(M)}$.

(3) Increased RE deposition above $_{(M)}$ results in a metallic phase which eventually leads to a thin film displaying a photoemission similar to that of clean RE.

A multistep RE/Si interface formation process (with increasing RE coverage) is schematically shown in FIGURE 4. However, studies of the RE/silicon interface formation

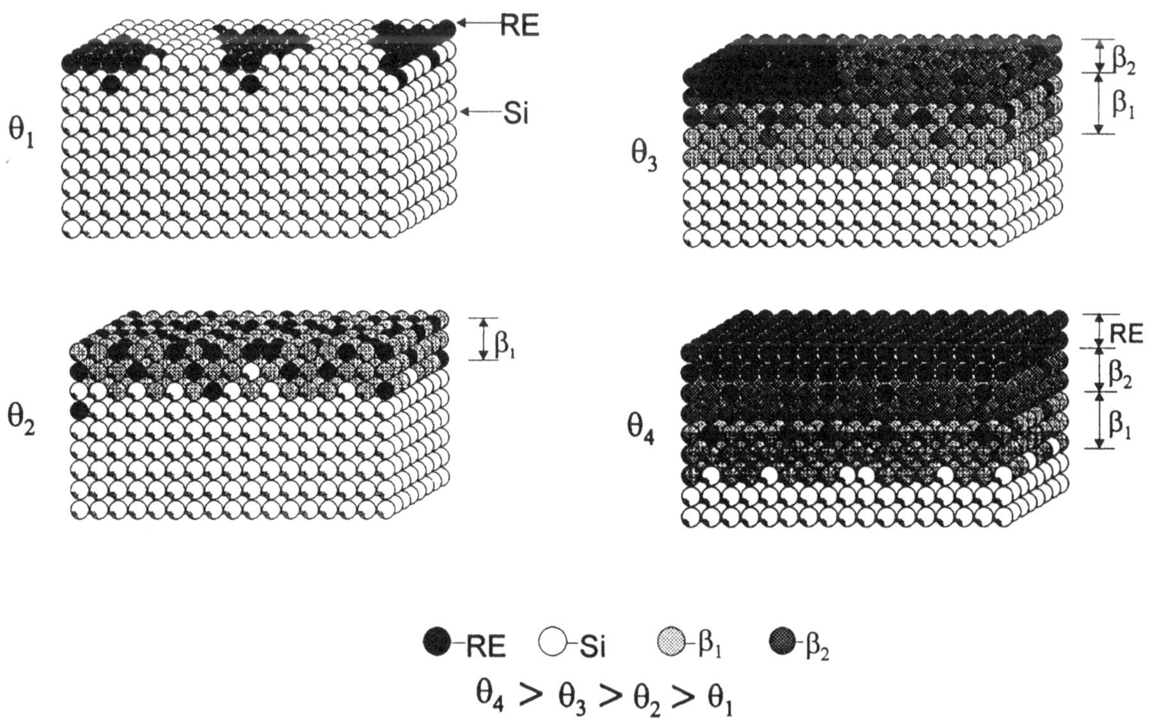


FIGURE 4 Schematic representation of a RE/Si multistep interface formation with increasing RE coverage.

show that the details of interactions between RE and Si differ from system to system. Therefore, it is difficult to put forward a general growth model based on the very limited available experimental results of the RE/Si interface formation. Up to the present time, a complex multistep interface growth has been proposed with a detailed model that accounts for the spectroscopic results of V/Ge(111), Ce/Si(111) and Ce/Ge(111) reactive metal/ semiconductor interfaces [2,3,26].

## REFERENCES


[1] E. Puppin, H. Guyot, Z.X. Shen, J. Hwang, I. Lindau [ *Solid State Commun. (USA)* vol.67 (1988) p.23-7 ]

[2] M. Grioni, J. Joyce, M. del Giudice, D.G. O'Neill, J.H. Weaver [ *Phys. Rev. B (USA)* vol.30 (1984) p.7370-3 ]

[3] M. Grioni, J. Joyce, S.A. Chambers, D.G. O'Neill, M. de Gindice, J.H. Weaver [ *Phys. Rev. Lett. (USA)* vol.53 (1984) p.2331-4 ]

[4] A. Franciosi, J.H. Weaver, P. Perfetti, A.D. Katnani, G. Margaritondo [ *Solid State Commun. (USA)* vol.47 (1983) p.427-30 ]

[5] A. Franciosi, P. Perfetti, A.D. Katnani, J.H. Weaver, G. Margaritondo [ *Phys. Rev. B (USA)* vol.29 (1984) p.5611-6 ]

[6] A. Fäldt, H.P. Myers [ *Phys. Rev. B (USA)* vol.33 (1986) p.1424-6 ]

[7] J. Onsgaard, J. Ghijsen, R.L. Johnson, F. Orskov, I. Chorkendorff, F. Grey [ *J. Electron Spectrosc. Relat. Phenom. (Netherlands)* vol.52 (1990) p.67 ]

[8] J. Onsgaard, M. Christiansen, F. Orskov, P.J. Godowski [ *Surf. Sci. (Netherlands)* vol.247 (1991) p.208-13 ]

[9] P.J. Godowski, F. Orskov, M. Christiansen, J. Onsgaard [ *Thin Solid Films (Switzerland)* vol.25 (1991) p.35-8 ]

[10] G. Rossi, J. Nogami, I. Lindau, J.J. Yeh [ *Surf. Sci. (Netherlands)* vol.152/153 (1985) p.1247-54 ]

[11] J. Nogami, C. Caarbone, J.J. Yeh, I.L. Lindau, S. Nannarone [ *Proc. 17th Int. Conf. on the Physics of Semiconductors* Eds J. Chadi, W. Harrison (Springer, New York, 1985) p.201 ]

[12] W.A. Henle, M.G. Ramsey, F.P. Netzer, K. Horn [ *Surf. Sci. (Netherlands)* vol.254 (1991) p.182-90 ]

[13] C. Carbone, J. Nagami, I. Lindau [ *J. Vac. Sci. Technol. A (USA)* vol.3 (1985) p.927 ]

[14] E. Puppin et al [ *J. Vac. Sci. Technol. B (USA)* vol.5 (1987) p.1083-6 ]

[15] W.A. Henle, F.P. Netzer, R. Cimino, W. Braun [ *Surf. Sci. (Netherlands)* vol.221 (1989) p.131-43 ]

[16] W.A. Henle, M.G. Ramsey, F.P. Netzer, R. Cimino, W. Braun [ *Solid State Commun. (USA)* vol.71 (1989) p.657-60 ]

[17] W.A. Henle, M.G. Ramsey, F.P. Netzer, S. Witzel, W. Braun [ *Surf. Sci. (Netherlands)* vol.243 (1991) p.141-50 ]

[18] S. Kennou, J.Y. Veuillen, T.A. Nguyen Tan [ *Appl. Surf. Sci. (Netherlands)* vol.56-58 (1992) p.520-4 ]

[19] Y.X. Wang, B.Q. Li, Y.R. Xing, C.C. Hsu [ *J. Vac. Sci. Technol. A (USA)* vol.7 (1989) p.2604-8 ]

[20] S. Cokhale, N. Ahmed, S. Mahamuni, V.J. Rao, A.S. Nigavekar, S.K. Kulkarni [ *Surf. Sci. (Netherlands)* vol.210 (1989) p.85-98 ]

[21] D.B.B. Lollman, T.A. Nguyen Tan, J.Y. Veuillen [ *Surf. Sci. (Netherlands)* vol.269/270 (1992) p.959-63 ]

[22] P. Paki et al [ *Surf. Sci. (Netherlands)* vol.269/270 (1992) p.970-4 ]

[23] G. Rossi et al [ *J. Vac. Sci. Technol. A (USA)* vol.1 (1983) p.781-4 ]

[24] G. Rossi, J. Nogami, J.J. Yeh, I. Lindau [ *J. Vac. Sci. Technol. B (USA)* vol.1 (1983) p.530-2 ]

[25] L. Braicovich, I. Abbati, C. Carbone, J. Nogami, I. Lindau [ *Surf. Sci. (Netherlands)* vol.168 (1986) p.193-203 ]

[26] R.A. Butera, M. del Giudice, J.H. Weaver [ *Phys. Rev. B (USA)* vol.33 (1986) p.5435-49 ]

CHAPTER 3

# THERMOCHEMICAL PROPERTIES

# 3.1 General remarks on ternary phase diagrams for metal silicides

R. Sinclair

April 1995

As pointed out by Beyers and co-workers [1-3], there are many situations in which three elements co-exist at important silicide interfaces in microelectronic circuits. For instance, in their example in FIGURE 1, Ti-Si-N are present at a $TiSi_2$-TiN diffusion barrier, while Al-Ti-N are at the upper aluminium-TiN interface. During manufacture or processing, high-temperature annealing or oxidation steps take place during which the three elements will attempt to equilibrate. Understanding the phases which are likely to be produced requires reference to the ternary, or three-element, phase diagrams. Very few such diagrams have been established experimentally for silicides, one notable exception being the Ti-Al-Si diagram of Raman and Schubert [4]. Datareviews by this author contain known phase diagrams and those calculated by the method described by Beyers [1], together with some examples of how they might be employed in typical processing situations.

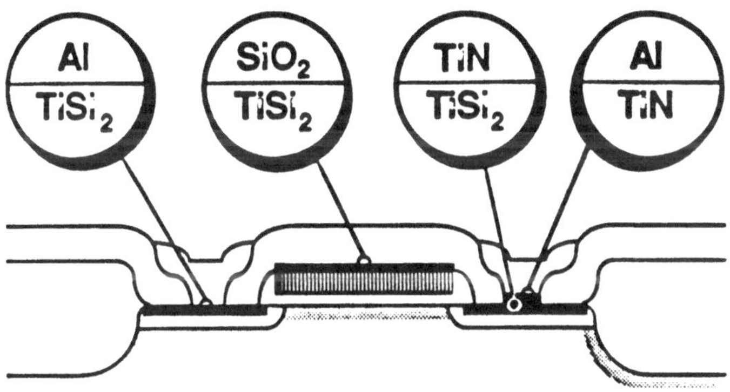


FIGURE 1 Cross-section of a transistor after salicide processing. The right contact has a TiN barrier layer between $TiSi_2$ and Al while the left contact does not. Note that three elements are typically present at each interface [3].

The isothermal section of a three-element phase diagram is conventionally illustrated as an equilateral triangle [5,6]. The elemental compositions are at the corners, binary phases along the edges and ternary compounds at the appropriate compositional point in the interior. According to the Gibbs phase rule there is a maximum of three phases in mutual equilibrium for an arbitrary composition. Regions of three phase equilibrium form triangles (tie-triangles), the corners of which identify the chemically compatible phases. The 'tie-lines' bounding the tie-triangles connect two phases which are in equilibrium with each other. Two tie-lines cannot cross or intersect since this would violate the Gibbs phase rule. Thus the disposition of the tie-lines, as well as that of the tie-triangles, is an important indication of phases which are mutually stable when in contact. Conversely, absence of such a tie-line demonstrates thermodynamic instability at the relevant temperature.

The existence of stable tie-lines can be established by calculating the Gibbs free energy of reaction at a composition point where two possible but 'competing' tie-lines would cross.

For instance in the hypothetical ternary system A-B-C in which there are only three binary phases AB, BC, CA, one elemental phase might be in equilibrium with the complementary binary compound (e.g. A-BC), or the two A-related binary compounds might be in equilibrium (see FIGURE 2). Calculation of the free energy change of the balanced chemical reaction,

$$2A + BC \rightarrow AB + AC$$

from the free energies of the participating phases, decides the issue. If the energy change (i.e. $G_{AB} + G_{AC} - 2G_A - G_{BC}$) is negative then the products are stable with respect to the reactants and a tie-line can be drawn between AB and AC. Other tie-lines can be determined in a similar manner. It should be noted that it is the free energy change of the reaction which is important, not the individual free energies (or heats of formation) of the individual compounds, although of course the latter have bearing on the former.

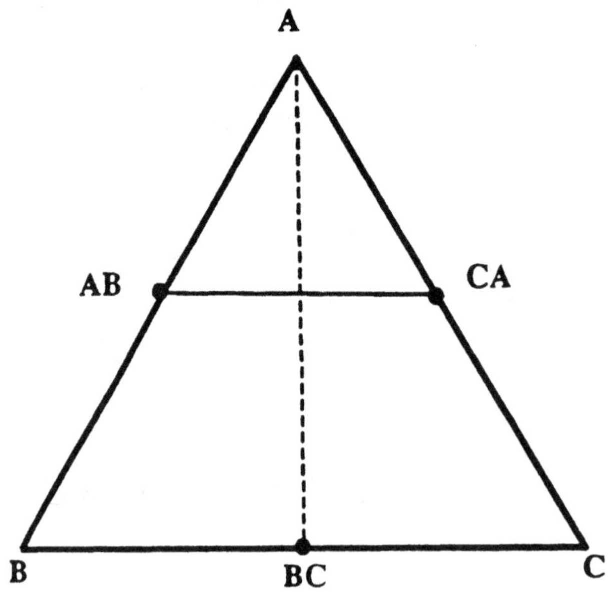


FIGURE 2 Hypothetical ternary system A-B-C in which there are only three binary phases AB, BC, CA. One elemental phase might be in equilibrium with the complementary binary compound (e.g. A-BC), or the two A-related binary compounds might be in equilibrium.

Further consideration of the present example brings to light a point of some real consequence. The existence of a key tie-line in a system can automatically preclude several others and so many phase equilibria can be inferred from one combination alone. Thus FIGURE 3 shows the four possible phase diagrams for our A-B-C system. If the A-BC tie-line exists, then not only is AB-AC stability prevented, but also that for B-AC and C-AB (since these tie-lines also cannot cross the one that is present). Accordingly, the phase diagram would automatically be that shown in FIGURE 3 (top left). This immediately indicates how one can proceed in determining the tie-line distribution experimentally, and hence the 'skeleton' of the phase diagram, without recourse to extensive repetitious work. Heating a thin film of A, which has been deposited on a substrate of the compound BC, to equilibrium followed by standard phase characterisation (e.g. by X-ray or electron diffraction) would decide the mutually stable phases. Because of the relative simplicity of modern deposition techniques, and the fact that equilibrium can be arrived at quite rapidly in thin-film reactions (since the diffusion distances are small and there is a high defect density), such an experiment can be carried out in a short period of time.

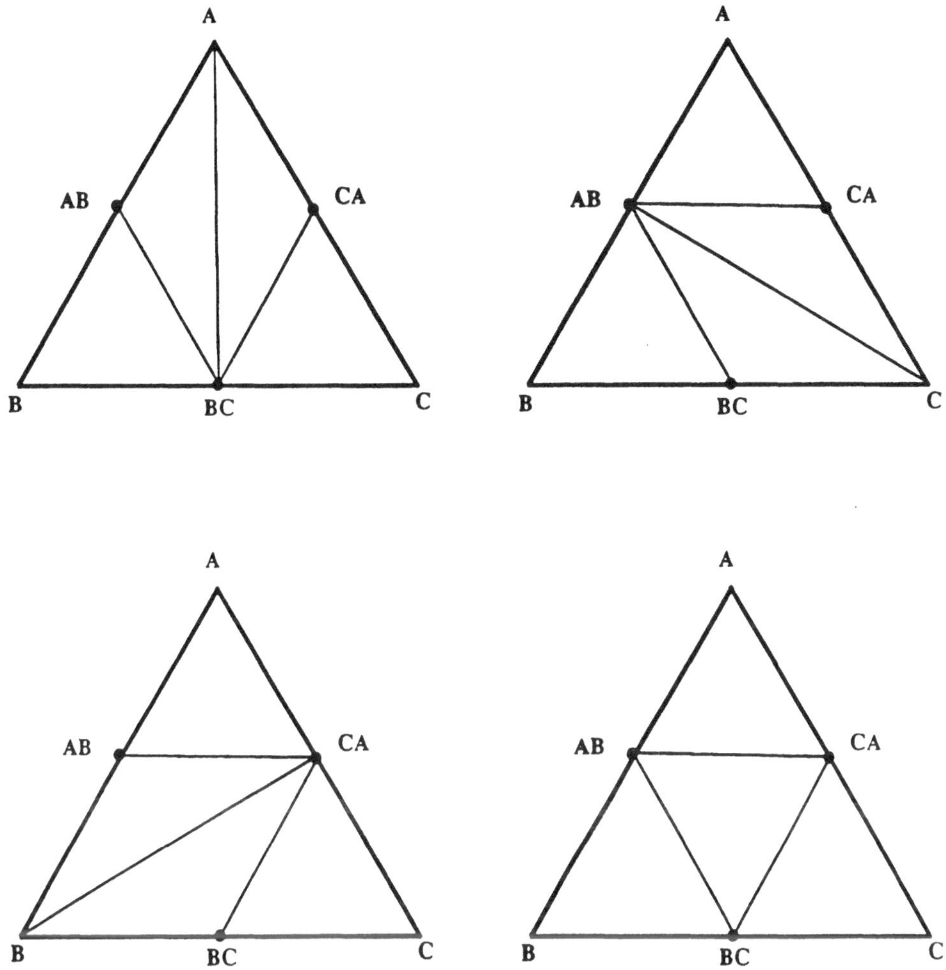


FIGURE 3 Four possible phase diagrams for the A-B-C system with three binary compounds.

A computer program to perform such calculations has been developed, the algorithm of which is shown in FIGURE 4. The existence of compound phases in a system is obtained from the ASTM X-ray powder data files and from computer literature searches. Their free energies are given in standard tables, the most extensive of which, in our experience, is that of Barin and Knacke [7,8]. Generation of the diagram now takes only a few minutes. Of course, since some thermochemical data are not reliable, reasonable uncertainties can be applied to the free energies to determine whether the diagram is altered. Whenever this is the case, careful choice of the appropriate experimental thin-film reaction is used to identify the correct tie-lines.

FIGURE 5 shows a real example for the Ti-Si-O system at 700-1000°C, which is important for understanding the oxidation of titanium disilicide (as might occur for instance during a salicide type process). The competing tie-lines in the appropriate region of the phase diagram, and their associated chemical reactions, are shown. It can be seen that, as written, the free energy change is negative in all cases and so the product phases $TiSi_2$ and $SiO_2$ are stable with respect to titanium oxides and silicon. Thus a tie-line in the diagram can be drawn between the silicide and silicon oxide. This then predicts that when a thin $TiSi_2$ layer on a silicon substrate is oxidised, so long as the silicon mobility is high, $SiO_2$ will be formed since the system is effectively in the Si-$TiSi_2$-$SiO_2$ composition region. If the alternative tie-lines had been present, a titanium oxide would be predicted. It is well-known that $SiO_2$ is produced under these circumstances [9]. The complete Ti-Si-O phase diagram is generated by considering all possible tie-lines and the result is shown in FIGURE 6(d). Prior to this calculation [1] this diagram had never been determined.

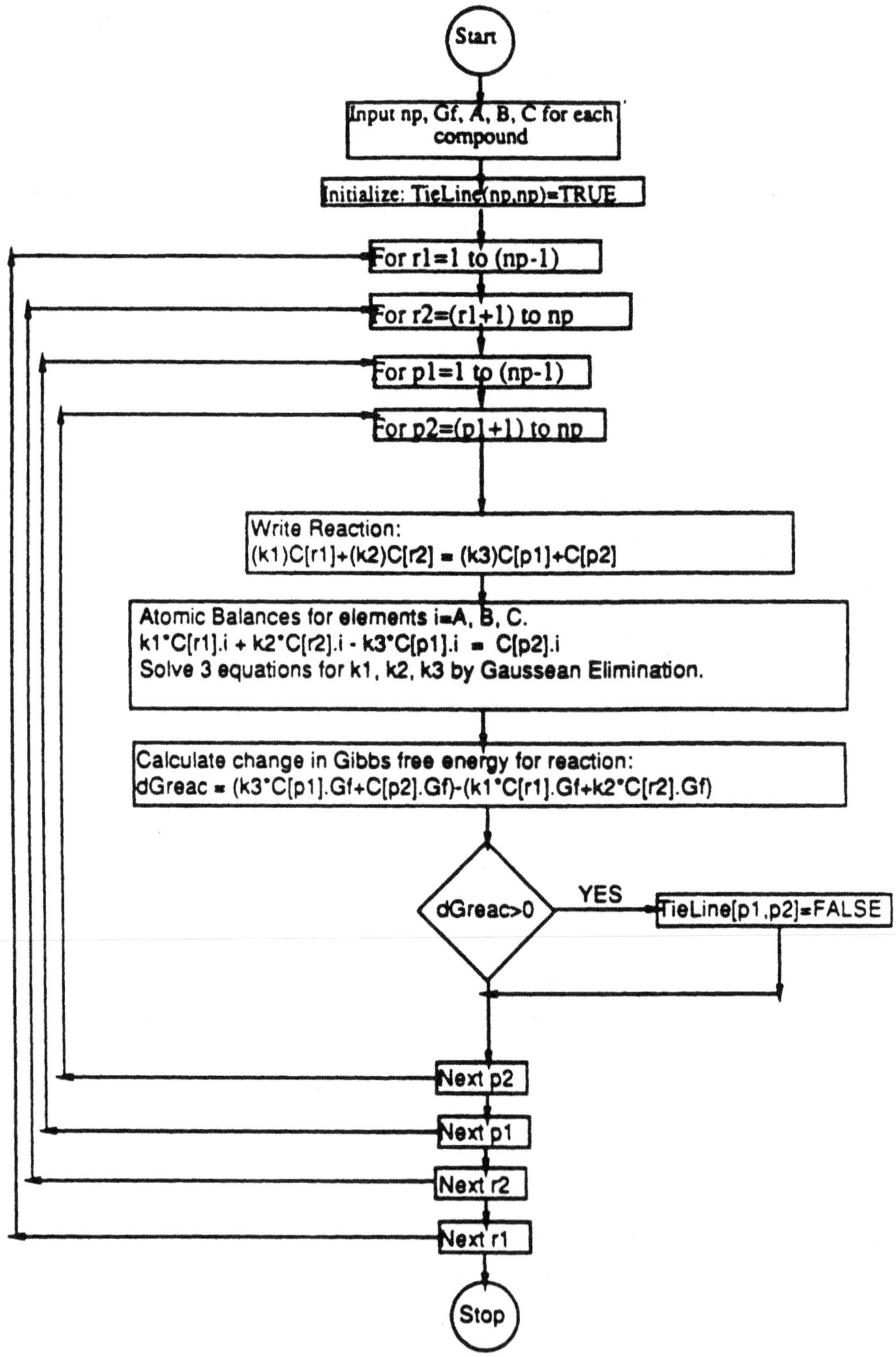


FIGURE 4 Algorithm of ternary phase diagram program.

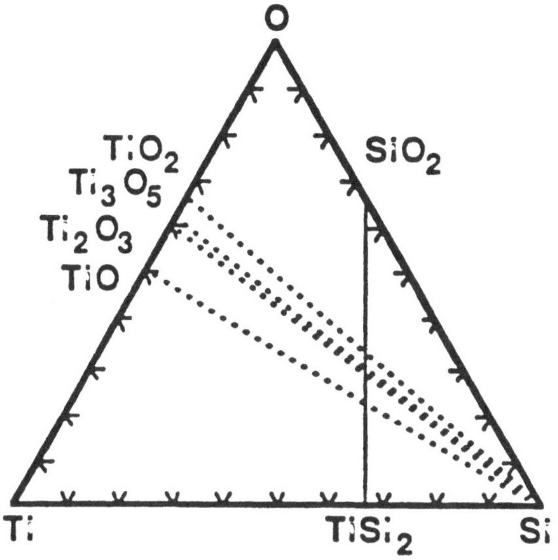


| Tie line reactions | | ΔG, kcal/mole |
|---|---|---|
| 2 TiO + 5 Si | = 2 $TiSi_2$ + 1 $SiO_2$ | -25 |
| 2 $Ti_2O_3$ + 11 Si | = 4 $TiSi_2$ + 3 $SiO_2$ | -51 |
| 2 $Ti_3O_5$ + 17 Si | = 6 $TiSi_2$ + 5 $SiO_2$ | -94 |
| 1 $TiO_2$ + 3 Si | = 1 $TiSi_2$ + 1 $SiO_2$ | -23 |

FIGURE 5 Some competing hypothetical tie-lines in the Ti-Si-O system. The existence of the $TiSi_2$-$SiO_2$ tie-line is established by the free-energy change of the balanced associated reaction [3].

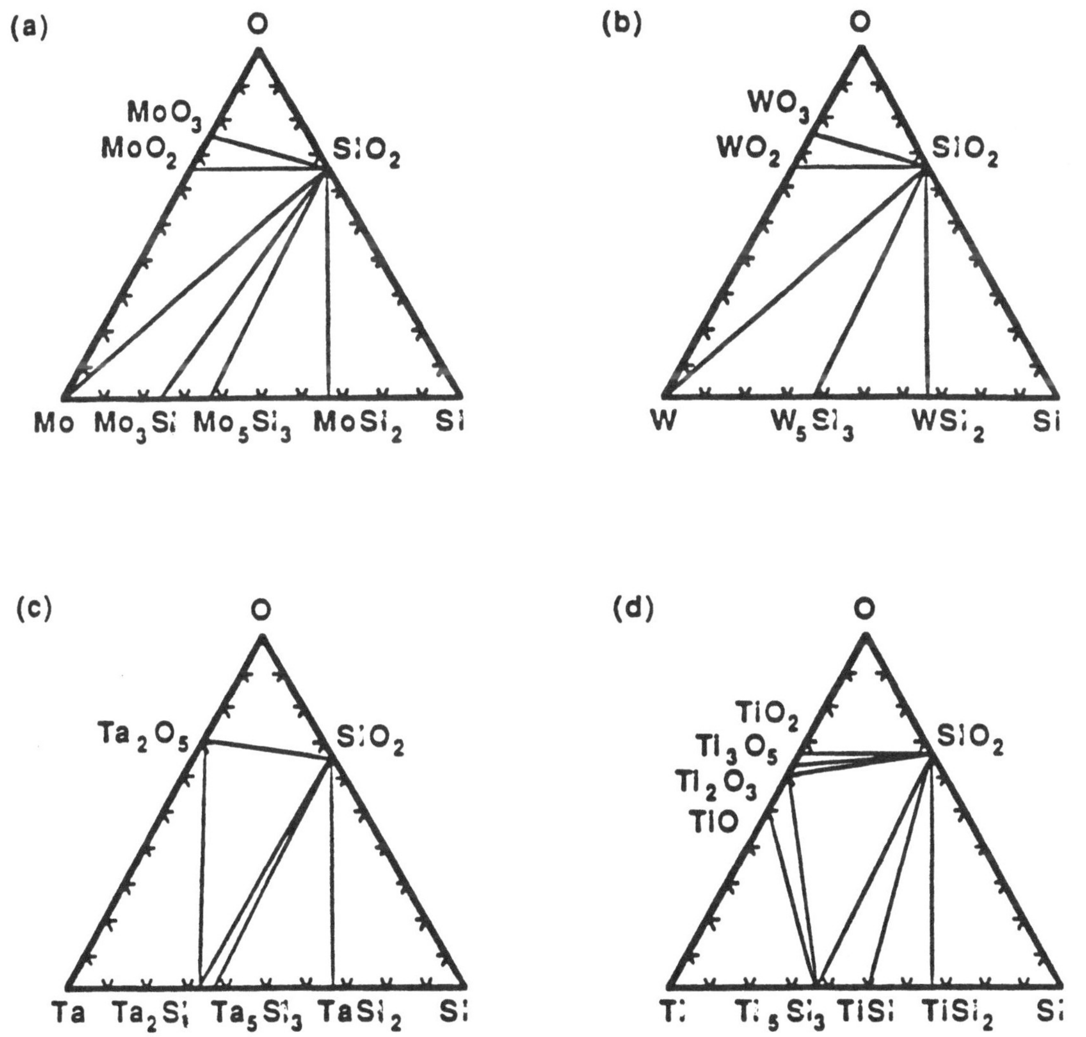


FIGURE 6 Calculated isothermal ternary phase diagrams for refractory metal-silicon-oxygen systems at 700-1000°C. The tie-lines between phases indicate mutual thermodynamic stability and partition the system into composition regimes in which three phases are mutually in equilibrium [3].

As alluded to earlier, the $TiSi_2$-$SiO_2$ tie-line can also be tested by an appropriate, independent thin-film reaction. A layer of (Ti + Si), with an atomic ratio Si:Ti = 1.5, was co-deposited on a $SiO_2$ substrate and heated in an inert atmosphere to 1000°C. Electron microscopy analysis shows that the metal layer is composed of a mixture of TiSi and $TiSi_2$, in clear thermochemical stability with the $SiO_2$ phase [10]. If this were not the case, a titanium oxide would have been produced. Other aspects of the phase diagram can similarly be tested. Thus, heating an elemental titanium film on a $SiO_2$ substrate results in the formation of titanium monoxide [3] confirming that there is no Ti-$SiO_2$ tie-line as predicted in FIGURE 6(d) and that titanium will reduce silica.

## ACKNOWLEDGEMENT

Clearly this and Datareviews 3.2, 3.3, 3.6 and 3.8 rely heavily on the original, pioneering work by R. Beyers, subsequently extended by A.S. Bhansali. I am very grateful for their excellent contributions to this field. Much assistance in putting together these reviews was given by M.R. Visokay.

## REFERENCES

[1] R. Beyers [ *J. Appl. Phys. (USA)* vol.56 (1984) p.147 ]
[2] R. Beyers, R. Sinclair, M.E. Thomas [ *J. Vac. Sci. Technol. B (USA)* vol.2 (1994) p.781 ]
[3] R. Beyers [ Ph.D. Thesis, Stanford University, USA (1989) ]
[4] A. Raman, K. Schubert [ *Z. Met.kd. (Germany)* vol.56 (1965) p.44 ]
[5] F.N. Rhines [ *Phase Diagrams in Metallurgy* (McGraw-Hill, New York, 1956) ]
[6] F.A. Hummel [ *Phase Equilibria in Ceramic Systems* (Dekker, New York, 1984) ]
[7] I. Barin, O. Knacke [ *Thermochemical Properties of Inorganic Substances* (Springer, Berlin, 1973 and 1977) ]
[8] I. Barin [ *Thermochemical Data of Pure Substances, Parts I and II* (VCH, Weinheim, 1989) ]
[9] S.P. Murarka [ *Silicides for VLSI Applications* (Academic, New York, 1983) p.72 ]
[10] R. Beyers [ *Mater. Res. Soc. Symp. Proc. (USA)* vol.47 (1985) p.143 ]

# 3.2 Ternary phase diagrams for TM-Si-O systems

R. Sinclair

April 1995

## A CLASSIFICATION SCHEME

While there are many possible combinations of tie-line distributions, Beyers [1] showed that there are essentially three types of diagram as illustrated schematically in FIGURE 1. If the free energy of formation of one phase is so much greater than that of the others in the system, then the tie-lines tend to emanate from that phase. In FIGURE 1(a) $SiO_2$ is the dominant phase and in such a case is stable in contact with the elemental metal. Such a metal (e.g. W) could not reduce native oxide on silicon for instance. In FIGURE 1(b) neither oxide nor the silicide phases are particularly dominant over the other phases and so the possible reactions are critically determined by the overall composition of the local system. For silicide contacts on silicon, the silicon-rich composition corner of the diagram is most important, and as also in FIGURE 1(a), oxidation of the silicide on silicon results in formation of $SiO_2$ (not metal oxide) so long as the silicon mobility is sufficiently high (e.g. Ti). In the final example, the metal oxide is dominant: now oxidation of the silicide on silicon first results in formation of the metal oxide (e.g. Zr).

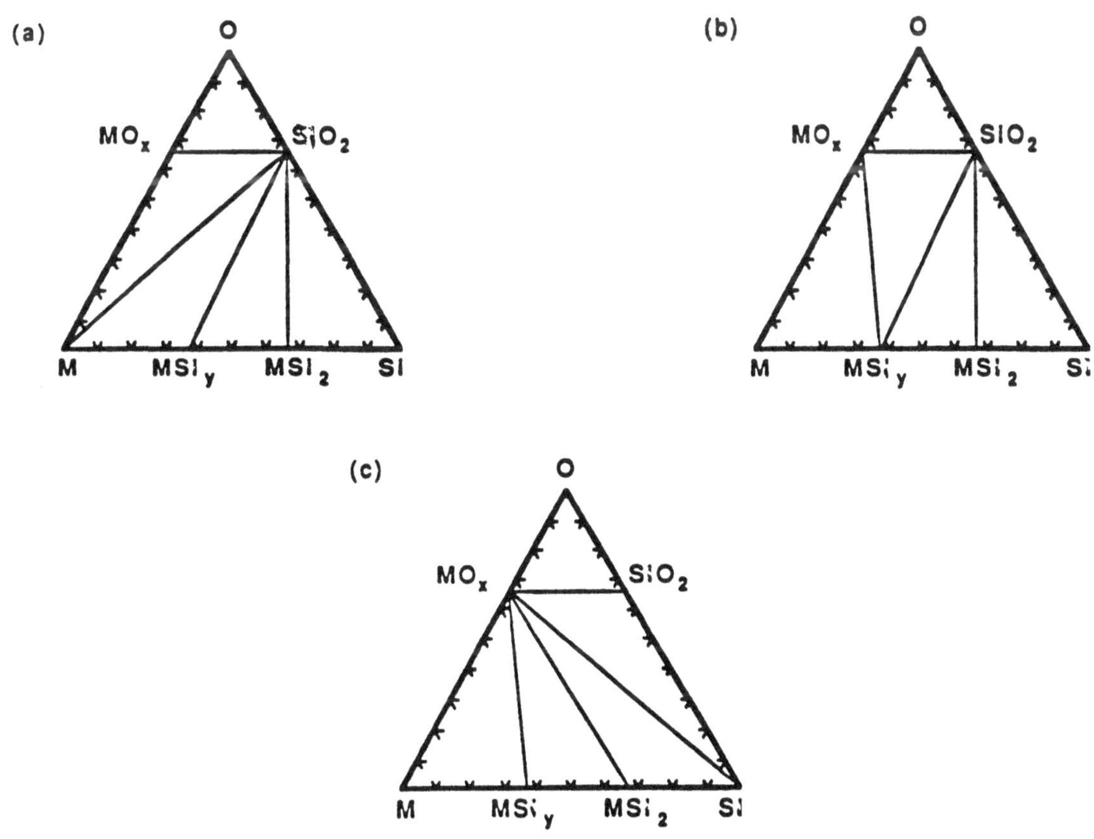


FIGURE 1 Beyers' categorisation of M-Si-O phase diagrams according to the 'dominant' phase: (a) $SiO_2$ dominant; (b) no phase dominant; (c) $MO_x$ dominant.

## B REFRACTORY METAL SILICIDES

Calculated tie-line diagrams for Mo, W, Ta and Ti are shown in FIGURE 2.

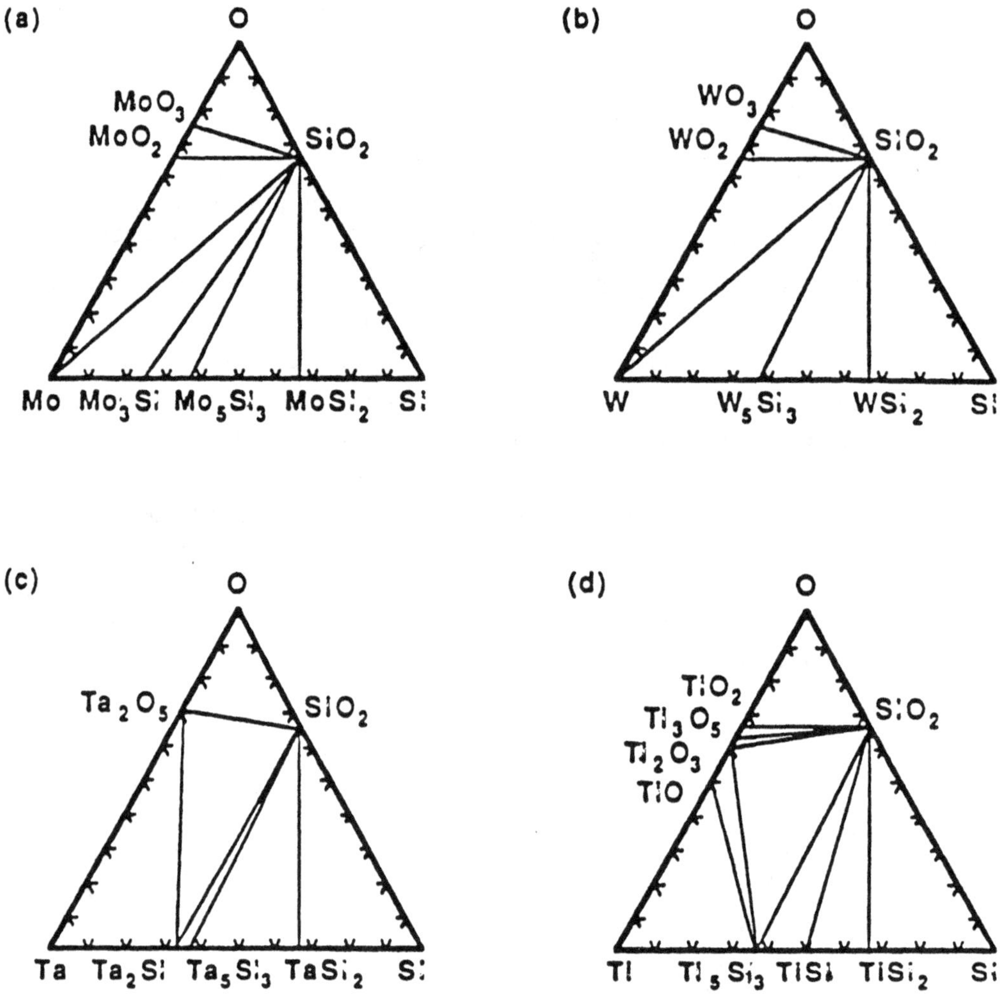


FIGURE 2 Calculated isothermal ternary phase diagrams for refractory metal-silicon-oxygen systems at 700-1000°C. The tie-lines between phases indicate mutual thermodynamic stability and partition the system into composition regimes in which three phases are mutually in equilibrium [1].

Tungsten and molybdenum therefore belong to the $SiO_2$-dominated category in which the metal does not reduce silicon oxide.

Titanium (FIGURE 2(d)) is one of the more important metals owing to the ubiquitous application of titanium silicide contacts. Its type is 'no-phase-dominant'. However, because of the degree of uncertainty in the exact free-energies of all the oxides and silicides, some of the interior tie-lines may not be reliable. Of major interest is the $TiSi_2$-Si-$SiO_2$ tie-triangle (i.e. $TiSi_2$ is stable with $SiO_2$ and with silicon itself), and the lack of a tie-line between Ti and $SiO_2$ (i.e. titanium reduces native silicon oxide, allowing ready titanium-silicon reaction in a salicide process). Tantalum (FIGURE 2(c)) also belongs to this category.

Indirect proof of these tie-line dispositions is also given by the result of oxidation of disilicide films on a $SiO_2$ substrate. For $WSi_2$ oxidation, elemental tungsten is sometimes observed as a reaction product [2]. A similar result is found for $MoSi_2$ oxidation [3]. On the other hand, for silicides such as $TaSi_2$, the elemental metal is never observed as a reaction product [4]. So, why is the elemental metal a possible reaction product for $WSi_2$ and $MoSi_2$, but not for $TaSi_2$ or $TiSi_2$? The ternary diagrams can predict which phases are possible oxidation products, although kinetics will determine which of the possible products actually form. Any phase having a tie-line that intersects the line drawn between oxygen and the disilicide is a possible oxidation product. Hence, elemental molybdenum and tungsten are possible, but elemental tantalum and titanium are not.

Zirconium (FIGURE 3) belongs to the metal oxide dominant category according to Beyers' calculations [1]. When a co-sputtered $ZrSi_2$ film containing excess silicon is oxidised at 900°C, the metal oxide $ZrO_2$ is identified by TEM analysis [1] as the product, not $SiO_2$. This is a consequence of the $ZrO_2$-Si tie-line in the phase diagram. Because $HfSi_2$ reportedly displays similar behaviour during oxidation [3], hafnium probably also has a ternary diagram similar to that of Zr-Si-O.

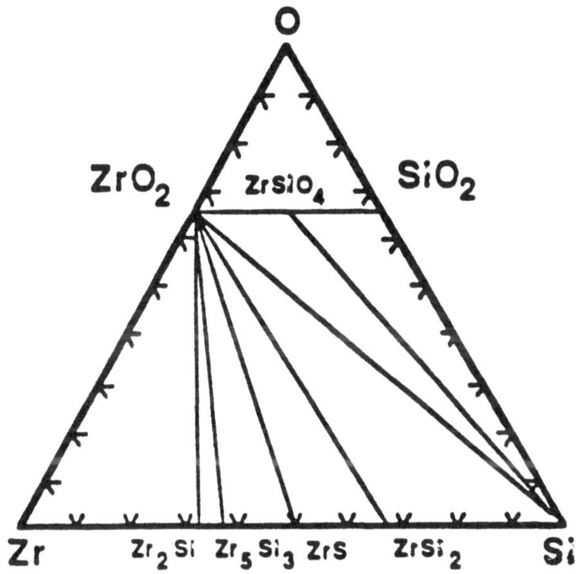


FIGURE 3 Calculated Zr-Si-O ternary phase diagram, showing that it belongs to the 'metal-oxide dominant' category [1].

## C NON-REFRACTORY METALS

Neither experimental nor calculated ternary phase diagrams have been determined for these systems, despite the importance of cobalt, nickel and iron silicides. Because the free energy of formation of their oxides is not as large as those of Ti and Zr, it would be expected that most such metals belong to $SiO_2$-dominant or no-phase dominant categories. Not all thermodynamic data are available for Co and Ni silicides, but the suggested Fe-Si-O diagram is shown in FIGURE 4. It should be emphasised that this still needs experimental testing.

Although the non-silicide forming metals are not entirely relevant to this Datareview, their phase diagrams can be readily suggested. Tie-lines must exist between the metal and silicon oxide, and the final important tie-line is then determined primarily by the relative free energy of formation of these two phases. In Al-Si-O a tie-line would be expected from $Al_2O_3$ to Si (metal oxide dominated), consistent with the chemical stability implied by silicon-on-sapphire technology, whereas for Ag-Si-O, oxide domination (i.e. the $SiO_2$-Ag tie-line) is anticipated (FIGURE 5).

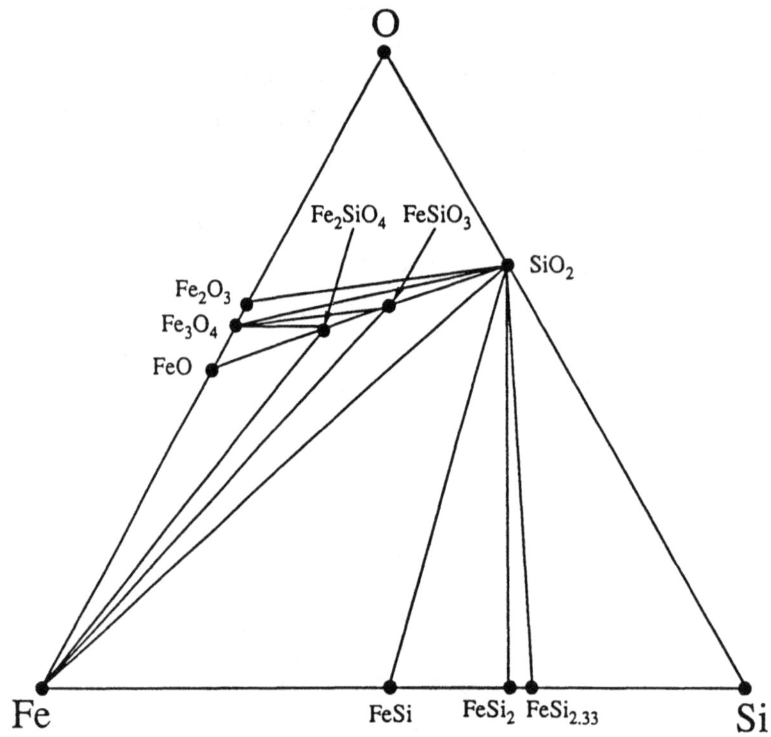


FIGURE 4 The proposed ternary phase diagram for Fe-Si-O at 700 K.

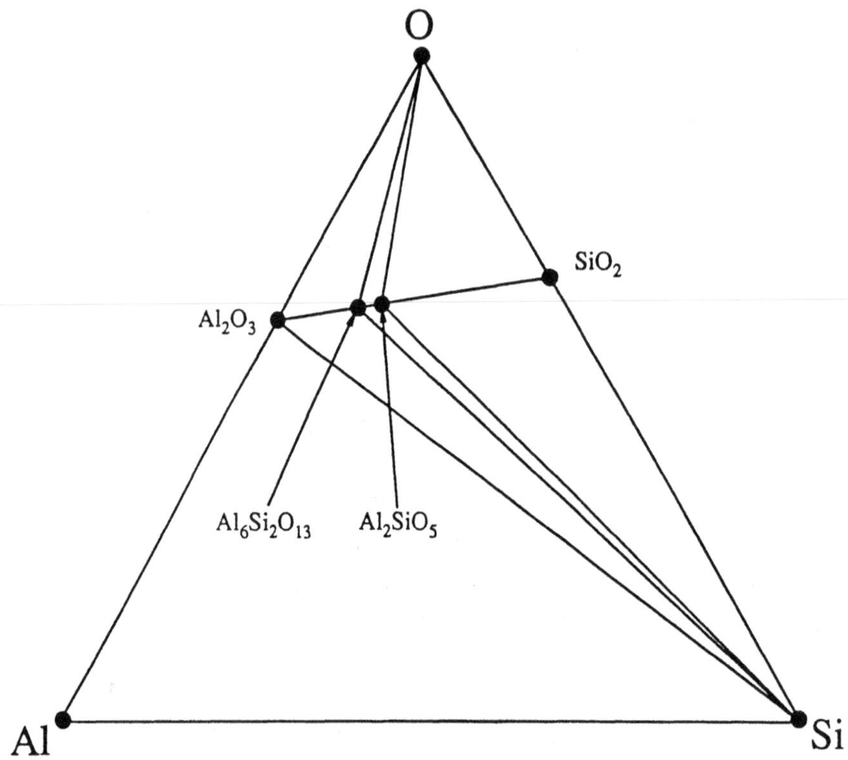


FIGURE 5 (a) The simplified Al-Si-O phase diagram at 700-900 K. Note that the binary metal-silicon system is a eutectic (no silicides).

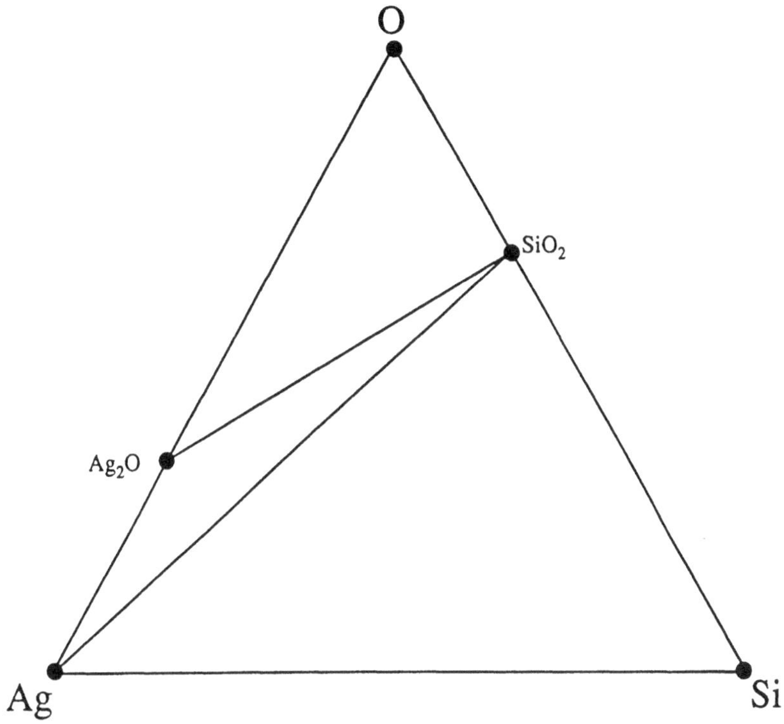


FIGURE 5 (b) The Ag-Si-O phase diagram at 500 K showing that it is oxide dominated. Note that the binary metal-silicon system is a eutectic (no silicides).

## REFERENCES

[1] R. Beyers [ Ph.D. Thesis, Stanford University, USA (1989) ]

[2] S. Zirinsky, W. Hammer, F. d'Heurle, J. Baglin [ *Appl. Phys. Lett. (USA)* vol.33 (1978) p.76 ]

[3] T. Mochizuki, M. Kashiwagi [ *J. Electrochem. Soc. (USA)* vol.127 (1980) p.1128 ]

[4] K.C. Saraswat, R. Nowicki, J.F. Moulder [ *Appl. Phys. Lett. (USA)* vol.41 (1982) p.1127 ]

## 3.3 Ternary phase diagrams for TM-Si-N systems

R. Sinclair

April 1995

By far the most important system is Ti-Si-N, the simplified version of which is shown in FIGURE 1. It belongs to the metal nitride dominant category. Accordingly, $TiSi_2$ is in equilibrium with TiN, so the latter is a thermodynamically stable diffusion barrier in conjunction with $TiSi_2$-silicon contacts. However, the absence of a $TiSi_2$-Si-$Si_3N_4$ tie-triangle precludes the insulating nitride being formed as the first phase during high temperature nitridation of $TiSi_2$ or Si (conducting TiN is produced instead) [2].

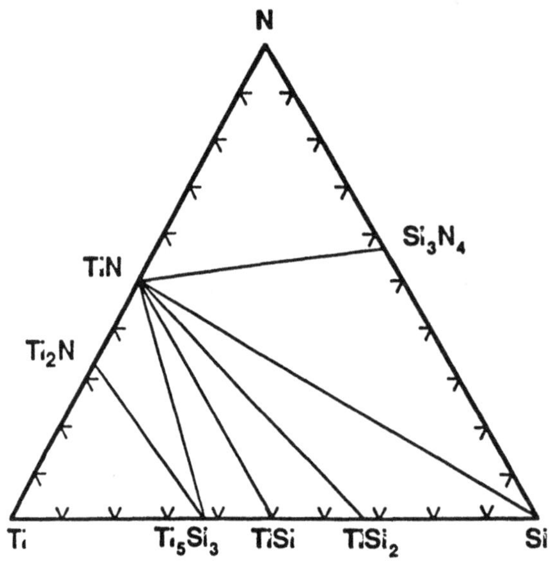


FIGURE 1 The calculated Ti-Si-N diagram, for T = 700-1000°C. It is metal nitride dominated [1].

One limitation of the present approach is the restriction of each phase to a single composition point. This is justified in many cases (e.g. $SiO_2$, $TiSi_2$ etc.). However, for titanium nitride it is not, as this phase has a wide range of stoichiometry. Of critical importance is the TiN composition at which the tie-lines no longer connect with $TiSi_2$ but directly with Si. Thus titanium-rich TiN will be stable with increasingly titanium-rich silicides (and not with Si) whereas nitrogen-rich TiN will not be stable with the silicides but will be with silicon. In the latter case, reaction of the nitride with $TiSi_2$ will result in some Si formation and decomposition of some of the silicide. Therefore, the composition of TiN used in diffusion barriers is important. Some efforts to establish this composition are currently under way.

Other TM-Si-N phase diagrams should be determined if alternative nitride diffusion barriers are to be considered.

### REFERENCES

[1] R. Beyers [ Ph.D. Thesis, Stanford University, USA (1989) ]

[2] P.J. Rosser, G.J. Tomkins [ *Mater. Res. Soc. Symp. Proc. (USA)* vol.37 (1985) p.607 ]

# 3.4 Phase diagrams and crystallographic data for TM-Si-dopant systems

R. Madar

January 1995

## A INTRODUCTION

Silicides play a central role in the metallisation of integrated circuit (IC) devices in silicon ULSI technology. One of the most widespread techniques of forming a silicide both on the highly doped polycrystalline silicon gate and directly on the doped silicon in the source and drain areas is called the self-aligned (SALICIDE) process.

During this process, the dopant will redistribute and this may have important consequences on the contact resistance leading to unexpected and sometimes misinterpreted phenomena. The understanding of this redistribution process requires an equilibrium approach based on the knowledge of the corresponding thermodynamic ternary phase diagram in addition to the determination of the kinetic aspect, as given by diffusion experiments on these materials. The dopants which have been taken into consideration are those usually used in VLSI technology: phosphorus, arsenic and boron.

This Datareview deals with the first aspect of this process, i.e. the thermodynamic equilibrium, while the diffusion of dopants in the silicides of interest will be treated in Datareview 7.2. The Datareview is based on a literature survey of such TM-Si-X (X = P, As, B) ternary phase diagrams and is presented as isothermal (293 K) sections of the three-element phase diagrams together with a review of the crystallographic data for the existing ternary phases.

## B NOBLE AND NEAR NOBLE METALS (Cu, Pd, Pt)

### B1 Crystallographic Data for Ternary TM-Si-Dopant Phases

TABLE 1 gives the relevant data for the Cu-Si-P system.

TABLE 1 Crystallographic data for the Cu-Si-P system.

| Composition | Symmetry | Space group | Unit cell parameters (nm) | Ref |
|---|---|---|---|---|
| $CuP_3Si_2$ | Cubic | F43m | a = 1.009 | [1] |
| $CuP_3Si_4$ | Cubic | F43m | a = 0.52853 | [2] |

There are no similar data for the TM-Si-As and TM-Si-B systems.

## B2 Ternary Phase Diagrams for TM-Si-Dopant Systems

FIGURES 1-3 present the ternary phase diagrams for the TM-Si-P, TM-Si-As and TM-Si-B systems, where TM = Cu, Pd and Pt respectively for FIGURES 1 to 3.

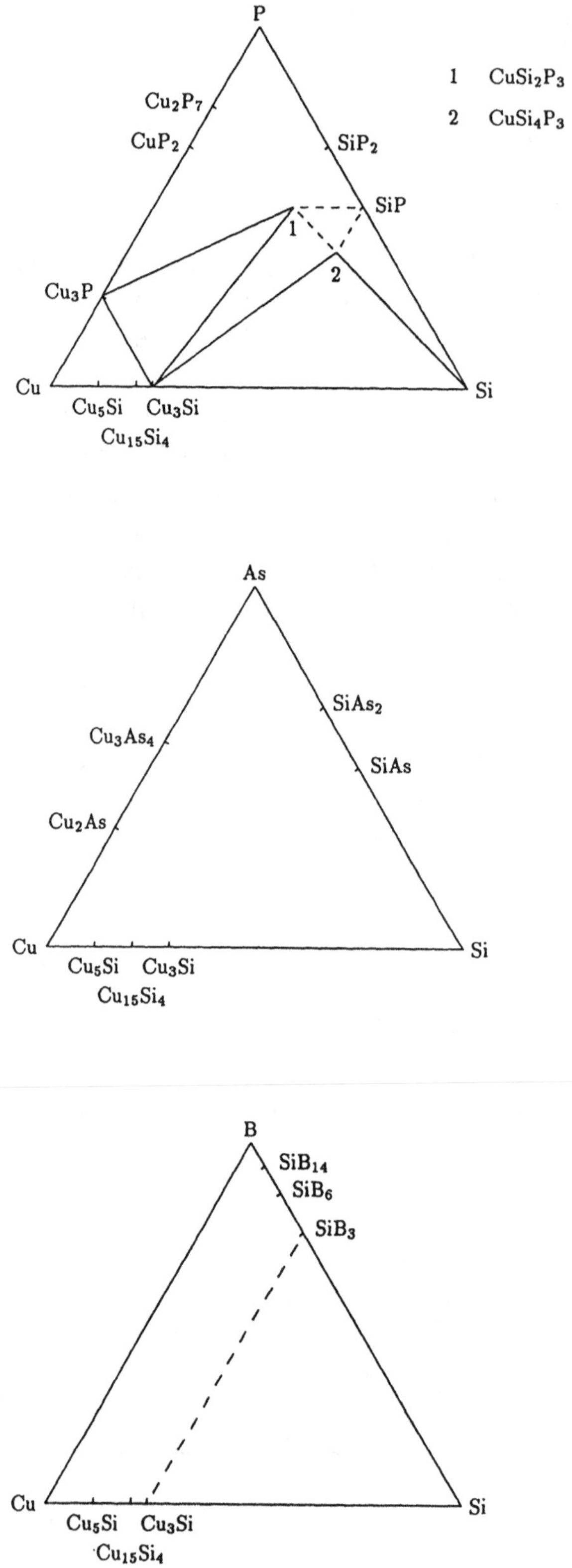


FIGURE 1.

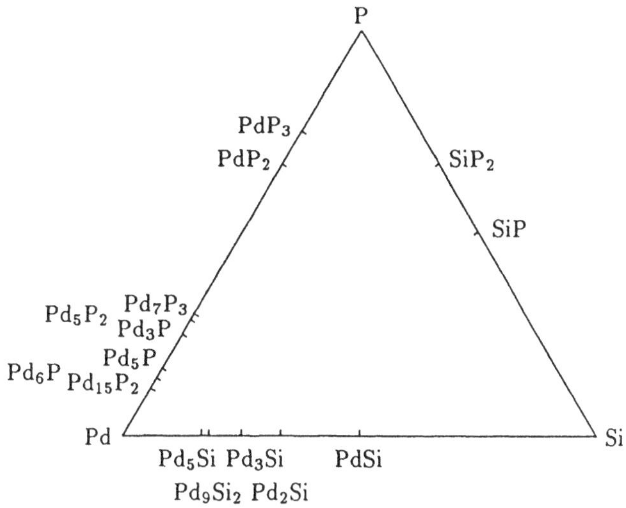
P
PdP3
PdP2
SiP2
SiP
Pd5P2
Pd7P3
Pd3P
Pd5P
Pd6P
Pd15P2
Pd
Si
Pd5Si Pd3Si
PdSi
Pd9Si2 Pd2Si

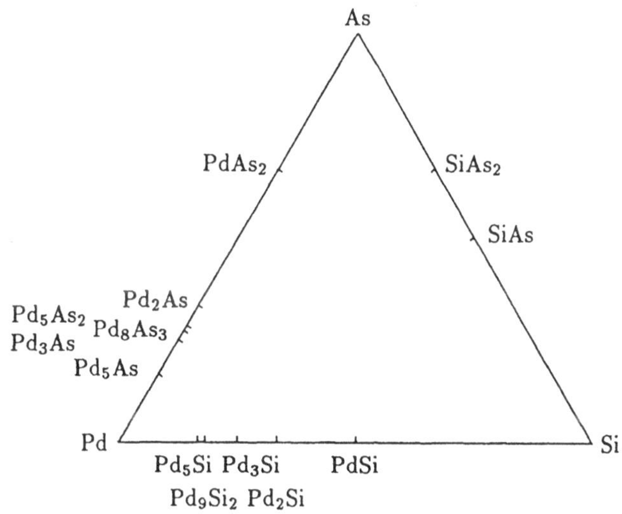
As
PdAs2
SiAs2
SiAs
Pd2As
Pd5As2
Pd8As3
Pd3As
Pd5As
Pd
Si
Pd5Si Pd3Si
PdSi
Pd9Si2 Pd2Si

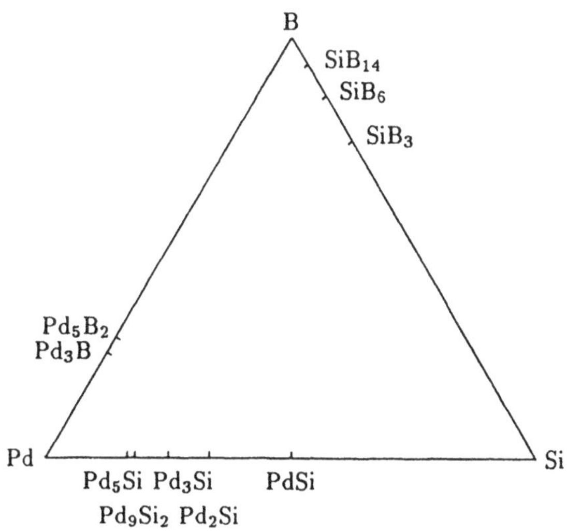
B
SiB14
SiB6
SiB3
Pd5B2
Pd3B
Pd
Si
Pd5Si Pd3Si
PdSi
Pd9Si2 Pd2Si

FIGURE 2.

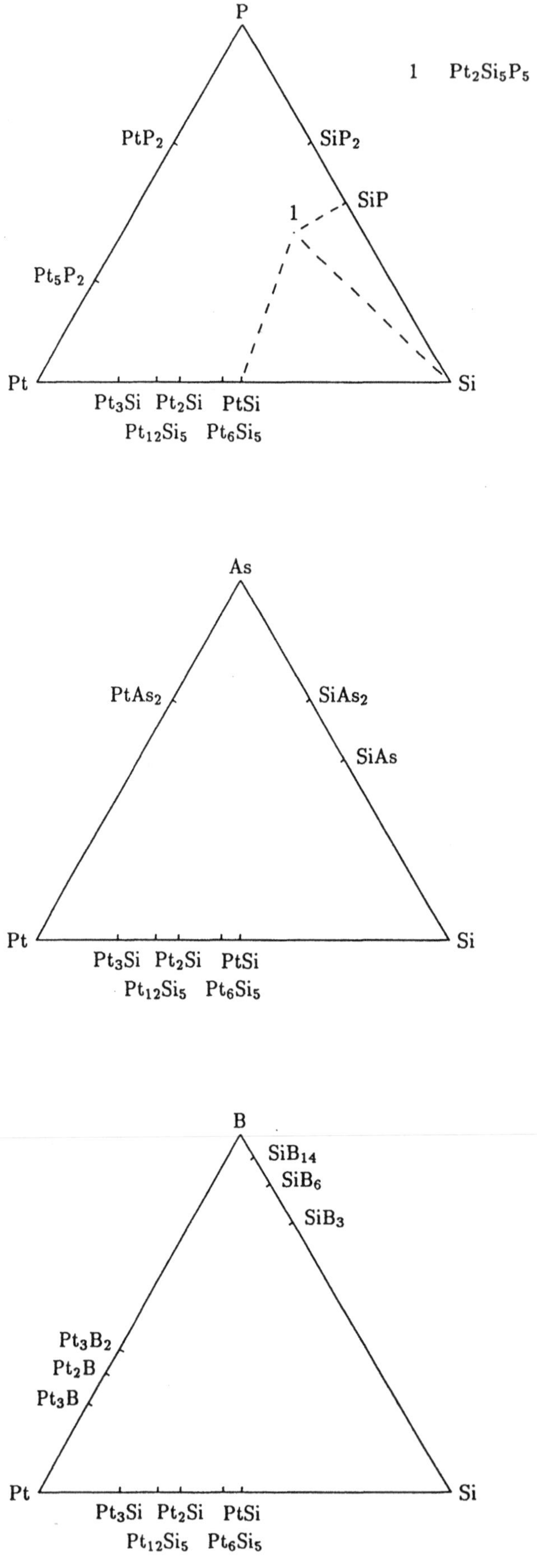
P
1 $Pt_2Si_5P_5$
$PtP_2$
$SiP_2$
SiP
1
$Pt_5P_2$
Pt
Si
$Pt_3Si$ $Pt_2Si$ PtSi
$Pt_{12}Si_5$ $Pt_6Si_5$
As
$PtAs_2$
$SiAs_2$
SiAs
Pt
Si
$Pt_3Si$ $Pt_2Si$ PtSi
$Pt_{12}Si_5$ $Pt_6Si_5$
B
$SiB_{14}$
$SiB_6$
$SiB_3$
$Pt_3B_2$
$Pt_2B$
$Pt_3B$
Pt
Si
$Pt_3Si$ $Pt_2Si$ PtSi
$Pt_{12}Si_5$ $Pt_6Si_5$

FIGURE 3.

# C GROUP VIII METALS (Co, Fe, Ni)

## C1 Crystallographic Data for Ternary TM-Si-Dopant Phases

TABLES 2 and 3 give the relevant data for the TM-Si-P and TM-Si-B systems respectively.

TABLE 2 Crystallographic data for TM-Si-P systems.

| Composition | Symmetry | Space group | Unit cell parameters (nm) | Ref |
|---|---|---|---|---|
| $Co_5P_3Si_2$ | Orthorhombic | Pnma | a = 0.5060<br>b = 0.3264<br>c = 0.5653 | [3] |
| $CoP_3Si_3$ | Orthorhombic | $P2_12_12_1$ | a = 0.5848<br>b = 0.5702<br>c = 1.2750 | |
| $Fe_2PSi$ | Orthorhombic | Pnma | a = 0.5242<br>b = 0.3059<br>c = 0.5853 | [3] |
| $FeP_4Si_4$ | Triclinic | $P_1$ | a = 0.4876<br>b = 0.5545<br>c = 0.6064<br>$\alpha = 85.33°$<br>$\beta = 68.40°$<br>$\gamma = 70.43°$ | |
| $Fe_{20}P_9Si$ | Hexagonal | P62m | a = 0.5927<br>b = 0.4980<br>c = 1.3633 | [4] |
| $Ni_2PSi$ | Orthorhombic | Pbca | a = 0.5927<br>b = 0.4980<br>c = 1.3633 | [5] |
| $Ni_3P_6Si_2$ | Orthorhombic | Imm2 | a = 0.35058<br>b = 1.10868<br>c = 0.53130 | [6] |
| $NiSi_3P_4$ | Hexagonal | I42m | a = 0.51598<br>c = 1.0350 | [7] |
| $Ni_5Si_2P_3$ | Orthorhombic | Pcba | a = 0.4932<br>b = 3.414<br>c = 0.5941 | [8] |

TABLE 3 Crystallographic data for TM-Si-B systems.

| Composition | Symmetry | Space group | Unit cell parameters (nm) | Ref |
|---|---|---|---|---|
| $Co_5Si_2B$ | Tetragonal | I4/mcm | a = 0.862<br>c = 0.425 | [9] |
| $Fe_5SiB_2$ | Tetragonal | I4/mcm | a = 0.554<br>c = 1.032 | [10] |
| $Ni_4Si_2B$ | Tetragonal | I4/mcm | a = 0.8669<br>c = 0.4275 | [11] |

Table 3 continued

| Composition | Symmetry | Space group | Unit cell parameters (nm) | Ref |
|---|---|---|---|---|
| $Ni_6Si_2B$ | Hexagonal | P62m | a = 0.6105<br>c = 0.2895 | [12] |
| $Ni_9Si_4B_2$ | Tetragonal | I4/mcm | a = 0.8632<br>c = 0.4290 | [13] |

There are no corresponding data for the TM-Si-As systems.

## C2 Ternary Phase Diagrams for TM-Si-Dopant Systems

FIGURES 4-6 present the ternary phase diagrams for the TM-Si-P, TM-Si-As and TM-Si-B systems, where TM = Co, Fe and Ni respectively for FIGURES 4 to 6.

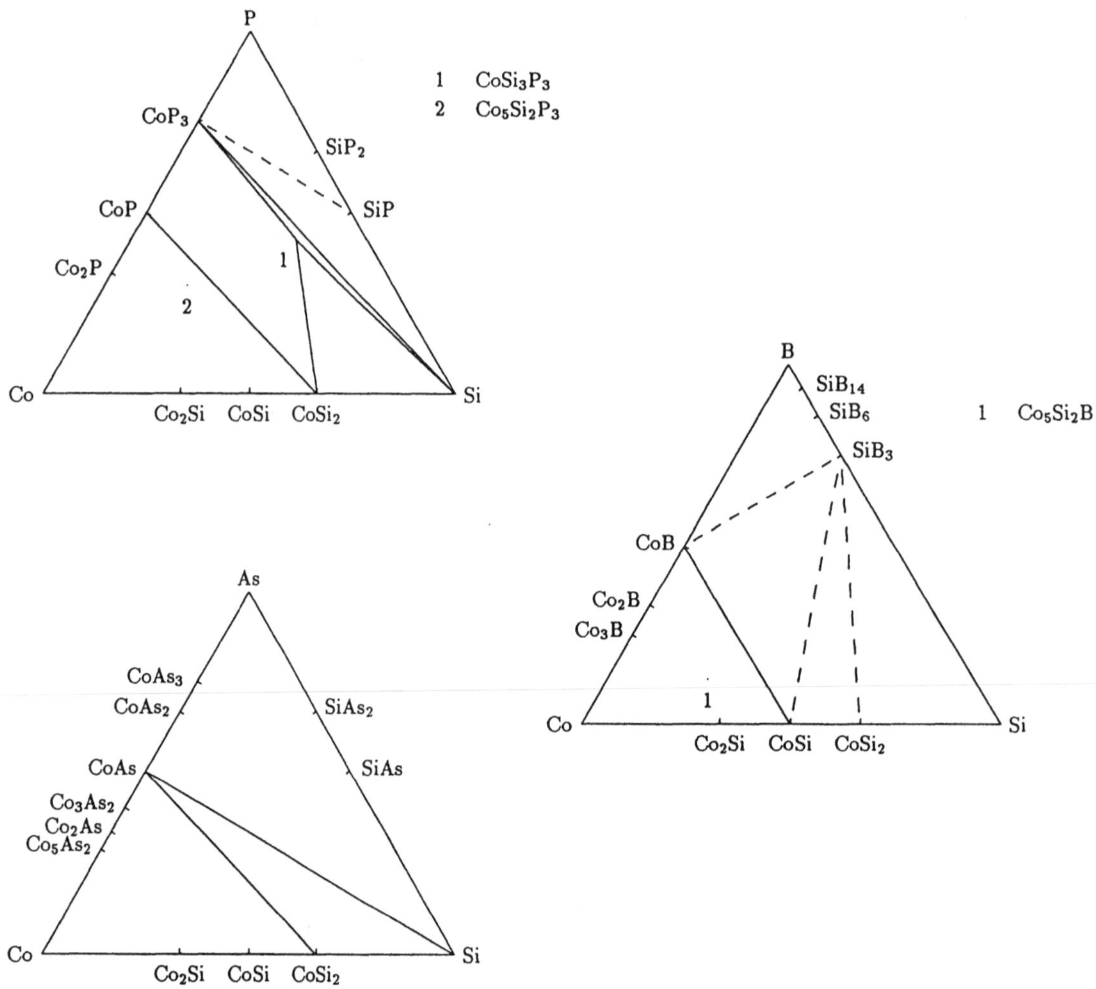


FIGURE 4.

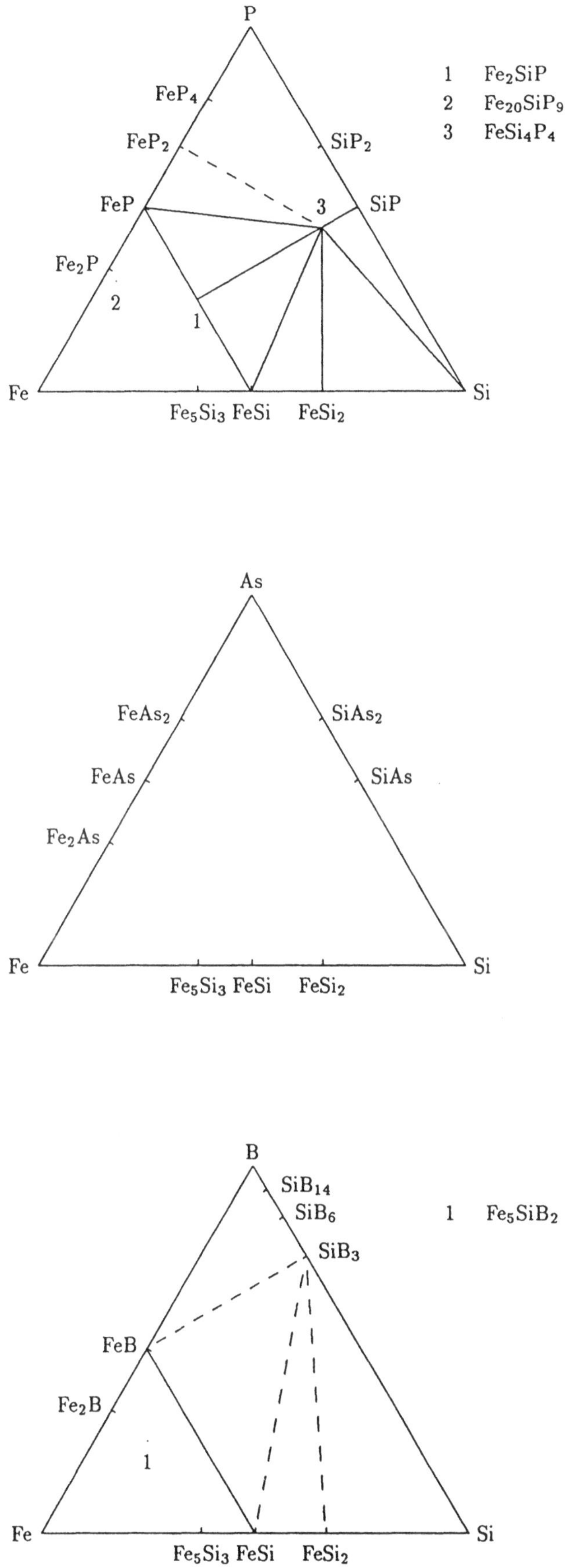
P
FeP$_4$
FeP$_2$
SiP$_2$
FeP
SiP
3
Fe$_2$P
2
1
Fe
Si
Fe$_5$Si$_3$ FeSi FeSi$_2$
1 Fe$_2$SiP
2 Fe$_{20}$SiP$_9$
3 FeSi$_4$P$_4$
As
FeAs$_2$
SiAs$_2$
FeAs
SiAs
Fe$_2$As
Fe
Si
Fe$_5$Si$_3$ FeSi FeSi$_2$
B
SiB$_{14}$
SiB$_6$
SiB$_3$
FeB
Fe$_2$B
1
Fe
Si
Fe$_5$Si$_3$ FeSi FeSi$_2$
1 Fe$_5$SiB$_2$

FIGURE 5.

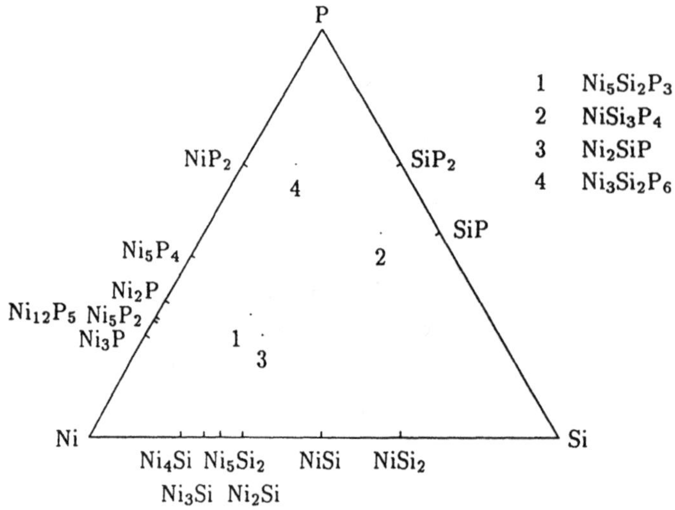
P
NiP2
SiP2
SiP
Ni5P4
Ni2P
Ni12P5 Ni5P2
Ni3P
Ni
Si
Ni4Si Ni5Si2 NiSi NiSi2
Ni3Si Ni2Si
1 Ni5Si2P3
2 NiSi3P4
3 Ni2SiP
4 Ni3Si2P6

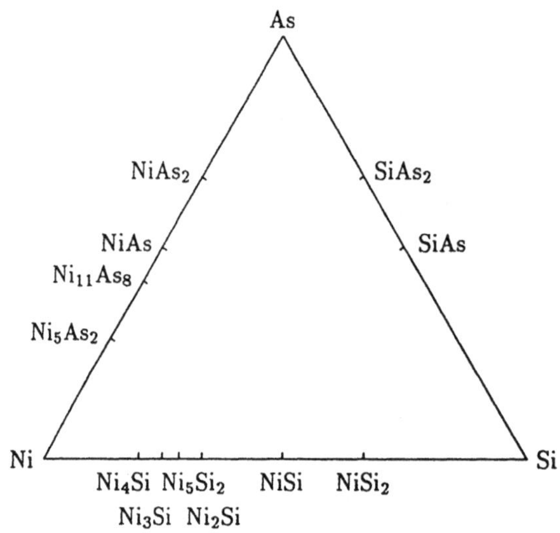
As
NiAs2
SiAs2
NiAs
SiAs
Ni11As8
Ni5As2
Ni
Si
Ni4Si Ni5Si2 NiSi NiSi2
Ni3Si Ni2Si

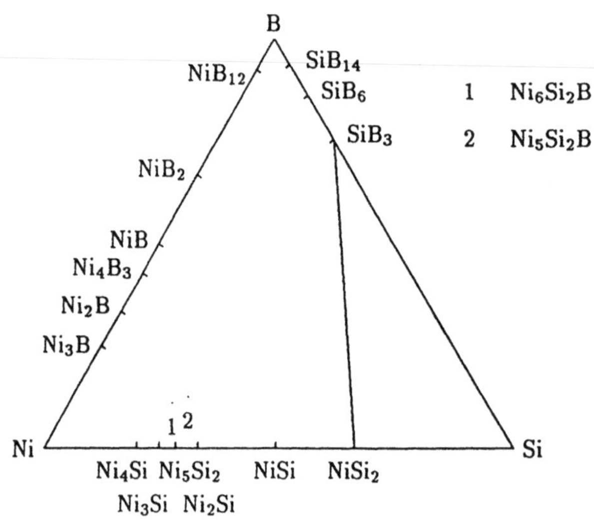
B
NiB12
SiB14
SiB6
SiB3
NiB2
NiB
Ni4B3
Ni2B
Ni3B
Ni
Si
Ni4Si Ni5Si2 NiSi NiSi2
Ni3Si Ni2Si
1 Ni6Si2B
2 Ni5Si2B

FIGURE 6.

# D REFRACTORY METALS (Cr, Ti, Mo, W, Ta)

## D1 Crystallographic Data for Ternary TM-Si-Dopant Phases

TABLES 4-6 give the relevant data for the TM-Si-P, TM-Si-As and TM-Si-B systems respectively.

TABLE 4 Crystallographic data for TM-Si-P systems.

| Composition | Symmetry | Space group | Unit cell parameters (nm) | Ref |
|---|---|---|---|---|
| $Cr_{25}P_8Si_7$ | Tetragonal | I4/mcm | a = 0.8960<br>c = 0.4685 | [14] |
| $Mo_{10}PSi_5$ | Tetragonal | I4/mcm | a = 0.9850<br>c = 0.4817 | [14] |

TABLE 5 Crystallographic data for TM-Si-As systems.

| Composition | Symmetry | Space group | Unit cell parameters (nm) | Ref |
|---|---|---|---|---|
| $Cr_3Si_{1-x}As_x$ | Cubic | Pm3m | a = 0.4588 | [15] |
| TaSiAs | Tetragonal | P4/nmm | a = 0.3499<br>c = 0.7857 | [16] |

TABLE 6 Crystallographic data for TM-Si-B systems.

| Composition | Symmetry | Space group | Unit cell parameters (nm) | Ref |
|---|---|---|---|---|
| $Cr_5Si_3B$ | Hexagonal | $P6_3/mcm$ | a = 0.8960<br>c = 0.4685 | [17] |
| $Mo_5SiB_2$ | Tetragonal | I4/mcm | a = 0.5998<br>c = 1.1027 | [18] |
| $W_2SiB$ | Tetragonal | I4/mcm | a = 0.6047<br>c = 1.099 | [19] |

## D2 Ternary Phase Diagrams for TM-Si-Dopant Systems

FIGURES 7-11 present the ternary phase diagrams for the TM-Si-P, TM-Si-As and TM-Si-B systems, where TM = Cr, Ti, Mo, W and Ta for FIGURES 7 to 11.

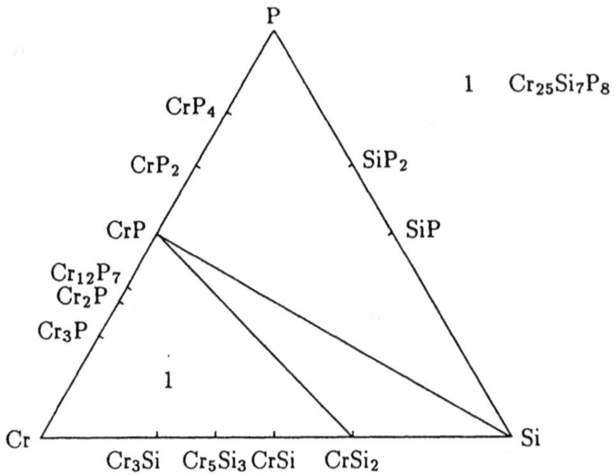
P
1 $Cr_{25}Si_7P_8$
$CrP_4$
$CrP_2$
$SiP_2$
CrP
SiP
$Cr_{12}P_7$
$Cr_2P$
$Cr_3P$
1
Cr
Si
$Cr_3Si$ $Cr_5Si_3$ CrSi $CrSi_2$

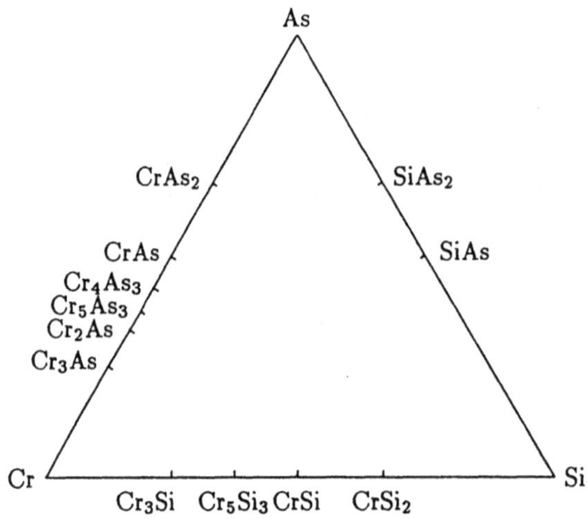
As
$CrAs_2$
$SiAs_2$
CrAs
SiAs
$Cr_4As_3$
$Cr_5As_3$
$Cr_2As$
$Cr_3As$
Cr
Si
$Cr_3Si$ $Cr_5Si_3$ CrSi $CrSi_2$

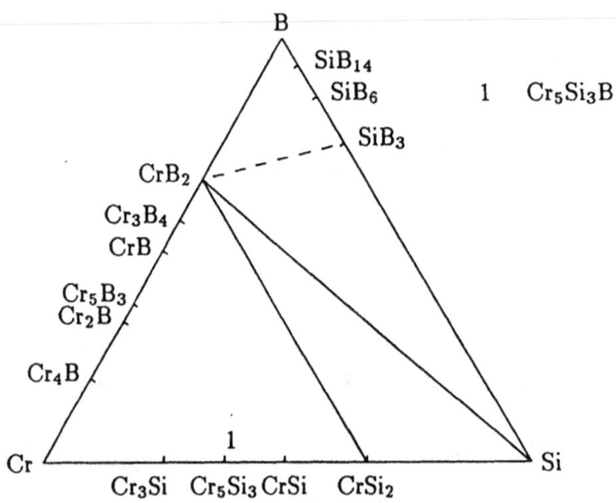
B
$SiB_{14}$
$SiB_6$
1 $Cr_5Si_3B$
$SiB_3$
$CrB_2$
$Cr_3B_4$
CrB
$Cr_5B_3$
$Cr_2B$
$Cr_4B$
1
Cr
Si
$Cr_3Si$ $Cr_5Si_3$ CrSi $CrSi_2$

FIGURE 7.

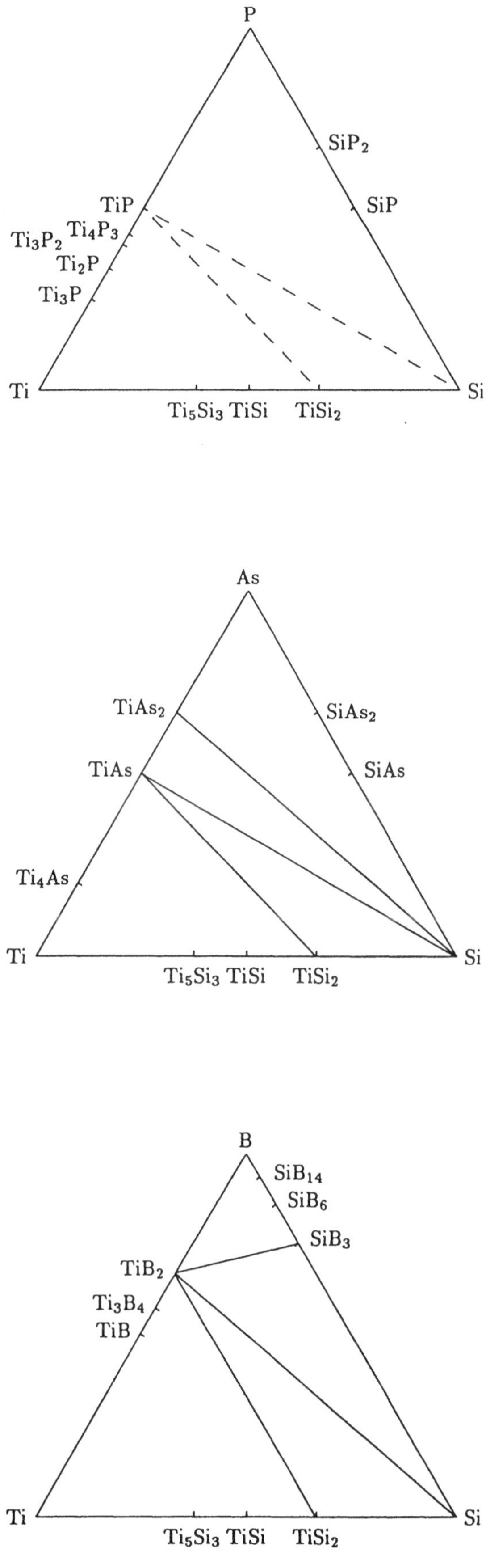
P
$SiP_2$
$SiP$
$TiP$
$Ti_3P_2$ $Ti_4P_3$
$Ti_2P$
$Ti_3P$
Ti
Si
$Ti_5Si_3$ TiSi $TiSi_2$
As
$TiAs_2$
$SiAs_2$
TiAs
SiAs
$Ti_4As$
Ti
Si
$Ti_5Si_3$ TiSi $TiSi_2$
B
$SiB_{14}$
$SiB_6$
$SiB_3$
$TiB_2$
$Ti_3B_4$
TiB
Ti
Si
$Ti_5Si_3$ TiSi $TiSi_2$

FIGURE 8.

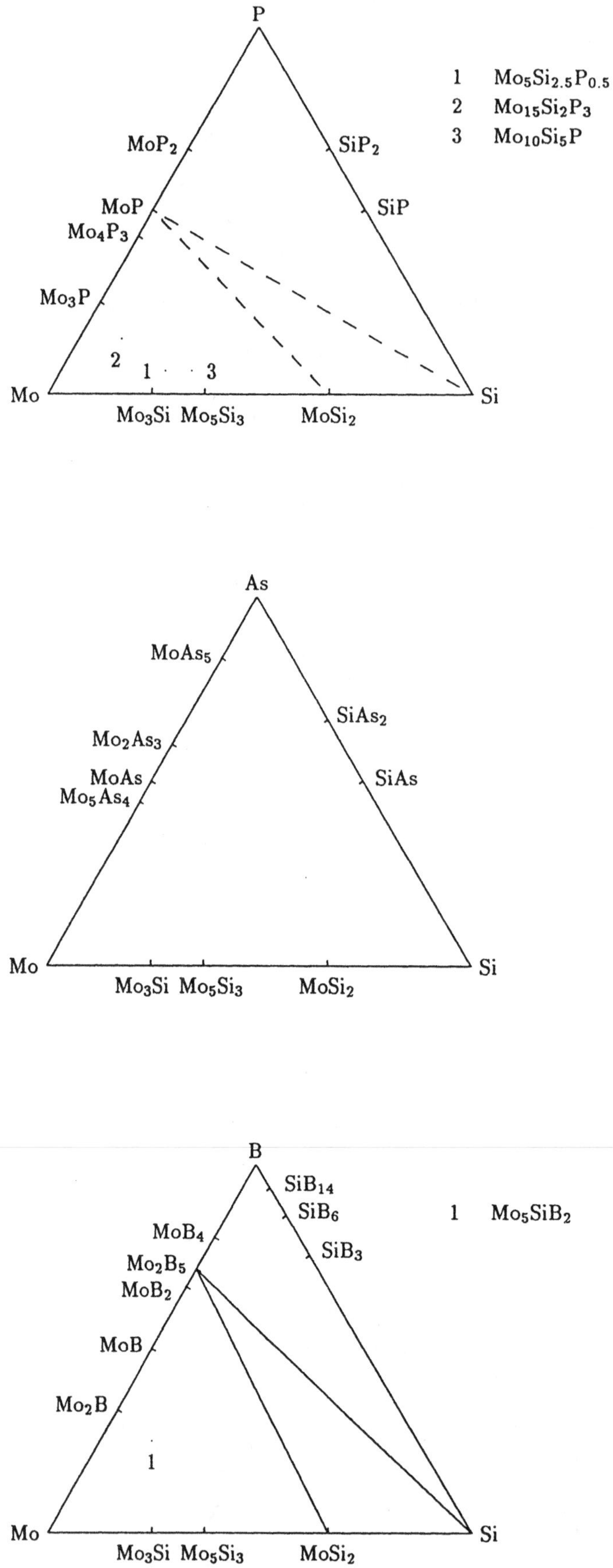
P
MoP2
MoP
Mo4P3
Mo3P
SiP2
SiP
1 Mo5Si2.5P0.5
2 Mo15Si2P3
3 Mo10Si5P
2
1
3
Mo
Si
Mo3Si Mo5Si3 MoSi2
As
MoAs5
Mo2As3
MoAs
Mo5As4
SiAs2
SiAs
Mo
Si
Mo3Si Mo5Si3 MoSi2
B
SiB14
SiB6
SiB3
1 Mo5SiB2
MoB4
Mo2B5
MoB2
MoB
Mo2B
1
Mo
Si
Mo3Si Mo5Si3 MoSi2

FIGURE 9.

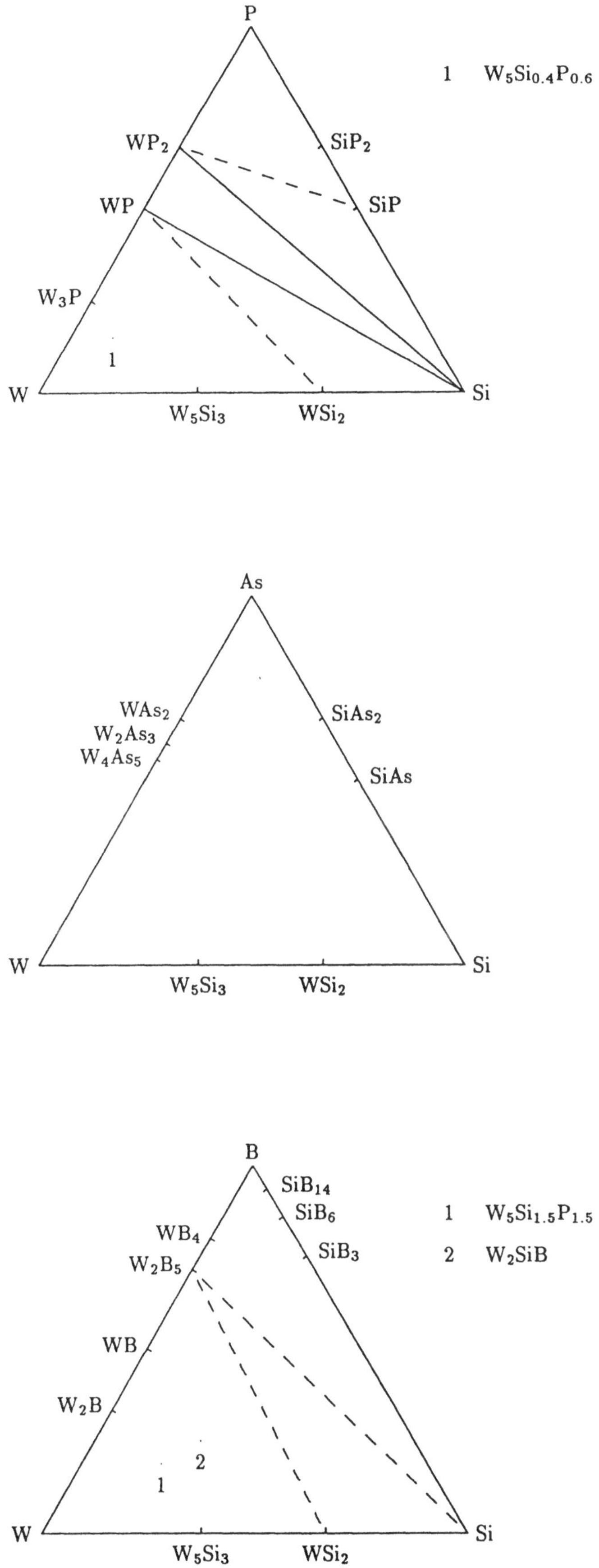
P
1 W5Si0.4P0.6
WP2
SiP2
WP
SiP
W3P
1
W
Si
W5Si3
WSi2
As
WAs2
SiAs2
W2As3
W4As5
SiAs
W
Si
W5Si3
WSi2
B
SiB14
SiB6
1 W5Si1.5P1.5
WB4
SiB3
2 W2SiB
W2B5
WB
W2B
2
1
W
Si
W5Si3
WSi2

FIGURE 10.

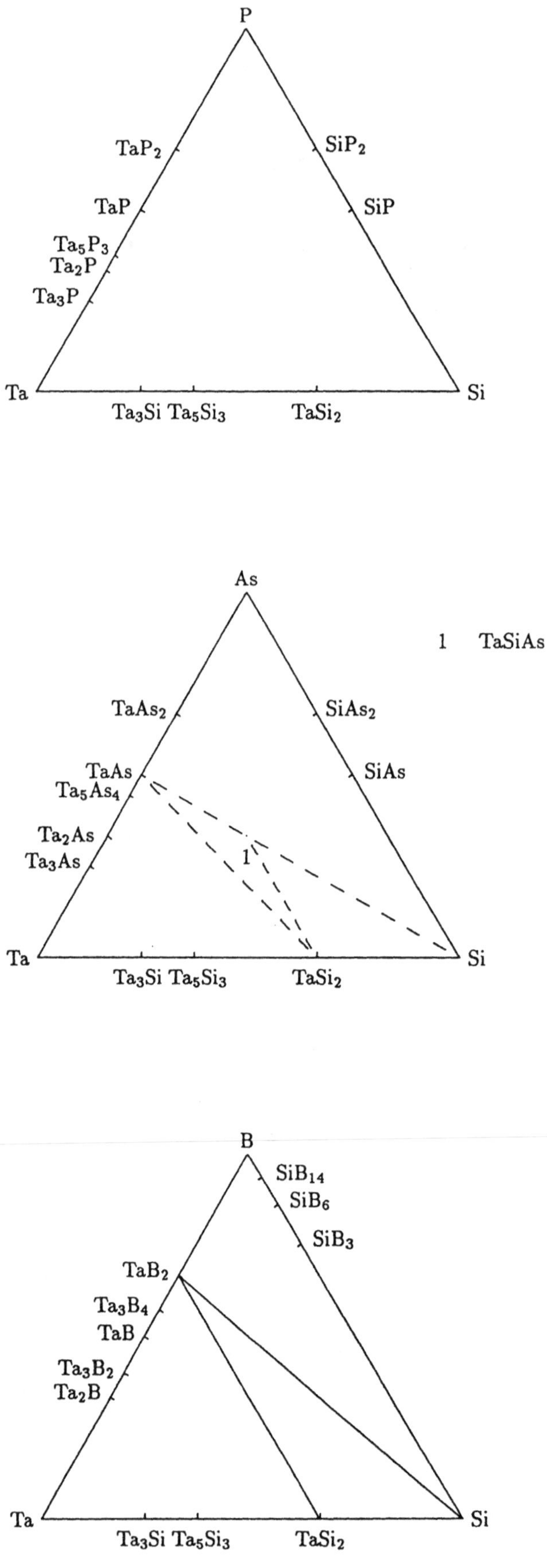
P
TaP2
TaP
Ta5P3
Ta2P
Ta3P
SiP2
SiP
Ta
Si
Ta3Si Ta5Si3
TaSi2
As
1 TaSiAs
TaAs2
TaAs
Ta5As4
Ta2As
Ta3As
SiAs2
SiAs
1
Ta
Si
Ta3Si Ta5Si3
TaSi2
B
SiB14
SiB6
SiB3
TaB2
Ta3B4
TaB
Ta3B2
Ta2B
Ta
Si
Ta3Si Ta5Si3
TaSi2

FIGURE 11.

## REFERENCES


[1] O.G. Folberth, H. Pfister [ *Acta. Crystallogr. (Denmark)* vol.14 (1961) p.325 ]

[2] G. Bhikshamaiah, S.V. Suryanarayana, M.S. Omar [ *J. Mater. Sci. Lett. (UK)* vol.7 (1988) p.1074 ]

[3] S. Rumdqvist [ *Acta. Chem. Scand. (Denmark)* vol.16 (1962) p.287 ]

[4] P. Jernberg, A.A. Yousif, L. Häggström, Y. Andersson [ *J. Solid State Chem. (USA)* vol.53 (1984) p.313 ]

[5] O.N. Ilnistskaya, Yu.B. Kuzma, V.S. Fundamensky [ *Dopov. Akad. Nauk. Ukr. 4 (Ukranian SSR)* vol.51 no.9 (1989) p.79 ]

[6] O.N. Ilnistskaya, P. Zavaly, Yu.B. Kuzma [ *Dopov. Akad. Nauk. Ukr. RSR (Ukranian SSR)* (1989) ]

[7] Yu.B. Kuzma, Ya.F. Lomnitskaya, O.N. Ilnitskaya, S.I. Chikhriy [ *Collected Abstracts XI Int. Conf. on Phosphorus Chemistry* vol.2 (1989) ]

[8] O.N. Ilnistskaya, Yu.N. Grin, Yu.B. Kuzma [ *Kristallografiya 1 (Russia)* vol.37 (1992) p.146 ]

[9] B. Aronsson [ *Acta. Chem. Scand. (Denmark)* vol.14 (1960) p.1414 ]

[10] B. Aronsson, G. Lundgre [ *Acta. Chem. Scand. (Denmark)* vol.13 (1959) p.433 ]

[11] E. Lugscheider, H. Reimann, O. Knotek [ *Monatsh. Chem.* vol.106 (1975) p.1155 ]

[12] S. Rundqvist, F. Jellinek [ *Acta. Chem. Scand. (Denmark)* vol.13 (1959) p.425 ]

[13] A.A. Uraz, S. Rundqvist [ *Acta. Chem. Scand. (Denmark)* vol.24 (1970) p.1843 ]

[14] O.N. Ilnistskaya, Yu.B. Kuzma [ *Russ. J. Inorg. Chem. (UK)* vol.27 no.10 (1982) p.1538 ]

[15] H. Boller, H. Wolfsgruber, H. Nowotny [ *Monatsh. Chem.* vol.98 (1967) p.2356 ]

[16] F. Hulliger [ *J. Less-Common Met. (Switzerland)* vol.30 (1973) p.397 ]

[17] H. Nowotny, E. Piegger, R. Kieffer, F. Benesovsky [ *Monatsh. Chem.* vol.89 (1958) p.611 ]

[18] B. Aronsson [ *Acta. Chem. Scand. (Denmark)* vol.12 (1958) p.31 ]

[19] H. Nowotny, E. Dimakopoulou, H. Kudielka [ *Monatsh. Chem.* vol.88 (1957) p.180-92 ]

# 3.5 Phase diagrams for TM-Si-Ge systems

R. Madar

January 1995

The reaction of metals with SiGe alloys is becoming more and more important since SiGe thin films are currently incorporated into silicon IC technologies in various ways. In order to understand the phenomena encountered in the metallisation processes of such films, an improved knowledge of the thermodynamic equilibria is needed.

This Datareview deals with the TM-Si-Ge ternary phase diagrams, taking into account the corresponding phase diagrams for TM-Si and TM-Ge.

Experimental phase diagrams for TM-Si-Ge (TM = Cr, Mn, Fe, Co) are given in FIGURE 1.

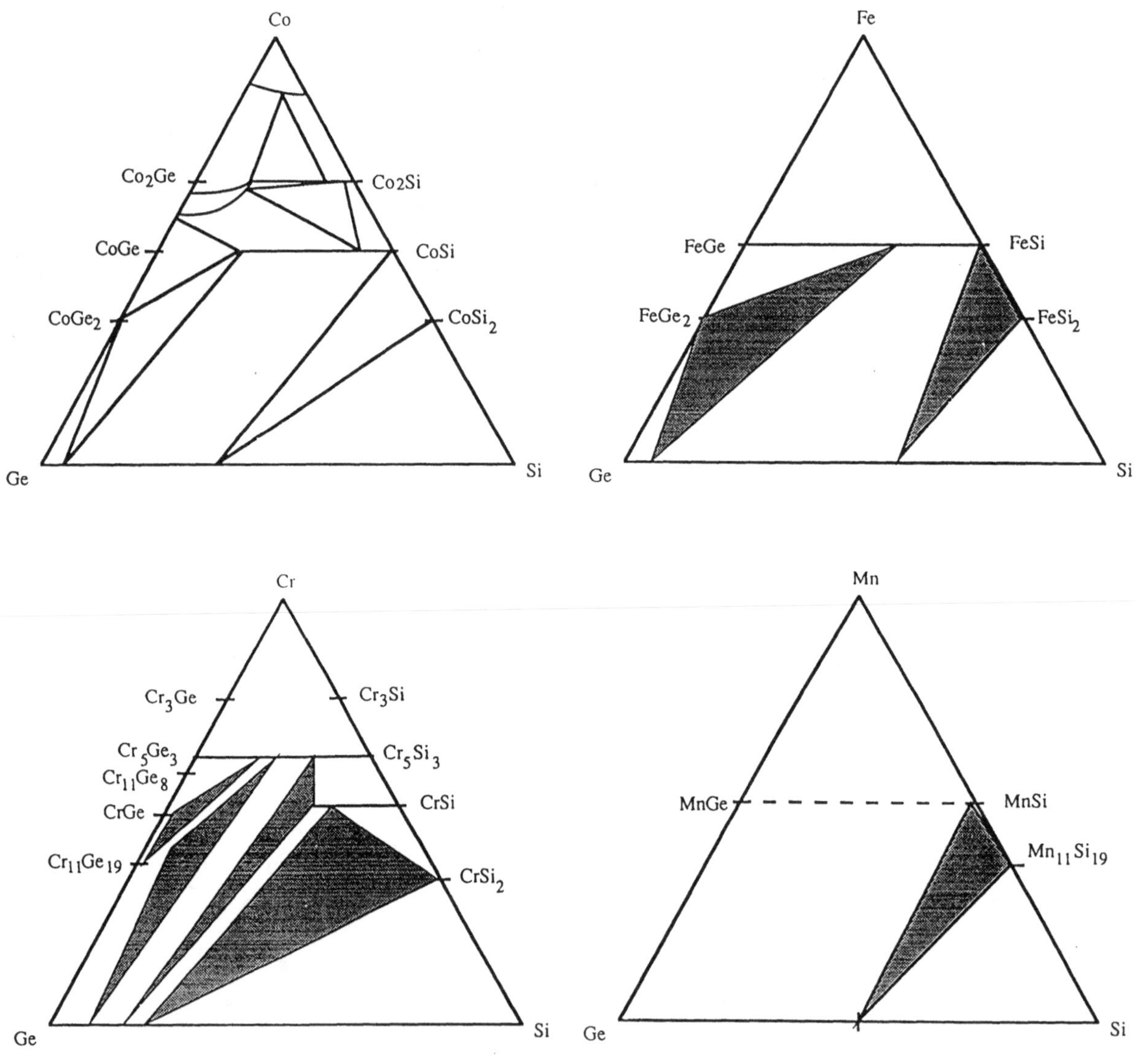


FIGURE 1 Experimental ternary phase diagrams TM-Si-Ge (TM = Cr, Mn, Fe, Co) [3].

TABLE 1 details the non-metal richest phases in the TM-Si and TM-Ge systems.

TABLE 1 Non-metal richest phases in both TM-Si and TM-Ge systems investigated together with their formation mode.

| Phase $M_nX_m$ | m/n | Reaction type | | Temperature (°C) |
|---|---|---|---|---|
| $TiSi_2$ | 2 | Congruent | | 1500 |
| $TiGe_2$ | 2 | Peritectic | $L + Ti_6Ge_5 \rightarrow TiGe_2$ | 1075 |
| $VSi_2$ | 2 | Congruent | | 1677 |
| $V_{17}Ge_{31}$ | 1.82 | Peritectic | $L + V_{11}Ge_8 \rightarrow V_{17}Ge_{32}$ | 960 |
| $CrSi_2$ | 2 | Congruent | | 1490 |
| $Cr_{11}Ge_{19}$ | 1.72 | Peritectic | $L + CrGe \rightarrow Cr_{11}Ge_{19}$ | 928 |
| $Mn_{11}Si_{19}$ | 1.72 | Peritectic | $L + MnSi \rightarrow Mn_{11}Si_{19}$ | 1152 |
| $\alpha$-$FeSi_2$ | 2.45 | Congruent | | 1220 |
| $\beta$-$FeSi_2$ | 2 | Peritectic | $\alpha$-$FeSi_2$ + FeSi $\rightarrow$ $\beta$-$FeSi_2$ | 982 |
| $FeGe_2$ | 2 | Peritectic | $L + Fe_4Ge_3 \rightarrow FeGe_2$ | 840 |
| $CoSi_2$ | 2 | Congruent | | 1326 |
| $Co_{0.9}Ge_2$ | 2.22 | Peritectic | $L + CoGe \rightarrow Co_{0.9}Ge_2$ | 832 |

TABLE 2 details the miscibility states for the TM-Si-Ge systems investigated, together with the annealing temperature.

TABLE 2 Systems investigated with the annealing temperature and the as-obtained equilibrium state.

| System | Annealing temperature (°C) | Equilibrium |
|---|---|---|
| $TiSi_2$-$TiGe_2$ | 900 | Total miscibility |
| $VSi_2$-$V_{17}Ge_{31}$ | 900 | Partial miscibility |
| $CrSi_2$-$Cr_{11}Ge_{19}$ | 900 | Immiscibility |
| $Mn_{11}S_{19-x}Ge_x$ | 900 | Immiscibility |
| $FeSi_2$-$FeGe_2$ | 800 | Immiscibility |
| $Fe_{29}Si_{71-x}Ge_x$ | 1050 | Partial miscibility |
| $CoSi_2$-$CoGe_2$ | 800 | Immiscibility |

## REFERENCES


[1] N. Boutarek [ Ph.D. Thesis, INPG, Grenoble (1993) ]

[2] N. Boutarek, R. Madar [ *Appl. Surf. Sci. (Netherlands)* vol.73 (1993) p.209 ]

[3] F. Wald, S.J. Michalik [ *J. Less-Common Met. (Switzerland)* vol.24 (1971) p.277 ]

# 3.6 Ternary phase diagrams for Al-TM-Si systems

R. Sinclair

April 1995

One such diagram, Ti-Al-Si has been examined in some detail. The original diagram [1] and its simplified representation [2] are shown in FIGURE 1. Aluminium is not thermodynamically stable with $TiSi_2$ for instance, and will react to produce the ternary compound, $Ti_7Al_5Si_{12}$. A diffusion barrier (TiN) is required to preserve $TiSi_2$ in contact with Al. The titanium aluminides can also dissolve a large amount of silicon (e.g. up to about 20% in $TiAl_3$). Thus the $TiAl_3$ particles formed by hot-sputter Al alloy deposition in Ti coated via holes will 'leach out' the Si from the metallisation (e.g. Al-Si-Cu) alloy. The composition determined experimentally in such circumstances by transmission electron microscopy [3] is in excellent agreement with the phase boundary found in the original paper [1].

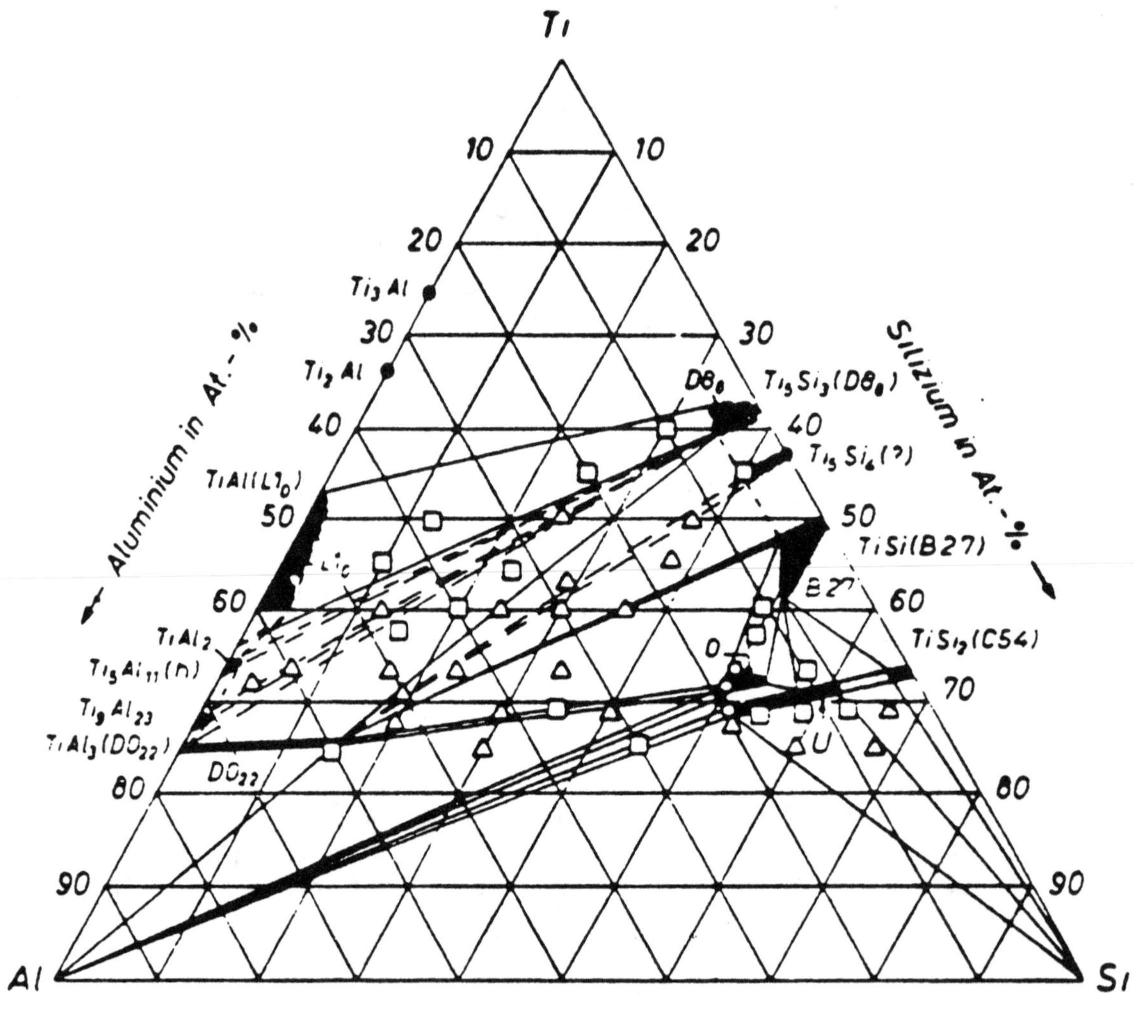


FIGURE 1 (a) The actual Ti-Al-Si phase diagram at 700°C as determined experimentally by Raman and Schubert [1].

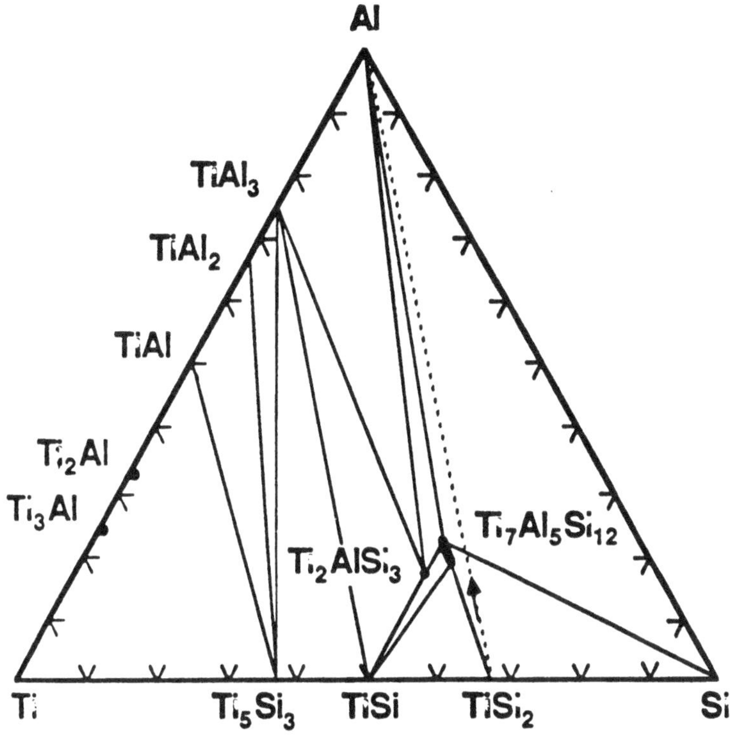


FIGURE 1 (b) The simplified form of the diagram, showing the major phases and tie-lines [2]. Note that the Ti-Si binary system is at the base of the diagram here, to be consistent with the other diagrams in Datareviews by this author.

Other Al-silicon diagrams should also be determined, because of the ubiquitous use of aluminium metallisation. The interested reader might also consult the Al-Ti-N diagram [4], for interpreting the breakdown of TiN diffusion barriers by reaction with Al at elevated temperatures (e.g. 550°C) [5]. This diagram is shown in FIGURE 2.

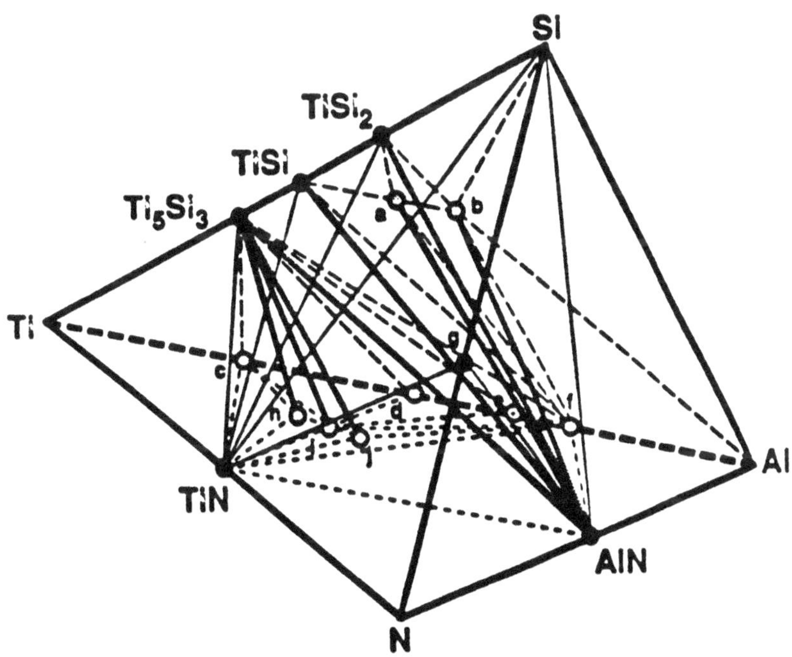


FIGURE 2 The complete Ti-Al-Si-N quaternary diagram (T = 900 K) with thick dashed lines as the interior (quaternary) tie-lines.

**REFERENCES**

[1] A. Raman, K. Schubert [ *Z. Met.kd. (Germany)* vol.56 (1965) p.44 ]
[2] R. Beyers [ Ph.D. Thesis, Stanford University, USA (1989) ]
[3] H. Nishimura, T. Yamada, R. Sinclair, S. Ogawa [ *Symp. on VLSI Technology*, Seattle, Washington, USA, 2-4 June 1992 (IEEE, New York, 1992) p.74-5 (digest) ]
[4] A.S. Bhansali [ Ph.D. Thesis, Stanford University, USA (1992) ]
[5] M. Okihara, N. Hirashita, K. Hashimoto, H. Onoda [ *Appl. Phys. Lett. (USA)* vol.66 no.11 (1995) p.1328 ]

# 3.7 Ternary TM-TM-Si reactions

M. Setton

June 1995

## A INTRODUCTION

In the quest for new materials, the study of silicides has provided a wealth of data regarding solid phase reactions. From a technological standpoint, the knowledge of key parameters needed for the introduction of silicides into semiconductor production is sufficient. Data such as formation or transformation temperatures, diffusing species, wet or dry etch chemistries and etch rates, oxidation behaviour, dopant behaviour and reactivity with $SiO_2$ are available for the processes including Ti, W, Ta or Co silicides [1]. However, because of their complexity, aspects of ternary Metal 1 - Metal 2 - Si systems are far less studied than binary ones even though they may have even greater technological impact. Applications of ternary systems to grow epitaxial films using diffusion through an interposed transport layer or to extend the thermal stability of silicide/Si contacts to higher temperatures are now being evaluated by several semiconductor device manufacturers.

In the literature, a growing number of papers dealing with the thermal evolution of metal alloys or metal bilayers on Si, Al-Metal-Si and Al/silicide can be found but important mechanisms such as moving species, dominance of nucleation or diffusion processes, phase segregation or solid solubility are still largely unknown.

This Datareview will present a general approach based on a review of some of the data available for both bulk and thin film studies. Experimental observations will be correlated with solid solubilities, crystal structures, electron affinities, identity of the moving species, atomic radii, kinetics and heats of formation to derive a qualitative model describing thin film reactions for ternary M1-M2-Si systems.

## B SYSTEM CLASSIFICATION

The evolution of metal bilayers or alloys on Si shows that ternary systems can be split into two basic categories:

(i) Systems without ternary compounds. This will lead to solid solutions or phase segregation accompanied in some cases by layer reversal.

(ii) Compound forming systems. Systems in this group will show some ternary phases either present with disilicides or stable in contact with Si.

TABLE 1 schematically illustrates the tendency of some selected systems to contain ternary silicides or solid solutions. It shows also that it should be possible to select chemical or physical parameters as indicators to place a system in one category or another.

TABLE 1 Schematic representation of the tendencies of systems to contain ternary phases or solid solutions. When available, the stability of Si-rich phases in contact with Si is indicated.

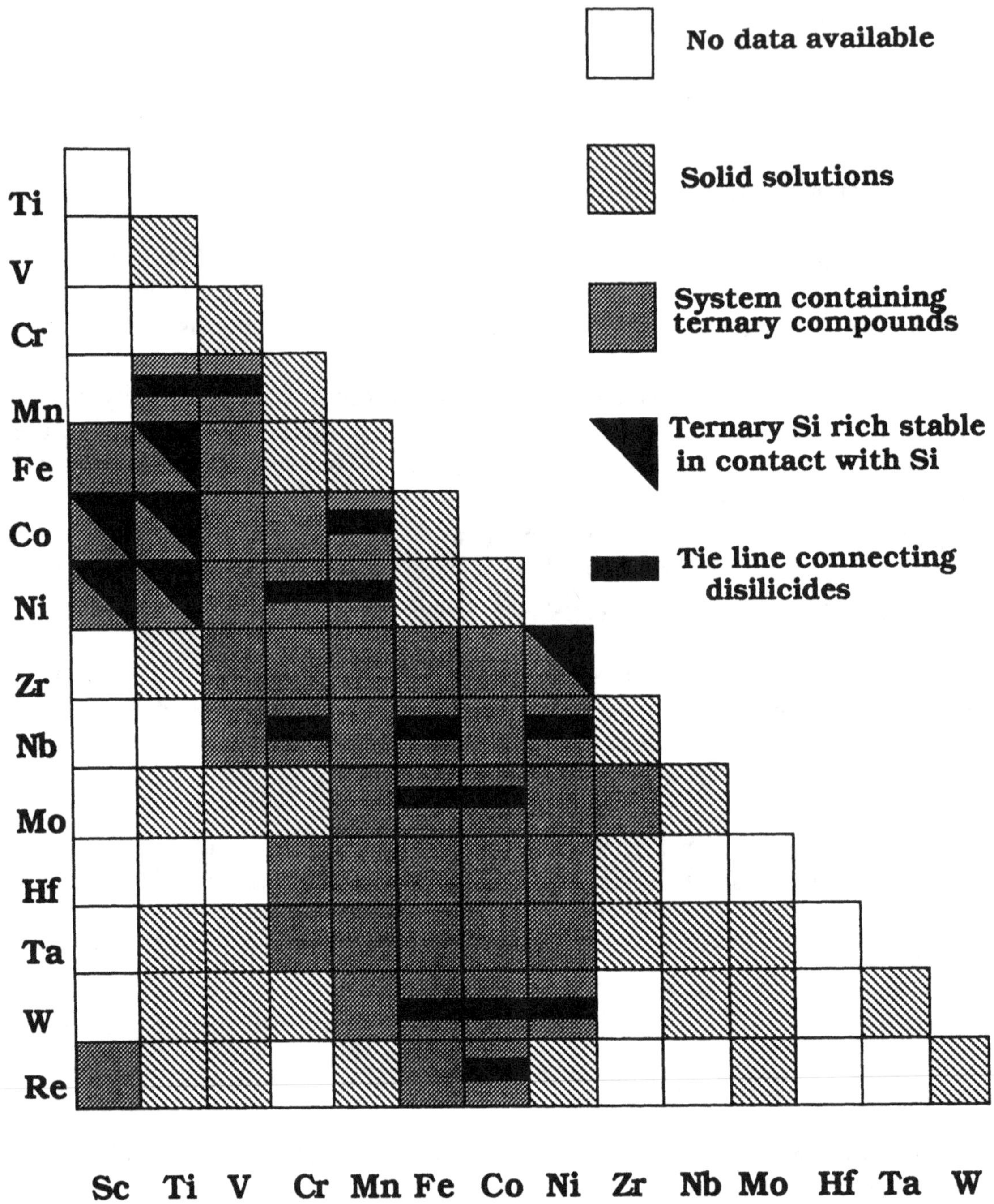


## B1 Solid Solutions

As expected from classical theory, the solubility of a third component in binary compounds depends on the closeness of the solute metal to one of the components of the solvent compound in the periodic table. Therefore, the parameters influencing the solubility are: (a) the dimensions of the substituting atom, (b) its chemical potential, and (c) the valency.

The solubility of phases in ternary systems has already been the object of a study by Tskai and Shaikina [2]. The dependence of the solubility on the electronegativity of species is attributed to the following explanation of the dissolution process. As with every equilibrium process it must be accompanied by a redistribution of the electronic density of the atoms. Because of this local equalisation of the potentials, the electron densities of each metal atom are modified. If the atom loses electrons, its valence shells will contract because of the increased effective charge of the nucleus and the electron density $\rho_{M1}$ increases, while simultaneously the orbital radius of M2 will increase and $\rho_{M2}$ decreases.

The absence of solid solubility occurs when the transfer of charge needed to equilibrate the potentials would cause an increase in the energy of the electrons in the solution compared to its value for the solid phases of pure components.

TABLE 2 presents bulk solubilities for selected systems. Only disilicides are mentioned but the solid solubility of intermediate phases is also important. For example, CoSi and NiSi

TABLE 2 Bulk solid solubility of silicides. (CSS = complete solid solubility.)

| System A-B-Si | Structure type of silicide A | Structure type of silicide B | Silicide solubility limit in mol % | | Ref |
|---|---|---|---|---|---|
| | | | B in A | A in B | |
| Ti-Zr-Si | $TiSi_2$ | $ZrSi_2$ | 5 | 55 | [3] |
| Ti-V-Si | $TiSi_2$ | $CrSi_2$ | 5 | 85 | [3] |
| Ti-Nb-Si | $TiSi_2$ | $CrSi_2$ | 3 | 45 | [4] |
| Ti-Ta-Si | $TiSi_2$ | $CrSi_2$ | 0 | 50 | [3] |
| Ti-Cr-Si | $TiSi_2$ | $CrSi_2$ | 7 | 88 | [3] |
| Zr-V-Si | $ZrSi_2$ | $CrSi_2$ | 0 | 0 | [3] |
| Zr-Nb-Si | $ZrSi_2$ | $CrSi_2$ | 6 | 21 | [4] |
| Zr-Ta-Si | $ZrSi_2$ | $CrSi_2$ | 0 | 32 | [4] |
| Zr-Cr-Si | $ZrSi_2$ | $CrSi_2$ | 0 | 0 | [3] |
| Zr-Mo-Si | $ZrSi_2$ | $MoSi_2$ | 5 | 0 | [3] |
| Zr-W-Si | $ZrSi_2$ | $MoSi_2$ | 12 | 3 | [5] |
| V-Nb-Si | $CrSi_2$ | $CrSi_2$ | CSS | CSS | [4] |
| V-Ta-Si | $CrSi_2$ | $CrSi_2$ | CSS | CSS | [4] |
| V-Cr-Si | $CrSi_2$ | $CrSi_2$ | CSS | CSS | [4] |
| V-Mo-Si | $CrSi_2$ | $MoSi_2$ | 65 | 4 | [3] |
| V-W-Si | $CrSi_2$ | $MoSi_2$ | 28 | 18 | [4] |
| Nb-Cr-Si | $CrSi_2$ | $CrSi_2$ | 0 | 0 | [4] |
| Nb-Mo-Si | $CrSi_2$ | $MoSi_2$ | 66 | 12 | [4] |
| Ta-Cr-Si | $CrSi_2$ | $CrSi_2$ | 40 | 12 | [3] |
| Ta-Mo-Si | $CrSi_2$ | $MoSi_2$ | 56 | 16 | [3] |
| Nb-W-Si | $CrSi_2$ | $MoSi_2$ | 30 | 15 | [5] |
| Ta-W-Si | $CrSi_2$ | $MoSi_2$ | 30 | 24 | [3] |
| Cr-Mo-Si | $CrSi_2$ | $MoSi_2$ | 32 | 46 | [3] |
| Cr-W-Si | $CrSi_2$ | $MoSi_2$ | 16 | 70 | [3] |
| Mo-W-Si | $CrSi_2$ | $MoSi_2$ | CSS | CSS | [3] |

have different crystal structures and during annealing of Co-Ni alloys segregation into distinct layers does occur [6]. At higher temperatures however, once the two cubic totally miscible disilicides are formed, a solid solution (Co-Ni)$Si_2$ constitutes the final state of the reaction.

Phase separation is illustrated in TABLE 3. Only results of thin film experiments are presented and the sequence of formation is not reported. We are dealing only with the final high temperature equilibrium products.

TABLE 3 Phase separation at high temperature for thin film experiments.

| System | Ref | Results |
|---|---|---|
| Ti-V-Si | [7] | $TiSi_2$ dissolved in $VSi_2$ lattice |
| Ti-W-Si | [9] | Solid solutions and phase separation |
| Ta-W-Si | [11] | See text for pseudo binary phase diagram reference |
| Ti-Pd-Si | [12] | PdSi + $VSi_2$ |
| Ti-Ta-Si | [13] | $TiSi_2$ + $TaSi_2$ (or (Ti,Ta)$Si_2$ see text) |
| Ta-Cr-Si | [14] | $TaSi_2$ + $CrSi_2$ |
| Ta-V-Si | [7] | $VSi_2$ dissolved in $TaSi_2$ lattice |
| Ir-V-Si | [15] | Intermetallics + phase separation |
| Ni-W-Si | [16,17] | Thermal: phase separation/ion beam: solid solutions |
| Ni-Cr-Si | [19,20] | $NiSi_2$ + $CrSi_2$ |
| Ni-Pt-Si | [22] | $Ni_{1-x}Pt_xSi \rightarrow PtSi + NiSi_2$ |
| Pt-Cr-Si | [20] | PtSi + $CrSi_2$ |
| Pt-V-Si | [23] | PtSi + $VSi_2$ |
| Pt-W-Si | [24] | PtSi + $WSi_2$ |
| Pt-Pd-Si | [25] | $Pd_{1-x}Pt_xSi$ stability ? |
| Pd-Cr-Si | [26] | $Pd_2Si$ + $CrSi_2$ |
| Pd-V-Si | [27] | $Pd_2Si$ + $VSi_2$ |
| Pd-W-Si | [24] | $Pd_2Si$ + $WSi_2$ |
| Gd-Ti-Si | [28] | $GdSi_2$ + $TiSi_2$ |
| Pt-Er-Si | [28] | Intermetallics + PtSi + $ErSi_2$ |
| Gd-V-Si | [29] | $GdSi_2$ + $VSi_2$ |
| Gd-Pt-Si | [29] | Intermetallics + PtSi + $GdSi_2$ |

## B1.1 Thin film results

TABLE 2 displays the bulk solubility of disilicides. The data presented always correspond to high temperature conditions (>1300°C) and thus represent limits. Since annealing of thin films is carried out at much lower temperatures, observed solubilities should be even less extensive.

Complete solid solubility is not limited to disilicides. Some monosilicides such as PdSi-NiSi or PdSi-PtSi also exhibit complete miscibility. Finstad and Nicolet [30] have studied the evolution of Pd/Ni/Si structures and observed between 750 and 800°C an orthorhombic compound labelled as $Pd_xNi_{1-x}Si$ with x = 0.35 and lattice parameters a = 5.79 Å, b = 5.34 Å and c = 3.21 Å. Bulk results of Wopersnow and Schubert [31] give a = 5.78 Å, b = 5.33 Å and c = 3.32 Å. In contrast with the other parameters, c does not follow a typical Vegard's law. It decreases up to x = 0.30 and then increases as more Pd is incorporated. This behaviour is the basis for a correlation between composition, binding type and valency electron concentration. For higher temperature annealing, the thin film solid solution [30] evolves towards phase separation of $NiSi_2$ and PdSi as expected from the phase diagram.

More recently, Long and Mahan [32] looked at the properties of semiconducting $Re_xMo_{1-x}Si_2$ and $Cr_xV_{1-x}Si_2$. Regarding the solubility, their results simply duplicate those of Verkhoglyadova [5]. For the electrical resistivity of $Cr_xV_{1-x}Si_2$, the thin film measurements of Long and Mahan are in conflict with those measured previously. The values obtained by Thomas et al [33] in an earlier study are in excellent agreement with the data of Sidorenko et al [34] and are therefore more reliable. This illustrates that in some cases a dual approach combining information for bulk samples with that from thin films helps in determining the validity of both data.

Because of its application as a diffusion barrier, the Ti-W alloy interactions with Si or with a silicide have been well documented. For bulk samples, Goldschmidt [3] indicates that solid solutions of C54 $TiSi_2$ and $WSi_2$ were found to be hexagonal for concentrations of $WSi_2$ less than 42%, tetragonal for concentrations above 96% and a mixture of tetragonal and hexagonal for intermediate concentrations. Harris et al [35] found the presence of a hexagonal $CrSi_2$-type structure for a $Ti_xW_{1-x}Si_2$ solid solution resulting from the annealing of the $Ti_{0.3}W_{0.7}$ alloy on Si.

Another work [36] with $Ti_{0.24}W_{0.76}$ alloy/PtSi/Si resulted in a solid solution with a hexagonal structure. For W (30 nm)/Ti (50 nm)/Si structures (60% $TiSi_2$/40% $WSi_2$), a pseudo-binary solid solution with a hexagonal structure was obtained as well [37]. For thin film structures one also has to remember that the formation of metastable Ti silicide with a C49 structure at low temperature may influence solid solubility as well as the crystal structure of the solid solution.

Gas et al [38] determined the phase diagram of the system $TiSi_2$-$WSi_2$. At the composition of interest for the last bilayer experiment mentioned (40% $WSi_2$) and for temperatures below 1000°C, their diagram shows that only the hexagonal phase is thermodynamically stable. The experimental results combined with those obtained for $TaSi_2$-$WSi_2$ [39] have been analysed in terms of the electron to atom ratio. If one attributes an electron concentration of 1, 2 and 3 respectively to Ti, Ta and W, then a 50% $TiSi_2$-50% $WSi_2$ solid solution has the same c/a ratio as $TaSi_2$ which is hexagonal and therefore is expected to have a similar structure.

For thinner Ti layers (5 to 10 nm) interposed [37], the reaction starts with the formation of $TiSi_2$ at 600°C. Up to 630°C, a pseudo-binary solid solution forms, followed by the appearance of $WSi_2$ at 650°C and silicidation completion at 850°C. Interestingly, there are no signs of hexagonal $WSi_x$.

The temperature of W silicidation is considerably reduced compared to that of W/Si samples. When dealing with W, the free energy of formation of $WO_3$ is smaller than that of silicon dioxide and the solubility limit of O in W is low; therefore as a W/Si sample is annealed, O distributed throughout the film migrates to the interface to form a diffusion barrier. The reaction starts at localised weak spots of the barrier. Interposing a Ti layer suppresses this effect. The formation of $TiO_2$ is actually energetically favourable ($\Delta G(SiO_2)$ = -80.2 kcal/g atom and $\Delta G(TiO_2)$ = 91.2 kcal/g atom at 1273°C) and the effect of the thin layer is not only to provide a transport medium for Si but, more importantly, to transform the interfacial layer into O dissolved into Ti, suppressing the diffusion barrier and lowering the silicidation temperature of W.

Another system that has been studied for thin film reactions is Ti-Ta-Si. A solid solution has been reported by Dahan and co-workers [40] and Choi et al [41]. The latter paper presents the formation of $TiSi_2/TaSi_2/Si$ at 850°C evolving to $(TiTa)Si_2$ at 1000°C. This solid solution, inappropriately called ternary 'phase', is thermally stable in contact with Si and has a fairly low resistivity. The observed results can be explained by the following factors:

(i) The absence of any binary silicide points towards a solid solution.

(ii) Identity of the moving species, Si, for both Ti-Si and Ta-Si systems.

(iii) Similarity of atomic radius (we will see later that this is unfavourable for true ternary compound formation), suggesting that Ti and Ta atoms probably statistically occupy the metal sites of $(M)Si_2$.

However, the different crystal structures leading to partial solid solubility for bulk samples (see TABLE 2) bring some interesting questions:

(i) Even if X-ray diffraction does not detect any phase separation, does this rule out demixing on a short range?

(ii) Since $TaSi_2$ seems to have negligible solubility in $TiSi_2$ for bulk samples, could the solid solution with low resistivity consist of Ti atoms substitutionally placed on Ta sites in the $TaSi_2$ lattice?

(iii) Should the electron to atom ratio be considered and lead to a hexagonal $CrSi_2$ type solid solution?

It is important to keep in mind that for all experiments on Si substrates, a low metal mobility could mask its solubility. Even though the metal might be totally miscible, silicidation due to Si diffusion can take place at temperatures too low to allow metal intermixing, causing layer separation due to silicide insolubility.

For alloys on Si, if one of the metals is the moving species during silicidation, it will tend to segregate at the alloy/Si interface. If, however, Si is the diffusing species for both silicides, either the resulting structure will be a pseudo-binary solid solution or macroscopic phase separation into two distinct layers will occur if the solid solubility of the two silicides is low.

Data and conclusions based on phase diagrams and bulk experiments are, of course, only valid for conditions associated with a steady state process, but the results presented here for solid solutions or phase separation for bulk samples and thin film reactions are usually in very good agreement.

### B1.2 Initial modelling of alloy reactions with Si

Thompson et al [42] developed a model to predict the outcome of alloy reactions with Si. In their approach, if for each subsystem M1-Si, M2-Si and M1-M2 the temperature of the reaction ($T_1$, $T_2$ and $T_m$ respectively) as well as the dominant diffusing species for the silicidation reaction are known, the behaviour of M1M2 alloys in contact with Si can be predicted. Based on these criteria, three categories were presented:

(I) Lowest reaction temperature is one of a silicide formation ($T_1$ or $T_2$) and metal atoms are the dominant diffusing species. For this case one expects the formation of a silicide layer between the alloy and Si. The remaining structure on top will evolve towards a depleted alloy followed by the growth of the silicide of the other metal.

(II) Lowest reaction temperature is the one of a silicide but Si is the diffusing species for that reaction. The alloy is transformed into a mixture of one silicide and unreacted metal. There is no redistribution of the metal atoms. Segregation into distinct layers is dependent on the temperature, solid solubility of the silicides and interfacial energy.

(III) Lowest reaction temperature is that of the intermetallic compound ($T_m$). Subsequent silicide formation is dependent on the phases, the morphology of the alloy and the identity of the moving species. The metal in excess with respect to the composition of the intermetallic compound will segregate at the Si interface to form a layer of pure silicide.

This model is adequate for systems exhibiting limited solid solubilities for which the end result at high temperature is phase separation. It also can be applied to isoelectronic or pseudo-isoelectronic metals in which silicides display complete solid solubility. It does not consider, however, the possibility of ternary compound formation, nor does it look at the dynamic aspect of thin film reactions.

### B1.3 Heat of mixing

From a thermodynamical point of view, we could say that the two metals will react with each other (case III) whenever their heat of mixing ΔHmix is greater than ΔHf of the first nucleating phase for both M1-Si and M2-Si. The threshold seems to be ΔHf = -11 kcal/g atom calculated using Miedema's theory [43] which takes into account both the difference in work function and the change in electron density at the boundary of the Wigner-Seitz atomic cells.

### B2 Ternary Compounds

Sprenger [44] divides the ternary phases into three categories depending on the Si concentration. They are (a) 0% < x(Si) < 30 at %, (b) 30% < x(Si) < 50%, (c) 50% < x(Si).

For the Si 'poor' phases, one of the classes is that of the Laves phases characterised by a formula T $(M_{1-x}Si_x)_2$, T being a transition metal. The electron concentration is critical for the formation of these phases. Bardos et al [45] showed that the number of electrons per atom outside closed shells should be less than 8 for binary phases to be stable. In ternary phases, the introduction of Si lowers the effective concentration by acting as an electron acceptor. The Si, playing the role of the smaller atom, has an effective radius much smaller than the metallic radius of Si for a coordination number of 12. Hume-Rothery [46] noticed that these estimated radii, ranging from 1.16 to 1.21 Å, were very close to the covalent radius of 1.174 Å. Therefore, it would seem that Si lowers the electron concentration by forming a covalent bond.

The ordered phases containing transition metals are usually $T_6M_{16}Si_7$ (G phases) with a $Th_6Mn_{23}$ type structure. TABLE 4 displays G phases involving transition metals. Their stability is related to an electronic factor (or, less likely, an elastic size effect since it increases as one moves from Fe to Co and finally Ni).

TABLE 4 Lattice spacings (Å) for G phases. Compiled from Pearson [ *Handbook of Lattice Spacings and Structures of Metals and Alloys* ].

| | Fe | Co | Ni |
|---|---|---|---|
| Sc | None | 11.438 | |
| Ti | None | 11.232 | 11.198 |
| V | None | None | 11.153 |
| Cr | None | None | None |
| Mn | None | None | 11.154 |
| Zr | None | 11.447 | 11.464 |
| Nb | None | 11.251 | 11.249 |
| Mo | None | None | None |
| Hf | 11.48 | 11.404 | 11.39 |
| Ta | None | 11.222 | 11.215 |
| W | None | None | None |

For compounds with intermediate Si contents 30% < x(Si) < 50%, the two common phase types are E phases M1M2Si (with a typical $Co_2Si$ Pnma structure) and V phases $M1_4M2_4Si_7$ or $M1M2Si_2$ (mostly of $Co_4Zr_4Ge_7$ symmetry). These two families (E and V) are shown in TABLES 5, 6 and 7. As we can see, V phases do not exist for the Cr group elements and a different structure $MgZn_2$ is common to E phases of Mo and W. This difference might be related to an electronic factor also, these refractory metals possessing an anomalously high electronegativity.

TABLE 5 Equiatomic E phases M1M2Si. Common structure is $Co_2Si$ unless indicated. (Lattice parameters a, b and c in Å.)

| | Fe | Co | Ni | Cu |
|---|---|---|---|---|
| Sc | None | a = 6.419<br>b = 3.953<br>c = 6.896 | a = 6.38<br>b = 4.02<br>c = 6.93 | a = 6.566<br>b = 3.976<br>c = 7.224 |
| Ti | hP186/mmm<br>Hexagonal | a = 6.107<br>b = 3.72<br>c = 6.936 | a = 6.1484<br>b = 3.67<br>c = 7.017 | a = 6.193<br>b = 3.746<br>c = 7.13 |
| V | Hexagonal<br>a = 6.21<br>c = 6.74 | a = 5.954<br>b = 3.584<br>c = 6.852 | a = 5.970<br>b = 3.580<br>c = 6.922 | None |
| Cr | None | a = 5.90<br>b = 3.70<br>c = 6.75 | None | No data |
| Mn | Hexagonal<br>a = 8.866<br>c = 8.655 | a = 5.819<br>b = 3.691<br>c = 6.853 | a = 5.897<br>b = 3.612<br>c = 6.916 | None |
| Zr | a = 6.405<br>b = 3.935<br>c = 7.199 | a = 6.452<br>b = 3.856<br>c = 7.170 | a = 6.470<br>b = 3.815<br>c = 7.263 | a = 6.506<br>b = 3.971<br>c = 7.278 |
| Nb | a = 6.231<br>b = 3.677<br>c = 7.190 | a = 6.231<br>b = 3.66<br>c = 7.070 | a = 6.223<br>b = 3.683<br>c = 7.088 | None |
| Mo | $MgZn_2$<br>a = 4.751<br>c = 7.662 | $MgZn_2$<br>a = 4.745<br>b = 7.570 | $MgZn_2$<br>a = 4.745<br>c = 7.578 | No data |
| Hf | a = 6.32<br>b = 3.91<br>c = 7.12 | a = 6.37<br>b = 3.83<br>c = 7.08 | a = 6.39<br>b = 3.89<br>c = 7.20 | a = 6.439<br>b = 3.679<br>c = 7.262 |
| Ta | a = 6.158<br>b = 3.712<br>c = 7.086 | a = 6.2<br>b = 3.65<br>c = 7.037 | a = 6.191<br>b = 3.678<br>c = 7.059 | No data |
| W | $MgZn_2$<br>a = 4.746<br>c = 7.630 | $MgZn_2$<br>a = 4.734<br>c = 7.722 | None | No data |

TABLE 6 E phases with elements other than Fe, Co, Ni or Cu.

| Compound | Structure | | Lattice parameters (Å) | | |
|---|---|---|---|---|---|
| PdTiSi | $Co_2Si$ | Pnma | a = 6.324 | b = 3.77 | c = 7.39 |
| PdZrSi | $Co_2Si$ | Pnma | a = 6.59 | b = 3.89 | c = 7.57 |
| PdScSi | $Co_2Si$ | Pnma | a = 6.543 | b = 4.069 | c = 7.396 |
| PtScSi | $Co_2Si$ | Pnma | a = 6.566 | b = 4.130 | c = 7.275 |
| CrHfSi | $Co_2Si$ | Pnma | a = 6.694 | b = 3.543 | c = 7.758 |
| CrMoSi | $MgZn_2$ | $P6_3/mmc$ | a = 5.001 | | c = 8.109 |
| CrNbSi | $Fe_2P$ | P62m | a = 6.597 | | c = 3.354 |

TABLE 7 V phases $M1_4M2_4Si_7$. Notice the absence of such phases for Cr group metals V, Mn and Mo, W. (Lattice parameters a, b and c in Å.)

| | Fe | Co | Ni |
|---|---|---|---|
| Sc | $MnTiSi_2$ Pbam<br>a = 8.960<br>b = 9.745<br>c = 7.639 | None | None |
| Ti | $MnTiSi_2$ Pbam<br>a = 8.6137<br>b = 9.5134<br>c = 7.639 | a = 12.513<br>c = 4.934 | a = 12.575<br>c = 4.945 |
| V | None | None | None |
| Cr | None | None | None |
| Mn | None | None | None |
| Zr | a = 13.056<br>c = 5.09 | a = 12.935<br>c = 5.09 | No |
| Nb | a = 12.652<br>c = 4.981 | a = 12.557<br>c = 4.981 | a = 12.571<br>c = 4.988 |
| Mo | None | None | None |
| Hf | $MnZrSi_2$ Immm<br>a = 17.05<br>b = 7.751<br>c = 5.124 | a = 12.73<br>c = 5.02 | a = 12.87<br>c = 5.09 |
| Ta | a = 12.591<br>c = 4.976 | a = 12.524<br>c = 4.962 | None |
| W | None | None | None |

Finally, for compounds with 50% < x(Si), Sprenger mentions, for example, $Ti_{0.75}Co_{0.25}Si_2$, $Zr_3Ni_2Si_8$ and $Ti_7Al_5Si_{12}$.

From TABLE 1 we notice that Mn, Fe, Co and Ni form ternary phases with almost all other transition elements. Just plotting the occurrence of a ternary compound as a function of the relative difference of atomic sizes gives an indication of the influence of the 'elastic' factor. The fact that large differences create a favourable condition for the formation of a compound has already been observed by Gladyshevskii [4]. It points in the same direction as a model established by Villars [47] in which the three important criteria for compound formation are:

(i) Difference between the Zunger pseudo-potential radii sum or, alternatively, metallic radii sum.

(ii) Difference between the number of valence electrons.

(iii) The sum of the ratios of the melting temperatures.

Using these criteria, Villars [47] was able to show that no compound formation occurred along the three axes of the three dimensional ($\Delta r$, $\Delta VE$, $\Delta T$ coordinates) stability diagram. This means that whenever one of the three parameters is small, no matter how large the others are, compound formation in unfavourable. Taking this idea further, Villars constructed structural stability domains related to the relative position of the atoms within the periodic table.

Even though ternary phases are reported for a given system, the chances that one of these will be observed for alloys or bilayers in contact with Si are rather small. Several factors contribute to this observation. First of all the free energy of formation of the ternary Si richest phases seems to be lower than that of binary disilicides. Only in a few cases does a tie line exist between the compound and Si, indicating a high thermal stability related to $\Delta G_{form}(tern) > \Delta G_{form}(M1Si_2)$ and $\Delta G_{form}(tern) > \Delta G_{form}(M2Si_2)$. The corresponding free energy diagram is shown in FIGURE 1. Secondly, if one of the metals M1 starts the silicidation process, the equilibrium at the interface will be moving along the Metal M1-Si line until the Si richest binary phase is obtained. Even if some unreacted metal M2 is left, it will react with Si before it can react with the silicide to form a ternary (a process occurring at higher temperatures).

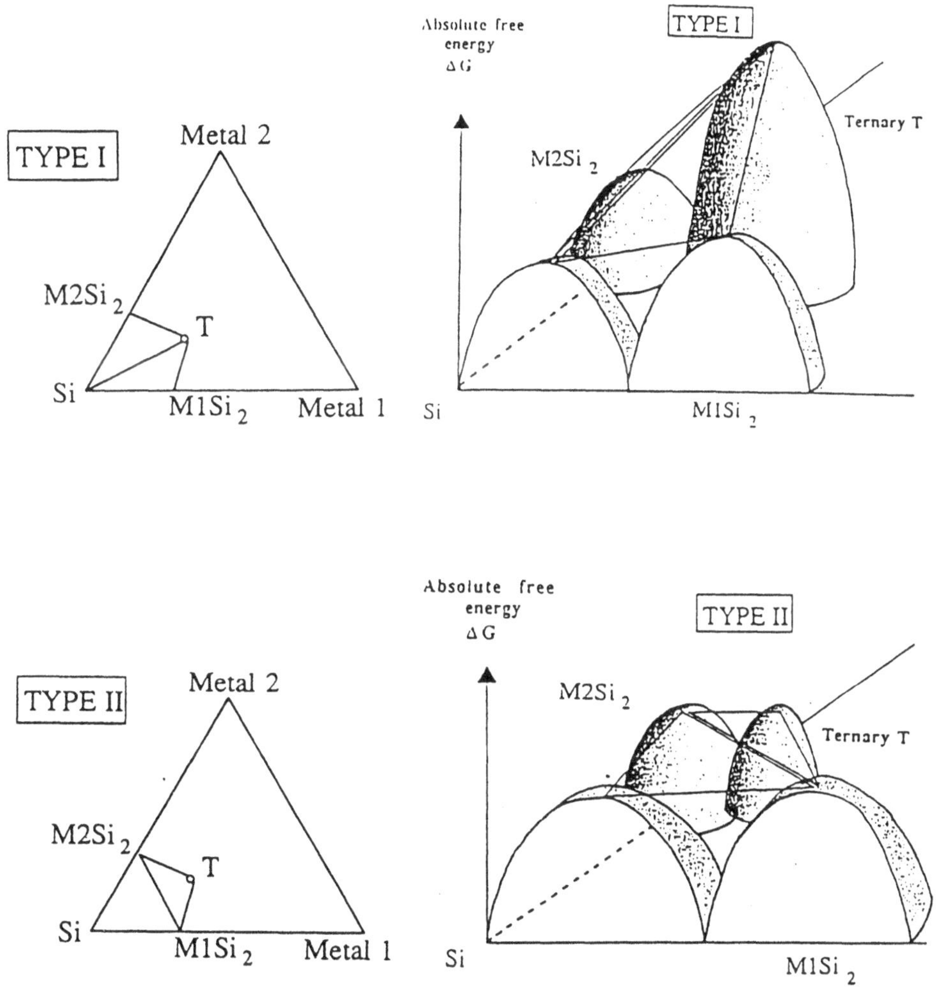


FIGURE 1 Schematic of the two types of phase diagram containing ternary compounds and the corresponding free energy plots.

Ternary compound growth is more likely if the two metals react first. Once the intermetallic phase is formed, Si can move into this phase (with a free energy of formation higher than the ΔG of both the first nucleated phases in the M1-Si and M2-Si systems) to cause its transformation to a ternary silicide either by a diffusion or by a nucleation controlled process.

Si might also react with excess metal to reach an equilibrium T + binary silicide + Si phase. In many cases the observation of a ternary phase follows the appearance of an intermetallic one.

A brief survey of the available phase diagrams for M1-M2-Si, M1 and M2 being transition metals, shows the absence of any tie lines between intermetallic compounds and Si. They will be stable only in contact with a ternary phase or with binary silicides.

Ternary phases can also be obtained in thin film experiments using two routes: (a) reaction of a pure metal with a silicide, and (b) reaction of Si with an intermetallic compound for systems with $\Delta G(M1\text{-}M2) > \Delta G_{\text{first nucleating phase}}$ in both M1-Si and M2-Si systems.

By limiting the amount of Si, it is also possible to obtain metal-rich silicides.

## C KINETIC ASPECTS

The selection of the first nucleating phase is not purely determined by minimisation of the free energy but also by kinetics.

The first phase to nucleate will be either M1-M2, M1-Si or M2-Si compound. As we shall see, ternary phase formation usually follows the reaction of metal with silicide or Si with an intermetallic compound. Miedema's theoretical heat of mixing has already been presented (see Section B1.3) as a method to predict whether intermetallic compound formation is more favourable than silicidation as the first stage of the reaction.

### C1 Effective Heat of Formation

Thermodynamics of transition metal silicides have been reviewed by Schlesinger [48].

The thermodynamical approach chosen in this Datareview, however, is that of Pretorius [49] who defined a concept of 'effective heat of formation' for which ΔH is a function of the concentration of the reactants, and therefore changes as the reaction progresses. This modification of ΔH links the thermodynamic approach to atomic availability. For the first phase, ΔH is calculated with respect to the eutectic composition, the point where the interfacial energy between the phases is the lowest. Let us suppose that the eutectic is x%A and y%B, x<y: for each phase $A_pB_q$, the effective heat of formation ΔH' is given by $\Delta H' = \Delta H\ (x/p)$, if x is the limiting element, or $\Delta H\ (y/q)$ otherwise. In this expression, ΔH is the conventional heat of formation in kcal/g atom. The first phase to form is the one with the lowest ΔH'. The second phase is the next congruent phase richer in the unreacted element, with the most negative ΔH'. We shall illustrate this methodology later on for the Ni-Ti system.

The concept of interfacial energy barrier was first taken into account by Gosele and Tu [50] in their elegant treatment of the growth kinetics of planar binary diffusion couples. They introduced two processes determining the evolution of the layers: (a) a diffusional flux of matter across the compound layer, slowing down as thickness increases, and (b) an interfacial reaction barrier related to the necessary rearrangement of atoms at the interfaces.

From a simple argument, the first silicide phase to form is the one with the lowest effective interface reaction barrier, and a second compound does not appear until this first phase has reached a critical thickness. This last point is of considerable importance since it means that because of kinetic instabilities (or nucleation difficulties), various equilibrium phases may only occur sequentially for thin films in contrast to simultaneously for bulk specimens.

Interfacial reactions play a decisive role in explaining why some layers which would thermodynamically be stable shrink and disappear. Using a similar reasoning, Dybkov [51] has also treated the growth of two compounds $A_pB_q$ and $A_rB_s$ at the interface of A and B by adding a chemical reaction term to the usual 'diffusional theory'.

### C2 Importance of Kinetics

We shall look at the binary Ti-Si system to illustrate the importance of kinetics. Using the silicide island growth model, Van Loenen et al [52] showed that the mechanism of Ti silicide formation was similar to that for near noble metals. The composition of the first phase, found by ion scattering, was close to TiSi. This composition was also observed for a TEM experiment on UHV films in which orthorhombic oriented monosilicide TiSi was detected for room temperature reaction between Ti and a-Si. There is no significant change in the structure until 550°C where all the Si is rapidly consumed. New phases identified as $Ti_5Si_3$ and $Ti_5Si_4$ are also present. An interesting observation is that if the sample is rapidly annealed to a temperature >600°C, the two Ti-rich compounds are bypassed and a heavily defective $TiSi_2$ grows at the expense of TiSi instead.

For the samples heated slowly, the structure evolves from Ti/TiSi/Si to Ti/$Ti_5Si_3$/Si because some Ti remains to react with TiSi. For samples processed at temperatures greater than 600°C, all the Ti is rapidly consumed by Si (at these temperatures Si is extremely mobile) and TiSi/Si is formed. Then the only possible reaction is $TiSi + Si \rightarrow TiSi_2$ and no Ti-rich silicides are involved in the sequence. This analysis is perfectly consistent with the work of Natan [53] who showed how the exhaustion of Ti prior to nucleation influenced the sequence. His study illustrated the fact that, as predicted by Gosele and Tu [50], the sequence is a function of the availability of the various components. Majni et al [54] also found that the amount of metal relative to Si determines the path followed for the formation of a final compound whose stoichiometry is determined by the initial relative concentrations.

To summarise what has been reviewed in the last sections, for ternary systems, the first phase to form at each interface can be predicted using Pretorius' methodology but the sequence is purely a function of atomic availability and kinetics. Ternary compound formation certainly adds to the complexity of the problem, but the diffusion of species through a growing compound and the rate of reaction at the interface should be included in any modelling of the 'dynamical' evolution of alloys or multilayers on Si.

## D USING INTERMEDIATE LAYERS AS TRANSPORT MEDIA AND EPITAXY PROMOTERS

Bilayers have been the subject of intense research because of the possibility of growing epitaxial silicides using an interposed layer as transport medium. This idea had been proposed by Setton [55] based on work for the Ni/Ti/Si system, the results from Yang and

Bene [56] and the O gettering effect of an interposed layer of Ti on the formation of W silicide.

Epitaxial NiSi formation using Ni/a-NiZr/Si structures had been reported by De Reus et al [57] and Rozgonyi et al [58] described the same result for NiCr semi-permeable amorphous membrane in contact with Si.

For many bilayers, the interposed layer has been observed to act as a selective diffusion barrier and to control intermetallic compound formation as well as silicidation.

We will illustrate intermetallic and ternary compound formation, layer reversal and solid state transport for the Ni/Ti/Si system as well as for the reactions of metal/silicides. The effect of promoting epitaxy will be discussed for Co/Metal (Ti,Ta,Zr)/Si experiments.

## D1 Experimental Observation of Solid State Transport Medium

For thick Ni and Ti metal bilayers (>30 nm) on Si, the reaction starts by the formation of an intermetallic compound [59]. The first phase detected by XRD is the Ni-rich phase $Ni_3Ti$. This can be explained using Pretorius' model. At the concentration of the lowest eutectic (24% Ni, 76% Ti), $Ti_2Ni$ has the most negative heat of formation. However, the first phase to nucleate has to be a congruent one according to the rules derived by Pretorius. This Ti-rich phase corresponds to a peritectic and it is therefore not formed while that of the closest congruently melting one, NiTi, is formed. As NiTi is unstable at low temperatures (it breaks down to $NiTi_2$ and $Ni_3Ti$ for temperatures below 800°C) we have to move to the next phase, namely $Ni_3Ti$ (also congruently melting), to reach the more thermodynamically likely intermetallic compound. This aspect of non-congruency of $Ti_2Ni$ could be at the root of the observation that $Ni_3Ti$ is the first compound to nucleate independently of the availability of metal atoms. Also, since Ni is much more mobile than Ti at low temperatures, one can infer that locally the interface is Ni-rich. In those terms, Ti would become the limiting element and $Ni_3Ti$ would be the dominant phase to form.

According to the ordered $Cu_3Au$ rule, Ni is the diffusing species during the growth of $Ni_3Ti$. This rule, pointed out by d'Heurle [60], says that the diffusing species constitutes the majority atoms in the compound considered. It seems fairly logical since there is a higher probability of 'atomic jumps' on the Ni sublattice which is more continuous than the Ti one in $Ni_3Ti$. Exchanges on different atomic sites are unlikely because of the increase in disorder and energy that would be associated with such a process.

The 'active' role of Ni compared to Ti is stressed when the deposition sequence is reversed. For Ti/Ni/Si films, AES data indicate that the intermediate layer of Ni is consumed in two different ways: Ni atoms diffuse towards the substrate to form $Ni_2Si$ and the complete sequence NiSi and $NiSi_2$ but diffusion into Ti also takes place. The second intermetallic compound to form is $Ti_2Ni$.

Above 550°C, Si becomes mobile and a ternary silicide indexed as tetragonal $Ti_4Ni_4Si_7$ appears at the intermetallic compound/Si interface. This V phase grows independently of the initial atomic ratios. No other ternary phases of the Ti-Ni-Si system could be formed by RTP of metal bilayers on a Si substrate because of the unlimited supply of Si leading to the richest Si phase in the ternary phase diagram.

When Ni is in excess with respect to the equi-atomic Ti/Ni ratio, a layer of pure Ni silicide forms between the substrate and the ternary on the surface. In contrast, because of their low mobility, excess Ti atoms remain within the ternary layer to form $TiSi_2$, but react with Si resulting in a mixture of phases $TiSi_2$ and V. Below 700°C, the C49 structure is observed whereas the low resistivity C54 forms at higher temperatures.

The Ni/$TiSi_2$ contact is highly unstable. Ni diffuses extremely rapidly through the disilicide, along grain boundaries. An activation energy of $1.25 \pm 0.2$ eV was found for this process. For Ni (30 nm)/$TiSi_2$ (175 nm), the same V phase grows starting at the metal/silicide interface, as shown by XRD and by a shift of the Ni $L_3VV$ Auger line. When looking closely at elemental profiles, one notices that as the diffusion front of Ni advances towards the substrate with increasing temperature, that of Si moves in the opposite direction at the same rate.

The reaction of Ni with Ti silicide is probably controlled by two different activation energies: (i) dissociation of the silicide, and (ii) diffusion of Ni in the silicide matrix (related to $D_0$ and the microstructure).

It is interesting to compare the evolution of the Ni/Ti/Si <100> structures with Ni/W/Si studied by Pai et al [61]. The sequence for the latter is the diffusion of Ni through W to react at the interface and form NiSi. There is no reaction of Ni with W. The depth distribution determined by RBS above 700°C is $WSi_2$ (some $NiSi_2$ mixed) /$NiSi_2$/Ni-Si/Si. We can see that the situation is very different from our case and applying Miedema's model presented earlier provides a clue. The theoretical heat of mixing of the two metals is much higher for Ni-Ti than for Ni-W, and therefore the energy balance pushes the reaction towards the reaction of the two elements first for Ni-Ti.

## D2 Metal/Silicide Reactions

The stability of metal layers on top of a silicide is also rich in information about solid state reactions.

Annealing of Ni/$TiSi_2$ leads to the formation of the ternary silicide $Ti_4Ni_4Si_7$ through the diffusion of Ni [59,62]. Ni atoms probably diffuse along high diffusivity paths and grains of ternary silicide nucleate at triple grain boundaries. The activation energy for Ni diffusion is found to be $1.25 \pm 0.2$ eV and corresponds fairly well to that determined for Al diffusing into $Pd_2Si$ and leading to $Al_3Pd_4Si$ formation [63,64]. When comparing the results with the outcome of Al/$CoSi_2$ and Al/$TiSi_2$, we see a fundamental difference: Al reacts first to form Co and Ti aluminides [63,65]. For two transition metals it is apparently unfavourable to free the silicided metal at low temperatures.

For Co/$TiSi_2$ samples annealed at 750°C, we observe $CoSi_2$/$Ti_{0.75}Co_{0.25}Si_2$, but increasing the temperature to 925°C gives homogeneous ternary silicide [66]. The observation of $CoSi_2$ at the surface indicates complex atomic redistribution with both Co and Si fluxes. The driving force for Si diffusion could be the reduction of free energy by Co silicidation at the interface. The same ternary phase is obtained by annealing Ti/$CoSi_2$ structures [66].

The diffusion of Cu through a silicide has also been studied [62,67] and the outcome of thin film reactions can be explained using ternary phase diagrams.

- $Cu/CoSi_2$: no ternary phases. $CoSi_2$ acts as a transport medium for the formation of $Cu_3Si$.

- $Cu/TiSi_2$: no ternary phase. Failure mechanism similar to $Cu/CoSi_2$, i.e. fast diffusion of Cu at low temperature (400°C).

- $Cu/ZrSi_2$: ternary phases present. It is then possible to form different phases $Zr_2CuSi_4$ and $ZrCuSi_2$ by changing the supply of Cu atoms. The excess Cu reacts with Si as well to form $Cu_3Si$ subsequently.

An interesting point for Cu/silicide structures is the oxygen diffusion barrier benefit provided by the silicide. For Cu/Si structures, the $Cu_3Si$ is highly unstable and conversion to $SiO_2$ and Cu can occur at room temperature. The metal silicide acts both as a transport medium and a layer protecting $Cu_3Si$ from oxidising because the O transport to the deeper layer is blocked.

## D3 Intermediate Layers as Epitaxy Promoters

The first experimental report of epitaxial growth of $CoSi_2$ on Si <100> for Co/Ti/Si <100> structures is the work of Dass et al [68]. This result has been obtained for other systems as well and we shall review the similarities and differences between them. Even though epitaxial $CoSi_2$ is always the end product, the sequence of formation seems to be different depending on the intermediate metal thickness and method of heating (RTP or conventional), and the mechanisms will be examined using the thermodynamic and kinetic models outlined previously.

For thick layers of Co and Ti [66], experimental results show that starting at 450°C Co atoms diffuse into the Ti layer to form cubic intermetallic compounds $Co_3Ti$ and $CoTi_2$. Further annealing induces the formation of $Co_2Ti$ at 550°C. Intermetallic compound formation has been confirmed by Liu et al [69].

Above 550°C (similar temperature threshold as for Ni-Ti-Si experiments), Si atoms become mobile and the movement of Si into the intermetallic compound and the opposite flux of Co atoms towards the substrate has several consequences: (a) growth of ternary silicides, and (b) nucleation and growth of Co silicides.

This phenomenon is perfectly illustrated by Barmak et al [70] who reported $Co/Co\text{-}Si\text{-}Ti/CoSi/CoSi_2/Si$ observed by XTEM from deposited 15 nm Co/5 nm Ti/Si annealed at 750°C for 30 s.

For thick Ti layers [66], at 600°C CoTi grows at the expense of $Co_2Ti$ and once the silicidation temperature of Ti is reached a structure $TiSi_2$ C49/$CoTi_2$/$CoSi_2$/Si is observed. The final equilibrium is $Ti_{0.75}Co_{0.25}Si_2/CoSi_2/Si$.

Hsia et al [71] have also reported the formation of a ternary layer Ti-Co-Si (with a composition close to 1:1:2) on top of epitaxial $CoSi_2$ for both alloys and bimetal source materials on Si.

The behaviour of Co/Zr bilayers [72] is very similar to that of Co/Ti/Si samples. At temperatures around 700°C, $Co_3Zr_6$ and $CoZr_2$ high resistivity phases are detected. As the temperature is increased $ZrSi_2$, $CoSi_2$ and ZrCoSi begin to appear.

For Co/Ta/Si structures, Byun et al [73] have observed the formation of an intermetallic compound close to $Co_3Ta$ and found XRD peaks corresponding to TaCoSi simultaneously with those of epitaxial $CoSi_2$.

To look at the different results, we will use the elegant treatment of the growth kinetics of planar binary diffusion couples by Gosele and Tu [50] mentioned in Section C1.

The Ti silicide sequence of formation for slow and fast heating rates from the work of Natan [53] showed how the exhaustion of Ti prior to nucleation influences the sequence. The experiment of Hsia et al [74] on $CoSi_2$ formation on SOI sheds more light on the importance of interfacial atomic availability.

For excess Co with respect to the amount of Si available on SOI, a phase reversal from $CoSi_2$ to CoSi is observed at high temperatures. However, $CoSi_2$ is stable for CoSi/Ti/Si structures. It is clear that we have to take atomic availability into account. This, along with Pretorius' model, helps us to answer the question as to whether the reaction proceeds through the entire $Co_2Si$, CoSi, $CoSi_2$ sequence or goes directly to $CoSi_2$.

$Co_2Si$ and CoSi may or may not be observed depending on the thickness of the intermediate transport layer. If this layer is thick (greater than 15 nm), Co availability will be low, Co atoms will reach the interface at a temperature high enough for $CoSi_2$ nucleation and the reaction will be controlled through Si diffusion. The Co tends to react with the other metal to form intermetallic compounds initially. In this case, because of low Co atomic concentration at the interface, Pretorius' model indicates that the effective heat of formation of Si-rich compounds is the largest and therefore Si-rich phases will tend to form. Co atoms adopt low energy positions in the $CoSi_2$ lattice with a coherent interface.

However, if this layer is thinner, the diffusion length approximately given by $(Dt)^{1/2}$ is sufficient for Co and Si fluxes to be equivalent at the interface and $Co_2Si$ or CoSi can form. Wei et al [75] presented the sequence $Co_2Si$, CoSi, $CoSi_2$ for Co (20 nm - 30 nm)/Ti 10 nm/Si <100> samples.

This reasoning of identity of the first nucleating phase as a function of the thickness of the intermediate layer is also supported by the sequence observed by Hong and Rozgonyi [76]. For Co/Ti/Co/Ti/Co/Ti (5 nm to 10 nm) /Si multilayered structures, initially Co availability is limited and $CoSi_2$ nucleates. However, upper layers maintain a Co supply and there is a shift from $CoSi_2$ to CoSi. The reversal of $CoSi_2$ to CoSi has been studied by Lau et al [77]. If, however, the thickness of the intermediate Ti layer is increased to 20 nm, a $TiSi_x$ amorphous layer forms initially. At 650°C interdiffusion increases leading to Ti-Co-Si/$CoSi_x$/Si. Finally $TiSi_2$ C54 forms. No intermediate Co silicide phases are observed.

Whenever taking thermodynamics into account, one should remember that the free energy term involves also the entropy of mixing which might not be negligible for alloys in contact with Si or for certain bilayer experiments.

## D4 Oxygen Gettering

Whether for W/Ti/Si, Co/Ti/Si or Co/Ta/Si, one of the effects of the interposed layer is to reduce the interfacial oxide. The capacity of a thin Ti layer to reduce $SiO_2$ and as a result lower $WSi_x$ formation temperature was presented previously.

Oxygen gettering is important also for Co since the presence of O leads to CoO formation and the development of voids under the silicide surface.

For epitaxial growth, and as d'Heurle has shown nucleation to be a competition between a chemical driving force and opposing interface effects [78], there could be a coherent non-glissile interface which is known to be sensitive to temperature because the energy necessary to free one atom on one side of the interface to join the growing compound is only provided via thermal activation. d'Heurle mentioned the superposition of rate control for $CoSi_2$ formation, by both diffusion and heterogeneous nucleation processes.

When there is some O present at the interface, the interfacial reaction rate is much smaller than the rate of diffusion and the reaction is governed by interface kinetics. On the other hand, when there is no oxide layer present, the reaction will proceed according to the rate of supply of atoms at the interface, i.e. it is diffusion-controlled.

## D5 Capping Layer Effect

For Ti (60 nm)/Co (30 nm)/Si <100> structures, Co diffuses both into Ti, to form intermetallic phases, and towards Si [66]. At 550°C, a structure $Ti/CoTi_2/CoTi/CoSi_2$ is observed. The formation of $CoSi_2$ at such low temperatures is quite surprising since for Co/Si <100> annealed in the same conditions, the majority phase at 550°C is CoSi.

This capping effect is supported by an experiment of Hong and Rozgonyi [76] using Co/Ti/Co/Ti/Co-Si <100> samples. At 550°C, the major Co silicide phase is $CoSi_2$. Because of Co flux from the upper layers and Si diffusion from the substrate, the layer evolves to CoSi at 650°C.

The mechanism of epitaxial $CoSi_2$ formation on Si <100> using an interposed layer as a transport medium has been described using thermodynamic and kinetic arguments such as the effective heat of formation introduced by Pretorius and the growth model of Gosele and Tu.

The thickness of the intermediate layer controls atomic availability at the interface and therefore influences the identity of the first nucleating phase as well as the phase sequence.

Possible effects of the intermediate layer are: (a) limitation of Co availability at the interface favouring $CoSi_2$ nucleation and growth, (b) reduction of interfacial oxide (because of higher oxidation potential than that of Si) usually hindering silicide nucleation, and (c) capping layer effect.

## E CONCLUSION

A review of experimental results for ternary systems Metal 1-Metal 2-Si has been presented. Based on the results of thin film experiments and on phase diagrams, systems were split into two categories, systems with solid solutions and systems containing ternary compounds.

When silicides exhibit limited solid solubilities, leading to phase separation for alloys, or when the metals are isoelectronic or pseudo-isoelectronic, a treatment of the ternary system as three separate ones is adequate. High temperature equilibrium results obtained for thin film alloys or bilayers on Si were found to agree with those of bulk experiments.

Solid solubility depends mostly on the dimensions of the substituting atom, its chemical potential and valency. The electron concentration has been shown to influence the crystalline structure of dimorphic solid solutions.

Ternary compound formation is usually associated with elements presenting large differences in valency and atomic size. Two routes can be used for ternary silicide growth: diffusion of Si into an intermetallic compound or reaction of a metal with a silicide. Free energy diagrams help to infer the high temperature stability of binary or ternary silicides in contact with Si.

Whether for alloys or bilayers on Si, the first phase to form as well as the sequence of formation can be analysed in terms of heat of formation and kinetics. Miedema's theoretical heat of mixing is useful in determining if intermetallic compound formation will be energetically favourable over silicidation at low temperatures.

Pretorius' concept of effective heat of formation helps to understand which phase will nucleate first and to explain the sequence of evolution of alloys or bilayers on Si. This model indirectly takes into account both the effects of diffusion through a growing compound and of atomic availability at the interface and explains differences observed for Co/Metal/Si experiments depending on the interposed metal thickness.

Three factors, namely reduction of an interfacial oxide, capping layer effect and modification of atomic fluxes at the interface, must be considered when studying ternary thin film experiments to explain lowering of the silicidation temperature as well as the promotion of epitaxy compared to metal/Si systems.

## REFERENCES


[1] S.P. Murarka [ *Silicides for VLSI Applications* (Academic Press, London, 1983) ]

[2] V.A. Tskai, O.I. Shaikina [ *Russ. Metall. (UK)* vol.5 (1986) p.190 ]

[3] H.J. Goldschmidt [ *Interstitial Alloys* (Plenum Press, 1967) ]

[4] E.I. Gladyshevskii, V.I. Lakh, R.V. Skolozdra, B.I. Stadnik [ *Russ. Powder Metall.* vol.4 no.22 (1964) p.278 ]

[5] T.S. Verkhoglyadova, O.I. Vivchar, E.I. Gladyshevskii [ *Russ. Powder Metall.* vol.4 (1966) p.316 ]

[6] T.G. Finstad et al [ *Thin Solid Films (Switzerland)* vol.135 no.2 (1985) p.229 ]

[7] A. Appelbaum, M. Eizenberg [ *J. Appl. Phys. (USA)* vol.56 no.8 (1984) p.2341 ]

[8] E.I. Gladyshevskii, V.Ya. Markiv, Yu.V. Efimov, E.M. Savitskii, V.V. Baron [ *Russ. J. Inorg. Mater.* vol.1 no.7 (1965) p.1023 ]

[9] P. Gas, F.J. Tardy, F.M. d'Heurle [ *J. Appl. Phys. (USA)* vol.60 no.1 (1986) p.193 ]

[10] V.A. Maksimov, F.I. Shamray [ *Russ. Metall.* (1970) p.197 ]

[11] P. Gas, J. Tardy, F.K. LeGoues, F.M. d'Heurle [ *J. Appl. Phys. (USA)* vol.61 no.6 (1987) p.2203 ]

[12] T.G. Finstad, M.A. Nicolet [ *Thin Solid Films (Switzerland)* vol.68 (1980) p.393 ]

[13] C.D. Capio, D.S. Williams, S.P. Murarka [ *J. Appl. Phys. (USA)* vol.62 no.4 (1987) p.1257 ]

[14] C.J. Palmstrom, J. Gyulai, J.W. Mayer [ *J. Vac. Sci. Technol. A (USA)* vol.1 no.2 (1983) p.452 ]

[15] M. Eizenberg, R. Brener [ *Thin Solid Films (Switzerland)* vol.89 (1982) p.355 ]

[16] M.F. Zhu, I. Suni, M.A. Nicolet, T. Sands [ *J. Appl. Phys. (USA)* vol.56 no.10 (1984) p.2740 ]

[17] C.S. Pai, S.S. Lau, D.B. Poker, L.S. Hung [ *J. Appl. Phys. (USA)* vol.58 no.11 (1985) p.4172 ]

[18] R.V. Skolozdra, Z.P. Shipka, E.I. Gladyshevskii [ *Russ. J. Inorg. Mater.* vol.5 no.8 (1969) p.1250 ]

[19] A. Appelbaum, M. Eizenberg, R. Brener [ *J. Appl. Phys. (USA)* vol.55 no.4 (1984) p.915 ]

[20] M.O. Naudé, R. Pretorius, D.J. Marais [ *Thin Solid Films (Switzerland)* vol.89 (1982) p.339 ]

[21] E.I. Gladyshevskii, L.K. Borusevich [ *Russ. J. Inorg. Chem. (UK)* vol.8 no.8 (1963) p.997 ]

[22] T.G. Finstad [ *Thin Solid Films (Switzerland)* vol.51 (1978) p.411 ]

[23] L.S. Hung, Q.Z. Hong, J.W. Mayer [ *J. Appl. Phys. (USA)* vol.63 no.5 (1988) p.1749 ]

[24] M. Eizenberg, K.N. Tu [ *J. Appl. Phys. (USA)* vol.53 no.3 (1982) p.1577 ]

[25] T.G. Finstad, M.A. Nicolet [ *J. Appl. Phys. (USA)* vol.50 no.1 (1979) p.303 ]

[26] J.O. Olowolafe, M.A. Nicolet, J.W. Mayer [ *J. Appl. Phys. (USA)* vol.47 (1976) p.5182 ]

[27] J.W. Mayer, S.S. Lau, K.N. Tu [ *J. Appl. Phys. (USA)* vol.50 no.9 (1979) p.5855 ]

[28] R.D. Thompson, K.N. Tu, G. Ottaviani [ *J. Appl. Phys. (USA)* vol.58 no.2 (1985) p.705 ]

[29] R. Thompson, M. Eizenberg, K.N. Tu [ *J. Appl. Phys. (USA)* vol.52 (1981) p.6763 ]

[30] T.G. Finstad, M.A. Nicolet [ *J. Appl. Phys. (USA)* vol.50 no.1 (1979) p.303 ]

[31] W. Wopersnow, K. Schubert [ *Z. Met.kd. (Germany)* vol.67 (1976) p.807 ]

[32] R.G. Long, E. Mahan [ *Appl. Phys. Lett. (USA)* vol.56 no.7 (1990) p.1655 ]

[33] O. Thomas et al [ *Appl. Surf. Sci. (Netherlands)* vol.38 (1989) p.94 ]

[34] F.A. Sidorenko, I.Z. Radovskii, L.P. Zelenin, P.V. Geld [ *Russ. Powder Metall.* vol.9 (1966) p.730 ]

[35] J.M. Harris, S.S. Lau, M.A. Nicolet, R.S. Nowicki [ *J. Electrochem. Soc. (USA)* vol.123 no.1 (1976) p.120 ]

[36] F. Nava, C. Nobili, G. Ferla, G. Ianuzzi, D. Queirolo, G. Celotti [ *J. Appl. Phys. (USA)* vol.54 no.5 (1983) p.2434 ]

[37] C.S. Wei, J. Van der Spiegel, M. Setton, J.J. Santiago, M. Tanielan, S. Blackstone [ *Mater. Res. Soc. Symp. Proc. (USA)* vol.52 (1986) p.297 ]

[38] P. Gas, F.J. Tardy, F.M. d'Heurle [ *J. Appl. Phys. (USA)* vol.60 no.1 (1986) p.193 ]

[39] P. Gas, F.J. Tardy, F.K. LeGoues, F.M. d'Heurle [ *J. Appl. Phys. (USA)* vol.61 no.6 (1987) p.2203 ]

[40] R. Dahan, J. Pelly, L. Zevin [ *J. Appl. Phys. (USA)* vol.67 (1990) p.2885 ]

[41] J.S. Choi et al [ *J. Appl. Phys. (USA)* vol.74 no.2 (1993) p.1456 ]

[42] R.D. Thompson, K.N. Tu, G. Ottaviani [ *J. Appl. Phys. (USA)* vol.8 no.2 (1985) p.705 ]

[43] A.R. Miedema, R. Boom, F.R. de Boer [ *J. Less-Common Met. (Switzerland)* vol.41 (1975) p.283 ]

[44] H. Sprenger [ *J. Less-Common Met. (Switzerland)* vol.34 (1974) p.39 ]

[45] A.M. Bardos, D.I.A. Beck [ *Trans. Am. Inst. Min. Eng. (USA)* vol.227 (1963) p.991 ]

[46] W. Hume Rothery [ *Trans. Am. Inst. Min. Eng. (USA)* vol.233 (1965) p.233 ]
[47] P. Villars [ *J. Less-Common Met. (Switzerland)* vol.119 (1986) p.175 ]
[48] M.E. Schlesinger [ *Chem. Rev.* vol.90 (1990) p.607 ]
[49] R. Pretorius [ *Mater. Res. Soc. Symp. Proc. (USA)* vol.25 (1984) p.15 ]
[50] U. Gosele, K.N. Tu [ *J. Appl. Phys. (USA)* vol.53 no.4 (1982) p.3252 ]
[51] V.I. Dybkov [ *J. Mater. Sci. (UK)* vol.21 (1986) p.3078 ]
[52] E.J. Van Loenen, A.E.M.J. Fischer, J.F. Van der Veen [ *Surf. Sci. (Netherlands)* vol.155 (1985) p.65 ]
[53] M. Natan [ *J. Appl. Phys. (USA)* vol.63 no.11 (1988) p.5534 ]
[54] G. Majni, M. Costato, F. Panini [ *Thin Solid Films (Switzerland)* vol.125 (1985) p.71 ]
[55] M. Setton [ Master's Thesis, University of Pennsylvania, USA (1987) ]
[56] H. Yang, R.W. Bene [ *J. Appl. Phys. (USA)* vol.59 no.5 (1986) p.1525 ]
[57] R. De Reus, H.C. Tissink, F.W. Saris [ *J. Mater. Res. (USA)* vol.5 (1990) p.341 ]
[58] G.A. Rozgonyi, J.H. Lee, D. Knoesen, D. Adams [ *Appl. Phys. Lett. (USA)* vol.58 (1991) p.729 ]
[59] M. Setton, J. Van der Spiegel, B. Rothman [ *J. Mater. Res. (USA)* vol.4 no.5 (1989) p.1218 ]
[60] F.M. d'Heurle, P. Gas [ *J. Mater. Res. (USA)* vol.1 no.1 (1986) p.205 ]
[61] C.S. Pai, S.S. Lau, D.B. Poker, L.S. Hung [ *J. Appl. Phys. (USA)* vol.58 no.11 (1985) p.4172 ]
[62] M. Setton [ Ph.D. Thesis, University of Pennsylvania, USA (1990) ]
[63] G.J. Van Gurp, J.L.C. Daams, A. von Ostroom, L.J.M. Augustus, Y. Tamminga [ *J. Appl. Phys. (USA)* vol.50 no.11 (1979) p.6915 ]
[64] U. Köster, P.S. Ho, J.E. Lewis [ *J. Appl. Phys. (USA)* vol.53 no.11 (1982) p.7445 ]
[65] C.Y. Ting, M. Wittmer [ *J. Appl. Phys. (USA)* vol.54 no.2 (1983) p.937 ]
[66] M. Setton, J. Van der Spiegel [ *Appl. Surf. Sci. (Netherlands)* vol.38 (1989) p.62 ]
[67] M. Setton, J. Van der Spiegel [ *J. Appl. Phys. (USA)* vol.70 (1991) p.193 ]
[68] M.L.A. Dass, D.B. Fraser, C.S. Wei [ *Appl. Phys. Lett. (USA)* vol.58 (1991) p.1308 ]
[69] P. Liu et al [ *J. Appl. Phys. (USA)* vol.74 no.3 (1993) p.1700 ]
[70] K. Barmak et al [ *Mater. Res. Soc. Symp. Proc. (USA)* vol.238 (1992) p.575 ]
[71] S.L. Hsia, T.Y. Tan, P. Smith, G.E. McGuire [ *J. Appl. Phys. (USA)* vol.70 no.12 (1991) p.7579 ]
[72] J.S. Byun, J.J. Kim, W.S. Kim [ *J. Electrochem. Soc. (USA)* vol.142 no.8 (1995) p.2805 ]
[73] J.S. Byun, S.B. Kang, H.Y. Kim, C.Y. Kim, K.H. Park [ *J. Appl. Phys. (USA)* vol.74 no.5 (1993) p.3156 ]
[74] S.L. Hsia, T.Y. Tan, P.L. Smith, G.E. McGuire [ *IEEE 1994 Symp. VLSI Tech. Dig. (USA)* p.123 ]
[75] C.S. Wei, D.B. Fraser, M.L.A. Dass, T. Brat [ *Proc. VLSI Multilevel Interconnection Conf.*, Santa Clara, CA, USA, June 1990 (VMIC, Tampa, USA) p.233 ]
[76] F. Hong, G.A. Rozgonyi [ *J. Electrochem. Soc. (USA)* vol.141 no.2 (1994) p.3480 ]
[77] S.S. Lau, J.W. Mayer, K.N. Tu [ *J. Appl. Phys. (USA)* vol.49 (1978) p.4005 ]
[78] F.M. d'Heurle [ *J. Mater. Res. (USA)* vol.3 no.1 (1988) p.167 ]

# 3.8 Higher-order metal silicide phase diagrams

R. Sinclair

April 1995

The ternary diagrams themselves are inadequate when four or more elements are present at a single interface (e.g. TiN diffusion barrier in contact with a $SiO_2$ insulating layer). Bhansali and co-workers [1,2] have demonstrated how to address this problem for the case of four elements. Such quaternary phase diagrams can be represented by a regular tetrahedron, with the binary systems as the edges, the ternaries as the faces and the quaternary compositions in the interior. Up to four phases can be present in equilibrium, but, as before, the tie-lines establishing two-phase compatibility are probably the most important aspect. Their calculation proceeds by the same method as previously, with the additional complexity of competing tie-lines over tie-planes (dual versus triple phase stability) as well as over each other.

The Ti-Si-O-N diagram is shown in FIGURE 1. The faces show the same tie-lines as seen earlier in the ternaries (see Datareviews 3.2-3.7). However, there are two interior tie-lines, emanating from TiN to $SiO_2$ and to 'SiON'. Accordingly both such interfaces are stable to heating. Application of the diagram to interpret the reaction of Ti with silicon oxynitride shows the power of the approach [1]: without it, any prediction of phase products would be merely guesswork.

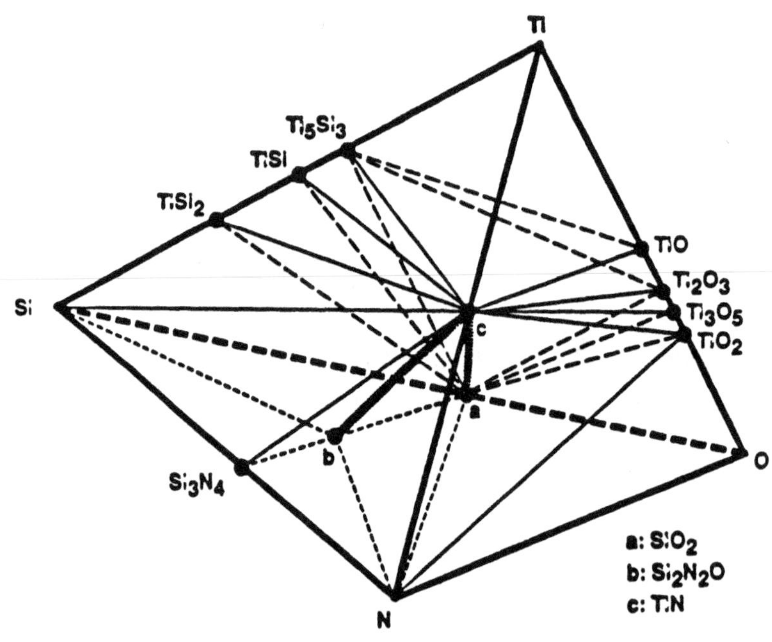


FIGURE 1 The calculated Ti-Si-O-N quaternary phase diagram (T = 700-1000°C). Thin solid or dashed lines represent the tie-lines on the faces of the tetrahedron (i.e. ternary systems) and the thick dashed lines are the interior (quaternary) tie-lines [1].

The Al-Si-O-N, Ti-Al-Si-N, Ti-Al-Si-O and Ni-Si-Al-C diagrams are shown in FIGURES 2 to 5 for completeness. Some ternaries not discussed earlier can be deduced from the faces of the tetrahedra.

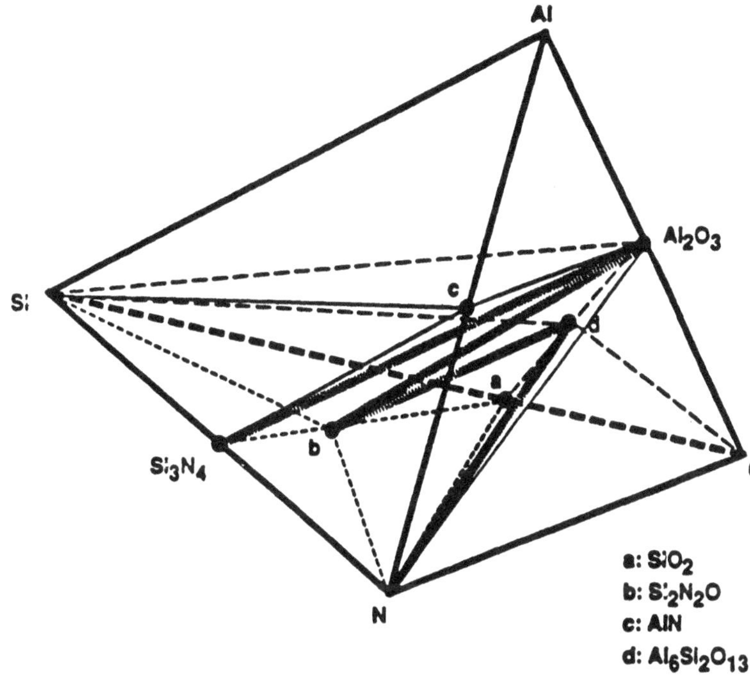


FIGURE 2 The Al-Si-N-O system represented in the form of a calculated quaternary phase diagram (T = 900 K). Thin solid or dashed lines represent the tie-lines on the faces of the tetrahedron (i.e. ternary systems) and the thick dashed lines are the interior (quaternary) tie-lines [1].

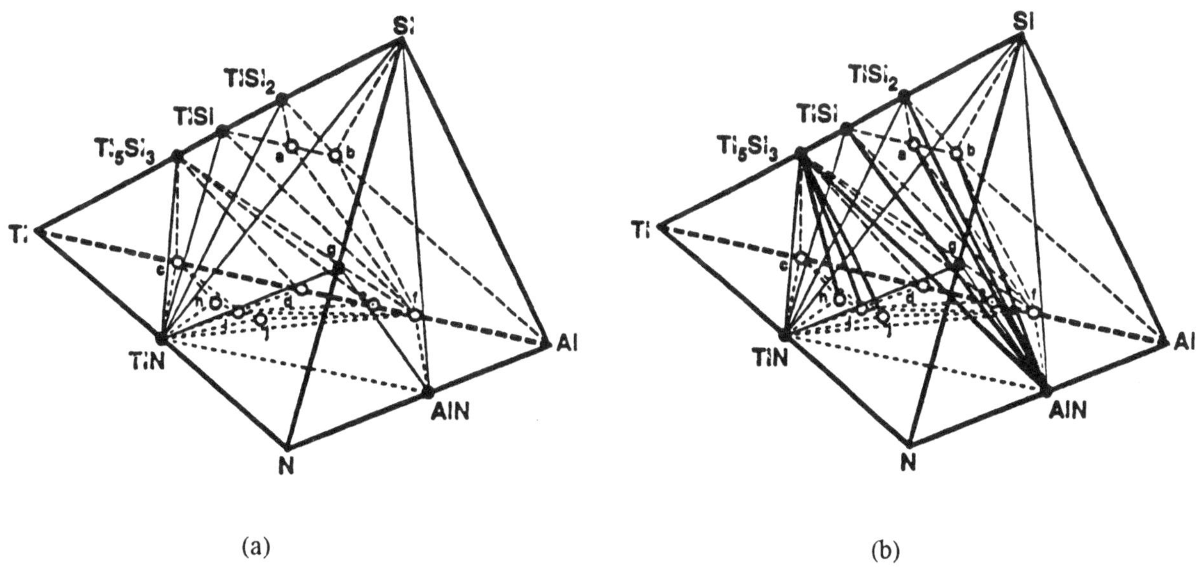


FIGURE 3 (a) Calculated ternary tie-lines (T = 900 K) forming the exterior structure of the Ti-Al-Si-N quaternary phase diagram (a: $Ti_2AlSi_3$; b: $Ti_7Al_5Si_{12}$; c: $Ti_3Al$; d: TiAl; e: $TiAl_2$; f: $TiAl_3$; g: $Si_3N_4$; h: $Ti_3AlN$; i: $Ti_2AlN$; j: $Ti_3Al_2N_2$) [1]. (b) The complete Ti-Al-Si-N quaternary diagram (T = 900 K) with thick dashed lines as the interior (quaternary) tie-lines.

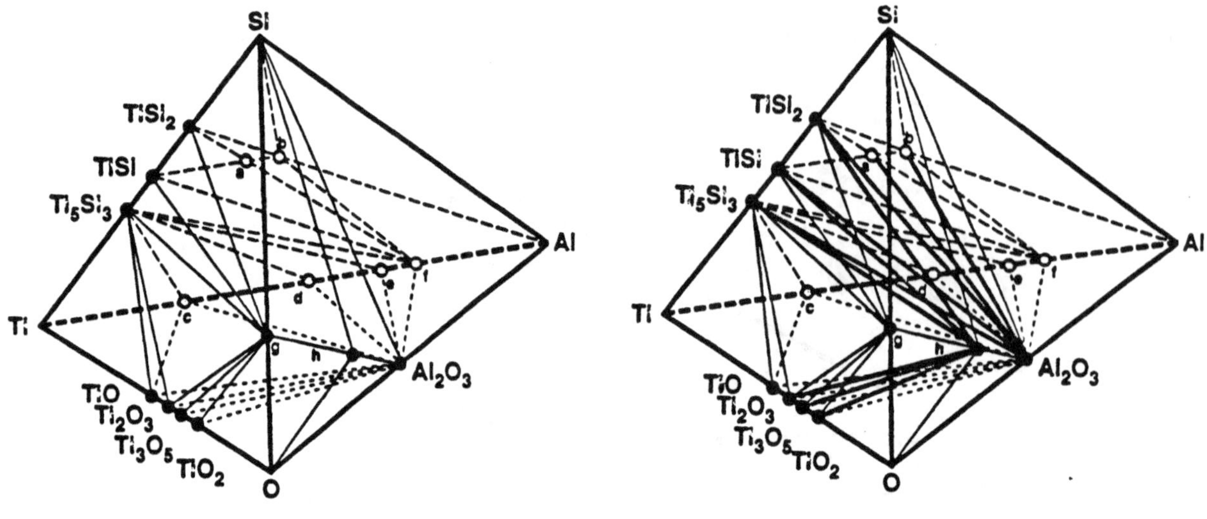


(a) (b)

FIGURE 4 (a) Calculated ternary tie-lines (T = 900 K) forming the exterior structure of the Ti-Al-Si-O quaternary phase diagram (a: $Ti_2AlSi_3$; b: $Ti_7Al_5Si_{12}$; c: $Ti_3Al$; d: TiAl; e: $TiAl_2$; f: $TiAl_3$; g: $SiO_2$; h: $Al_6Si_2O_{13}$) [1]. (b) The complete Ti-Al-Si-O quaternary phase diagram (T = 900 K) with thick dashed lines as the interior (quaternary) tie-lines.

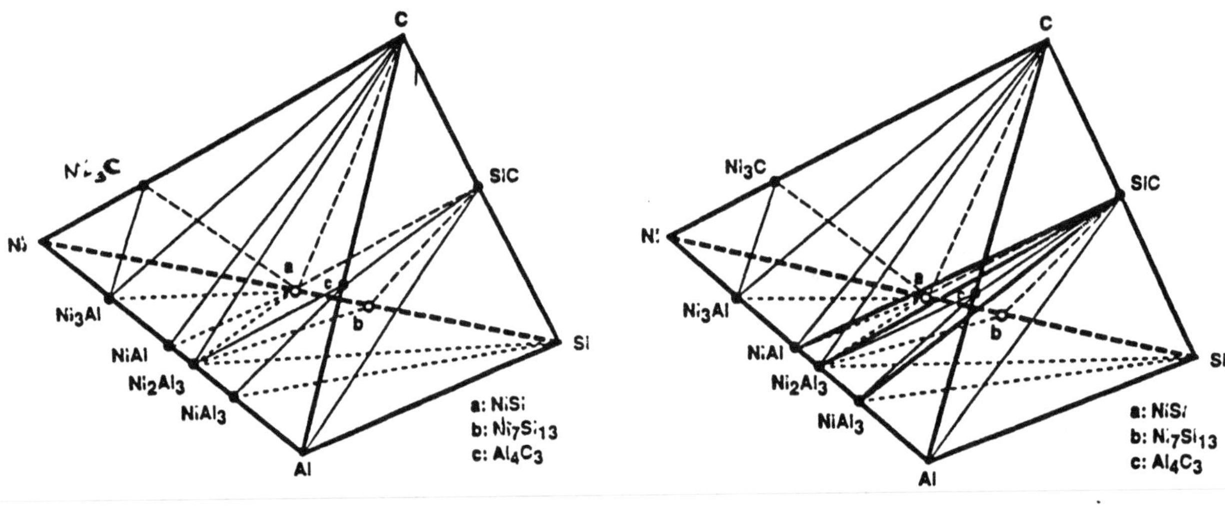


(a) (b)

FIGURE 5 (a) The exterior portion of the Ni-Al-Si-C quaternary phase diagram (T = 1300 K) [1]. (b) The complete Ni-Al-Si-C quaternary phase diagram (T = 1300 K) with thick dashed lines as the interior (quaternary) tie-lines. Note that $NiSi_2$ is omitted as no thermodynamic data are listed in [3,4].

## REFERENCES

[1] A.S. Bhansali [ Ph.D. Thesis, Stanford University, USA (1992) ]

[2] A.S. Bhansali, D.H. Ko, R. Sinclair [ *J. Electron. Mater. (USA)* vol.19 (1990) p.1171 ]

[3] I. Barin, O. Knacke [ *Thermochemical Properties of Inorganic Substances* (Springer, Berlin, 1973 and 1977) ]

[4] I. Barin [ *Thermochemical Data of Pure Substances, Parts I and II* (VCH, Weinheim, 1989) ]

CHAPTER 4

# ELECTRONIC PROPERTIES

# 4.1 Electronic band structure of TM silicides

J. Derrien

June 1994

## A INTRODUCTION

A very large number of silicide compounds can be formed by reaction between Si and transition metals. A detailed understanding of the electronic properties of each silicide is beyond the scope of this Datareview. We therefore will confine our discussion to a qualitative one based on first principles rather than on accurate, sophisticated, self-consistent but difficult calculations.

## B CHEMICAL BONDING IN SILICIDES

The formation of the electronic band structure of transition metal (TM) silicides can be approached within a simple molecular-orbital picture [1,2], considering the bonding between the metal d-orbitals and the silicon (Si) s- and p-electron states. Several years ago it was pointed out by Friedel [3] that the main contribution to the cohesive energies of transition metals is due to the formation of the d-band from atomic d-orbitals of the TM atoms which are brought together. The strength of the d-bonding, and hence that of the cohesive energy, depends upon the hopping integrals between d-orbitals on neighbouring TM atoms.

When silicon atoms are inserted into the TM matrix, the atomic d-orbitals are driven further from each other and the hopping integrals decrease and lead to a weakening of the cohesive energy. To balance this loss Si s- and p-states and TM d-orbitals should couple together so that bonding states can be formed. The strength of the chemical bonding between TM and Si atoms depends upon the filling of these sp-d bonding states and explains the stability of the compounds. On this basis, corresponding antibonding states are expected to be empty. Several total-energy calculations and a large body of experimental spectroscopic data have proven the qualitative validity of this physical picture of silicide formation. Moreover, they have delineated the role of the Si s-, the Si p- and the TM d-states in bonding and band structure properties of a variety of TM silicides. They confirm also the dependence of the band structure on d-band occupancy. The common characteristics of TM silicides are reflected in occupied non-bonding d-states near the Fermi level, $E_F$, the p-d bonding states extending to ~6 eV below $E_F$ and the s-derived states 10-14 eV below $E_F$. Most of the empty antibonding states are spread above $E_F$ [4].

## C BAND STRUCTURE OF 3d-TRANSITION METAL SILICIDES

The complete valence band (occupied) and conduction band (empty) electronic structures of materials are usually determined by photoemission spectroscopy (PES) techniques, both in an integrated mode (total density of states, DOS) or in an angle resolved mode (electronic states dispersion curve versus wave vector, E(**k**)).

The occupied states are probed by measuring the electron energy distribution N(E) of electrons ejected from the material into vacuum under photon irradiation. Ultraviolet photoelectron spectroscopy (UPS) uses exciting photons from gas discharge sources (Ar, Ne, He etc. exciting lines) whereas X-ray photoelectron spectroscopy (XPS) excites materials with soft X-ray photons from X-ray generators. With synchrotron radiation (SR) facilities nowadays, a rather complete continuous energy photon range can be achieved. It is very convenient since the photoionisation cross-sections of initial states vary with photon energy and can be monitored in order to distinguish the orbital symmetry of various probed states. Of particular interest is the photon energy ranging from 70 to 80 eV where TM d-states are favourably excited as compared to Si s- and p-states. For photoemission at lower excitation (hν ~ 20 eV) it seems that the weight of TM d-states with respect to Si s- and p-states does not change as drastically. The same behaviour is observed with soft X-ray photons (~ keV).

The empty states are probed using either inverse photoemission spectroscopy (IPS) or Bremsstrahlung isochromat spectroscopy (BIS). Both techniques are based on the injection of energetic electrons into empty states and the detection of photons emitted during the relaxation towards ground states of those hot electrons [5].

## C1 Disilicides

The dependence of the 3d-TM $Si_2$ band structure on the d-band occupancy is qualitatively demonstrated in FIGURE 1. The TM d-bands, schematically interpolated and drawn [4], dominate the DOS and move through $E_F$ with increasing atomic number Z. The TM d-derived states vary substantially within the series appearing first as a broad band of mostly empty states (at the beginning of the TM row) and displaying finally a narrow, fully-occupied d-band (at the end of the TM row) as observed in FIGURE 1(a) (solid line). The Si p-derived states are grouped into two energy regions straddling the Fermi level, one corresponding to occupied states, the other to the empty states (FIGURE 1(a), dotted lines). With increasing atomic number the centre of the occupied region moves to higher binding energies but the empty features stay relatively fixed at ~5 eV above $E_F$. TM d- and Si p-derived states hybridise together resulting in p-d bonding and antibonding combinations. The movement of the bonding states follows the shift of TM d-states to higher binding energy with increasing Z while corresponding antibonding states do not change so drastically in the silicide series. It is worth noting that FIGURE 1(a) displays theoretical predictions based on fictitious iso-structural calculations for the whole silicide series [4,15]. Detailed computations taking into account different crystal structures using a many-body treatment will modify slightly the density of states and its energy location. Nevertheless, the final results still suggest that the bonding in TM silicides always involves p-d hybridisation mechanisms, following qualitatively the above discussed trend.

In FIGURE 1(b) we report a compilation of UPS spectra recorded on disilicides fractured in-situ in order to avoid possible surface contamination. The disilicide DOS is mainly dominated by d-states whose movement towards higher binding energy with increasing Z is clearly measured. In order to outline this d-band movement through the whole 3d-TM series, photoemission spectroscopy recorded with the same photon energy range (hν ~ 70 eV) has been carried out to favour excitation of TM d-states as compared to Si s- and p-states which are not noticeably observed in those spectra. In order to distinguish also

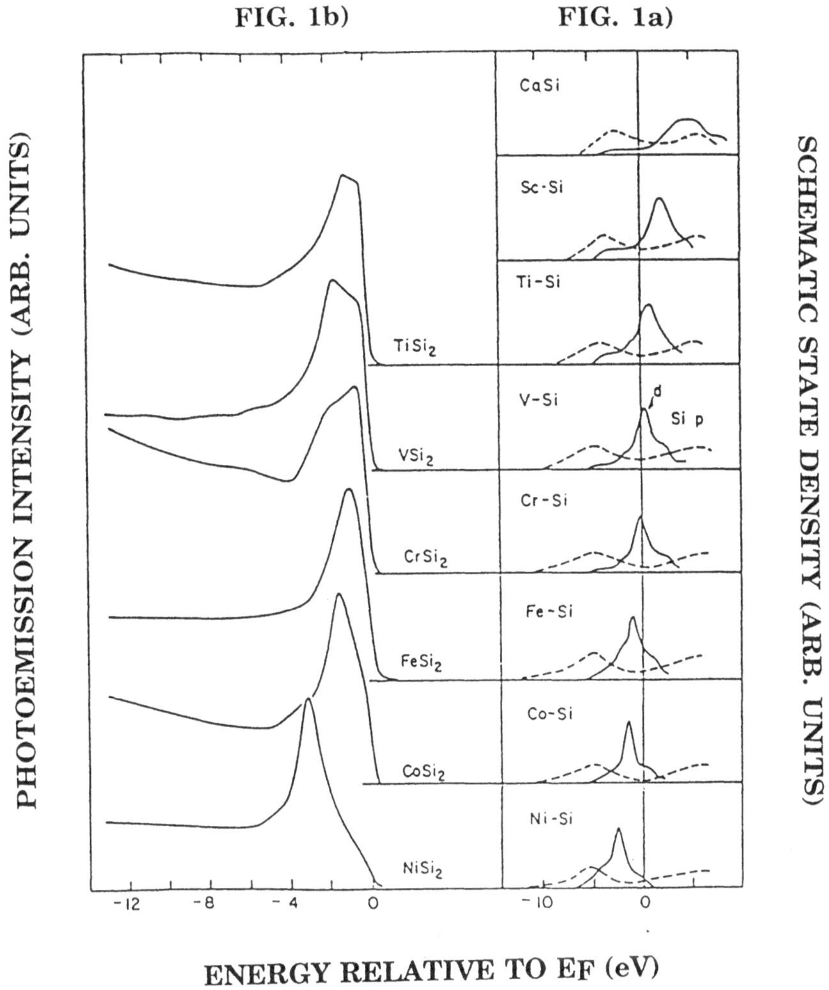


FIGURE 1 Experimental results (left, 1(b)) and interpolated state densities (right, 1(a)) for the metal d- (solid lines) and silicon p- (dashed lines) states for disilicides. The spectra have been obtained with photon energy hν = 70-80 eV. Therefore, the metal d-states are more strongly excited as compared to the Si s- and p-states. The spectra appear as mainly dominated by non-bonding d-derived states (from [4]).

those Si s- and p-states in disilicides, XPS (hν ~ 1487 eV) has been performed and the results are reported in FIGURE 2 for comparison with UPS spectra (hν ~ 70 eV) [6]. The comparison shows that XPS spectra carry information on both TM d- and Si s- and p-states whereas UPS spectra carry predominantly information on the TM d-states. In this way one can observe the mixing of those states and find information on the chemical bonding in disilicides. FIGURE 2 states that for the early 3d-metal disilicides the occupied electronic band structures reflect mainly the s- and p-states of Si. For the compounds in the middle of the row the main intensity arises from TM d-states. The empty states (antibonding states) have been measured with BIS and IPS [6-9] (FIGURE 3). Just above $E_F$ one observes qualitatively the symmetrical counterpart of occupied non-bonding and bonding states below $E_F$. For the early 3d-disilicides antibonding states are dominated by d-derived states (see also FIGURE 1(a)) whereas for the late 3d-disilicides the relative weight of Si s- and p-states slightly increases.

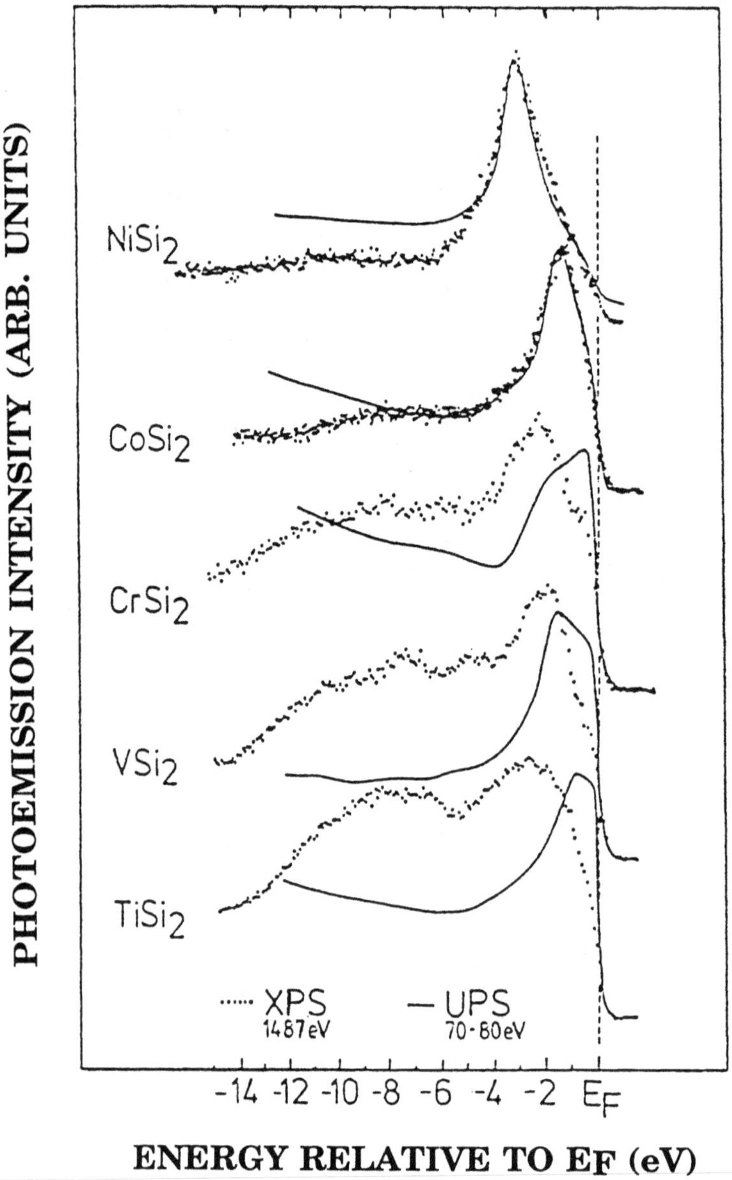


FIGURE 2 Comparison of XPS with photoemission spectra obtained with hν = 70-80 eV taken from FIGURE 1. With XPS (hν ~ 1487 eV) the Si s- and p-states are more clearly observed as compared to the metal d-states, especially for the early TM disilicides (from [6]).

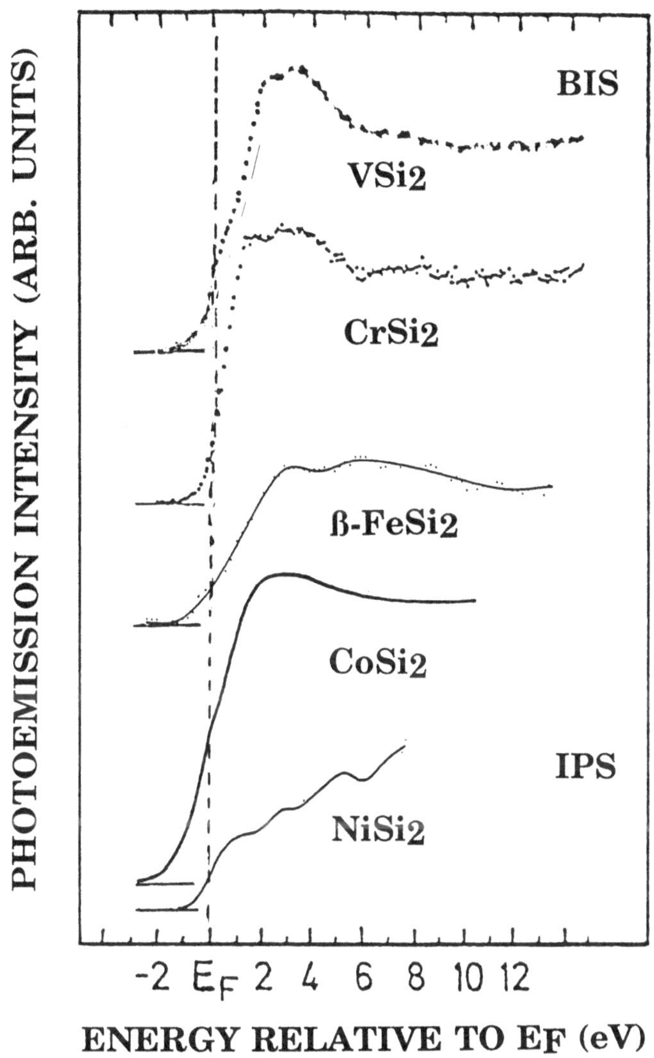


FIGURE 3 Antibonding states as measured with BIS and IPS techniques. The unoccupied states of the early 3d-disilicides are dominated by metal d-derived states whereas for the late disilicides the weight of Si s- and p-states slightly increases. The vertical scale is not normalised between spectra. All spectra have been recorded with the BIS technique except for the $NiSi_2$ spectrum (with IPS) (from [6-9]).

## C2 Other 3d-Silicides

Both experimental and theoretical results reflect the same trend already observed with disilicides. The electronic band structures of silicides whose compositions are different from the disilicide one are dominated by Si s-states (~10 eV below $E_F$), Si p-states centred at ~4-5 eV below $E_F$ and TM-d states extending down to ~5 eV below $E_F$. Slight differences can, however, be observed depending on the crystal structure and the atomic ratio of TM and Si. With increasing Si content the electronic band structures of silicides qualitatively vary from spectra dominated by strong TM non-bonding d-bands and weak Si s- and p-derived features towards spectra with strong d-sp hybridised bonding states. In particular the widths of the Si 3s- and Si 3p-derived bands increase with increasing Si content and they shift to somewhat higher binding energy.

FIGURE 4 shows photoemission spectra of a few well defined silicides of Ni, Co and Fe to illustrate the features [7,10]. It also clearly shows the relevant influence of the crystal structure (nearest neighbour interaction, consideration of the spatial extent of the wave function, energy position of the original bands) on the silicide electronic structure.

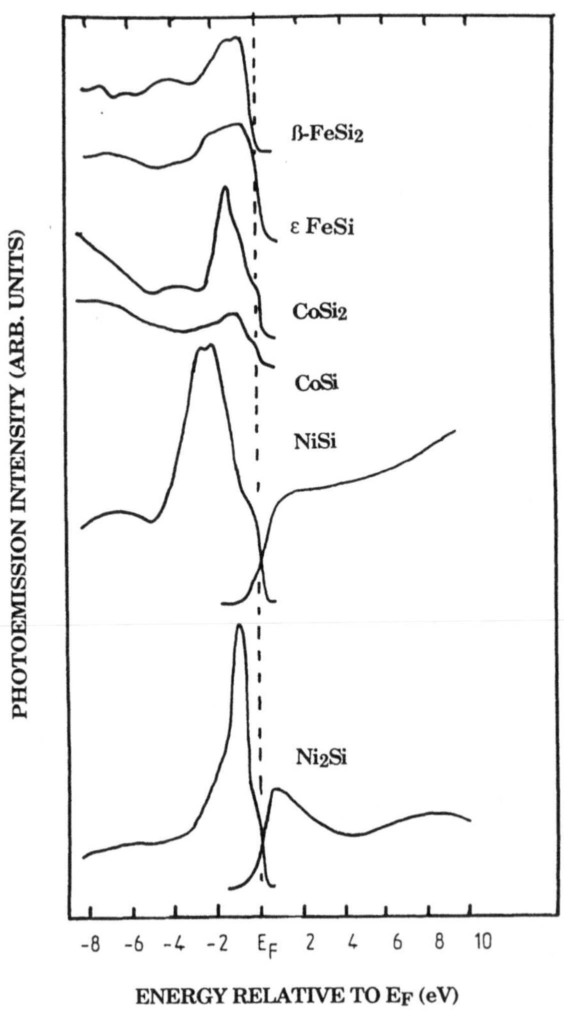


FIGURE 4 Comparison of the electronic structures for well-defined silicides (from [7,10]).

## D BAND STRUCTURE OF 4d- AND 5d-TRANSITION METAL SILICIDES

The substitution of a 3d metal by a 4d or 5d element results in a larger splitting of the bonding and antibonding TM d-Si p-states. This is attributed to an increased hybridisation between TM d- and Si p-states due to the larger expectation radius of the d-wave function [11]. FIGURES 5 and 6 show a compilation of available photoemission spectra of 4d- and 5d-TM silicides [12-14].

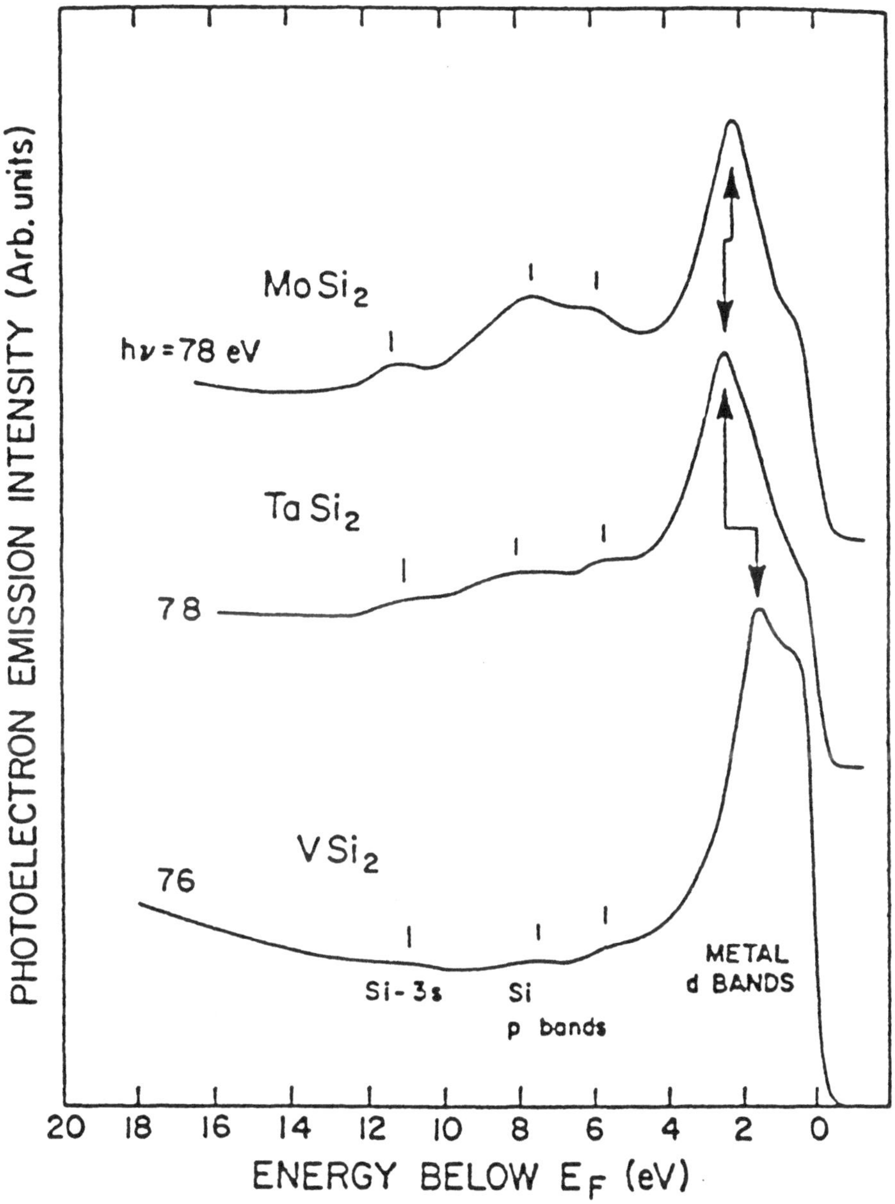


FIGURE 5 Comparison of photoemission spectra for $VSi_2$, $TaSi_2$ and $MoSi_2$. With the substitution of a 3d metal by a 4d or 5d element a larger d-band is observed, as well as a higher density of states at $E_F$ (from [12]).

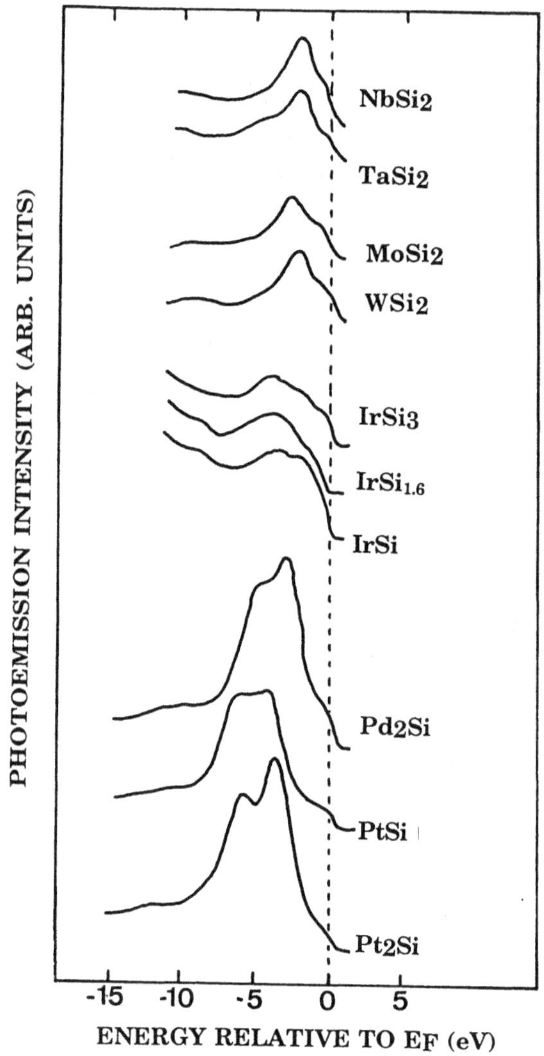


FIGURE 6 Compilation of 4d and 5d silicide spectra recorded with XPS. Only the Ir silicide spectra have been recorded with UPS (hν ~ 20 eV) (from [1,12-14]).

## E CONCLUSION

Reliable experimental electronic band structures of TM silicides are available. They can be understood qualitatively within a molecular orbital approach invoking hybridisation of TM d-orbitals and Si s- and p-states. More detailed but difficult calculations taking into account the real crystal structure in a many body approach can also now be undertaken in order to essentially reproduce electronic features observed by photoemission spectroscopy.

## REFERENCES

[1] C. Calendra, O. Bisi, G. Ottaviani [ *Surf. Sci. Rep. (Netherlands)* vol.4 (1985) p.271-364 ]
[2] C.D. Gelatt, A.R. Williams, V.L. Muruzzi [ *Phys. Rev. B (USA)* vol.27 (1983) p.2005 ]
[3] J. Friedel [ *The Physics of Metals* Ed. J.M. Ziman (Cambridge University Press, 1969) ch.8 ]
[4] J.H. Weaver, A. Franciosi, V.L. Moruzzi [ *Phys. Rev. B (USA)* vol.29 (1984) p.3293-302 ]
[5] M. Cardona, L. Ley (Eds) [ *Top. Appl. Phys. (Germany)* vol.26 and 27 (1978) ]

[6] W. Speier et al [ *Phys. Rev. B (USA)* vol.39 (1989) p.6008-16 ]

[7] M. Azizan [ Ph.D. Thesis, Université de Grenoble, France (1987) ]

[8] J.Y. Veuillen, T.T.A. Nguyen, R. Cinti, M. De Crescenzi, J. Derrien [ *Phys. Rev. B (USA)* vol.39 (1989) p.8015 ]

[9] M. De Crescenzi, G. Gaggiotti, N. Motta, F. Patella, A. Balzarotti, J. Derrien [ *Phys. Rev. B (USA)* vol.42 (1990) p.5871 ]

[10] C. Pirri [ Ph.D. Thesis, Université de Haute Alsace Mulhouse, France (1984) ]

[11] B.K. Bhattacharyya, D.M. Bylander, L. Kleinmann [ *Phys. Rev. B (USA)* vol.32 (1985) p.7973 ]

[12] J.H. Weaver, V.L. Moruzzi, F.A. Schmidt [ *Phys. Rev. B (USA)* vol.23 (1981) p.2916-2921 ]

[13] P.J. Grunthaner, F.J. Grunthaner, A. Madhurka [ *J. Vac. Sci. Technol. (USA)* vol.20 (1982) p.680 ]

[14] M. Wittmer, P. Oelhafen, K.N. Tu [ *Phys. Rev. B (USA)* vol.33 (1986) p.5391-400 ]

[15] N.E. Christensen [ *Phys. Rev. B (USA)* vol.42 (1990) p.7148-53 ]

# 4.2 Schottky barrier heights of TM silicides on Si and GaAs

J. Derrien

June 1994

## A INTRODUCTION

Since the first observations by Braun [1] and first explanations by Schottky [2] of rectifying properties of metal-semiconductor contacts, the mechanisms determining their Schottky barrier (SB) heights have been under discussion. (The barrier height $\Phi_B$ is defined as the energy distance from the Fermi level, $E_F$, to the top of the valence band or to the bottom of the conduction band at the interface for semiconductors doped p and n respectively.) The present Datareview does not intend to discuss extensively the physics of all the models proposed so far [3]. It will instead summarise general remarks and collect data available on the Schottky barrier height formation of transition metal silicides. From the technological point of view, TM silicide/semiconductor contacts are preferred to elemental metal/semiconductor junctions. Their thermal and chemical stability and their extreme hardness make them very attractive in device fabrication.

## B GENERAL REMARKS ON THE SCHOTTKY BARRIER HEIGHT

For reasons of charge neutrality at the interface between a metal and a semiconductor, the space charge of the depletion layer in the semiconductor is compensated by an equal amount of charge of opposite sign at the interface. According to the Schottky-Mott model this charge resides on the metal at the interface. Several data do not agree with this assumption and rather indicate the presence of additional electronic states at the interface which can absorb charges. The physical nature of these interface states has been the main aim of research in this field for the past half century and several models have been proposed [3]. However, the SB height mechanism is probably and primarily determined by the decay of the metal electron wave function into the semiconductor in the energy range between the top of the valence band and the Fermi level ($E_F$) where the metal conduction band overlaps with the semiconductor bandgap. This leads to metal-induced gap states (MIGS) which can carry charge. More accurate data on GaAs and Si still show slight deviations of SB heights from the MIGS predictions and it is recognised that a secondary mechanism attributed to fabrication-induced interface defects should be included. Indeed, thanks to spectroscopic surface studies it is now well known that the nature and the morphology of the very first interfacial layer of the metal/semiconductor contact depend drastically on chemical reactions, surface diffusion, interdiffusion etc. These phenomena will create defects at the interface and will govern the SB height value.

For barrier heights $\Phi_B$ between n-silicon and transition metals (TM) the question arises as to whether $\Phi_B$ is correlated to parameters related to silicon/metallic element interfaces or silicon/TM silicide interfaces. This fact should be kept in mind and defined prior to any $\Phi_B$ determination.

In principle for TM silicide/silicon contacts, during the interface formation by solid state reaction of TM and Si at high temperature, most of the contaminants are evacuated by outward diffusion from the interface leading to a clean, well-defined and stable silicide/silicon interface. The measured $\Phi_B$ on such contacts is then reliable. However, these junctions may still contain metallurgical defects induced by the solid state reaction. To improve this situation, lattice-matched epitaxial silicide/silicon interfaces [4], displaying quasi-ideal Schottky contacts, have been prepared. These studies confirm that the $\Phi_B$ values for these interfaces are higher for some orientations of the interfaces (~0.78 eV instead of ~0.65 eV for the $NiSi_2$/(n)Si(111) contact). The authors of [4] also stress the importance of interface strains on the SB heights.

The preceding discussions attempt to introduce several phenomena which govern the Schottky barrier. It appears that the stoichiometry of the semiconductor surface region upon contact formation is one relevant parameter to be controlled. This is the reason that reliable SB junctions can now be prepared with various 'recipes', passivating, for example, the semiconductor surface by exposure to reactive gases prior to metallisation or depositing a very thin interlayer between the metal and the semiconductor. Although clear understanding of the processes occurring at metal/semiconductor interfaces is still lacking it is not impossible to consider Schottky barrier engineering.

## C SCHOTTKY BARRIER VALUES FOR SILICIDE/SILICON CONTACTS

Most of the values reported in TABLE 1 [5-15] have been obtained with electrical measurements (current-voltage, capacitance-voltage or photoresponse-wavelength characteristics).

TABLE 1 Schottky barrier values $\Phi_B$ (eV) of metal silicides on n-silicon.

| Silicides | $\Phi_B$ (eV) | Ref | Remarks |
|---|---|---|---|
| $TiSi_2$ | 0.60 | [5] | |
| $VSi_2$ | 0.64 | [6] | |
| $CrSi_2$ | 0.57 | [5] | p-type semiconductor |
| MnSi | 0.65 | [7] | |
| $MnSi_{1.7}$ | 0.67 | [7] | |
| $FeSi_2$ | 0.67 | [6] | α phase is metallic; β phase is a p-type semiconductor |
| CoSi | 0.68 | [5] | |
| $CoSi_2$ | 0.64 | [5] | Mostly polycrystalline samples |
| $Ni_2Si$ | 0.66 | [8] | |
| NiSi | 0.65 | [9] | 0.78 if lattice-matched |
| $NiSi_2$ | 0.66 | [9,10] | 0.78 if lattice-matched type B |
| $YSi_{1.7}$ | 0.39 | [11] | 0.75 on (100) p-Si and 0.69 for Y metal |
| $ZrSi_2$ | 0.55 | [5] | |
| $NbSi_2$ | 0.63 | [6] | |
| $MoSi_2$ | 0.65 | [6] | Small gap semiconductor |
| $Ru_2Si_3$ | 0.72 | [12] | 0.79 for Ru metal |
| RhSi | 0.75 | [13] | 0.81 for Rh metal |
| $Pd_2Si$ | 0.71 | [8] | Depends drastically on heat treatment |
| HfSi | 0.53 | [5] | |
| $TaSi_2$ | 0.59 | [5] | |

Table 1 continued

| Silicides | $\Phi_B$ (eV) | Ref | Remarks |
|---|---|---|---|
| $WSi_2$ | 0.65 | [5] | |
| $ReSi_2$ | 0.77 | [6] | Small gap semiconductor, $\phi_B$ corresponds to Re metal |
| $OsSi_{1.8}$ | 0.85 | [6] | mixed phases |
| IrSi | 0.93 | [14] | |
| $Ir_2Si_3$ | 0.85 | [14] | |
| $IrSi_3$ | 0.94 | [14] | $p^+$-type semiconductor |
| $Pt_2Si$ | 0.85 | [15] | |
| PtSi | 0.88 | [15] | 0.21 for p-Si |

## D SCHOTTKY BARRIER VALUES FOR SILICIDE/GALLIUM ARSENIDE CONTACTS

Recently, in order to achieve high temperature stable contacts to GaAs in GaAs IC technology, a few studies have been undertaken. The barriers have been prepared either by solid state reaction of silicon and metals previously co-sputtered onto GaAs substrates [16] or by plasma enhanced chemical vapour deposition (PECVD) [17].

FIGURE 1 shows the SB height and the ideality factor n of a Ti-W-Si/GaAs junction during its annealing. The atomic ratio of Si to Ti-W is around 0.7. The barrier $\Phi_B$ remains very stable with increasing temperature. This kind of barrier might be formed with various TM-silicides ($TiSi_x$ and $WSi_y$) and a systematic study is preferred using only one kind of transition metal. FIGURE 2(a) shows the $\Phi_B$ variation versus the Si/W atomic ratio. Stable and achievable contacts have been found for x = 0.64 (FIGURE 2(b)) and $\Phi_B$ is ~0.75 eV. Similar behaviour can also be observed with Mo-Si/GaAs (FIGURE 2(c)).

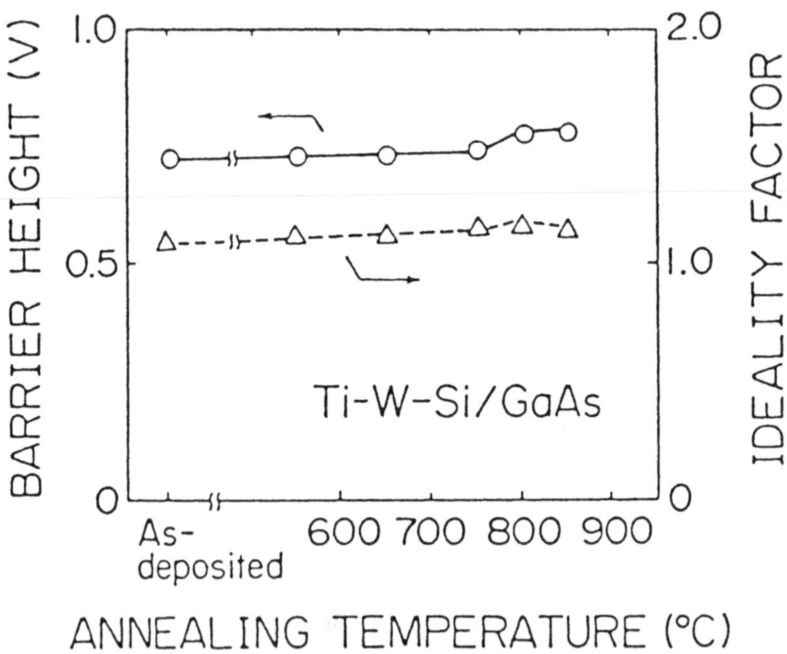


FIGURE 1 Schottky barrier height and ideality factor of Ti-W-Si/GaAs contacts (from [16]).

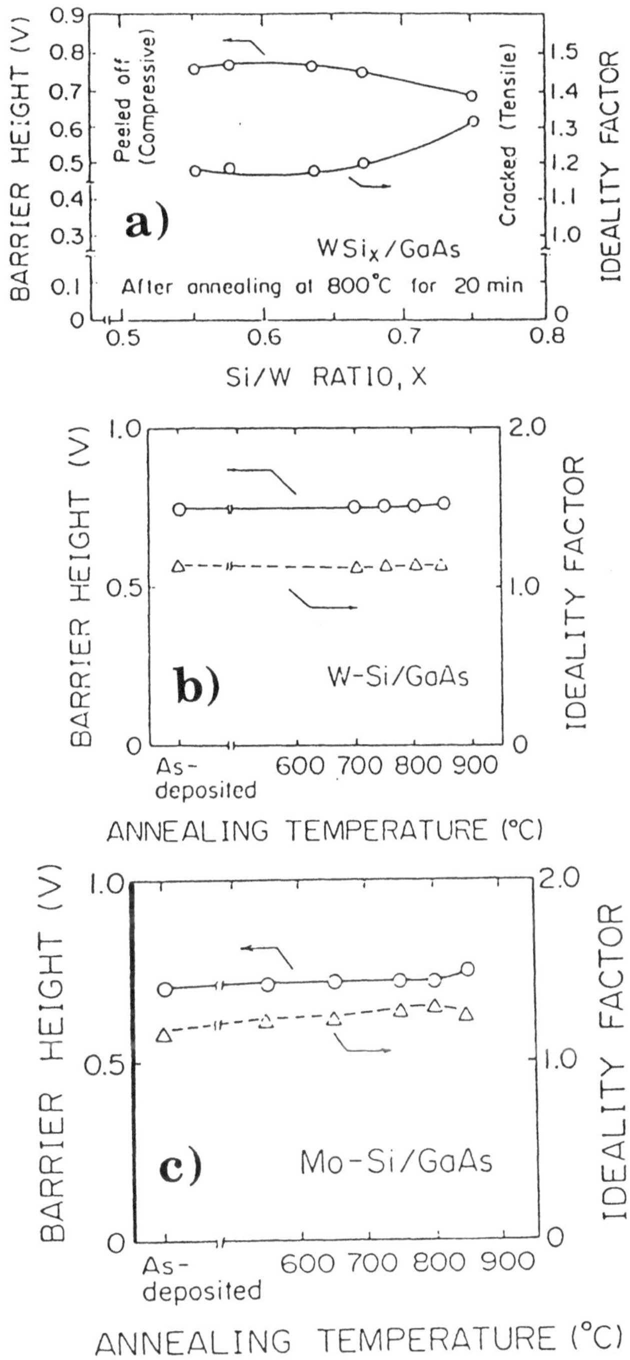


FIGURE 2 (a) Schottky barrier height and ideality factor of $WSi_x$/GaAs contacts versus Si/W ratio, x; (b) Schottky barrier height and ideality factor of $WSi_x$/GaAs contacts versus temperature (x = 0.64); (c) Schottky barrier height and ideality factor of $MoSi_x$/GaAs contacts versus temperature (x ~ 0.7) (from [16]).

## E CONCLUSION

From the fundamental point of view, the precise mechanisms which lead to SB formation at metal/semiconductor interfaces have not been definitively clarified. However, it appears that the metal-induced gap states and the fabrication-induced interface defect states are primarily responsible for the SB formation.

**REFERENCES**


[1] F. Braun [ *Pogg. Ann.* vol.153 (1924) p.556 ]

[2] W. Schottky [ *Naturwissenschaften (Germany)* vol.26 (1938) p.843 ]

[3] W. Mönch [ *Rep. Prog. Phys. (UK)* vol.53 (1990) p.221-78 ]

[4] E. Bucher, S. Schulz, M.Ch. Lux-Steiner, P. Munz, U. Gubler, R. Greuter [ *Appl. Phys. A (Germany)* vol.40 (1986) p.71-7 ]

[5] S.M. Sze [ *Physics of Semiconductor Devices 2nd edition* (Wiley, New York, 1981) ]

[6] R.D. Thompson, K.N. Tu [ *Thin Solid Films (Switzerland)* vol.93 (1982) p.265 ]

[7] M. Eizenberg, K.N. Tu [ *J. Appl. Phys. (USA)* vol.53 (1992) p.6885 ]

[8] P.S. Ho [ *J. Vac. Sci. Technol. A (USA)* vol.1 (1983) p.745 ]

[9] M. Liehr, P.E. Schmid, F.K. Le Gones, P.S. Ho [ *Phys. Rev. Lett. (USA)* vol.54 (1985) p.2139 ]

[10] R.T. Tung, J.M. Gibson [ *J. Vac. Sci. Technol. A (USA)* vol.3 (1985) p.987 ]

[11] K.N. Tu, R.D. Thomson, B.Y. Tsaur [ *Appl. Phys. Lett. (USA)* vol.38 (1981) p.626 ]

[12] D. Donoval, L. Stolt, H. Norde, J. de Sousa Pires, P.A. Tove, C.S. Petersson [ *J. Appl. Phys. (USA)* vol.53 (1982) p.5352 ]

[13] P.A. Psaras, R.D. Thomson, S.R. Herd, K.N. Tu [ *J. Appl. Phys. (USA)* vol.55 (1984) p.3536 ]

[14] S.P. Murarka [ *J. Vac. Sci. Technol. (USA)* vol.17 (1980) p.775 ]

[15] G. Ottaviani, K.N. Tu, J.W. Mayer [ *Phys. Rev. B (USA)* vol.34 (1981) p.3354 ]

[16] N. Yokoyama, T. Ohnishi, T. Nakamura, H. Nishi [ *Mater. Res. Soc. Symp. Proc. (USA)* vol.54 (1986) p.401-8 ]

[17] S. McClatchie, H. Thomas, D.V. Morgan [ *Appl. Surf. Sci. (Netherlands)* vol.73 (1993) p.58-63 ]

# 4.3 Schottky barrier heights of RE silicides on Si

Xu Zhenjia (C.C. Hsu)

June 1994

## A INTRODUCTION

Almost all metals can form silicides, but most of them result in high Schottky barrier heights on n-type silicon and low barrier heights on p-type silicon. For example, platinum has a barrier height of 0.84 eV on n-type silicon and 0.24 eV on p-type silicon, while palladium silicide gives values of 0.72 eV and 0.35 eV for n- and p-type silicon respectively.

Tu et al [1] and Norde et al [2] showed that the RE silicides, unlike most other silicides, produce a high Schottky barrier height of about 0.7 eV on p-type Si and a low barrier height of about 0.4 eV on n-type Si. TABLE 1 lists the most recent values of barrier heights $\Phi_B^n$ and $\Phi_B^p$ of some of the RE metals and RE silicides.

## B SCHOTTKY BARRIER OF $ErSi_{2-x}$ ON Si

### B1 Schottky Barrier Height

Recently, special attention has been devoted to $ErSi_{2-x}$, which appears to be an ideal candidate for the investigation of the fabrication of photodetectors operating in the infrared region (0.5-1 eV). There are two reasons for this consideration. Firstly, from a metallurgical point of view, the silicide layer can be epitaxially grown on silicon with a high crystallographic quality. Secondly, $ErSi_2$ like other RE silicides forms a high Schottky barrier on p-type silicon (~0.8 eV) and low Schottky barrier on n-type silicon (~0.28 eV).

$ErSi_2$/Si diodes prepared from 300 Å thick $ErSi_{2-x}$ MBE films grown on n-type and p-type (111) Si have been carefully studied [4-6]. The authors observed an excellent rectifying behaviour of the $ErSi_{2-x}$/p-type Si diodes. The diode leakage current at RT was less than $10^{-8}$ A at 1 V, while the saturation current approached $10^{-2}$ A for a diameter of 0.25 mm. The ideality factor, measured for five orders of magnitude of the forward current, was found to be 1.03, indicating a high perfection of the diode. Using a Richardson constant of 80 A $cm^{-2}$ $K^{-2}$, a Schottky barrier height of $ErSi_{2-x}$ on p-type Si, $\Phi_B^p = 0.73 \pm 0.01$ eV at 300 K, was calculated [4]. By photoemission methods, a barrier height $\Phi_B^p = 0.74 \pm 0.002$ eV at RT was measured [4]. The n-type junction is ohmic at room temperature and rectifying at low temperatures. Capacitance-voltage (C-V) measurements gave $\Phi_B^n = 0.29 \pm 0.01$ eV at 77 K [4], while internal photoemission measurements yield $\Phi_B^n = 0.280 \pm 0.004$ eV [4]. The accuracy of optical barrier height determination (less than 5 meV) was always better than that of electrical determinations (10 meV).

TABLE 1 Schottky barrier heights, $\Phi_B$, of RE metals and silicides.

| | Diode fabrication technique | $\Phi_B^n$ (eV) | Methods of measurement | Ref | Diode fabrication technique | $\Phi_B^p$ (eV) | Methods of measurement | Ref |
|---|---|---|---|---|---|---|---|---|
| $GdSi_2$ | TE | $0.37 \pm 0.02$ | (I-V) | [1] | TE | $0.71 \pm 0.03$ | (I-V)<br>n = 1.05 | [1] |
| Tb | | | | | | 0.65<br>0.63 | Photo<br>(I-V) | [2] |
| $DySi_2$ | TE | $0.37 \pm 0.02$ | (I-V) | [1] | TE | $0.73 \pm 0.03$ | (I-V)<br>n = 1.05 | [1] |
| $HoSi_2$ | TE | $0.37 \pm 0.02$ | (I-V) | [1] | | | | |
| Er | | | | | | 0.68<br>0.68 | (I-V)<br>Photo | [2] |
| $ErSi_2$ | TE | $0.39 \pm 0.02$ | (I-V) | [1] | TE | $0.70 \pm 0.02$ | (I-V)<br>n = 1.1 | [1] |
| | | | | | TE | 0.77<br>0.76 | (I-V)<br>Photo | [2] |
| | TE | $0.277 \pm 0.002$<br>$0.276 \pm 0.002$<br>$0.273 \pm 0.005$ | (I-V)<br><br>Photo | [3] | | | | |
| | MBE<br>(Epitaxy) | $0.29 \pm 0.01$<br>$0.29 \pm 0.01$<br>$0.280 \pm 0.004$ | (I-V, 77 K)<br>(C-V, 77 K)<br>Photo, 77 K | [6] | MBE | $0.74 \pm 0.01$<br>$0.740 \pm 0.005$ | (I-V)<br>n = 1.03<br>Photo | [6] |
| Yb | | | | | TE | 0.77<br>0.76 | (I-V)<br>Photo (77 K) | [2] |

TE: thermal evaporation; MBE: molecular beam epitaxy; Photo: photoemission method; I-V: current-voltage.

It should be noted that the sum of $\Phi_B^n + \Phi_B^p$ is equal to $1.020 \pm 0.006$, a value about 100 meV smaller than the silicon bandgap $E_g = 1.115$ eV. The given values $\Phi_B^n$ and $\Phi_B^p$ were not corrected for the image force lowering which is about 15 meV. This slight discrepancy is not yet explained [4-6].

As thermal evaporation is a cheaper approach and more compatible with large scale structures, Schottky diodes fabricated from $ErSi_{2-x}$ on n-type silicon by thermal evaporation have been studied to examine their suitability as an infrared detector [3]. I-V measurements gave a barrier height of $(0.276 \pm 0.002)$ eV and an ideality of 1.05. This is as low as for epitaxially grown diodes [4]. Photoresponse measurements gave a $\Phi_B^n = (0.273 \pm 0.005)$ eV, which is in close agreement with the barrier heights measured by other methods [3]. A plot of quantum efficiency against wavelength is given in FIGURE 1 [3].

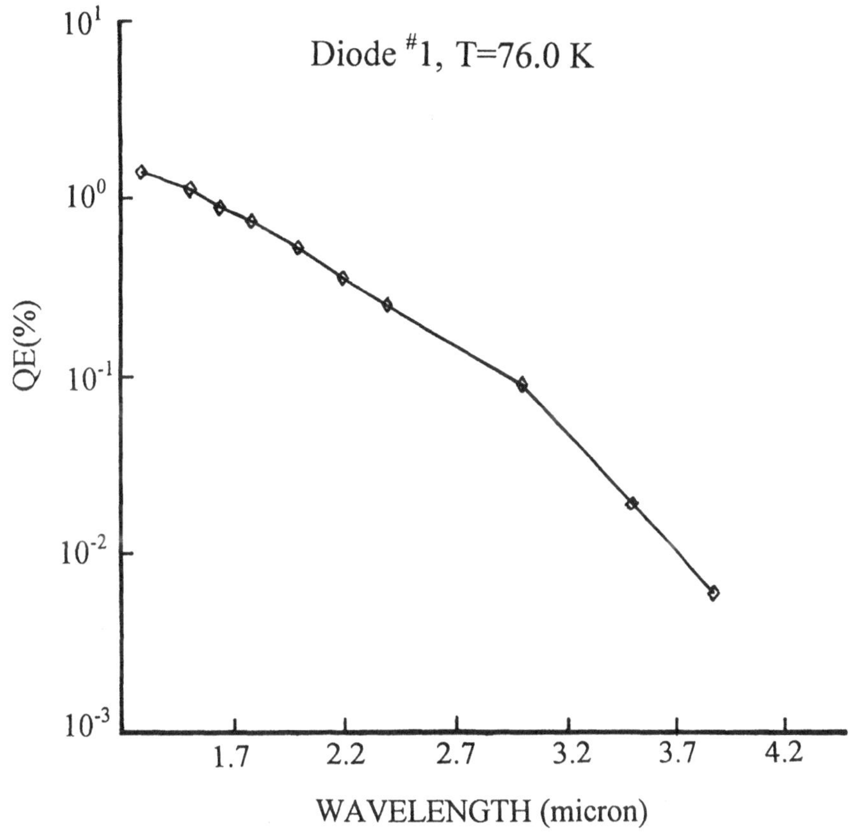


FIGURE 1 Plot of log (quantum efficiency) against wavelength (after [3]).

As can be seen from FIGURE 1 the quantum efficiency at 2.0 μm is about 0.52%, which is comparable to that obtained with platinum silicide diodes made in a similar fashion, while the dark current is less than $10^{-7}$ A $cm^{-1}$ at 77 K. This opens up the possibility of a whole new generation of infrared detector array architectures with n-type silicon based multiplexes.

## B2 The Effect of an Interfacial Oxide Layer

Since RE metals are very chemically reactive the problem of oxidation is always of great importance. Wu et al [7] have studied the effect of an interfacial oxide layer on the Schottky barrier height of the Er and $ErSi_2$ contacts on p-type silicon. A correlation was found between the barrier height and the presence of an interfacial oxide layer. Samples with a thin interfacial oxide (~3 Å) exhibit a barrier height of 0.68 eV. Samples without a detectable interfacial oxide, or chemically cleaned samples after annealing, show a higher barrier height of 0.8 eV.

## REFERENCES

[1] K.N. Tu, R.D. Thompson, B.Y. Tsau [ *Appl. Phys. Lett. (USA)* vol.38 (1981) p.626-8 ]

[2] H. Norde, J. de Sousa Pries, F. d'Heurle, F. Pesavento, P.A. Tove [ *Appl. Phys. Lett. (USA)* vol.38 (1981) p.865-6 ]

[3] M.H. Unewisse, J.W.V. Stroey [ *Appl. Phys. Lett. (USA)* vol.72 (1992) p.2367-71 ]

[4] J.Y. Duboz, P.A. Badoz, F. Arnaud, J.A. Chroboczek [ *Appl. Phys. Lett. (USA)* vol.55 (1989) p.84-6 ]

[5] J.Y. Duboz et al [ *Appl. Surf. Sci. (Netherlands)* vol.38 (1989) p.171-7 ]

[6] F. Arnaud d'Avitaya et al [ *Thin Solid Films (Switzerland)* vol.184 (1990) p.283-93 ]

[7] C.S. Wu, D.M. Scott, S.S. Lau [ *J. Appl. Phys. (USA)* vol.58 (1985) p.1330-4 ]

# 4.4 Electronic properties of epitaxial TM silicide/Si heterojunctions

J. Derrien

June 1994

## A INTRODUCTION

Transition metal (TM) silicides have been widely used in microelectronics as Schottky barriers, gate electrodes, interconnects and contacts. Among these materials, silicides epitaxially grown on silicon possess several attractive advantages such as a higher thermal and chemical stability, a well ordered interface etc. Moreover, it is possible to re-epitaxy silicon material on top of these epitaxial silicides opening the way to novel classes of devices based on silicon/silicide/silicon multilayer structures.

A silicide is expected to grow epitaxially on silicon if the crystal structures are similar and the lattice mismatch between them is small. It now appears that almost all transition metal silicides can be grown epitaxially to a certain extent on silicon. However, due to a large misfit and different crystal structure from silicon substrates, these TM silicides form small size grains (a few thousand Å) leaving large areas of bare Si substrate between them. We will not discuss those cases in this Datareview and the reader is referred to available review papers [1,2]. So far, uniform lateral epitaxy leading to continuous, high quality layers has only been achieved for $NiSi_2$ and $CoSi_2$ which display the cubic $CaF_2$ structure and small misfit with respect to Si (about 0.4% and 1.2%, respectively). Both disilicides are metallic [3]. Their growth and characterisation have been extensively reviewed elsewhere [4-7]. In Section B we will confine our discussion to their electronic properties and potential applications. Very recently, epitaxial semiconducting silicides have been suggested to be well-suited to combine the benefits of silicon-based digital-technology with new optoelectronic devices, optical fibre links, infrared detectors etc. [8]. Although work in this field is still in progress and many results on these semiconducting silicides are expected in the near future we will introduce in Section C a discussion on the electronic properties of $\beta$-$FeSi_2$ epitaxially grown on silicon.

## B EPITAXIAL $NiSi_2$ AND $CoSi_2$/Si HETEROJUNCTIONS

Both $NiSi_2$ and $CoSi_2$ crystallise in the cubic $CaF_2$ structure (FIGURE 1) and the lattice mismatches to Si, defined as

$$\Delta = \frac{a_{silicide} - a_{Si}}{a_{Si}}$$

where $a_{silicide}$ and $a_{Si}$ are the lattice parameters of silicide and Si, respectively, are -0.46 and -1.2%, at room temperature, respectively.

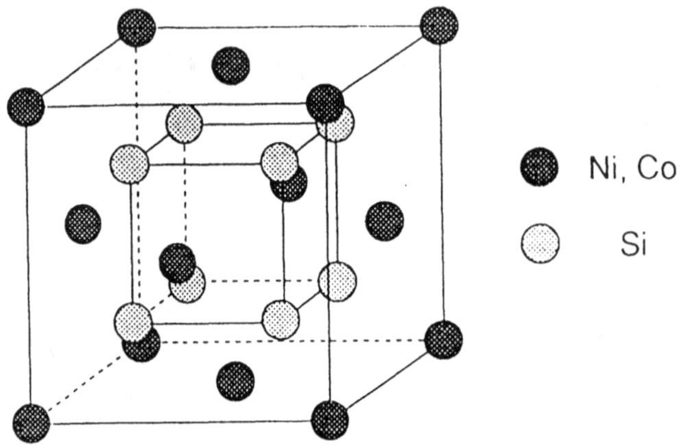


FIGURE 1 Cubic unit cell of the fluorite structure.

Epitaxial growth of $NiSi_2$ and $CoSi_2$ on silicon is expected and has been achieved both on Si(111) and (100) with various growth methods and surface characterisation techniques [4-7]. It appears that better epitaxial growth has been observed on Si(111) due to the lower interfacial and surface energies [4]. Basically there are two possible orientations (A and B) of the silicides on top of the Si(111) surface. They are related by a 180° rotation around the [111] axis. Concerning the interface structure various possibilities can be expected and they differ in whether the silicide is terminated by a Si plane or a metal plane (FIGURE 2). There is general agreement that $NiSi_2$ displays a sevenfold coordinated interface structure whereas the situation for $CoSi_2$ still remains somewhat ambiguous (5 or 8 fold coordinated). Of course the exact interface structure, in those quasi-ideal cases, is the most relevant parameter (interface states) for the Schottky barrier determination.

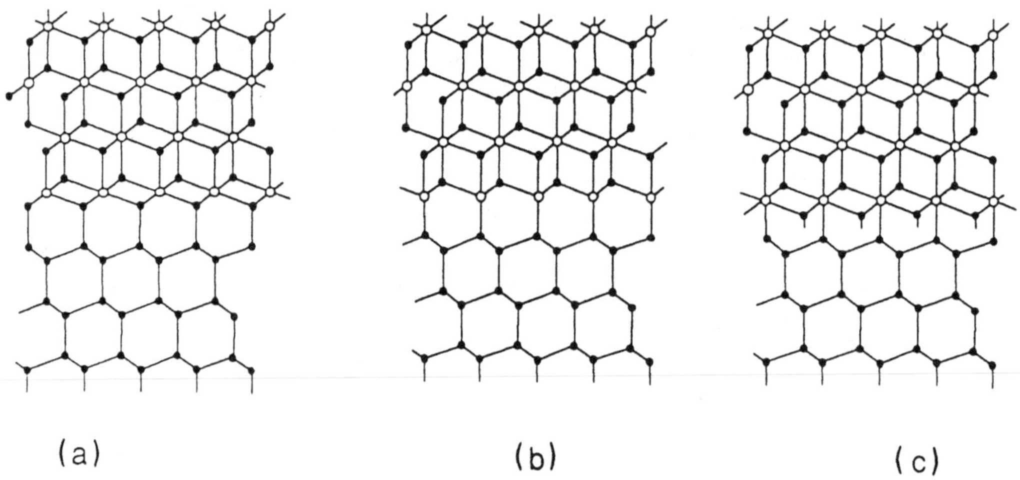


FIGURE 2 Possible interface structures for a type-B interface ([110] projection). (a) Seven-fold, (b) five-fold and (c) eight-fold coordination of the metal atoms (from [4]).

The Si plane or metal plane termination occurs also at the topmost surface of silicides and may affect their electronic band structures, with additional surface state features possible [9,10].

Since the crystal structure and atomic positions in the unit cell of $NiSi_2$ and $CoSi_2$ are well-defined it has been possible to compute self-consistent electronic band structures [11]. FIGURE 3 shows the density of states (DOS) calculated for $NiSi_2$ and $CoSi_2$ [12]. As already discussed in [3] the occupied DOS is dominated by non-bonding TM d-bands

(0-2 eV below $E_F$) and bonding states between TM d- and Si s- and p-derived states (~4 eV below $E_F$) and Si s-states (~8-12 eV below $E_F$). These data agree with those collected from photoemission experiments performed with synchrotron radiation [9]. FIGURE 4 shows such a comparison. We notice that besides measured bulk band features surface features attributed to various terminations of the topmost silicide surface [9,10] are observed, as discussed previously.

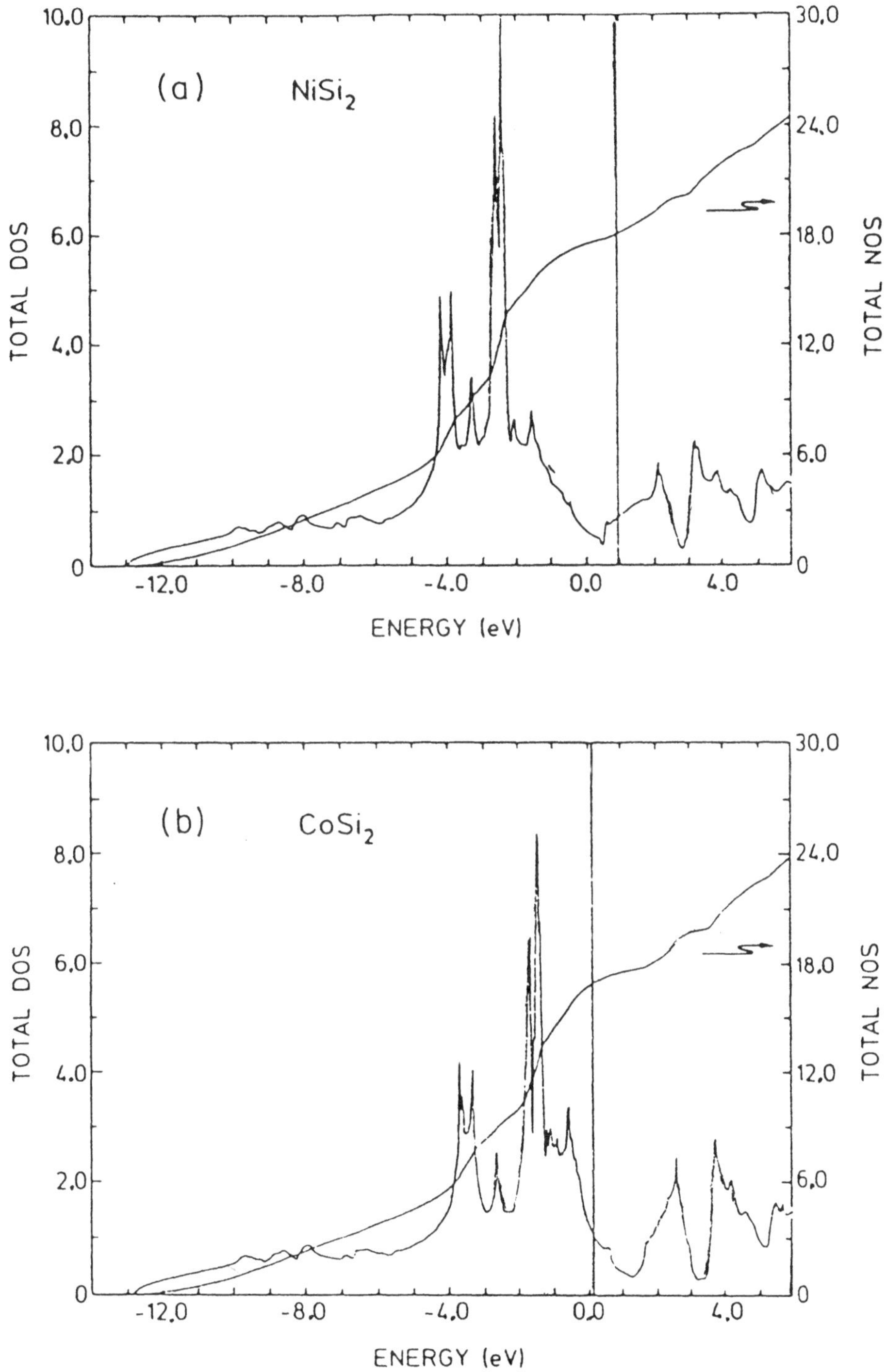


FIGURE 3 Density of states functions calculated for (a) $NiSi_2$, (b) $CoSi_2$ in the fluorite structure. Density of states (DOS) and number of states (NOS) are in units of (electrons/eV) per formula unit (f.u.) and electrons/f.u., respectively. The Fermi level is indicated by a vertical line. (From [12].)

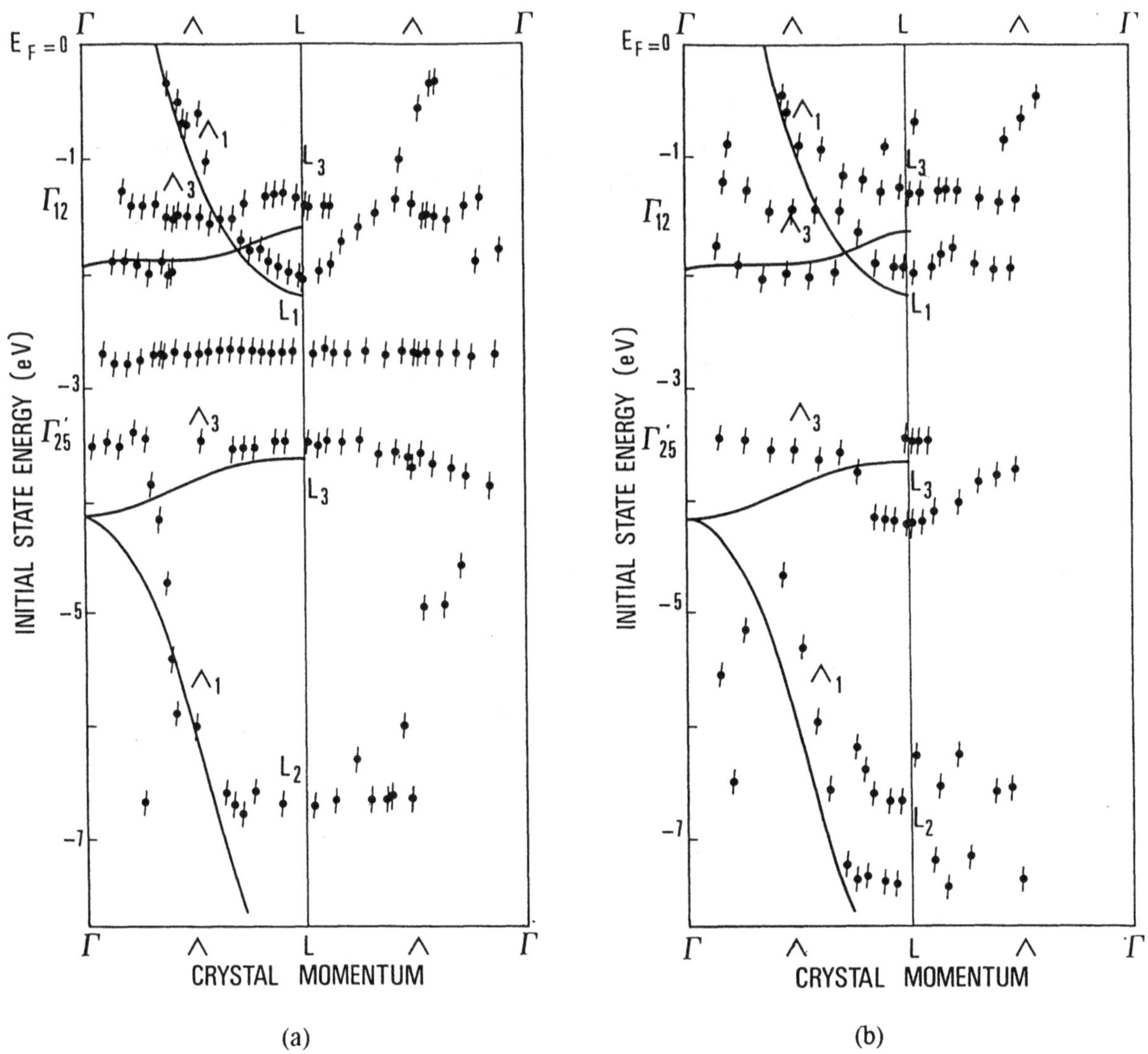


FIGURE 4 Energy band structure deduced from synchrotron radiation angle-resolved photoemission spectroscopy: (a) as recorded on a Co-rich $CoSi_2$ surface; (b) as recorded on a Si-rich $CoSi_2$ surface. Full circles are from data. Solid lines are from calculations on bulk $CoSi_2$ according to [11]. Note the surface state band observed in (a) at ~2.7 eV below $E_F$. (From [6].)

Thanks to the advances in Molecular Beam Epitaxy (MBE) growth techniques, single crystals of $NiSi_2$ and $CoSi_2$ can be grown with thickness ranging from 10 to ~ 1000 Å. Their electrical properties have been determined [4,7]. Furthermore, a silicon layer re-epitaxially grown on top of the silicide film should provide us with semiconductor/metal/semiconductor heterojunctions for potential applications in high-speed devices. Examples are the metal base transistor (MBT) consisting of two Schottky diodes, arranged face-to-face. The common metal ($CoSi_2$) acting as the base of the transistor has to be thin enough to be 'transparent' to ballistically injected electrons [2,4,7,13]. An alternative is the so-called permeable base transistor (PBT) in which the base is not continuous but rather consists of a grid fabricated by lithography and etching, or some other means [4,7].

## C EPITAXIAL $FeSi_2$/Si HETEROJUNCTIONS

In the equilibrium bulk Fe-Si phase diagram two $FeSi_2$ compounds exist. One, called α-phase, is stable at high temperature (≥940°C). It displays a tetragonal structure and a metallic behaviour. Another one, called β-phase, stable at lower temperature (<940°C), adopts an orthorhombic structure and is semiconducting in nature. Its optical gap is about 0.87 eV (~1.5 μm) and places the β-phase in a class of optoelectronic materials suitable for optical fibre links, infrared detectors, etc. [14]. This is why epitaxial growth of β-$FeSi_2$ has been extensively studied recently [6,8]. Due to a large misfit to silicon substrates (up to -5.5% depending on the orientation) and different crystal structures, the β-$FeSi_2$ single crystal heteroepitaxy has not been realised. The silicide layer is composed mostly of several epitaxial variants [15,16].

The semiconducting nature of β-$FeSi_2$ is a puzzling problem as compared to the metallic behaviour of the neighbouring TM silicides such as $CoSi_2$ and $NiSi_2$. Indeed, if one calculates the electronic properties of $FeSi_2$ assuming the same fluorite structure as in the case of $CoSi_2$ and $NiSi_2$, then $FeSi_2$ would be metallic (FIGURE 5(a)). However, the Fermi level falls, for this hypothetical fluorite structure of $FeSi_2$ (called γ-phase), in a very high DOS peak. It indicates that the fluorite form of $FeSi_2$ is highly unstable. The system lowers its energy by undergoing a transition from the fluorite γ-metallic phase towards the orthorhombic β-semiconducting phase. This transition occurs through a distortion of the fluorite γ lattice to achieve the orthorhombic β lattice [17], similar to a solid state Jahn-Teller phenomenon. FIGURE 5(b) shows the opening of the gap due to this lattice distortion, from γ to β phase [12,18]. The nature of the gap in the band structure (FIGURE 5(c)) is indirect although the difference between the calculated direct and indirect gaps is very small. These values (~0.8 eV) agree with experimental results [14,19]. It is worth mentioning that besides the stable β-$FeSi_2$ phase, usually found to grow on silicon with techniques close to equilibrium conditions, several other unexpected phases have been observed during $FeSi_2$ heteroepitaxy on silicon. They are stabilised by kinetics conditions, by the strain field exerted by the silicon substrate and by chemical bonding at the interface in the heteroepitaxy mechanisms. So far, ε-FeSi, CsCl-$FeSi_x$, γ-$FeSi_2$ and α-$FeSi_2$ have been epitaxially grown on silicon [4,7,20]. Most of these metastable phases have different electronic properties and transform, with increasing temperature and/or thickness, to the β-$FeSi_2$ phase.

Up to the present time there is no realised optoelectronic device with semiconducting $FeSi_2$. However, due to a high thermoelectric coefficient, β-$FeSi_2$ has been used for thermoelectric power generators [21].

## D CONCLUSION

Epitaxial silicides provide interesting models for a better understanding of both interface and bulk properties. Their epitaxy on silicon has triggered several metastable, strained interfacial phases bearing unexpected physical properties. The investigation of their solid-state transitions towards stable phases appears very promising.

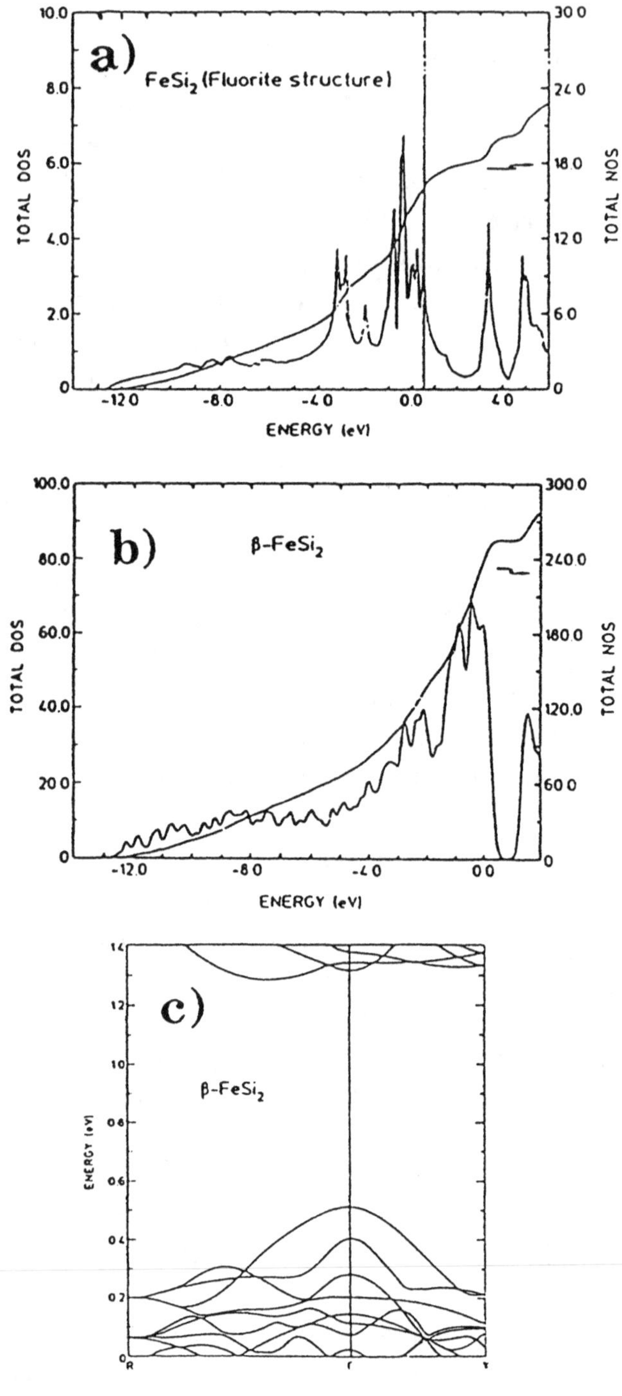


FIGURE 5 Density of states calculated for (a) hypothetical fluorite structure $FeSi_2$; (b) β-$FeSi_2$ in its orthorhombic structure; (c) band structure of β-$FeSi_2$ (from [12]).

## REFERENCES


[1] L.J. Chen, K.N. Tu [ *Mater. Sci. Rep. (Netherlands)* vol.6 (1991) p.53-140 ]

[2] J. Derrien, F. Arnaud d'Avitaya [ *J. Vac. Sci. Technol. A (USA)* vol.5 (1987) p.2111 ]

[3] J. Derrien [ Datareview in this book: 4.1 Electronic band structure of TM silicides ]

[4] H.J. Von Känel [ *Mater. Sci. Rep. (Netherlands)* vol.5 (1992) p.193-269 ]

[5] J. Derrien, J. Chevrier, A. Younsi, V. Le Thanh, J.P. Dussaulcy, N. Cherief [ *Phys. Scr. (Sweden)* vol.T35 (1991) p.251 ]

[6] N. Cherief, R. Cinti, M. de Crescenzi, J. Derrien, T.A. Nguyen Tan, J.Y. Veuillen [ *Appl. Surf. Sci. (Netherlands)* vol.41/42 (1989) p.241 ]

[7] S. Mantl [ *Mater. Sci. Rep. (Netherlands)* vol.8 (1992) p.1-95 ]

[8] J. Derrien, J. Chevrier, V. Le Thanh, T.E. Crumbaker, J.Y. Natoli, I. Berbezier [ *Appl. Surf. Sci. (Netherlands)* vol.70/71 (1993) p.546 ]

[9] G. Gewinner, C. Pirri, J.C. Peruchetti, D. Bolmont, J. Derrien, P. Thiry [ *Phys. Rev. B (USA)* vol.38 (1988) p.1879 ]

[10] L. Haderbaahe, P. Wetzel, C. Pirri, J.C. Peruchetti, D. Bolmont, G. Gewinner [ *Appl. Surf. Sci. (Netherlands)* vol.41/42 (1989) p.257 ]

[11] W.R.L. Lambrecht, N.E. Christensen, P. Blöchl [ *Phys. Rev. B (USA)* vol.36 (1987) p.2493 ]

[12] N.E. Christensen [ *Phys. Rev. B (USA)* vol.42 (1990) p.7148 ]

[13] E. Rosencher, S. Delage, Y. Campidelli, F. Arnaud d'Avitaya [ *Electron. Lett. (UK)* vol.20 (1984) p.762 ]

[14] M.C. Bost, J.F. Mahan [ *J. Appl. Phys. (USA)* vol.58 (1985) p.2696 ]

[15] J.E. Mahan, V. Le Thanh, J. Chevrier, I. Berbezier, J. Derrien [ *J. Appl. Phys. (USA)* vol.74 (1993) p.1747 ]

[16] I. Bezbezier, J.L. Regolini, C. d'Anterroches [ *Microsc. Microanal. Microstruct. (France)* vol.4 (1993) p.5-21 ]

[17] Y. Dusausoy, J. Protas, R. Wandji, B. Roques [ *Acta Crystallogr. B (Denmark)* vol.27 (1971) p.1209 ]

[18] R. Eppenga [ *J. Appl. Phys. (USA)* vol.68 (1990) p.3027 ]

[19] C. Giannini, S. Lagomarsino, F. Scarinsi, P. Castrucci [ *Phys. Rev. B (USA)* vol.45 (1992) p.8222-4 ]

[20] J. Derrien et al [ *Appl. Surf. Sci. (Netherlands)* vol.73 (1993) p.90-101 ]

[21] U. Stöhrer, R. Voggesberger, G. Wagner, U. Birkholz [ *Energy Convers. Manage. (UK)* vol.30 (1990) p.143-7 ]

# 4.5 Electronic properties of RE/Si interfaces

Xu Zhenjia (C.C. Hsu)

June 1994

## A THEORETICAL INVESTIGATIONS

The fundamental study of electronic properties should be based on the theoretical calculations of density of states (DOS) of RE/Si systems. Unfortunately, very little theoretical information is available to date on the study of RE silicides.

Ye and co-workers [1,2] studied the chemisorption of Ce, Sm and Yb on Si(111) by theoretical calculations with local density formalism (LDF), the discrete vibrational method (DVM) and the semi-empirical EHT (extended Huckel theory) method. Cluster models are used to simulate two possible sites, the top site and the threefold site, where the latter is found to be more stable than the former [1,2]. Ye and co-workers [1,2] also found the charge transfer from the RE atom to silicon due to the difference of electronegativity between RE and Si, which results in the energy shift of Si3p. From the calculated TDOS curves, it can be shown that no charge transfer is observed for the Si3s peaks after adsorption of Ce, Sm or Yb on Si(111). Basically, these peaks keep their silicon characteristics. This result indicates that the Si3s states are not involved in chemical bonding. However, hybridisation between Si3p and the valence orbitals of RE occurs, and results in a broadening and a shift of the energy level.

In the case of Ce/Si(111), there is an unfilled peak just below the $E_F$. This peak is attributed to the (d-f) hybridisation. Upon adsorption of Ce on the Si(111) surface, the Si3p peak located at -2.2 eV shifts 0.4 eV due to the hybridisation with (d-f) electrons and a shift of 0.7 eV of surface states of Si was also observed [1,2].

There is a large difference between Yb and Sm after adsorption on Si(111), although they have the same valence electron configuration ($5d^0 6s^1$) [1,2]. For Yb/Si(111), a very sharp peak, attributed to 4f, is located at -3.4 eV below $E_F$. Two peaks attributed to the hybridisation between Si3p and the 5d, 6s valence electrons are observed. In the case of Sm/Si(111), the level just below $E_F$ is mainly due to the contribution of Sm(5d). The Si3p peak becomes stronger, but the energy shift is very small (0.1 eV) after the hybridisation between Si(3p) and Sm(5d) [1,2].

## B ELECTRONIC STRUCTURE OF RE SILICIDES

In the field of RE silicides very little information is available on photoemission from the valence states since, in most cases, the 4f signal dominates the spectra and the valence states are not directly accessible. In addition, these silicides are characterised by a low value of the d occupancy and therefore the traditional model of the chemical bond, which has been generally recognised for transition metal silicides, cannot be taken a priori for granted. In this section, we choose some experimental results as the prototype of the electronic structure of RE silicides for our discussion, and we shall not attempt to give a systematic review on the electronic structure of all RE silicides.

## B1 Erbium Silicide

Up to now, the most studied material among the RE silicides is probably erbium silicide. The Er-metal ground-state configuration is $4f^{11}$ and the photoemission processes lead to the $4f^{10}$ configuration giving rise to 4f multiplet states which are localised and core-like even though their binding energies are shallow and degenerate with valence-band states. Taking advantage of the large photoionisation cross-section and favourable location in the 4-11 eV binding energy range of 4f electrons, He II photoemission has been used to investigate the change in their binding energy versus Er coverage with a very good energy resolution. There remains a window at least 4 eV wide in which the electron states directly involved in the Er-Si chemical bonds (Er5d-Si3ps) are easily observed without interference with Er4f atomic-like states.

### B1.1 $Er_xSi_{1-x}$ valence bands

Wetzel et al [3] reported the valence spectra recorded with He II (hν = 40.8 eV) radiation for three stoichiometries, x = 0.45, 0.55 and 0.8. They also showed the valence-band spectra measured on a 30 Å thick epitaxial $ErSi_{1.7}$ layer grown on Si(111) and on a 50 Å thick polycrystalline erbium film. The prominent emission close to the Fermi level is associated with the Er5d derived states while the emission in the 4-11 eV binding energy window reflects the various 4f multiplet states. For crystalline $ErSi_{1.7}$, the Er5d-Si3sp hybridised states are reflected by two broad structures at ~2.25 and 0.4 eV. The 4f related multiplet structure is located in the 4-11 eV binding-energy window. It reveals a main well-resolved peak at 4.95 eV dominating the low binding energy part of the 4f structure while the high binding energy region comprises a series of incompletely resolved peaks. The 4f related multiplet structure shape for $ErSi_{1.7}$ looks different from that observed on a pure Er layer which involves both surface and bulk related 4f multiplet structures. This suggests that in crystalline $ErSi_{1.7}$ all Er atoms are in the same bulk-like atomic environment. The 5d related features are strongly affected by the silicide stoichiometry.

For $Er_{0.45}Si_{0.55}$ the valence-band spectrum displays only a broad feature at 1.8 eV below $E_F$. When x increases, the relative density of states increases at the Fermi level, and the broad structure at 1.8 eV below $E_F$ becomes more structured showing two components at 1.2 and 2.3 eV for $Er_{0.55}Si_{0.45}$. These features shift towards the Fermi level with increasing Er content and are located at ~0.30 and 1.8 eV for $Er_{0.8}Si_{0.2}$. In the higher binding energy region of the UPS spectra, one can observe that the 4f emission is much less structured than that for crystalline $ErSi_{1.7}$ or even for pure Er. This is clearly visible for the low-binding 4f peak, located at 4.95 eV in crystalline $ErSi_{1.7}$ while in all amorphous compounds it evolves into a very broad structure as wide as 1.5 eV. It appears that the Er5d valence states hybridise with the Si3ps states, while the Er4f states, although located in the valence-band region, behave like atomic-like states.

### B1.2 Valence band for the RT Er/Si(111) interface

Wetzel et al [3] also reported the valence-band spectra measured after deposition of Er onto a clean Si(111) substrate versus Er coverage. The reported spectrum is essentially dominated by intrinsic bulk related features at ~3.4 and 7.5 eV below $E_F$ along with surface states at 0.2, 0.85 and 1.8 eV. The upper-most spectrum shows the emission from a 50 Å-thick polycrystalline Er film. At submonolayer coverages (= 0.5 ML), it shows that both bulk and surface Si(111) structures are strongly attenuated whereas three new structures appear: two structures at 5.8 and 9.0 eV binding energy associated with Er4f states and a broad one at ~1.5-2 eV arising from Er5d-Si3sp hybridised states. The binding energy shift versus Er coverage shows that this chemical shift is about 0.5 eV for 0.3 ML [3]. This indicates the chemical interaction between the incoming Er atoms and the Si(111) substrates with a charge transfer from Er to Si species. Furthermore, the UPS spectrum recorded for 0.5 ML shows no distinct Fermi edge and only poorly resolved 4f structures.

Upon further Er deposits the 4f peaks increase more rapidly towards lower binding energies at 5.50 eV for 1 ML. At this stage, both bulk and surface Si(111) states are swept out and only a broad structure is observed at ~2 eV. Concomitantly the broad 4f structure centred at ~0.9 eV for 0.5 ML evolves into a more structured feature. Between 1 and 2 ML, the UPS spectrum remains quite unchanged and is very similar to the one recorded on $Er_xSi_{1-x}$ (x = 0.45). Higher Er deposition up to ~6 ML results essentially in a general broadening of the 4f line shape for the lower as well as higher binding energy parts. To a large extent this broadening results from the evolution of the interfacial silicide stoichiometry towards a more Er-rich silicide with a larger distribution of local surface and bulk Er environments.

### B2 Gadolinium Silicide

Gd is also a good candidate for investigation of the electronic states at the interface and in the bulk silicides since the 4f emission has a peak around 7-8 eV below the Fermi level so that there is a window at least 5 eV wide in which the electronic states involved in the chemical bond are directly observable at all photon energies, without the complication due to the superposition with the 4f emission.

Puppin et al [4] compared the synchrotron-radiation photoemission data from the Si(111)/Gd interface

$$\frac{\text{Si2p}}{\text{Gd4f}}$$

(core levels and valence states) with those from all of the silicides in the Si-Gd phase diagram ($Gd_3Si_5$, GdSi, $Gd_5Si_3$) measured under strictly comparable conditions. From the observed experimental results, Puppin et al [4] concluded that the first silicide-like reaction product is formed for a Gd coverage of about 2 ML and that this interface silicide-like product is correlated with GdSi.

By exploiting the energy dependence of the photoemission cross-section and the valence resonance at the crossing of the Gd 4d-4f threshold, Braicovich et al [5] investigated the valence photoelectron spectroscopy of Gd silicides $Gd_3Si_5$, GdSi and $Gd_5Si_3$ with photoemission spectroscopy in the phonon energy range 40.8-149 eV, where the 4f emission leaves a window of about 6 eV below $E_F$ totally available for observation of the valence states. Taking advantage of the photon energy dependence of the photoemission cross-sections, it is possible to understand the main aspects of the chemical bond and to recognise the most relevant orbital contribution to the density of occupied states, because the relative Gd and Si contributions to the spectra in this energy range between 40.8 and 149 eV change very strongly. The modification of the spectra versus photon energy, along with their stoichiometry dependence, shows the relevance of covalent mixed Gd5d-Si3sp states in the formation of the chemical bond [5]. However, these experimental results cannot distinguish the Si3s from the Si3p contribution, although it is necessary to account for a more extended basis (typically by including Si s-states) in more complete models.

## C THE VALENCE FLUCTUATION

A new type of ground state, the so-called mixed- or intermediate-valence state, has now been observed in RE silicide systems, of which ytterbium silicide probably is the best studied example. The fundamental origin of this state is that two configurations $4f^{n+1}V^2$ and $4f^nV^3$ of the rare earth atom (where V stands for the valence electrons), depending on the chemical surroundings, may be brought so close in energy that the ground state has to be described quantum mechanically as a superposition of these configurations. Photoemission has proved to be a valuable tool for the investigation of the electronic properties of mixed valence compounds. Especially accessible are mixed valence systems involving rare-earth metals. This is due to their unfilled and localised 4f orbital. The spectra so obtained show peaks originating from both divalent ($f^{n+1}$) and trivalent ($f^n$) initial configurations and the intensities of these peaks are often used to derive a value for the degree of valence mixing for the material in question. A photoemission spectrum of the 4f level always shows a multiplet structure characteristic of the initial occupation number. The multiplet structures from different initial occupancies are generally separated in binding energy, which facilitates the detection of a mixed valence behaviour. However, recent photoemission studies of several mixed-valence RE compounds have shown that the surface is divalent, in contrast to the mixed valence bulk. The preference for a divalent state at the surface can be understood from the decrease in cohesive energy due to the reduced coordination for a surface atom. This means that the surface core levels will shift to higher binding energies with respect to the bulk core level, whereby the divalent configuration will be stabilised. In order accurately to determine the bulk mean valence, the experiment must be performed with a resolution high enough to distinguish between the surface and bulk structures.

### C1 Samarium Valence Change at the Sm/Si Interface

Samarium is an excellent test system with which to study the interaction between chemisorbed metallic overlayers and different kinds of substrates because the 4f configuration is sensitive to the Sm valence state. Photoemission spectra exhibit only divalent Sm features at coverages below 2-3 Å and a mixture of trivalent and divalent Sm features at high coverages. This transition was accompanied by strong adatom-substrate chemical interactions as revealed by large core-level shifts.

Franciosi et al [6] have studied the samarium valence change at the Sm/Si(111) interface. Fäldt et al [7] and Onsgaard et al [8] have studied the valence change at the Sm/Si(100) interface. Peaks within 5 eV of $E_F$ in a series of valence band photoelectron energy distribution curves are attributed to the $4f^5$ final state multiplet produced by ionisation of $Sm^{+2}$; peaks at higher energies are attributed to the $4f^4$ final state multiplet of $Sm^{+3}$ [6]. For metal coverages below 3 Å, Sm atoms appear only in a divalent state, but a shift to lower binding energy ($0.2 \pm 0.1$ eV) for the divalent feature is observed as the coverage increases

(0.11 Å). When the coverage is increased above 3 Å, the $Sm^{+3}$ $4f^4$ final state multiplets emerge, and are rigidly shifted 1 eV to higher binding energy with respect to Sm metal [6].

## C2 Ytterbium Valence Change at the Yb/Si Interface

Yb is divalent, and its complete $4f^{14}$ subshell lies at very shallow energies (~1.2 eV below $E_F$), overlapping the $(6s5d)^2$ delocalised states. Solid Yb gains energy by delocalising one atomic 4f electron, becoming trivalent. As a consequence, the valence of Yb becomes unstable for relatively small perturbations of the local potential, which may occur under pressure (fcc 2+ hcp 3+ transition) as in most intermetallics where Yb shows mixed valency with a temperature-dependent mean valency.

### C2.1 Substrate-dependent valency of Yb chemisorbed onto Si surfaces

Rossi et al [9] have studied the substrate-dependent valence of Yb chemisorbed onto Si(111)77, Si(100)21 and a-Si surfaces using X-ray absorption resonance spectroscopy (XARS), since this technique is a very sensitive probe of valence on rare-earth $L_2$ and $L_3$ edges. On the basis of XARS measurements for 1 ML Yb absorbed on different Si surfaces, it was shown that the Yb valence is 2 for Si(111)77, 2.09 for Si(100)21 and 2.15 for amorphised Si. This means that at the early stage of the Yb/Si interface formation the Yb-Si chemisorption on different Si substrates induces mixed valence in the Yb atoms chemisorbed on the a-Si surface sites and a lower mixed valence in those on the Si(100) sites, but leaves purely divalent Yb ions on the Si(111) surface.

### C2.2 Mixed valence in $Yb_3Si_5$, YbSi and $Yb_5Si_3$

In the whole Yb-Si phase diagram, there are three compounds. Two compounds ($Yb_3Si_5$ with $Th_3Pd_5$ structure and YbSi with CrB structure) have all equivalent bulk Yb sites while $Yb_5Si_3$ ($Mn_5Si_3$ structure) has two inequivalent bulk Yb sites. We note that $Yb_3Si_5$ is the only stoichiometric compound on the Si-rich side of the phase diagram. The other compounds are defect disilicides ($YbSi_{2-x}$) with a statistical distribution of vacant sites in the $AlB_2$ structure [10]. Thus the spectroscopy of the mixed valence in these compounds demonstrates the effect of local environment on the valence fluctuations.

Braicovich et al [11] and Abbati and co-workers [12-14] have studied the mixed valence in $Yb_3Si_5$, YbSi and $Yb_5Si_3$ with photoemission spectroscopy. The determination of the Yb valence requires a detailed knowledge of the shape of the $4f^{2+}$ emission in order to separate the bulk contribution from the emission due to the surface region. In fact the latter has a different valence due to the lower atomic coordination. For Yb metal, the decomposition can be done with just two components: the bulk (B) and surface (S) contributions. However, in $Yb_3Si_5$ and YbSi the decomposition cannot be carried out with just two components; three components must be used. The other components are due to the emission from the surface region (subsurface) [14]. This gives the bulk valence of $2.52 \pm 0.02$ for $Yb_3Si_5$ and $2.72 \pm 0.02$ for YbSi. In both cases the surface and subsurface 4f emission is well below $E_F$ and thus the surface and subsurface Yb is divalent. In $Yb_5Si_3$ there are two Yb families, one containing 4 and the other 6 atoms over each 10 Yb atoms. The inequivalent sites produce the increase of the total width $4f^{2+}$ spectrum and a correspondence of the peak position with the other silicides. This makes the decomposition of the spectrum very difficult. Abbati et al [14] have thus followed the opposite approach, that is the synthesis of the spectrum by using as much of the experimental information from the other two silicides as possible in order to eliminate any ambiguity. In $Yb_5Si_3$, which has inequivalent Yb sites, they confirmed that one Yb species is divalent and the other has mixed valence. In the surface region, Yb is divalent in all compounds.

### C2.3 The effect of growth conditions

Braicovich et al [11], Abbati et al [12] and Wigren et al [15] have studied the effect of growth conditions (different coverages and temperatures) on the Yb atomic valence during interface formation. From the photoemission spectra from Yb 4f states as a function of the coverage (= 2, 1, 0.8, 0.5, 0.2 ML) in the Si(111)/Yb interfaces at room temperature and after treatment at 300 and 450°C [12], it can be shown that the (3+) compound is very weak in the interface at RT where it appears only at the higher coverages, while the thermal treatment strongly enhances the (3+) component. The appearance of the (3+) component is due to the fact that a fraction of the Yb atoms show mixed valence. The relation between the intensity of the emission from the (3+) configuration and from the Yb in the (2+) configuration for Yb in a mixed valence situation is strictly linear, corresponding to the valence v = 2.37. This means that the mixed valence has the same value in all cases, independent of the coverage and of the annealing temperature [12]. This shows that the growth is chemically driven by the existence of a preferred interface site.

## REFERENCES


[1] Ye Ling, Xie Xi-de [ *Acta Phys. Sin. (China)* vol.37 (1988) p.1593-9 ]

[2] Wang Lei, Ye Ling [ *Chinese J. Semicond. (China)* vol.10 (1989) p.653-8 ]

[3] P. Wetzel, L. Hoderbache, C. Pirri, L.C. Peruchotti, D. Bolmont, G. Gewinner [ *Phys. Rev. B (USA)* vol.43 (1991) p.6620-6 ]

[4] E. Puppin et al [ *J. Vac. Sci. Technol. B (USA)* vol.5 (1987) p.1083-6 ]

[5] L. Braicovich, E. Puppin, I. Lindau, A. Iandelli, G.L. Olcese, A. Palenzona [ *Phys. Rev. B (USA)* vol.41 (1990) p.3123-7 ]

[6] A. Franciosi, J.H. Weaver, P. Perfetti, A.D. Katnani, G. Margaritondo [ *Solid State Commun. (USA)* vol.47 (1983) p.427-30 ]

[7] A. Fäldt, H.P. Myers [ *Phys. Rev. B (USA)* vol.33 (1986) p.1424-6 ]

[8] J. Onsgaard, J. Ghijsen, R.L. Johnson, F. Orskov, I. Chorkendorff, F. Grey [ *J. Electron Spectrosc. Relat. Phenom. (Netherlands)* vol.52 (1990) p.67 ]

[9] G. Rossi, D. Chanderris, P. Ronbin, J. Lecante [ *Phys. Rev. B (USA)* vol.33 (1986) p.2926-9 ]

[10] A. Iandelli, A. Palenzona, G.L. Olcese [ *J. Less-Common Met. (Switzerland)* vol.64 (1979) p.213-20 ]

[11] L. Braicovich, I. Abbati, C. Carbone, J. Nogami, I. Lindau [ *Surf. Sci. (Netherlands)* vol.168 (1986) p.193-203 ]

[12] I. Abbati et al [ *Solid State Commun. (USA)* vol.60 (1986) p.595-8 ]

[13] I. Abbati et al [ *Phys. Rev. B (USA)* vol.34 (1986) p.4150-4 ]

[14] I. Abbati et al [ *Solid State Commun. (USA)* vol.62 (1987) p.35-9 ]

[15] C. Wigren, J.N. Andersen, R. Nyholm, U.O. Karlssen [ *J. Vac. Sci. Technol. A (USA)* vol.9 (1991) p.1942-5 ]

CHAPTER 5

# ELECTRICAL TRANSPORT

5.1 Electrical transport in metallic TM silicides
5.2 Electrical transport in semiconducting TM silicides
5.3 Electrical transport in RE silicides

# 5.1 Electrical transport in metallic TM silicides

U. Gottlieb, F. Nava, M. Affronte, O. Laborde and R. Madar

January 1995

## A INTRODUCTION

From an electrical viewpoint, silicides can be classified on the basis of resistivity behaviour. The resistivity of metals ranges between μΩ cm and mΩ cm at room temperature. Moreover, their resistivity decreases as the temperature decreases. Most of the silicides are metals. The silicides are classified into metallic and semiconducting silicides in TABLE 1. Some silicides are superconductors and their transition temperatures are listed in TABLE 2. Semiconducting behaviour is observed in silicides formed with metals of the VIA, VIIA and VIIIA group of the periodic table [1-3] (see Datareview 5.2).

TABLE 1 Electrical classification of transition metal silicides (**bold** is used for metals; [ ] for semiconductors; ( ) for silicides non-characterised; underlined for superconductors; for transition temperatures refer to TABLE 2).

| IIIA | IVA | VA | VIA | VIIA | VIIA | | | IB |
|---|---|---|---|---|---|---|---|---|
| | **$TiSi_2$** | **$V_3Si$**<br>**$V_3Si_2$**<br>**$V_5Si_3$**<br>**$VSi_2$** | **$Cr_3Si$**<br>**$Cr_5Si_3$**<br>[$CrSi_2$] | [$MnSi_{1.7}$] | **α-$FeSi_2$**<br>[β-$FeSi_2$] | **$Co_2Si$**<br>**CoSi**<br>**$CoSi_2$** | **$Ni_2Si$**<br>**NiSi**<br>**$NiSi_2$** | **$Cu_3Si$** |
| **$YSi_2$** | ($ZrSi_2$) | **$Nb_3Si$**<br>**$NbSi_2$** | **$Mo_3Si$**<br>**$MoSi_2$** | | [$Ru_2Si_3$] | ($RhSi_2$) | **$Pd_2Si$** | |
| | **$HfSi_2$** | **$TaSi_2$** | **$WSi_2$** | [$ReSi_{1.75}$] | ($OsSi_2$) | (IrSi)<br>[$IrSi_3$] | **PtSi** | |

TABLE 2 Transition temperatures of superconducting silicides.

| Silicide | $T_c$ (K) |
|---|---|
| $V_3Si$ | ~17.1 |
| $Nb_3Si$ | ~19 (?) |
| $Mo_3Si$ | ~1.7 |
| $CoSi_2$ | ~1 |
| $NbSi_2$ | 0.130 |
| $TaSi_2$ | 0.353 |

The metallic silicides present some peculiarities especially at high temperatures: in some silicides, the resistivity increases linearly with temperature, in others it increases more than linearly with temperature, the so-called superlinear behaviour, and in others the resistivity increases less than linearly, the so-called sublinear behaviour.

Some silicides exhibit magneto-effects [4] and some are superconductors with transition temperatures which vary by orders of magnitude [5,6]. The non-cubic single crystal silicides, characterised by a metallic behaviour, show anisotropy in resistivity. This anisotropy is particularly large for the silicides which crystallise in the hexagonal structure [7,8].

## B RESISTIVITY OF METALLIC SILICIDES

We refer to Datareview 1.1 for the crystal structures of the transition metal silicides of TABLE 1. One can distinguish two families of crystal structures among these materials. The silicides of the near noble metals (the VIIIB columns) crystallise generally in high symmetry structures such as the cubic structures of $CoSi_2$ and $NiSi_2$. On the other hand, the disilicides of the refractory metals of the IVB, VB and VIB groups form another family of crystal structures characterised by hexagonal $MSi_2$-type layers which form the structures by different stacking sequences. The orthorhombic structure of $TiSi_2$ is formed by an ABCDA... sequence, the hexagonal structure of $VSi_2$, $NbSi_2$, $TaSi_2$ and $CrSi_2$ is formed by an ABCA... sequence, while the tetragonal structure of $MoSi_2$ and $WSi_2$ is formed by an ABA... sequence.

Experimentally, it has been shown that the resistivity $\rho(T)$ for a great variety of metals and alloys can be approximated by Matthiessen's rule [9,10]:

$$\rho(T) = \rho_0 + \rho_{ep}(T) + \rho_{ee}(T) + \mathrm{DMR}(T) \qquad (1)$$

where $\rho_0$ is the electron-impurity component (which is independent of temperature), $\rho_{ep}$ is the contribution due to electron-phonon interactions, and $\rho_{ee}$ is the contribution due to electron-electron interactions. DMR(T), the deviation from Matthiessen's rule, takes into account all deviations from simple additivity. Several sources are possible for DMR: changes in the Debye temperature $\theta_D$, the difference in the angular dependence of the distribution of electrons in k space for electron-phonon and electron-impurity scattering at temperatures below $\theta_D$ and a possible saturation behaviour of the resistivity.

The resistivity $\rho(T)$ of metallic silicides generally follows a classical metal behaviour, i.e. it is composed of the sum of a temperature independent term $\rho_0$ and a temperature dependent term $\rho_i(T)$. FIGURE 1 shows some examples of the resistivity temperature dependence observed in thin films of metallic silicides. The resistivity of $TiSi_2$ and $CoSi_2$ increases almost linearly with temperature, following a classical metal behaviour [11]. For other metallic silicides a deviation from this linearity is observed at high temperatures: the $\rho(T)$-curve shows an upwards curvature (positive $d^2\rho/dT^2$) for $WSi_2$ and $MoSi_2$ and a downwards curvature (negative $d^2\rho/dT^2$) for $TaSi_2$. $NbSi_2$ and $VSi_2$ also show this downwards curvature [7]. This variety of behaviour, due to different scattering mechanisms in the silicides, will briefly be discussed in this section.

The low temperature resistivity contribution $\rho_0$ comes from charge carrier scattering by crystal defects and impurities. The amounts of defects and impurities in a crystal generally do not depend on temperature. $\rho_0$ is also called the residual resistivity and increases with increasing concentration of impurities and defects. In contrast, the temperature dependent contribution $\rho_i(T)$ depends only on the material characteristics. Hence, a good measure of the sample quality is the so-called residual resistance ratio RRR = $\rho$(300 K)/$\rho$(4.2 K), under the condition that $\rho_i$(300 K) is greater than $\rho_0$.

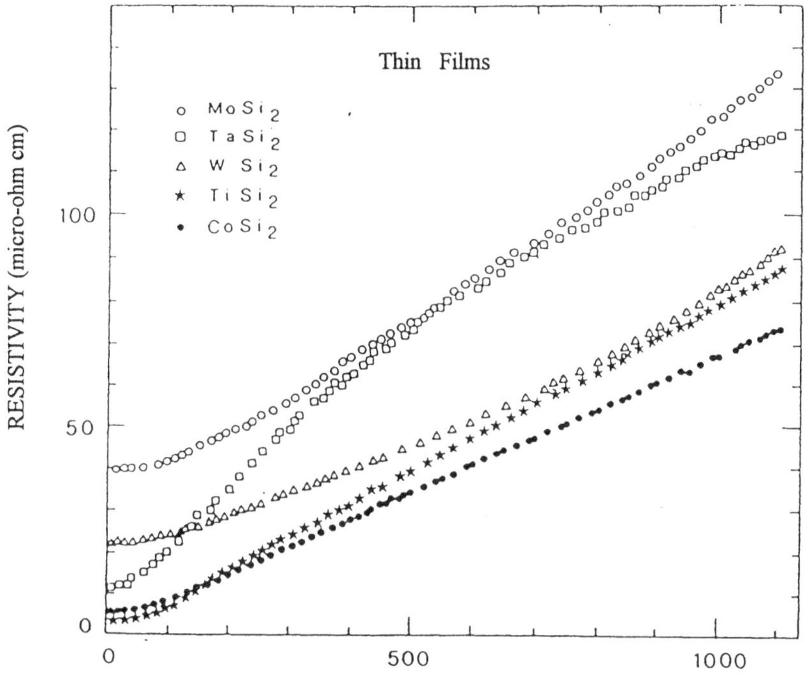


FIGURE 1 Temperature dependence of the resistivity of $MoSi_2$, $TaSi_2$, $WSi_2$, $TiSi_2$ and $CoSi_2$ thin films.

TABLE 3 gives the resistivity measured at room temperature and at 4.2 K of several metallic silicides. For materials with the same composition, those having the highest $\rho$(300 K)/$\rho$(4.2 K) ratio probably have the lowest concentration of impurities or defects. Note also that the RRR values are much higher for single crystals than for thin films. This is due to the presence of defects such as grain boundaries in thin films.

TABLE 3 Main features of the resistivity measured in metallic silicides. Note that the ratio $\rho$(293 K)/$\rho$(4.2 K) is much larger for single crystals (s.c.) than for thin films due to the presence of defects such as grain boundaries in thin films.

| Compound | Type of sample and orientation | $\rho$(293 K) ($\mu\Omega$ cm) | $\rho$(4.2 K) ($\mu\Omega$ cm) | $\rho$(293 K)/ $\rho$(4.2 K) | Ref |
|---|---|---|---|---|---|
| $YSi_2$ | Thin film | 50 | 18 | 2.8 | [47] |
| $TiSi_2$ | s.c. [100]<br>s.c. [010]<br>s.c. [001] | 11.0<br>10.0<br>12.0 | 0.177<br>0.196<br>0.138 | 62<br>51<br>87 | |
| $HfSi_2$ | Thin film | 70 | 38 | 1.8 | [16] |
| $V_5Si_3$ | Thin film | 92 | 15 | 6.1 | [17] |
| $V_3Si$ | Thin film | 65 | 6.2 | 10.5 | |
| $VSi_2$ | s.c. [0001]<br>s.c. [$10\bar{1}0$] | 34.13<br>59.64 | 1.61<br>2.80 | 21.2<br>21.3 | |
| $NbSi_2$ | s.c. [0001]<br>s.c. [$10\bar{1}0$] | 22.01<br>39.15 | 0.085<br>0.174 | 259<br>225 | |

TABLE 3 continued

| Compound | Type of sample and orientation | ρ(293 K) (μΩ cm) | ρ(4.2 K) (μΩ cm) | ρ(293 K)/ ρ(4.2 K) | Ref |
|---|---|---|---|---|---|
| $TaSi_2$ | s.c. [0001]<br>s.c. [10$\bar{1}$0] | 20.2<br>40.1 | 0.05<br>0.04 | 404<br>1003 | |
| $Cr_3Si$ | Thin film | 32 | 10.9 | 2.9 | [17] |
| $Cr_5Si_3$ | Thin film | 80 | 25.9 | 3.1 | |
| $MoSi_2$ | s.c. [001]<br>s.c. [110] | 12.6<br>16.9 | 0.0092<br>0.0096 | 1380<br>1760 | |
| $WSi_2$ | s.c. [001]<br>s.c. [110] | 12.9<br>11.9 | 0.536<br>0.064 | 24<br>186 | |
| $CoSi_2$ | Thin film | 15 | 3 | 5 | [48] |
| CoSi | Thin film | 147 | | | [49] |
| $Co_2Si$ | Thin film | 110 | | | |
| $Co_4Si$ (?) | Thin film | 90 | 71 | 1.3 | [50] |
| $Ni_2Si$ | Thin film | 24 | | | [51] |
| NiSi | Thin film | 10.5 | 2 | 5.2 | |
| $NiSi_2$ | Thin film | 34 | 19.8 | 1.7 | [3] |
| $Pd_2Si$ | s.c. I ∥ c-<br>s.c. I⊥c- | 13.73<br>18.91 | 0.066<br>0.097 | 224<br>209 | |
| $Cu_3Si$ | Thin film | 53 | 38 | 1.4 | [52] |

The temperature dependent part $\rho_i(T)$ of the resistivity generally comes from scattering of the charge carriers by lattice vibrations. For metallic silicides with relatively simple electronic structures, $\rho_i(T)$ is found to vary approximately linearly with T at high temperatures (T > 300 K) [8,12,13] and gradually varies with a higher T-power law at low temperatures. The simplest analytical expression for $\rho_i(T)$ due to electron-phonon scattering is given by the Bloch-Grüneisen model [14,15]

$$\rho_{ep}(T) = \rho' T \left(\frac{T}{\theta_D}\right)^4 \int_0^{\frac{\theta_D}{T}} \frac{x^5 dx}{(e^x-1)(1-e^{-x})} \qquad (2)$$

where ρ' is a constant and $\theta_D$ the Debye temperature. This expression can be obtained by the solution of the Boltzmann equation [15] with the following assumptions:

- spherical Fermi surface,
- no Umklapp processes,
- phonon system in thermal equilibrium,
- electron-phonon coupling only with longitudinal phonons,
- Debye phonon spectrum.

This model predicts a linear variation of $\rho_i(T)$ with T at high temperature and a $T^5$ variation for $\theta_D/T \geq 20$. In FIGURE 2 we report the ρ(T) curves measured in $TiSi_2$ and $CoSi_2$ single crystals: above ~100 K the ρ(T) follows an almost perfect linear behaviour.

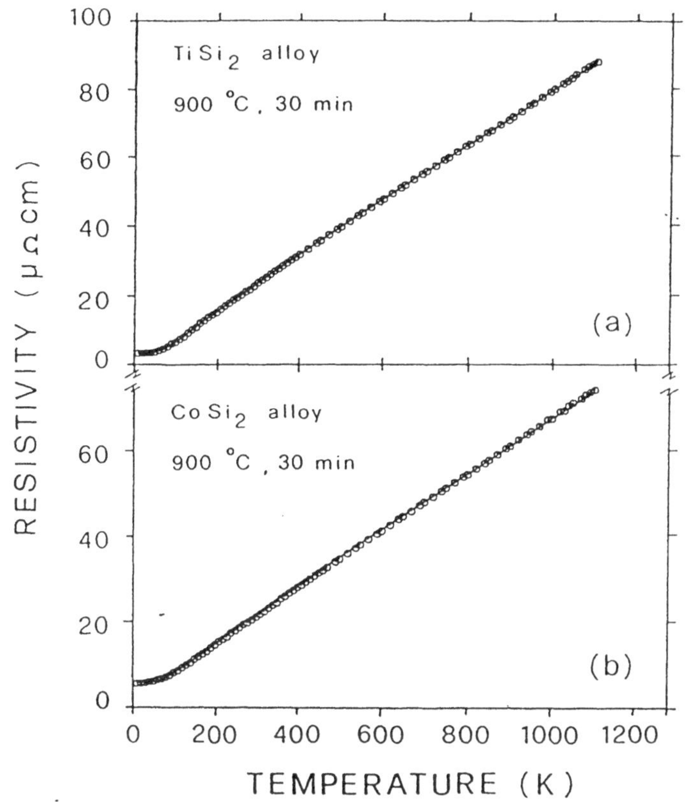


FIGURE 2 Resistivity of $TiSi_2$ and $CoSi_2$ thin films. Note the almost linear T dependence above about 100 K.

Another effect, typical of A15 superconductors, is found for the high-temperature resistivity of the majority of the metallic silicides [12,16-18]. When electron-phonon scattering is the dominant temperature-dependent scattering mechanism, the temperature dependence of resistivity exhibits negative curvature ($d^2\rho(T)/dT^2 < 0$) and it shows a tendency to saturate. The materials showing saturation generally exhibit strong deviations from Matthiessen's rule. The resistivity saturation is pronounced in systems where the room temperature resistivity is large (50-100 μΩ cm) [19]. In FIGURE 3 we show the experimental resistivity data of $V_3Si$, $V_5Si_3$ and $VSi_2$ thin films [16]. This behaviour of resistivity saturation has been interpreted in terms of the Ioffe-Regel criterion [20]: the electron mean free path cannot be shorter than the interatomic distance. This leads to the parallel resistor formula [20],

$$\frac{1}{\rho(T)} = \frac{1}{\rho_{ideal}(T)} + \frac{1}{\rho_{sat}} \tag{3}$$

which is quite successful in interpreting the resistivity data of most of the metallic silicides.

$\rho_{ideal}(T)$ is given by:

$$\rho_{ideal}(T) = \rho_0 + \rho_{ep}(T) \tag{4}$$

and $\rho_{sat}$ (saturation value of the resistivity) is characteristic for a given system not depending on disorder (i.e. on the residual resistance $\rho_0$) and has to be measured. Note that the parallel resistor formula disagrees with Matthiessen's rule.

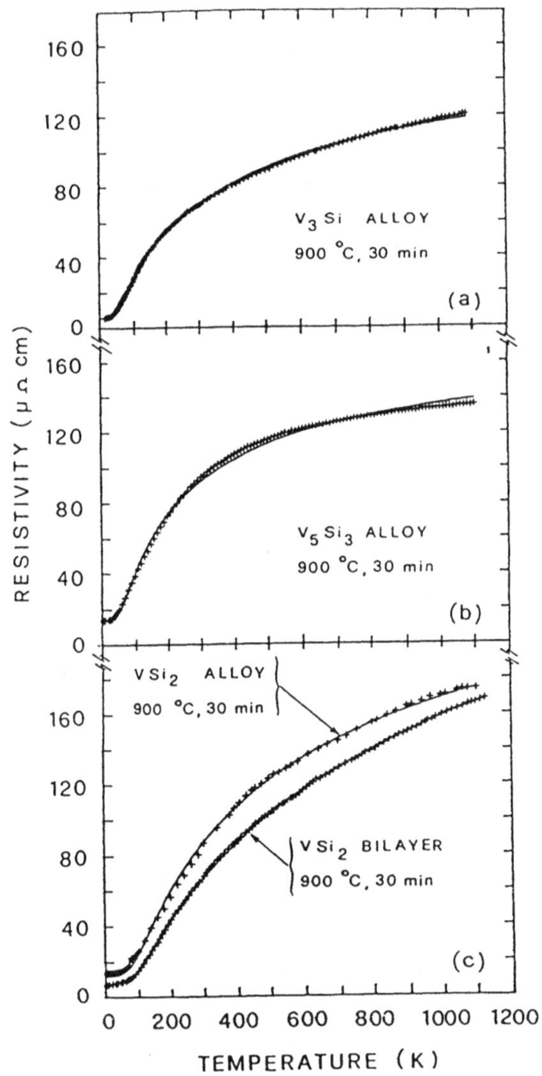


FIGURE 3 Comparison between the experimental resistivity data (crosses) and the parallel resistor formula (solid line) with the parameters listed in TABLE 4 for (a) $V_3Si$ alloy, (b) $V_5Si_3$ alloy and (c) $VSi_2$ alloy and bilayer.

In TABLE 4, we have listed the fitting parameters $\rho_0$, $\rho'$, $\theta_D$ and $\rho_{sat}$ obtained by fitting the experimental data measured in several silicides with the parallel resistor formula. $\rho_0$ is the temperature independent low temperature resistivity, $\rho'$ is a proportionality factor, $\theta_D$ is the Debye temperature and $\rho_{sat}$ is the saturation value of the resistivity.

TABLE 4 Fitting parameters evaluated from resistivity data of metallic silicides (s.c. = single crystal). Fitting curves given in the text.

| Compound | | $\rho_0$ (μΩ cm) | $\rho'$ (μΩ cm) | $\theta_D$ (K) | $\rho_{sat}$ (μΩ cm) |
|---|---|---|---|---|---|
| $TiSi_2$ s.c. | [010] | 0.020 | 0.040 | 520 | |
| $VSi_2$ alloy | | 14.84 | 0.474 | 550 | 289 |
| $VSi_2$ bilayer | | 7.92 | 0.320 | 555 | 301 |

TABLE 4 continued

| Compound | | $\rho_0$ ($\mu\Omega$ cm) | $\rho'$ ($\mu\Omega$ cm) | $\theta_D$ (K) | $\rho_{sat}$ ($\mu\Omega$ cm) |
|---|---|---|---|---|---|
| $V_5Si_3$ alloy | | 15.36 | 0.677 | 345 | 169 |
| $V_3Si$ alloy | | 6.19 | 0.424 | 299 | 158 |
| $NbSi_2$ alloy | | 17.56 | 0.297 | 545 | 345 |
| $TaSi_2$ | [10$\bar{1}$0]<br>[0001] | 0.039<br>0.048 | 0.170<br>0.074 | 375<br>366 | 451<br>400 |
| $Cr_3Si$ bilayer | | 10.88 | 0.078 | 439 | 240 |
| $Cr_5Si_3$ bilayer | | 25.87 | 0.554 | 502 | 238 |
| $MoSi_2$ s.c. | [001]<br>[110] | 10.0<br>10.6 | 0.059<br>0.063 | 600<br>580 | |
| $WSi_2$ s.c. | [001]<br>[110] | 0.057<br>0.066 | 0.056<br>0.052 | 613<br>568 | |
| NiSi bilayer | | 1.34 | 0.04 | 408 | 339 |
| $NiSi_2$ bilayer | | 23.1 | 0.044 | 260 | 406 |
| $Pd_2Si$ s.c. | I \|\| a<br>I \|\| c | 0.01<br>0.092 | 0.072<br>0.044 | 168<br>200 | 1604<br>936 |

Concerning the values given in TABLE 4, a few points must be emphasised. The Debye temperatures found for $NbSi_2$ and $VSi_2$ thin films are equal within the experimental error ($\Delta\theta_D \sim 10$ K). The mean values are 507 K and 485 K for $VSi_2$ and $NbSi_2$, respectively. The Debye temperatures of $VSi_2$ and $NbSi_2$ determined from low temperature specific heat measurements are quite different: $\theta_D = 665$ K and 579 K for $VSi_2$ and $NbSi_2$ respectively [6]. There is, in general, good agreement between the fitting values of $\theta_D$ obtained in thin film and single crystal samples [3,7,21,22]. A higher value of $\rho'$ is found for those silicides which already show a deviation from linearity in the resistivity curves at room temperature. Finally, we note that $\rho_{sat}$ varies within a wide range of values.

In some silicides one can observe that $\rho(T)$ increases more rapidly than T ($d^2\rho/dT^2 > 0$) in the high temperature region [23]. This may arise from several causes, which are in competition with those mentioned above. The most plausible explanation of the positive deviation from linearity observed in silicides seems to be related to the variation of the Debye temperature due to the effect of thermal expansion [24]. A second mechanism [25,26] could be that the structure of silicides is such that two or more bands overlap near the Fermi energy and that the only function which exhibits a rapid energy dependence near the Fermi energy is the electronic density of states. The temperature dependence of the resistivity may actually show an upward curvature if the Fermi energy falls near a minimum of the density of states. At high temperatures an increasing superlinear behaviour of $\rho(T)$ is observed for $WSi_2$ in single crystal samples [8], for both the [001] and [110] crystallographic directions as can be seen in FIGURE 4, and in polycrystalline films [12]. Similar results have been found for $MoSi_2$ [8], which is isoelectronic with $WSi_2$.

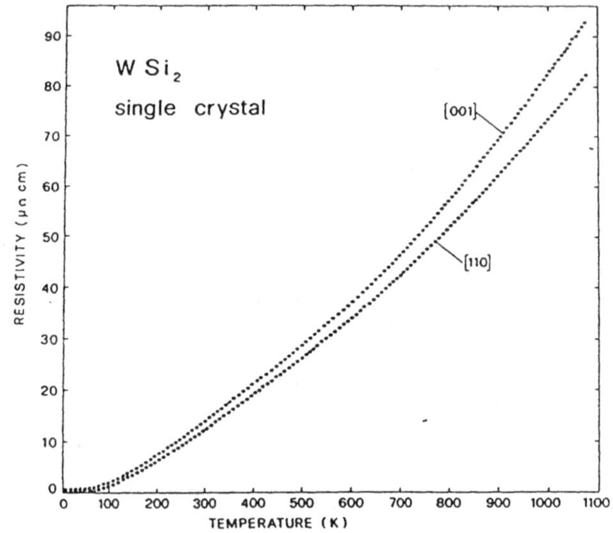


FIGURE 4 Resistivity of $WSi_2$ single crystals for two different current flow directions. Note the anisotropy and the evident superlinear behaviour.

Low-temperature resistivity measurements can be useful not only to clarify the effects of defects and impurities but also to determine the dominant contribution to the total resistivity by best-fit procedures. At temperatures much lower than the Debye temperature, the Bloch-Grüneisen theory predicts $\rho_i(T) = \rho(T) - \rho_0 \propto T^5$. On the other hand, Wilson [14,27] predicts a temperature range where $\rho_i(T) \propto T^3$ if a particular type of electron-phonon scattering dominates. A combination of these two extremes presumably applies and the low temperature resistivity data are often fitted by a curve: $\rho(T) = \rho_0 + cT^z$ where z is between 3 and 5. A value of 4.8 can be attributed to z for T < 20 K for nickel silicide thin films, while at higher temperatures z is almost equal to 1 in agreement with the Bloch-Grüneisen theory. For $TaSi_2$ $[10\bar{1}0]$ indications are that z is equal to 5 for T < 28 K, then continuously decreases with increasing T. For T > 185 K z becomes less than unity, according to the marked departure from linearity, and quasi-saturation is observed in ρ(T) at high temperature [8]. From the analysis performed on the low-temperature resistivity data of single crystal $MoSi_2$ [110], it can be inferred that z is equal to 3.7 in the temperature range 15 < T < 40 K. A similar result was obtained for single crystal $WSi_2$ [001] indicating that $MoSi_2$ and $WSi_2$, in both the [001] and [110] crystallographic directions, behave like a transition metal. For $VSi_2$ and $NbSi_2$ [28] the thermal variation of ρ(T) is well accounted for by a $T^5$ law at low temperatures.

Generally, the resistivity of a crystal is a tensor which, except for cubic lattices, has different elements. All single crystals of non-cubic silicides actually show anisotropy in resistivity. In FIGURES 5 and 6 we report the resistivity measured in different directions on $TaSi_2$ and $VSi_2$ and $NbSi_2$ single crystals, respectively. For silicides with hexagonal crystal structures like $Pd_2Si$, $VSi_2$, $NbSi_2$ and $TaSi_2$, the ratio between ρ measured along two different directions is of the order of 2. The same ratio is almost undetectable for the silicides with an orthorhombic crystal structure, such as $TiSi_2$ [8,29].

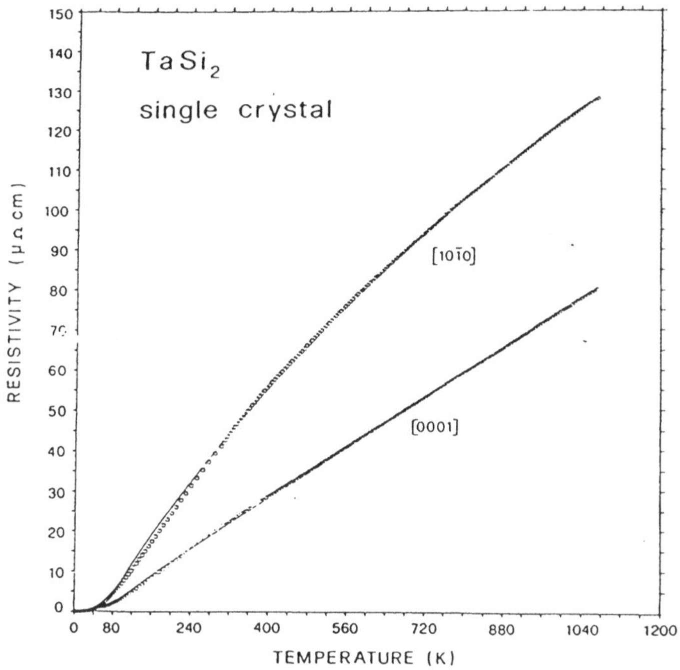


FIGURE 5 Temperature dependence of the resistivity of $TaSi_2$ single crystals. The anisotropy at room temperature is about 2. The solid lines correspond to a fit with parameters in TABLE 4.

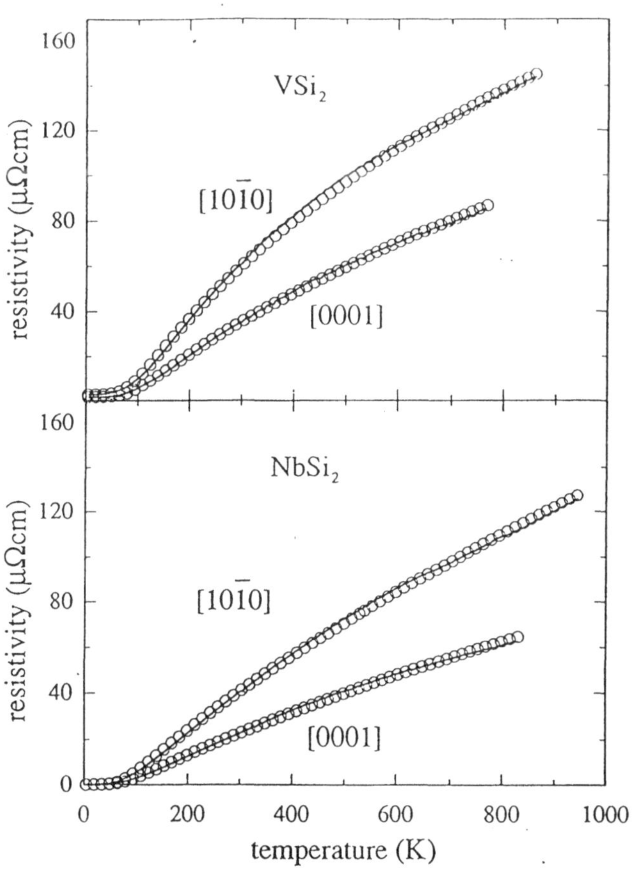


FIGURE 6 Temperature dependence of the resistivity of $VSi_2$ and $NbSi_2$ single crystals. The anisotropy at room temperature is about 2.

## C HALL EFFECT AND MAGNETORESISTANCE OF METALLIC SILICIDES

The Hall effect is the appearance of a transverse voltage $V_H$ when both a longitudinal current I and a perpendicular magnetic field of strength B are applied. The Hall voltage $V_H$ is proportional to I, B and 1/t (t = thickness of the sample), i.e.:

$$V_H = R_H \frac{IB}{t} \tag{5}$$

where $R_H$ is the so-called Hall coefficient. For the free electron case and for parabolic bands:

$$R_H = \pm \frac{1}{ne} \tag{6}$$

The sign of $R_H$ indicates the type of the charge carriers, and $R_H$ is often used to determine n, the number of carriers.

The high intrinsic resistivity of several silicides (see TABLE 3) raises the question as to whether it is due to low charge carrier density or to a low charge carrier mobility. Electronic transport properties of metallic compounds are known to be controlled by the properties of the Fermi surface [15]. Galvanomagnetic effects, namely Hall effect and magnetoresistance, can provide information about the Fermi surface and about the conduction mechanisms involved [11,30].

In FIGURE 7, we present the Hall coefficient measured on several refractory metal silicide thin films. The $R_H$ temperature dependence in FIGURE 7, observed also in many other

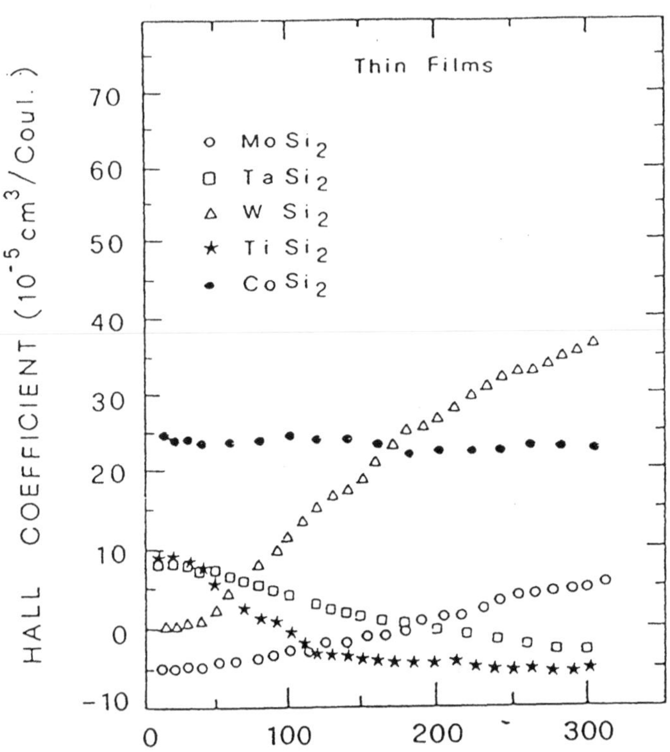


FIGURE 7 Temperature dependence of the Hall coefficient of $MoSi_2$, $TaSi_2$, $WSi_2$, $TiSi_2$ and $CoSi_2$ thin films.

silicides [16,17,31-34], suggests the presence of multi-carrier effects. In this case the use of Eqn (6) may overestimate the carrier density and underestimate the carrier mobility [14,15].

As can be seen in FIGURE 7, the Hall coefficient of $MoSi_2$, $TiSi_2$ and $TaSi_2$ changes sign. However, care is needed in interpreting the data especially for polycrystalline thin films. When a silicon excess is present in the thin film sample, the results for the Hall coefficient can arise from the silicon formed at the grain boundaries as well as from the silicide, and completely change the observed value of $R_H$.

Another possible effect that can lead to misinterpretation of the Hall effect data, when one tries to estimate the charge carrier density, is the possibility of the appearance of a low magnetic field - high field transition. This effect can be observed not only in pure metals, such as copper or aluminium [35] (especially in Al, $R_H$ changes sign with the applied magnetic field), but also in $NbSi_2$ [36,37] as can be seen in FIGURE 8, where we have plotted the Hall coefficient as a function of the reduced parameter B/ρ also compared to the magnetoresistance of this compound. For this case, the sample used was a high quality single crystal and one can see that $R_H$ varies from -3.5 to -4.5 x $10^{-10}$ $m^3/C$ in the low field region to -10.5 x $10^{-10}$ $m^3/C$ in the high field region. It is worth noting that these results are in good agreement with results obtained by other authors [38] on good quality thin films, but they observed only the low field limit.

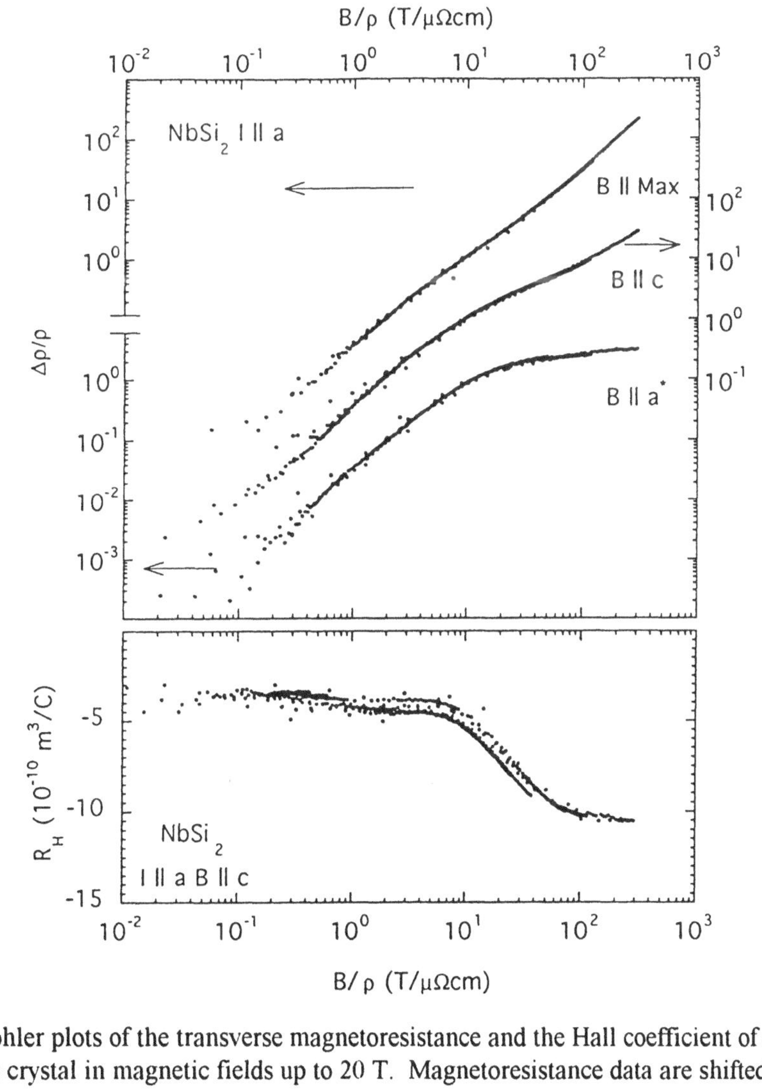


FIGURE 8 Kohler plots of the transverse magnetoresistance and the Hall coefficient of a $NbSi_2$ high quality single crystal in magnetic fields up to 20 T. Magnetoresistance data are shifted for clarity.

Transverse magnetoresistance measurements may also allow the evaluation of the charge carrier density (but not the sign of the dominant carrier type!) and the mobility. The theory of magnetoresistance is very complex (for a detailed review see for example [39]), but it appears that the magnetoresistance of nearly all non-magnetic metals varies only as a function of the reduced parameter $B/\rho$ [40], where $\rho$ is the zero field resistivity. As $\rho = m^*/(ne^2\tau)$ ($m^*$ = effective mass, n = charge carrier density, $\tau$ = scattering time) and $B = m'\omega_c/e$ ($m'$ = cyclotron effective mass, $\omega_c$ = cyclotron frequency) we have $B/\rho \sim \omega_c\tau$. $\omega_c\tau$ corresponds to the number of orbits that charge carriers can complete on the Fermi surface before being scattered. This is the so called Kohler rule. The low field magnetoresistance always varies as $\alpha(B/\rho)^2$. When a compensated metal is assumed, $\alpha$ is found to be proportional to $(1/n)^2$.

The transverse magnetoresistance coefficients of single crystal silicides have been measured as a function of temperature [7,12,41,42]. Some examples demonstrating the applicability of the Kohler rule are shown in FIGURE 9, where the transverse magnetoresistivity of $MoSi_2$, $Pd_2Si$, $TaSi_2$ and $VSi_2$ is plotted for different temperatures and for characteristic orientations of the magnetic field and current along the main crystallographic axes. Kohler's law is well satisfied for a given compound, and for given field current configuration over a wide range of temperature and magnetic field.

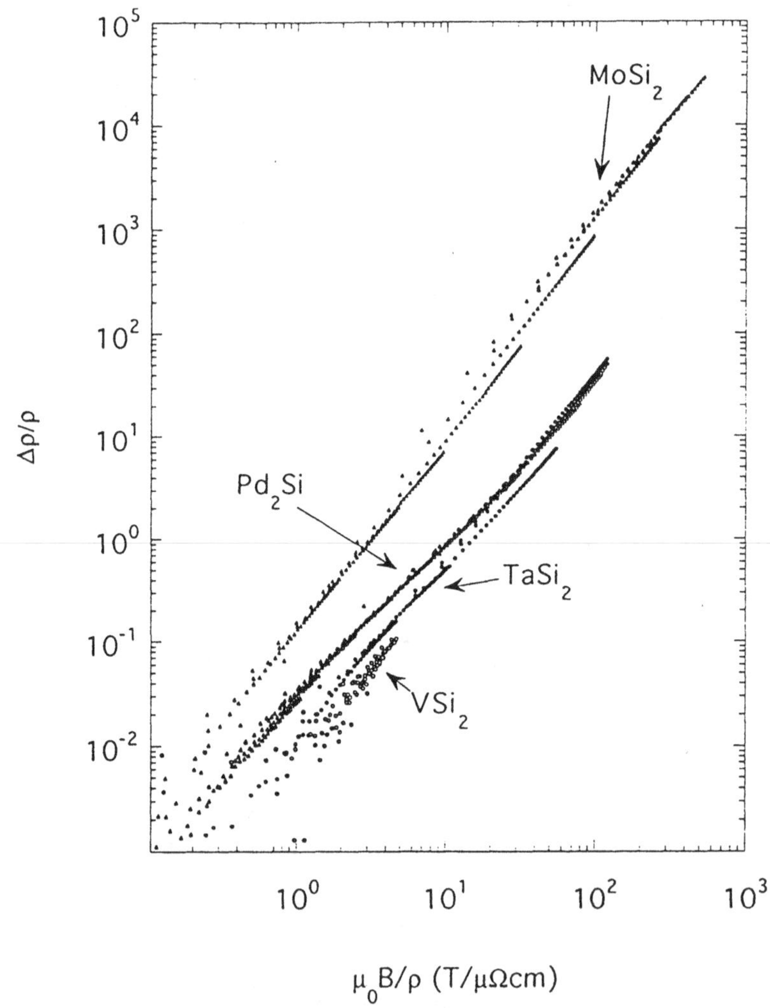


FIGURE 9 Kohler plots of the transverse magnetoresistance of $TaSi_2$ with I || c and B || a, $VSi_2$ with I || c and B || a, $MoSi_2$ with I || c and B || a* and $Pd_2Si$ with I || c and B || a*.

To a first approximation, these results suggest that the examined silicides behave like metals whose charge carriers are electrons and holes in equal number having the same characteristics, i.e. mobility and carrier density. Such a simple model for $MoSi_2$ is also supported by electronic density of state calculations, which indicate that two bands, one electron-like and one hole-like, cross the Fermi level [43]. The charge carrier densities, evaluated from Kohler's rule and the Hall effect, are quoted in TABLE 5 for $MoSi_2$, $Pd_2Si$, $TaSi_2$, $NbSi_2$ and $VSi_2$.

TABLE 5 Carrier concentration n evaluated by Hall effect measurements and magnetoresistance (s.c. = single crystal). Note that the formula $R_H = 1/en$ is generally used to evaluate n from the $R_H$: this can lead to an overestimation of n if two or more bands contribute to electron transport.

| Compound | | Carrier density by Hall measurements ($\times 10^{21}$ $cm^{-3}$) | Carrier density by magnetoresistance ($\times 10^{21}$ $cm^{-3}$) | Ref |
|---|---|---|---|---|
| $TiSi_2$ thin film | | -400 | | [44] |
| $HfSi_2$ thin film | | -4 | | |
| $VSi_2$ | s.c. [0001] | | 6.8 | |
| $NbSi_2$ | s.c. $[10\bar{1}0]$ | 5.9 | 5.3 | |
| $TaSi_2$ thin film | | -65 | | [31] |
| $TaSi_2$ | s.c. $[10\bar{1}0]$<br>s.c. [0001] | | 4.7<br>5.7 | |
| $MoSi_2$ | s.c. [001]<br>s.c. [110] | | 2.6<br>1.2 | |
| $WSi_2$ thin film | | +10 | | [53] |
| $Co_2Si$ thin film | | +2.0 | | |
| $CoSi_2$ thin film | | +33 | | |
| CoSi thin film | | -1.0 | | |
| $Ni_2Si$ thin film | | +90 | | |
| $NiSi_2$ thin film | | +20 | | |
| NiSi thin film | | -70 | | |
| $Pd_2Si$ thin film | | -4 | | [54] |
| $Pd_2Si$ | s.c. $[11\bar{2}0]$<br>s.c. [0001] | | 3.4<br>4.3 | |

In FIGURE 8, we have plotted the transverse magnetoresistance of $NbSi_2$ for different temperatures, for the same current direction but for different magnetic field directions relative to the crystallographic axes. Kohler's rule is very well obeyed for low values of $B/\rho$. In the high field region however, at the same values of $B/\rho$ where the Hall coefficient changes, important differences appear in the magnetoresistance depending on the magnetic field direction. This can be more clearly seen in FIGURE 10 where we have plotted the transverse magnetoresistance as a function of the field angle relative to the crystallographic axis at 4.2 K for various magnetic fields. The occurrence of the important structures in the magnetoresistance can be related to the topology of the Fermi surface of $NbSi_2$. The deep minimum when $B \parallel a^*$ indicates an open orbit along the hexagonal c direction of $NbSi_2$. Similar experiments have been carried out on $Pd_2Si$ [42].

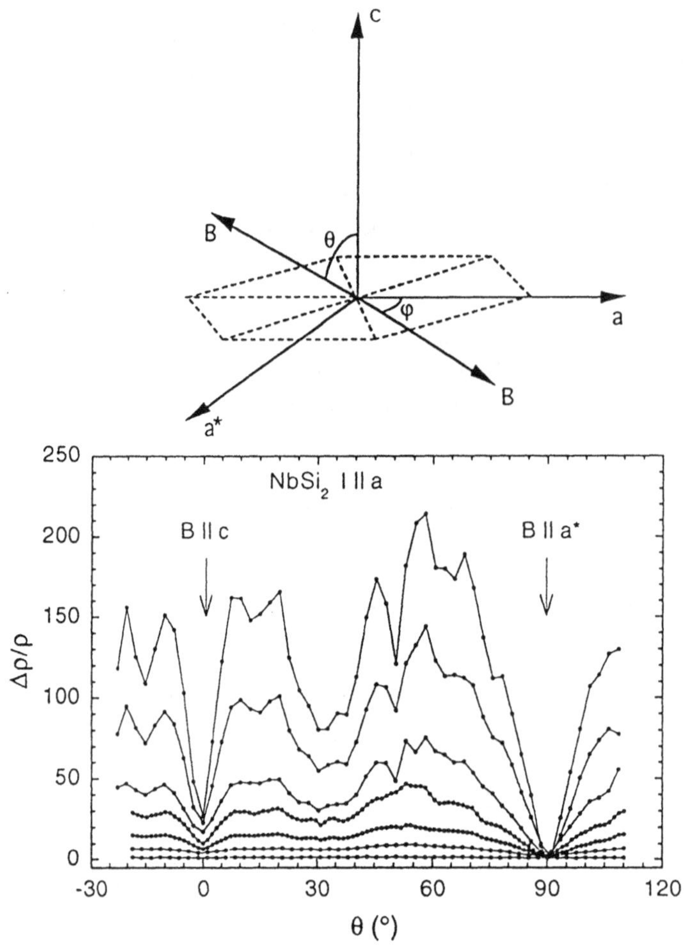


FIGURE 10 Transverse magnetoresistance of $NbSi_2$ at 4.2 K for I || a and B = 1, 3, 5, 7.5, 10, 15 and 20 T as a function of magnetic field direction relative to the crystal axis.

The magnetoresistance mobility μ can also be estimated from the magnetoresistance data in the framework of a single band of charge carriers which dominates the conduction process. For some silicides, like $TiSi_2$ [44], the correlation between the temperature dependence of the resistivity and the mobility going from liquid nitrogen to room temperature is very good. In contrast, it is not perfect for $MoSi_2$ [41] since the estimated mobility decreases by a factor of about 8 from 4.2 K to 80 K, while ρ increases only by a factor of about 5. Similar results [45] have been obtained for the isoelectronic compound $WSi_2$. For silicides like $MoSi_2$ [46], characterised by a complex Fermi surface, μ is a non-linear, density-weighted average for all the charge carriers participating in the conduction process. The high-mobility carriers give a stronger contribution. Therefore, poor correlation in the temperature dependence of ρ and μ may be ascribed to a variation of dominant carriers with temperature.

## D CONCLUSION

We have presented and reviewed the electrical properties, namely resistivity, Hall effect and magnetoresistance, of a large variety of metallic silicides measured on both thin films and single crystals. Most of the metallic silicides show a non-classical behaviour of the resistivity. A negative deviation from linearity has mostly been observed with the exception of $MoSi_2$ and $WSi_2$, which show a superlinear behaviour. Hall effect data on thin films of metallic silicides have to be interpreted very carefully as it was shown that these materials may show a transition between a low magnetic field and a high field regime in the galvanomagnetic effects. The magnetoresistance measured on single crystalline silicides almost follows Kohler's rule and also shows a transition between the low field and high field regions.

## ACKNOWLEDGEMENT

The work reported in this and the subsequent Datareview was financially supported by the European Community in the framework of the programme Capital and Mobility, contract number ERBCHRXTCT930318.

## REFERENCES

[1] F. Nava et al [ *Mater. Sci. Rep. (Netherlands)* vol.9 (1993) p.141 ]
[2] M.C. Bost, J.E. Mahan [ *J. Appl. Phys. (USA)* vol.58 (1985) p.2696 ]
[3] J.C. Hensel, R.T. Tung, J.M. Poate, F.C. Unterwald [ *Appl. Phys. Lett. (USA)* vol.44 (1984) p.913 ]
[4] S. Auffret et al [ *Physica B (Netherlands)* vol.173 (1991) p.265 ]
[5] V.A. Marchencko [ *Sov. Phys. - Solid State (USA)* vol.15 (1973) p.126 ]
[6] J.C. Lasjaunias, O. Laborde, U. Gottlieb, R. Madar, O. Thomas [ *J. Low Temp. Phys. (USA)* vol.92 (1993) p.335 ]
[7] U. Gottlieb, O. Laborde, O. Thomas, A. Rouault, R. Madar [ *Appl. Surf. Sci. (Netherlands)* vol.53 (1991) p.247 ]
[8] F. Nava et al [ *J. Appl. Phys. (USA)* vol.65 (1989) p.1584 ]
[9] A. Matthiessen, C. Vogt [ *Ann. Phys. (Germany)* vol.129 (1964) p.19 ]
[10] M.R. Cimberle, G. Bobel, C. Rizzuto [ *Adv. Phys. (UK)* vol.23 (1974) p.639 ]
[11] N.W. Ashcroft, N.D. Mermin [ *Solid State Physics* (Saunders College, Philadelphia, 1976) ]; F.J. Blatt [ *Physics of Electronic Conduction in Solid* (McGraw Hill, New York, 1968) ]
[12] F. Nava, K.N. Tu, E. Mazzega, M. Michelini, G. Queirolo [ *J. Appl. Phys. (USA)* vol.61 (1987) p.1085 ]
[13] O. Thomas, O. Laborde, R. Madar, J.P. Senateur [ *J. Less-Common Met. (Switzerland)* vol.136 (1987) p.175 ]
[14] A.H. Wilson [ *The Theory of Metals* 2nd edition (Cambridge Univ. Press, London and New York, 1956) ]
[15] J.M. Ziman [ *Electrons and Phonons* (Clarendon Press, Oxford, 1960) ]
[16] F. Nava, O. Bisi, K.N. Tu [ *Phys. Rev. B (USA)* vol.34 (1986) p.6143 ]
[17] E. Mazzega, M. Michelini, F. Nava [ *J. Phys. F, Met. Phys. (UK)* vol.17 (1987) p.1135 ]
[18] F. Nava, P. Psaras, H. Takay, K.N. Tu, S. Valeri, O. Bisi [ *J. Mater. Res. (USA)* vol.1 (1986) p.327 ]
[19] H.M. Milchberg, R.R. Freeman, S.C. Davey, R.M. Moore [ *Phys. Rev. Lett. (USA)* vol.61 (1988) p.2364 ]
[20] M. Gurvitch [ *Phys. Rev. B (USA)* vol.24 (1981) p.7404 ]
[21] J.C. Hensel, R.T. Tung, J.M. Poate, F.C. Unterwald, D.C. Jacobson [ *Mater. Res. Soc. Symp. Proc. (USA)* vol.25 (1984) p.575 ]
[22] J.W.C. de Vries, A.H. van Ommen [ *J. Appl. Phys. (USA)* vol.64 (1988) p.749 ]
[23] G. Grüneisen [ *Handb. der Phys.* vol.13 (1928) p.16 ]
[24] N.F. Mott, H. Jones [ *The Theory of the Properties of Metals and Alloys* (Dover, New York, 1958) ]
[25] H. Jones [ *Handb. der Phys.* vol.19 (1956) p.227 ]

[26] F.J. Blatt [ *J. Phys. Chem. Solids (UK)* vol.17 (1961) p.177 ]

[27] A.H. Wilson [ *Proc. R. Soc. Lond. A (UK)* vol.167 (1938) p.331 ]

[28] U. Gottlieb et al [ *Surf. Coat. Technol. (Switzerland)* vol.45 (1991) p.237 ]

[29] R. Marani, F. Nava, A. Rouault, R. Madar, J.P. Senateur [ *J. Phys., Condens. Matter (UK)* vol.1 (1989) p.5887 ]

[30] F.J. Blatt [ *Physics of Electronic Conduction in Solids* (McGraw-Hill, New York, 1968) ]

[31] M.T. Huang, T.L. Martin, V. Malhotra, J.E. Mahan [ *J. Vac. Sci. Technol. B (USA)* vol.3 (1985) p.836 ]

[32] R.E. Weinmeister, J. Mahan [ *Appl. Phys. Lett. (USA)* vol.39 (1981) p.977 ]

[33] A.H. van Ommen, A.H. Reader, J.W.C. de Vries [ *J. Appl. Phys. (USA)* vol.64 (1988) p.3754 ]

[34] P.H. Woerlee, P.M.Th.M. van Attekum, A.A.M. Hoeden, G.A.M. Hurk, P.A.M. Wolters [ *Appl. Phys. Lett. (USA)* vol.44 (1984) p.876 ]

[35] C.M. Hurd [ *Hall Effect in Metals and Alloys* (Plenum Press, New York, 1972) ]

[36] U. Gottlieb [ PhD Thesis, Institut National Polytechnique de Grenoble, France (1994) ]

[37] U. Gottlieb, O. Laborde, R. Madar [ *Mater. Res. Soc. Symp. Proc. (USA)* vol.320 (1994) p.473 ]

[38] K. Pomoni, Ch. Krontiras, J. Salmi [ *J. Phys. D, Appl. Phys. (UK)* vol.23 (1990) p.354 ]

[39] A.B. Pippard [ *Magnetoresistance in Metals* (Cambridge University Press, London, 1989) ]

[40] E. Fawcett [ *Adv. Phys. (UK)* vol.13 (1964) p.139 ]

[41] O. Laborde, O. Thomas, J.P. Senateur, R. Madar [ *J. Phys. F, Met. Phys. (UK)* vol.16 (1986) p.1745 ]

[42] O. Laborde, U. Gottlieb, R. Madar [ *J. Low Temp. Phys. (USA)* vol.95 (1994) p.835 ]

[43] B.K. Bhattacharyya, D.M. Bylander, L. Kleiman [ *Phys. Rev. B (USA)* vol.32 (1985) p.7973 ]

[44] V. Malhotra, T.L. Martin, J.E. Mahan [ *J. Vac. Sci. Technol. B (USA)* vol.2 (1984) p.10 ]

[45] T.L. Martin, V. Malhotra, J.E. Mahan [ *J. Electron. Mater. (USA)* vol.13 (1984) p.309 ]

[46] J.M. van Ruitenbeek, W. Joss, R. Pauthenet, O. Thomas, J.P. Senateur, R. Madar [ *Phys. Rev. B (USA)* vol.35 (1987) p.7936 ]

[47] M. Gurvitch, A.F.J. Levi, R.T. Tung, S. Nakahara [ *Mater. Res. Soc. Symp. Proc. (USA)* vol.91 (1987) p.457 ]

[48] J.C. Hensel [ *Mater. Res. Soc. Symp. Proc. (USA)* vol.54 (1986) p.499 ]

[49] C.D. Lien, M. Finetti, A.M. Nicolet, S.S. Lau [ *J. Electron. Mater. (USA)* vol.13 (1984) p.95 ]

[50] G. Aprilesi, E. Mazzega, M. Michelini, F. Nava, G. Queirolo, L. Meda [ *J. Appl. Phys. (USA)* vol.60 (1986) p.310 ]

[51] E.G. Colgan, M. Maenpaa, F. Finetti, M.A. Nicolet [ *J. Electron. Mater. (USA)* vol.12 (1984) p.413 ]

[52] M.O. Aboelfotoh, L. Krusin-Elbaum [ *J. Appl. Phys. (USA)* vol.70 (1991) p.3382 ]

[53] B.Z. Li, R.G. Aitken [ *Appl. Phys. Lett. (USA)* vol.46 (1985) p.401 ]

[54] M. Wittmer, D.L. Smith, P.W. Lew, M.A. Nicolet [ *Solid-State Electron. (UK)* vol.21 (1978) p.573 ]

# 5.2 Electrical transport in semiconducting TM silicides

U. Gottlieb, F. Nava, M. Affronte, O. Laborde and R. Madar

January 1995

## A INTRODUCTION

In contrast to metallic silicides, semiconducting silicides have resistivities ranging from mΩ cm up to hundreds of Ω cm at room temperature. The resistivity of a semiconductor increases as the temperature decreases contrary to the behaviour in metals.

The family of semiconducting silicides presents some attractive features for application in microelectronics devices: the wide range of energy gaps (from about 0.12 eV to 1.1 eV) [1,2], the ability to grow silicon dioxide as a native oxide [3] and the prospect of relatively simple device processing in microelectronics technology [4]. Despite the work already done, results are still often contradictory, depending mainly on the quality of the examined material. In our review we will limit discussion to the semiconducting silicides $CrSi_2$, $ReSi_{1.75}$, β-$FeSi_2$ and $Ru_2Si_3$.

## B RESISTIVITY AND HALL EFFECT IN SEMICONDUCTING SILICIDES

In TABLE 1 we present the crystal structures of the semiconducting silicides $CrSi_2$, $ReSi_{2-\delta}$, β-$FeSi_2$ and $Ru_2Si_3$, as reported by several authors. It is important to note that semiconducting silicides only appear at the boundary between the two structure families: $CrSi_2$ from the VIB column, $ReSi_{2-\delta}$ from the VIIB column, and β-$FeSi_2$ and $Ru_2Si_3$ from the VIIIB column.

$CrSi_2$ crystallises in the same hexagonal structure as the metallic silicides $VSi_2$, $NbSi_2$ and $TaSi_2$. This material offers the opportunity of epitaxial growth on silicon as the lattice mismatch is only 0.14% in atomic spacings between (0001) $CrSi_2$ and (111) Si [5].

In the case of $ReSi_{2-\delta}$, the literature on its structure is not clear. The older work describes it as having the C11b structure of $MoSi_2$ and $WSi_2$ [6-8]. More recently it seemed to be established that the real structure is distorted C11b which is orthorhombic [9,10]. Also, the composition of this compound is reported to be different: most authors describe it as $ReSi_2$ but other workers found that the real composition is $ReSi_{1.8}$ [11] or $ReSi_{1.75}$ [12]. The most recent result indicates that the real structure of $ReSi_{1.75}$ is triclinic with the appearance of disordered Si vacancies in the structure [12]. The triclinic structure is still a distorted C11b structure, so it is possible that the gap of $ReSi_{1.75}$ comes from the deformation of this structure relative to $MoSi_2$.

β-$FeSi_2$ [13,14] crystallises in an orthorhombic structure which is a distorted version of the cubic $CoSi_2$ structure. The distortion of the structure seems to be responsible for the opening of the gap. $FeSi_2$ also exists as a high temperature, cubic, non-stoichiometric metallic phase, the so-called $\alpha$-$FeSi_2$ phase.

TABLE 1 Crystal structures of the semiconducting silicides.

| Silicide | Structure type | Space group | Lattice parameters (Å) | Ref |
|---|---|---|---|---|
| $CrSi_2$ | Hexagonal C40 | P6222 | a = 4.431<br>c = 6.364 | |
| $ReSi_2$ | Tetragonal C11b | I4/mmm | a = 3.131<br>c = 7.676 | [6-8] |
| $ReSi_2$ | Orthorhombic | Immm | a = 3.128<br>b = 3.144<br>c = 7.677 | [9,10] |
| $ReSi_{1.8}$ | Tetragonal C11b | I4/mmm | a = 3.132<br>c = 7.681 | [11] |
| $ReSi_{1.75}$ | Triclinic | P1 | a = 3.139<br>b = 3.121<br>c = 7.670<br>$\alpha$ = 89.87° | [12] |
| $\beta$-$FeSi_2$ | Orthorhombic | Cmca | a = 9.863<br>b = 7.791<br>c = 7.833 | [13,14] |
| $Ru_2Si_3$ | Orthorhombic | Pbcn | a = 11.057<br>b = 8.934<br>c = 5.533 | [17,18] |

$Ru_2Si_3$ belongs to a large family of transition metal compounds, known as 'Nowotny chimney-ladder' [15] or defect-$TiSi_2$ compounds [16]. It crystallises in an orthorhombic structure [17,18] with 40 atoms per unit cell, the transition metal atoms forming a sublattice of the $\beta$Sn type. Above about 1000°C $Ru_2Si_3$ is reported to undergo a diffusionless phase transition to a tetragonal structure [17].

In TABLE 2, we present a compilation of some transport measurement results for the semiconducting silicides $Ru_2Si_3$, $ReSi_{2-\delta}$, $CrSi_2$ and $\beta$-$FeSi_2$. These results were obtained on both oriented and non-oriented single crystals and on thin films. As one can see from the values given in TABLE 2, experimental results show a great divergence.

The room temperature resistivity is quite different for the different silicides. It ranges from 0.3 to 0.8 m$\Omega$ cm for $CrSi_2$ [19-25], from 3.5 to 18 m$\Omega$ cm for $ReSi_{2-\delta}$ [9,10,12,23,26], from 0.3 to 1.6 $\Omega$ cm for $\beta$-$FeSi_2$ [27-31] and from 1 to 5 $\Omega$ cm for $Ru_2Si_3$ [19,32-35]. Also, the obtained gap values show a large variation: from 0.12 to 0.33 eV for $ReSi_{2-\delta}$ [9,10,12,26], from 0.27 to 0.35 eV for $CrSi_2$ [20,21,25], from 0.8 to 1.0 eV for $\beta$-$FeSi_2$ [28-31] and from 0.7 to 1.04 eV for $Ru_2Si_3$ [32,34,35]. We will now discuss some characteristic properties for each of these silicides in detail.

TABLE 2 Electrical properties of semiconducting silicides (s.c. = single crystal).

| Material | Sample type | ρ (300 K) (Ω cm) | $n_H$ (300 K) ($cm^{-3}$) | $\mu_H$ (300 K) ($cm^2$/V s) | $E_g$ (eV) | Ref |
|---|---|---|---|---|---|---|
| $Ru_2Si_3$ | Bulk | | | | 0.7, 0.4 | [32] |
| | sc [100] | 2 | | | | [19,33] |
| | sc [010] | 1 | | | | |
| | sc [001] | 1.6 | $+1.0 \times 10^{18}$ | 3.5 | | |
| | sc [010] | 4 | $+1.0 \times 10^{18}$ | ~2 | 1.04 | [34,35] |
| | sc [001] | 5 | $+1.0 \times 10^{19}$ | ~1 | | |
| $ReSi_2$ | Bulk | $3\text{-}10 \times 10^{-3}$ | | | 0.12 - 0.22 | [9] |
| | Thin film | $18 \times 10^{-3}$ | $+1.6 \times 10^{19}$ | 4.6 | 0.18 | [10] |
| | Thin film | $\sim 18 \times 10^{-3}$ | | | 0.09 - 0.33 | [26] |
| | Thin film | $\sim 3.5 \times 10^{-3}$ | | | | [23] |
| $ReSi_{1.75}$ | sc [100] | $6 \times 10^{-3}$ | $+3.7 \times 10^{18}$ | 370 | 0.16 - 0.32 | [12] |
| | sc [001] | $4 \times 10^{-3}$ | | | | |
| $CrSi_2$ | sc [11$\bar{2}$0] | $0.74 \times 10^{-3}$ | | | | [19] |
| | sc [0001] | $0.35 \times 10^{-3}$ | | | | |
| | Bulk | $\sim 0.7 \times 10^{-3}$ | $+7.7 \times 10^{20}$ | ~12 | 0.29 - 0.32 | [20] |
| | sc | $0.8 \times 10^{-3}$ | $+4.0 \times 10^{20}$ | ~15 | 0.35 | [21] |
| | sc [11$\bar{2}$0] | $0.8 \times 10^{-3}$ | | | | [22] |
| | sc [0001] | $0.5 \times 10^{-3}$ | | | | |
| | Thin film | $\sim 9 \times 10^{-3}$ | | | | [23] |
| | Thin film | $5.6 \times 10^{-3}$ to $7.3 \times 10^{-3}$ | $+1.1 \times 10^{20}$ to $+0.5 \times 10^{20}$ | 10.3 - 17.6 | | [24] |
| | Thin film | $\sim 9 \times 10^{-3}$ | $+4 \times 10^{19}$ | ~20 | 0.27 | [25] |
| β-$FeSi_2$ | Thin film | 1.4 - 1.6 | | | | [27] |
| | Bulk | 1.2 | $\sim +10^{18}$ | ~9 | 1.0 | [28] |
| | Thin film | ~0.3 | $+1.4 \times 10^{18}$ | 4 | 0.8 | [29] |
| | Thin film | 0.3 | $+7.8 \times 10^{18}$ | 2.3 | 0.85 | [30,31] |

## B1 $Ru_2Si_3$

Below ~1000°C, $Ru_2Si_3$ is reported to be a semiconductor with a bandgap of the order of about 1 eV [32]. Band structure calculations [36] predict a gap with a width in the same order of magnitude as found by experiments. The gap seems to be related not only to the Si p - Ru d hybridisation, but also to the 5p ruthenium orbitals. $Ru_2Si_3$ seems to be a promising thermoelectric material [34]. Iridium was found to be a possible n dopant and manganese seems to be a good p dopant [37].

In FIGURE 1, we present the resistivity as a function of temperature of $Ru_2Si_3$ single crystals. The resistivity of $Ru_2Si_3$ is anisotropic and the anisotropy is roughly constant in the low temperature range: $\rho_a/\rho_b \approx 2$ and $\rho_c/\rho_b \approx 1.5$. The variation of the resistivity of $Ru_2Si_3$ with temperature is typical for a doped semiconductor [38] with an extrinsic regime below ~300 K and the intrinsic regime at higher temperatures. At low temperatures ($T \leq 20$ K), one would expect that all free carriers have been trapped by impurities. In this case, conduction is only possible in an impurity band (localisation state) or by some kind of hopping mechanism [39]. For temperatures lower than 10 K, the conductivity of $Ru_2Si_3$ follows a $\sigma(T) = 1/\rho = \sigma(T = 0\text{ K}) + mT^{1/2}$ law [19], indicating a conduction mechanism in the localisation state.

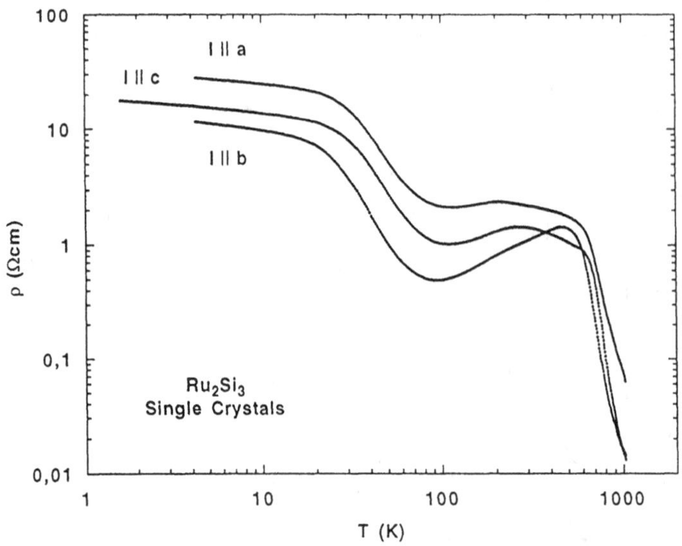


FIGURE 1 Resistivity vs. temperature of $Ru_2Si_3$. Note the anisotropy of the resistivity in the low temperature range.

Between ~20 K and ~100 K, the resistivity rapidly decreases with temperature. This results from the rapidly increasing number of free charge carriers as the impurities are progressively ionised. $\rho$ varies as $\exp(-E_0/k_BT)$, where $E_0$ is an activation energy. Here the $E_0$ values are between 19 and 24 meV, which are typical values of ionisation energies of most impurities in semiconductors.

Above 100 K, $\rho$ increases with T. If we assume a roughly constant charge carrier concentration n, the variation of $\rho$ results from the mobility variation. Charge carrier scattering by phonons leads generally to a $T^{-3/2}$ dependence of the mobility [40], i.e. $\rho \sim T^{3/2}$ when n is constant. At high temperatures (T > 300 K), thermal energy becomes sufficient to activate carriers across the bandgap. The resistivity follows an activation law $\rho \sim \exp(-E_g/2k_BT)$ with an energy gap varying between 0.95 eV and 1.1 eV depending on sample orientation.

In FIGURE 2, we present the Hall mobility of $Ru_2Si_3$ taken from [19,34]. The values of the mobility below 6 K are absolute values as the Hall effect changes sign at about 10 K. One can clearly distinguish three regions: (i) below 10 K, a conduction mechanism in the localisation state, which can be assigned to heavy electrons with a conductivity $\sigma_e$, a Hall coefficient $R_e$ and a mobility $\mu_e = R_e\sigma_e$; (ii) the extrinsic regime between about 10 K and about 400 K, where charge carriers are holes created by impurity ionisation with a conductivity $\sigma_{te} = n(T)e\mu_{te}$; and (iii) at high temperatures the intrinsic regime where thermally activated holes are the predominant charge carriers. The change between two successive regimes can be understood within the framework of the two band model, when one assumes that the mobilities of the two types of carrier involved are very different [38].

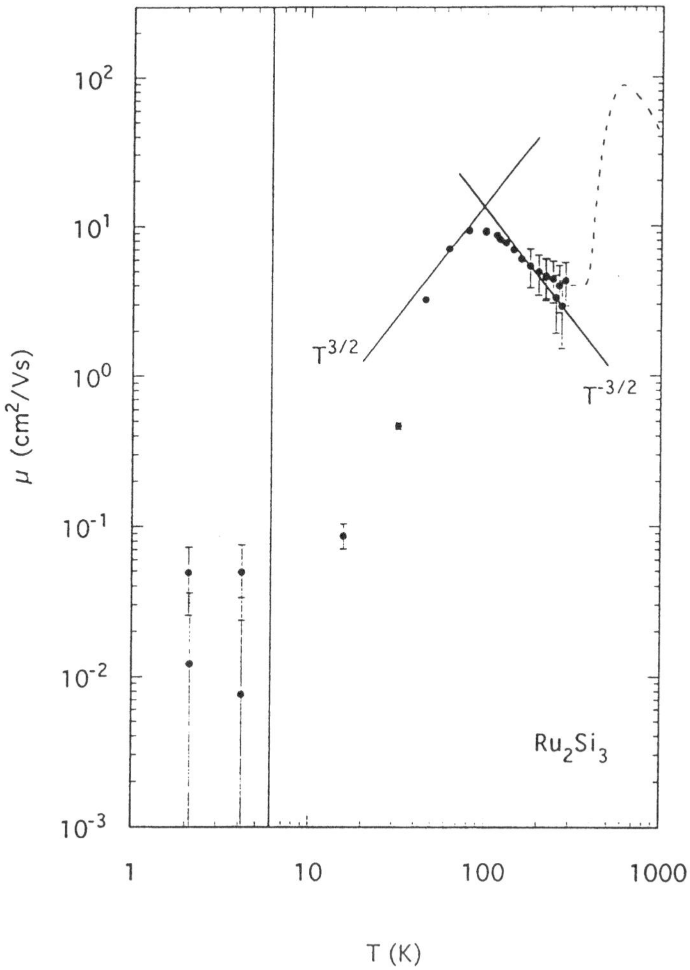


FIGURE 2 Hall mobility vs. T of $Ru_2Si_3$ (high temperature data from [34]). Mobility values below 6 K are absolute values since the Hall coefficient changes sign below around 10 K.

Between 20 K and 100 K, where the impurities are progressively ionised, charge carrier scattering by the ionised impurities should lead to a $T^{3/2}$ law [40]. This behaviour can be observed only within a very small range of temperature. For higher temperatures ($100\ K \leq T \leq 300\ K$), the mobility is determined by electron-phonon scattering, leading to a $(m^*)^{5/2}T^{-3/2}$ law [40], where m* is the effective mass. Effectively, the mobility of $Ru_2Si_3$ follows this behaviour between ~100 K and 300 K in the extrinsic region and above ~500 K in the intrinsic region. One can therefore estimate for $Ru_2Si_3$ the ratio between the effective mass in the intrinsic region $m_i$ and the effective mass in the extrinsic region $m_e$: $m_e/m_i \approx 2.8$.

### B2 $ReSi_{2-\delta}$

As mentioned above, there is great inconsistency in the literature regarding the crystal structure and the composition of $ReSi_{2-\delta}$. Nevertheless, all authors observe semiconducting behaviour of this material by transport and optical measurements [9,10,26], which is in contradiction with band structure calculations [41,42]. It is also difficult to see how $ReSi_2$ can have only filled bands as the number of valence electrons in the formula unit is odd. Recent results [12] resolve this inconsistency, as the real composition is determined to be $ReSi_{1.75}$ in a triclinic crystal structure. The structure modification leads to a doubling of the primitive unit cell, the unit formula is now $Re_2Si_{3.5}$ (= $ReSi_{1.75}$), and so the number of valence electrons would be even.

In FIGURE 3 we present the resistivity of $ReSi_{1.75}$ single crystals for two different current directions as a function of temperature. For the two directions, the resistivity behaviour is quite similar, but the resistivity for I || a is always higher than for I || c. As for $Ru_2Si_3$, the conductivity of $ReSi_{1.75}$ below 10 K indicates a conduction mechanism in a localisation state [19].

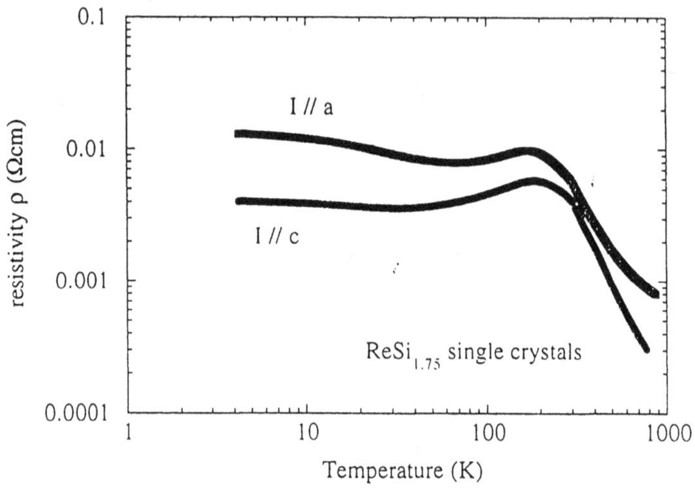


FIGURE 3 Resistivity vs. T for two different crystallographic directions of $ReSi_{1.75}$.

Above 300 K, the resistivity follows an activation law as expected for a semiconductor in the intrinsic region. In the extrinsic region ($10\ K \leq T \leq 300\ K$), however, it is difficult to determine the different effects as the resistivity is nearly constant. It may be that deviations from perfect Si stoichiometry could act as a dopant for this material, but more work is needed in order to understand these effects.

In FIGURE 4, we present the high temperature resistivity in an Arrhenius plot. For I || a, ρ follows a simple linear behaviour with a well-defined slope corresponding to an energy gap of $E_g = 0.16$ eV. For I || c, however, one can observe two resistivity slopes for different

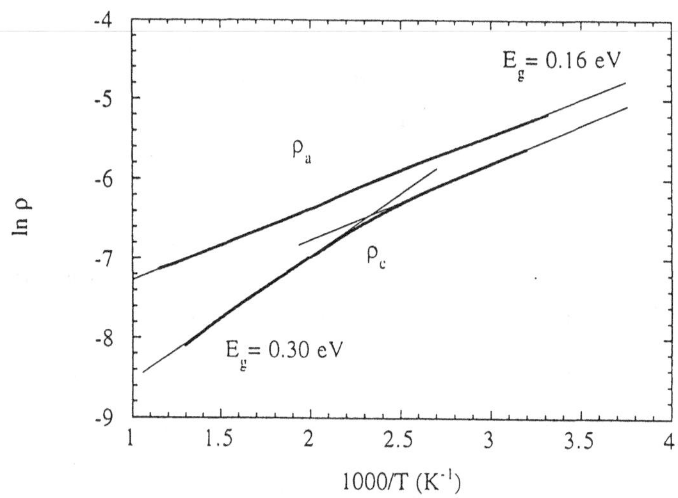


FIGURE 4 ln(ρ) vs. 1000/T for high temperature resistivity of $ReSi_{1.75}$ for two different current directions.

temperature ranges. Close to room temperature an energy gap $E_g = 0.16$ eV can be observed, identical to I || a. For T > ~400 K one can observe $E_g = 0.30$ eV. These values are quite close to those reported by Long et al [26]. Optical measurements on polycrystalline films show a direct energy gap of 0.36 eV and an indirect energy gap of 0.12 eV.

In FIGURE 5, we present the Hall mobility of $ReSi_{1.75}$. Above 300 K, it follows a $T^{-3/2}$ law originating from the scattering of charge carriers by phonons [40]. Between 100 K and 200 K the mobility seems to follow a $T^{3/2}$ law via scattering of the carriers by ionised impurities. It is worth noting that the Hall mobility of $ReSi_{1.75}$ single crystals at room temperature is quite high: $\mu_H = 370$ cm$^2$/V s.

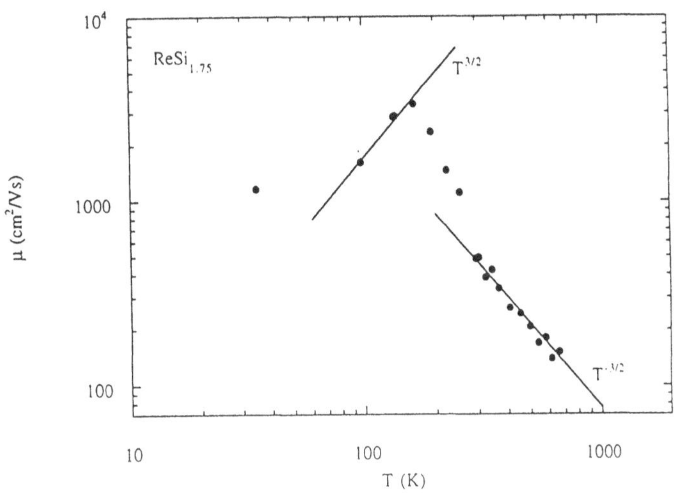


FIGURE 5 Hall mobility vs. T for $ReSi_{1.75}$.

## B3 $CrSi_2$

A great number of papers on $CrSi_2$ have been published in the recent literature. We will limit our discussion to some significant results and discuss its main properties without being complete. It seems now to be established that $CrSi_2$ is a semiconductor with an indirect bandgap on the order of 0.3 eV, as determined from optical and transport measurements and also obtained by band structure calculations.

Bost et al [43] measured the optical properties of $CrSi_2$ polycrystalline thin films and obtained an indirect bandgap of 0.35 eV. These values are in good agreement with the results from Bellani et al [44] who also obtained an indirect bandgap of 0.36 eV on single crystal samples. They also calculated an indirect bandgap of 0.38 eV [44]. Krijn et al [45] calculated an indirect bandgap of 0.21 eV and the band structure calculations of Mattheiss [46,47] also give an indirect energy gap $E_g = 0.30$ eV.

In FIGURE 6, we present the high temperature resistivity of both $CrSi_2$ thin films and a single crystal sample. One can see that the gap values determined from the two curves are slightly different: $E_g = 0.27$ eV for the thin film data and $E_g = 0.30$ eV for the single crystal. We must, however, emphasise that low temperature resistivity measurements of single crystals of $CrSi_2$ have given a metallic behaviour [19,22].

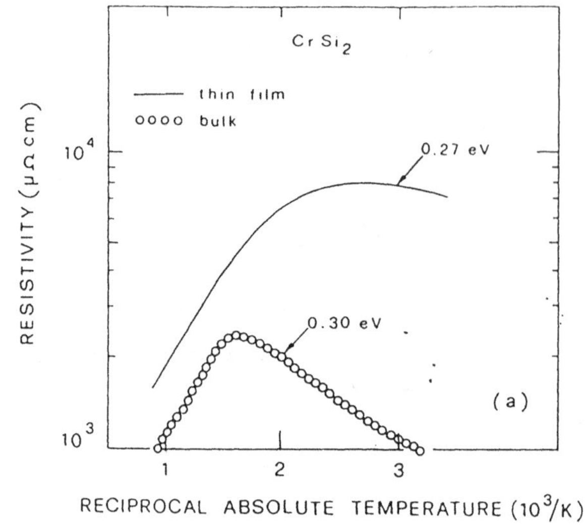


FIGURE 6 High temperature resistivity vs. 1000/T of $CrSi_2$ thin film and single crystal.

The electrical properties of $CrSi_2$ seem to be strongly dependent on the silicon stoichiometry as can be seen in FIGURE 7 where we present the resistivity of $CrSi_2$, $CrSi_{2.01}$, $CrSi_{2.02}$ and $Cr_{0.88}Mn_{0.12}Si_2$. Increasing the Si content seems to give better semiconducting behaviour. A lack of Si seems to lead to a metallic behaviour, as found by low temperature specific heat measurements [48].

In FIGURE 7, it can also be seen that manganese is a good n dopant of $CrSi_2$. Similar results were obtained by several authors [20,21]. On the other hand light doping with vanadium will result in a metallic behaviour [24]. This behaviour could be explained in the framework of a rigid band model when one assumes that the band structures of $VSi_2$ and $CrSi_2$ are nearly the same and the Fermi level in between the bands would only be determined by the number of valence electrons in the compound. This behaviour is predicted by some band structure calculations [46], where the Fermi level for $CrSi_2$ lies just above the upper valence band edge. Doping with Mn or higher Si content would increase the number of valence electrons and so push the Fermi level into the gap, whereas doping with V or a lower Si content would decrease the valence electron number and the Fermi level would fall into the valence band.

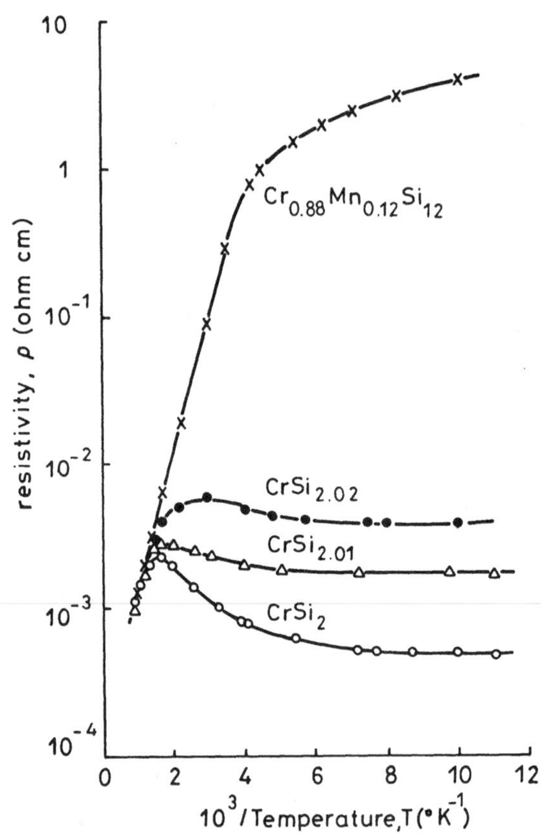


FIGURE 7 Resistivity vs. 1000/T for $CrSi_2$, $CrSi_{2.01}$, $CrSi_{2.02}$ and $Cr_{0.88}Mn_{0.12}Si_2$ crystals (data taken from [21]).

## B4 $\beta$-FeSi$_2$

Recently, $\beta$-FeSi$_2$ has received a great deal of interest due to its semiconducting nature and its relatively high thermoelectric power [49]. It is reported to have a direct energy gap of about 0.9 eV (see also Chapter 4). It was shown that p doping could be achieved with manganese and aluminium, and cobalt was found to be an appropriate n dopant [28,49].

Band structure calculations have been done by several authors. Eppenga [50] predicted that $\beta$-FeSi$_2$ is a semiconductor with similar indirect and direct bandgaps ($E_g$ = 0.44 eV for the indirect and $E_g$ = 0.46 eV for the direct transition). This result of similar values was confirmed by Christensen [51], who predicted a direct transition of 0.80 eV. Optical measurements [1] also found a direct bandgap of 0.87 eV, in good agreement with the value determined by photothermal deflection spectroscopy of 0.85 eV [31].

In FIGURE 8(a), we present the low temperature resistivity of $\beta$-FeSi$_2$ thin films in the extrinsic region compared to the expected activation law behaviour in the intrinsic region for a bandgap of 0.80 eV. In FIGURE 8(b) we present the comparison of thin film and single crystal high temperature conductivities. From these curves, one can deduce a gap value of 0.8 eV.

The Hall voltage of both p- and n-type $\beta$-FeSi$_2$ shows considerable nonlinearities at low temperatures [30,31,52,53]. As an example we have plotted $R_H$ versus applied magnetic field in FIGURE 9. The interpretation of this behaviour is not clear. Some authors explain it by supposing a ferromagnetic ordering in $\beta$-FeSi$_2$ for temperatures below 100 K [30,31] together with a two valence band model and an impurity level. Other authors explain the Hall coefficient behaviour of $\beta$-FeSi$_2$ with a two band model excluding any possibility of ferromagnetic ordering [52,53].

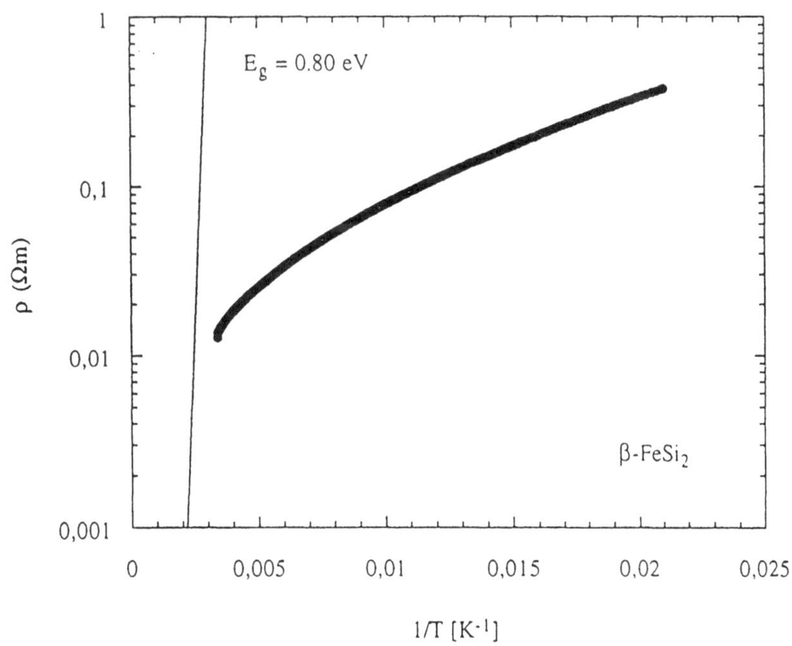


FIGURE 8 (a) log(ρ) vs. 1000/T of low temperature resistivity of $\beta$-FeSi$_2$ thin film compared with an activation law for $E_g$ = 0.80 eV.

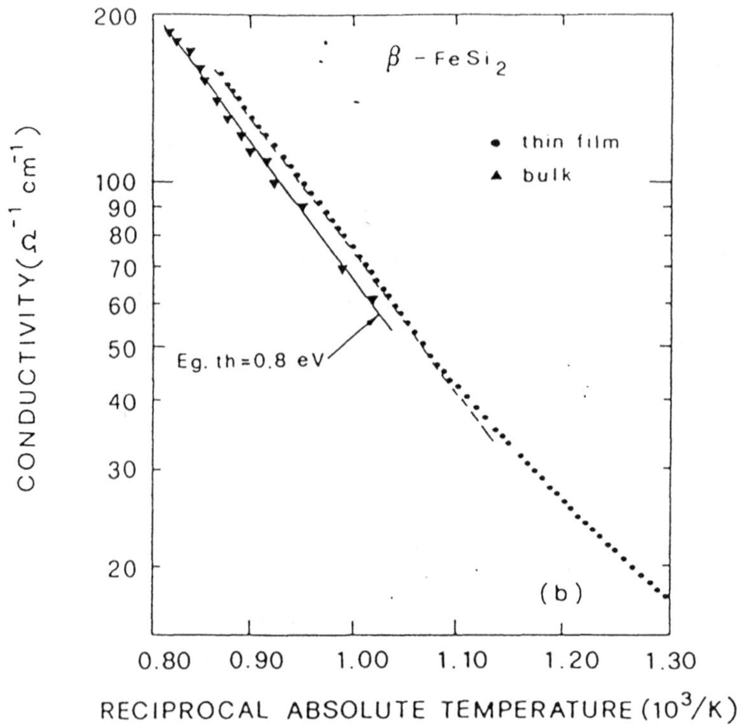


FIGURE 8 (b) High temperature conductivity vs. 1000/T of $\beta$-$FeSi_2$ thin film and single crystal.

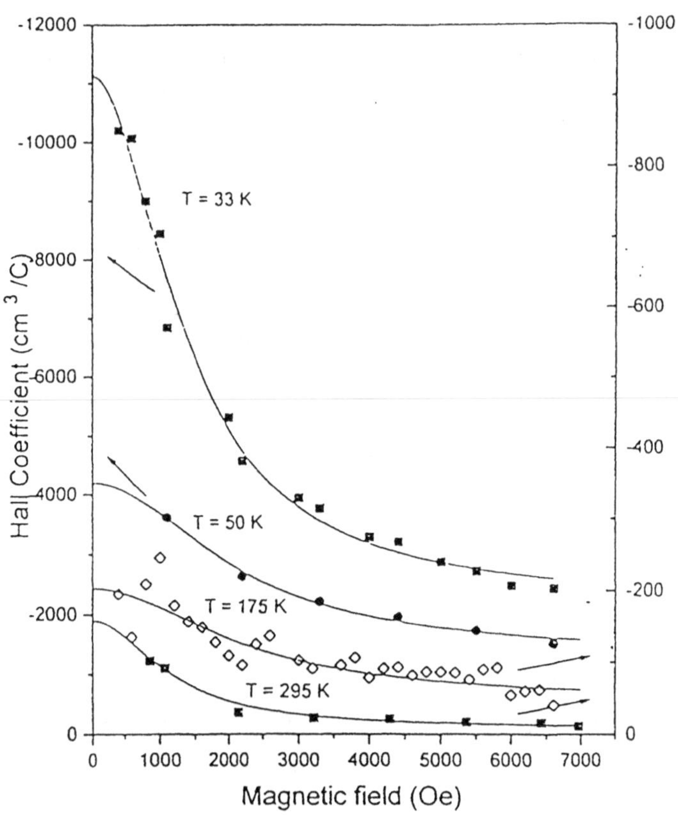


FIGURE 9 Hall coefficient vs. magnetic field of $\beta$-$FeSi_2$ single crystal at various temperatures (data taken from [52]).

## C CONCLUSION

We have presented and reviewed the electrical properties, namely resistivity and Hall effect, of some characteristic silicides measured on both thin films and single crystals. The results for the resistivity and Hall coefficient obtained and presented in the literature show a large divergence, indicating that the material fabrication processes are still not very well controlled. Therefore, the physical interpretation of the experimental results is ambiguous leading to many contradictions appearing in the literature. Recently, studies on single crystal materials have been appearing and the situation is becoming clearer.

## REFERENCES

[1] M.C. Bost, J.E. Mahan [ *J. Appl. Phys. (USA)* vol.58 (1985) p.2696 ]
[2] M.C. Bost, J.E. Mahan [ *J. Appl. Phys. (USA)* vol.64 (1988) p.2034 ]
[3] M. Batur, M.-A. Nicolet [ *J. Electrochem. Soc. (USA)* vol.131 (1984) p.371 ]
[4] S.P. Murarka [ *J. Vac. Sci. Technol. (USA)* vol.17 (1980) p.775 ]
[5] F.Y. Shiau, H.C. Sheng, L.J. Chen [ *Appl. Phys. Lett. (USA)* vol.45 (1984) p.524 ]
[6] H.J. Wallbaum [ *Z. Met.kd. (Germany)* vol.33 (1941) p.378 ]
[7] V.S. Nesphor, G.V. Samsonov [ *Sov. Phys.-Solid State (USA)* vol.2 (1960) p.1966 ]
[8] V.S. Nesphor, G.V. Samsonov [ *Phys. Met. Metallogr. (UK)* vol.11 (1960) p.146 ]
[9] T. Siegrist, F. Hullinger, G. Travaglini [ *J. Less-Common Met. (Switzerland)* vol.92 (1983) p.119 ]
[10] C. Krontiras et al [ *Thin Solid Films (Switzerland)* vol.161 (1988) p.197 ]
[11] J.L. Jorda, M. Ishikawa, J. Muller [ *J. Less-Common Met. (Switzerland)* vol.85 (1982) p.27 ]
[12] U. Gottlieb et al [ to be published ]
[13] R. Wandji, Y. Dusausoy, J. Protas, B. Roques [ *C. R. Hebd. Seances Acad. Sci. C, Sci. Chim. (France)* vol.267 (1968) p.1587 ]
[14] P.Y. Dusausoy, J. Protas, R. Wandji, B. Roques [ *Acta Crystallogr. B (Denmark)* vol.27 (1971) p.1209 ]
[15] W.B. Pearson [ *Acta Crystallogr. B (Denmark)* vol.26 (1970) p.1044 ]
[16] H. Nowotny [ *The Chemistry of Extended Defects in Non-Metallic Solids* Eds L.R. Eyring, M. O'Keefe (North-Holland, Amsterdam, 1970) p.223 ]
[17] D.J. Poutcharovsky, K. Yvon, E. Parthé [ *J. Less-Common Met. (Switzerland)* vol.40 (1975) p.139 ]
[18] D.J. Poutcharovsky, E. Parthé [ *Acta. Crystallogr. B (Denmark)* vol.30 (1974) p.2692 ]
[19] U. Gottlieb [ Ph.D. Thesis, Institut National Polytechnique de Grenoble, France (1994) ]
[20] I. Nishida, T. Sakata [ *J. Phys. Chem. Solids (UK)* vol.39 (1978) p.499 ]
[21] D. Shinoda, S. Asanabe, Y. Sasaki [ *J. Phys. Soc. Jpn. (Japan)* vol.19 (1964) p.269 ]
[22] T. Hirano, M. Kaise [ *J. Appl. Phys. (USA)* vol.68 (1990) p.627 ]
[23] R.G. Long, J.E. Mahan [ *Appl. Phys. Lett. (USA)* vol.56 (1990) p.1655 ]
[24] O. Thomas et al [ *Appl. Surf. Sci. (Netherlands)* vol.38 (1989) p.94 ]
[25] F. Nava, T. Tien, K.N. Tu [ *J. Appl. Phys. (USA)* vol.57 (1985) p.2018 ]
[26] R.G. Long, M.C. Bost, J.E. Mahan [ *Thin Solid Films (Switzerland)* vol.162 (1988) p.29 ]
[27] U. Gottlieb, Y. Morand, R. Madar [ unpublished ]
[28] T. Kojima [ *Phys. Status Solidi A (Germany)* vol.111 (1989) p.233 ]
[29] M. Michelini, F. Nava, E. Galli [ *J. Mater. Res. (USA)* vol.6 (1991) p.1655 ]
[30] O. Valassiades, C.A. Dimitriadis, J.H. Werner [ *J. Appl. Phys. (USA)* vol.70 (1991) p.890 ]
[31] C.A. Dimitriadis, J.H. Werner, S. Logothetidis, M. Stutzmann, J. Weber, R. Nesper [ *J. Appl. Phys. (USA)* vol.68 (1990) p.1726 ]

[32] C.P. Susz, J. Muller, K. Yvon, E. Parthé [ *J. Less-Common Met. (Switzerland)* vol.71 (1980) p.1 ]

[33] U. Gottlieb, O. Laborde, A. Rouault, R. Madar [ *Appl. Surf. Sci. (Netherlands)* vol.73 (1993) p.243 ]

[34] C.B. Vining, C.E. Allevato [ Xth Int. Conf. on Thermoelectrics, Cardiff, Wales, UK, 10-12 Sept. 1991, Ed. D.M. Rowe (Barbrow Press, Cardiff, 1991) p.167 ]

[35] C.B. Vining [ Space Nuclear Power Systems, AIP Conf. Proc. no.246 (1992) p.338 ]

[36] P. Pécheur, G. Toussaint [ *Phys. Lett. A (Netherlands)* vol.160 (1991) p.193 ]

[37] C.B. Vining, C.E. Allevato [ *Proc. 27th Intersociety Energy Conversion Engineering Conf.*, San Diego, California, USA, 3-7 Aug. 1992 (Soc. Automotive Eng., Warrendale, Pennsylvania, USA, 1992) p.3.489-3.492 (vol.3) ]

[38] B.I. Shklovskii, A.L. Elfros [ *Electronic Properties of Doped Semiconductors* (Springer Verlag, Berlin, 1984) ]

[39] N. Mott [ *Conduction in Non-Crystalline Materials* (Clarendon Press, Oxford, 1987) ]

[40] H.M. Rosenberg [ *Low Temperature Solid State Physics* (Clarendon Press, Oxford, 1965) ]

[41] B.K. Bhattacharyya, D.M. Bylander, L. Kleinman [ *Phys. Rev. B (USA)* vol.33 (1986) p.3947 ]

[42] S. Itoh [ *Mater. Sci. Eng. B (Switzerland)* vol.6 (1990) p.37 ]

[43] M.C. Bost, J. Mahan [ *J. Appl. Phys. (USA)* vol.63 (1988) p.839 ]

[44] V. Bellani et al [ *Phys. Rev. B (USA)* vol.46 (1992) p.9380 ]

[45] M.P.C.M. Krijn, R. Eppenga [ *Phys. Rev. B (USA)* vol.44 (1991) p.9042 ]

[46] L.F. Mattheiss [ *Phys. Rev. B (USA)* vol.43 (1991) p.12549 ]

[47] L.F. Mattheiss [ *Phys. Rev. B (USA)* vol.45 (1992) p.3252 ]

[48] U. Gottlieb, O. Thomas, J.C. Lasjaunias [ unpublished ]

[49] U. Birkholz, J. Schelm [ *Phys. Status Solidi (Germany)* vol.27 (1968) p.413 ]

[50] R. Eppenga [ *J. Appl. Phys. (USA)* vol.68 (1990) p.3027 ]

[51] N.E. Christensen [ *Phys. Rev. B (USA)* vol.42 (1990) p.7148 ]

[52] E. Arushanov, Ch. Kloc, H. Hohl, E. Bucher [ *J. Appl. Phys. (USA)* vol.75 (1994) p.5106 ]

[53] E. Arushanov, Ch. Kloc, E. Bucher [ to appear in Phys. Rev. B (USA) ]

# 5.3 Electrical transport in RE silicides

Xu Zhenjia (C.C. Hsu)

June 1994

## A INTRODUCTION

Our knowledge of the electrical bulk properties of RE silicides is very limited and fragmentary. For instance, the list of silicides for which the resistivity and Hall coefficient have been measured to date is very short (TABLES 1 and 2), and only very few of these measurements have been performed on thin films prepared in accordance with the practice of semiconductor device technology. Because of the possible applications $ErSi_2$ has been studied by several authors and more is known about its properties than for most other RE silicides.

## B RESISTIVITY OF RE SILICIDES

TABLE 1 Resistivity of RE silicides (RT).

| Silicide | Resistivity ($10^{-6}$ μΩ cm) | Ref |
|---|---|---|
| LaSi | 340 | [1] |
| $La_5Si_3$ | 365 | [1] |
| $LaSi_2$ | 350, 274 | [1,2] |
| $CeSi_2$ | 408 | [2] |
| $PrSi_2$ | 202 | [2] |
| $NdSi_2$ | 349 | [2] |
| $GdSi_2$ | 263 | [2] |
| $DySi_2$ | 3020 | [2] |
| $TbSi_2$ (thin film) | 90 | [8] |
| $ErSi_2$ (thin film) | 30 | [1] |
| $LuSi_{2-x}$ (thin film) | 100 | [9] |

Duboz and co-workers [3,4], Arnaud et al [5] and Sagnes et al [6] have studied the resistivity of epitaxial $ErSi_{2-x}$ thin films with thickness ranging from 1000 to 50 Å as a function of temperature. We reproduce one of these results in FIGURE 1 [3]. The resistivity at 300 K was found to be 30 μΩ cm [3]. The resistivity decreases linearly with decreasing temperature and tends towards a limiting value of about 10 μΩ cm [3]. At 4.5 K, ρ(T) rapidly changes slope and decreases to reach 8.5 μΩ cm at 2 K [3]. The linear decrease of ρ(T) is typical of a metal and is consistent with Matthiessen's rule, which stipulates the additivity of the various scattering processes:

$$\rho(T) = \rho_0 + \rho_{ph}(T)$$

where $\rho_0$, the residual resistivity, results from defects, impurities, grain boundaries and interface scattering, and is essentially independent of temperature, and $\rho_{ph}(T)$ is the phonon scattering contribution which is linear with temperature at high temperature and vanishes at low temperature. The phonon contribution $\rho_{ph}(T)$ is about 24 μΩ cm, and the residual resistivity $\rho_0$ at 0 K due to crystallographic defects is less than 9 μΩ cm [3]. It seems unlikely that $ErSi_{2-x}$ films can be obtained with a resistivity lower than 25 μΩ cm at 300 K [3]. The abrupt decrease of ρ(T) below 4.5 K is attributed to magnetic ordering [3].

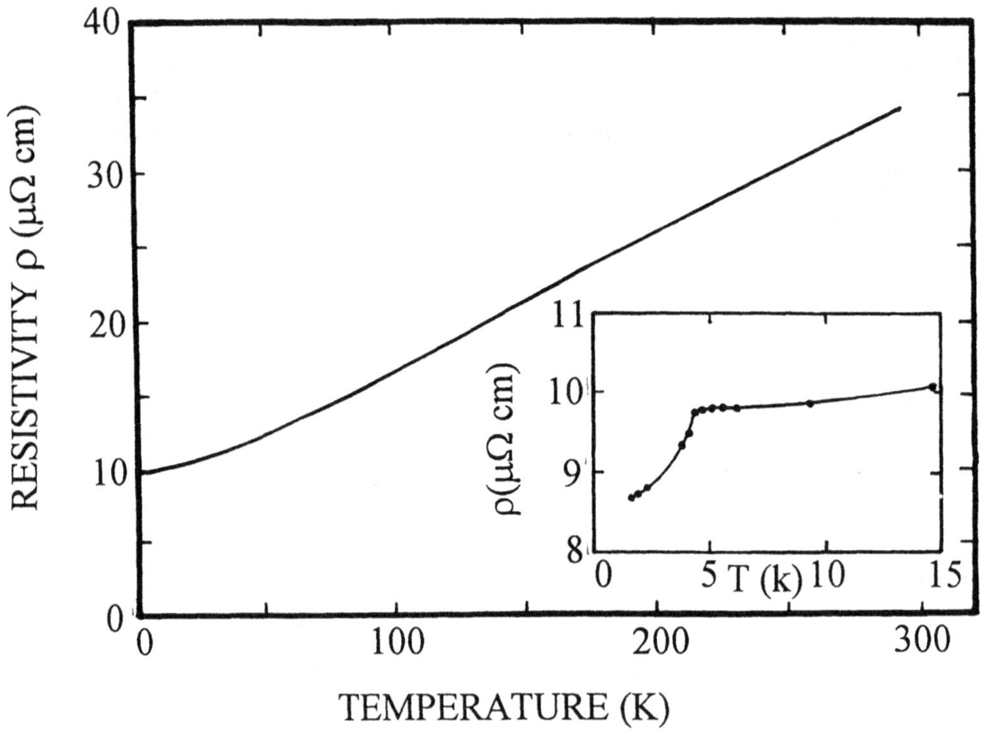


FIGURE 1 Resistivity of an erbium silicide film 300 Å thick, as a function of temperature. The inset shows the low temperature part of the curve and indicates the occurrence of a magnetic transition near 4.5 K (after [3]).

The resistivity versus temperature in thin $ErSi_{2-x}$ films with thickness ranging from 1000 to 50 Å is shown in FIGURE 2(a) [6]. A remarkable feature is that all the curves display virtually the same shape for temperature dependence. On the contrary, the residual resistivity at low temperature was found to depend strongly on the thickness. FIGURE 2(a) also shows that ρ(T) curves are translated as a whole towards higher resistivity when the thickness decreases. For all samples, $\rho_{ph}$ was found to be between 30 and 35 μΩ cm as the temperature varied from 15 K to room temperature [6]. This means that carrier-phonon interactions are not affected by thickness variations. On the contrary, the residual resistivity $\rho_0$ rapidly increases with decreasing thickness. FIGURE 2(b) shows the effect of thin film thickness on the residual resistivity at low temperature (18 K) [6]. At large thickness, $\rho_0$ tends to its bulk value. The low temperature resistivity for low thickness increases with thickness by around one order of magnitude. This increase is related to a surface scattering.

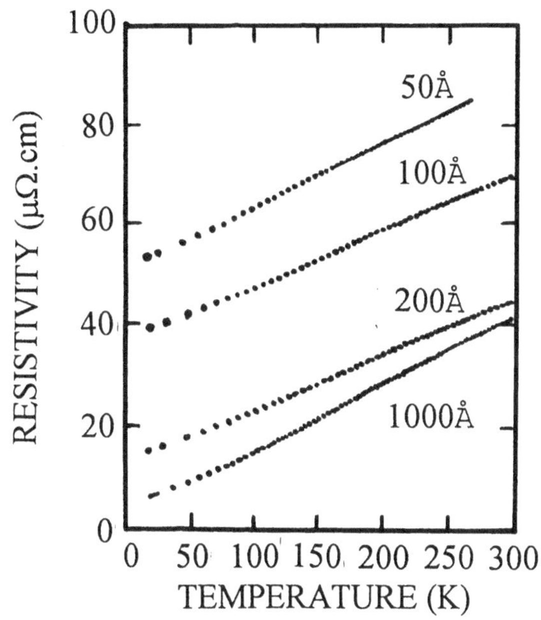


(a)

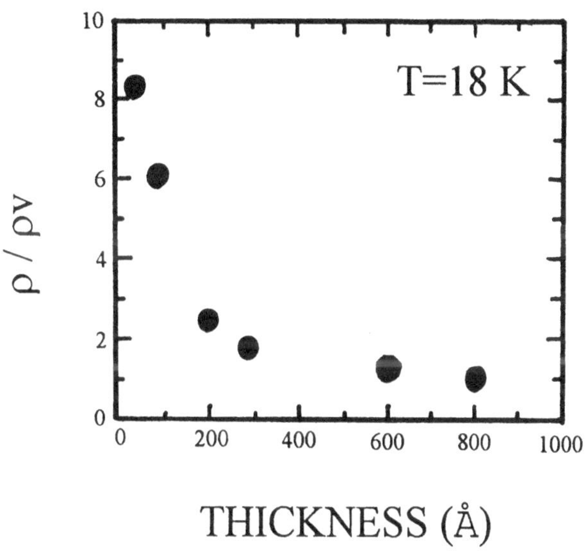


(b)

FIGURE 2 (a) Experimental resistivity (ρ) of $ErSi_2$ films vs. temperature for various film thicknesses. Data for 600 and 800 Å are superimposed on the 1000 Å curve (after [6]). (b) Normalised resistivity for $ErSi_2$ films at 18 K vs. thickness (circles). The bulk resistivity ρv has been taken as ρ (800 Å) (after [6]).

$ErSi_{2-x}$ ohmic contacts on n-type Si with resistivity as low as $3.7 \times 10^{-8}$ Ω cm were obtained by employing structures consisting of Pt, Si and Er layers [7]. These measured contact resistivities are the lowest among the known contacts to n-type Si. A low-leakage high-breakdown voltage Schottky diode on p-type Si was also realised [7]. Janega et al [7] showed no evidence of diffusion of Er into the substrate, within the sensitivity of the EDX technique, and $ErSi_{2-x}$ appeared to adhere very well to the silicon dioxide and the silicon during various process operations. Thus, the authors conclude that $ErSi_{2-x}$ contacts may prove to be extremely well suited for vertical as well as lateral integrated circuit scaling, and the Pt/Si/Er layer structures can be used to make good rectifying Schottky diodes as well as extremely low resistivity ohmic contacts.

As shown in FIGURE 3, Kaatz et al [8] found similar results for $TbSi_{2-x}$ to those for $ErSi_{2-x}$ in the study of resistivity in epitaxial $TbSi_{2-x}$ with a thickness of 120 Å. For T > 40 K, ρ(T) follows the Matthiessen dependence reaching 90 μΩ cm at room temperature [8]. At T = 38 K, an abrupt decrease in ρ(T) is observed, where the Tb ions order antiferromagnetically [8]. The phonon contribution to ρ(T) is about 22 μΩ cm [8].

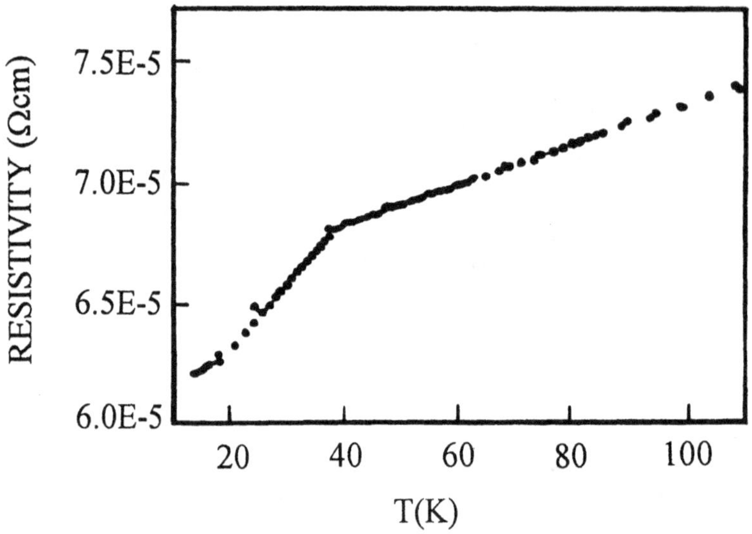


FIGURE 3 Resistivity of the 120 Å thick terbium silicide film as a function of temperature (after [8]).

Travlos et al [9] reported that the RT resistivity, $\rho_{300\,K}$, of the $LuSi_{2-x}$ layers prepared under different conditions was found to vary from 180 to 150 μΩ cm. The resistivity decreased with annealing time. FIGURE 4 shows the temperature variation of the resistivity of an epitaxial $LuSi_{1.7}$ layer grown on <100>Si at 400°C for 360 min which had the lowest value of $\rho_{300\,K}$. The resistivity initially decreases linearly with decreasing T and then levels off to the residual resistivity value $\rho_{4.6\,K}$ of about 75 μΩ cm [9].

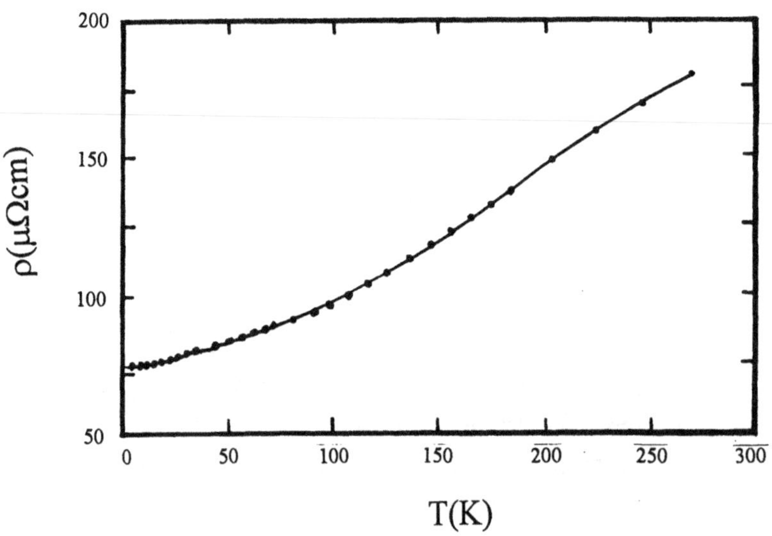


FIGURE 4 Temperature dependence of the resistivity of $LuSi_{2-x}$ layer grown on <100> Si at 400°C for 360 min (after [9]).

## C HALL COEFFICIENT $R_H$ OF RE SILICIDES

TABLE 2 Hall coefficient $R_H$ of RE silicides (RT).

| Silicide | | $R_H$ ($10^{-4}$ $cm^{-3}$/C) | Ref |
|---|---|---|---|
| $LaSi_2$ | | -17.5 | [2] |
| $CeSi_2$ | | 19.3 | [2] |
| $PrSi_2$ | | -10.7 | [2] |
| $ErSi_2$ | (1000 Å) | 4.8 | [6] |
| | (200 Å) | 1.4 | |
| | (100 Å) | 0.2 | |
| | (50 Å) | -2.2 | |

Duboz and co-workers [3,4], Arnaud et al [5] and Sagnes et al [6] have studied the Hall coefficient $R_H$ of epitaxial $ErSi_{2-x}$ thin films with thickness ranging from 1000 to 50 Å as a function of temperature. One of the results is reproduced in FIGURE 5 [3]. The $R_H$ is positive at 300 K, then decreases and becomes negative below 300 K. Subsequently, below 20 K, $R_H$ increases slightly and exhibits a cusp at 4.5 K, which is typical for the anomalous Hall effect and is attributed to magnetic ordering [4,5]. The sign of $R_H$ indicates that the majority carriers are holes at about 150 K and electrons below 150 K. In most Hall experiments, the electrical conduction is dominated by a single type of carrier. If both electrons and holes are involved in the electrical conduction, a two band model must be used. The value of $R_H$ depends on the concentrations and the mobilities of both electrons and holes; if the two concentrations are comparable, the variation of hole and electron mobilities with temperature leads to either positive or negative values of $R_H$ and the carrier densities cannot be calculated.

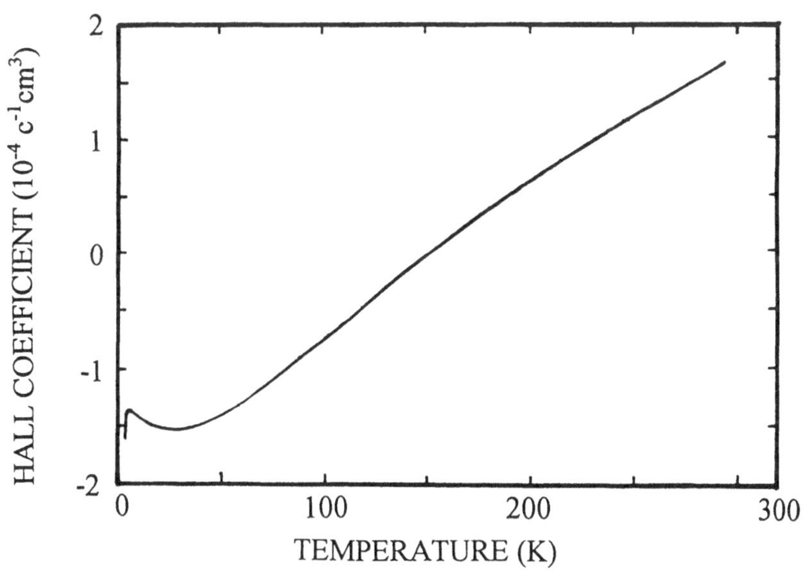


FIGURE 5 Hall coefficient, $R_H$, of an erbium silicide film 300 Å thick as a function of temperature, T. The evolution of $R_H(T)$ can be explained by a two-band conduction model, with hole and electron mobility varying with T (after [3]).

The variations in Hall coefficient with temperature are shown in FIGURE 6 for the set of $ErSi_{2-x}$ films [6]. The $R_H(T)$ curves, like $\rho(T)$, show the same translation properties. They can be obtained approximately from one another by a vertical shift. The $R_H(T)$ shape, which is positive at high temperatures, becomes negative as the temperature decreases. This phenomenon is attributed to the contribution of both electrons and holes to the conduction [6]. The $R_H(T)$ dependence varies significantly with the film thickness. However, the $R_H$ coefficient above 600 Å is no longer sensitive to the film thickness, as shown in the insert of FIGURE 6.

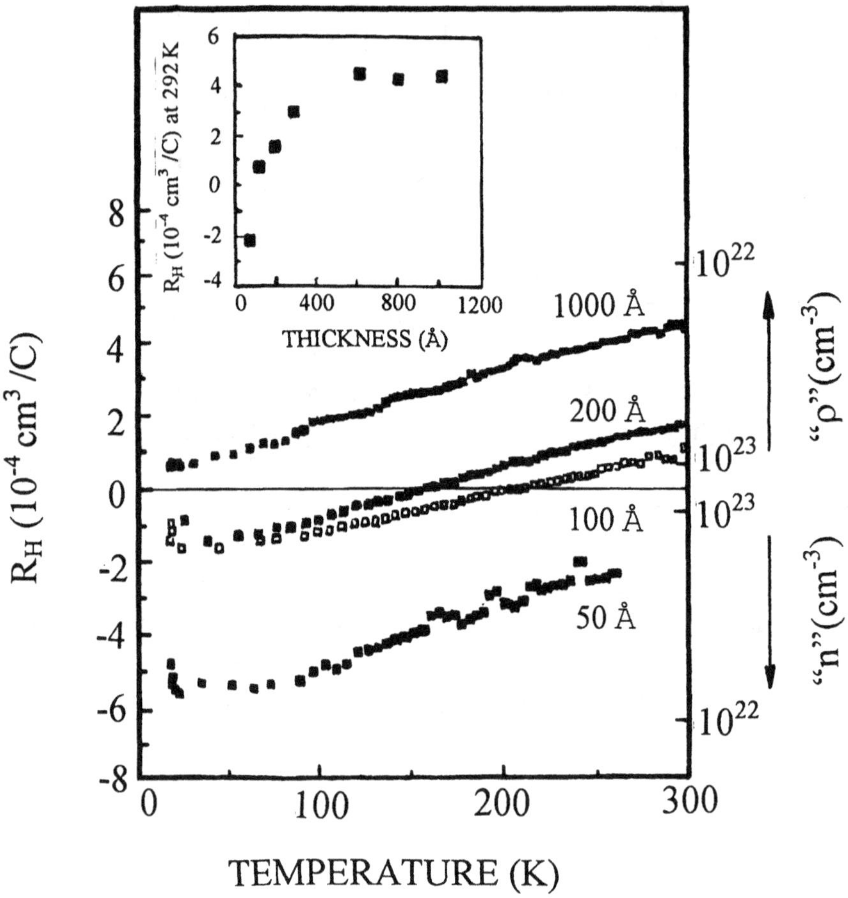


FIGURE 6 Experimental data for Hall coefficient $R_H$ of $ErSi_2$ films vs. temperature and thickness from 50 to 1000 Å. The insert represents the variation in the coefficient with thickness at room temperature (after [6]).

FIGURE 7 (data expressed in terms of the carrier density) shows the results of the Hall coefficient of epitaxial $TbSi_{2-x}$ as a function of temperature, reproduced from the data of Kaatz et al [8]. The Hall coefficient $R_H$ indicates electronic conduction and has minima at T = 38 K and ~16 K. At higher temperatures the carrier density saturates at the value of about $10^{22}$ cm$^{-3}$, indicating that $TbSi_{2-x}$ is a good metal. At T < 25 K, $R_H$ decreases strongly implying the presence of minority carriers (holes). The fact that n(T) peaks at 16 K indicates either the presence of the minority phase, or the occurrence of another magnetic phase transition [8].

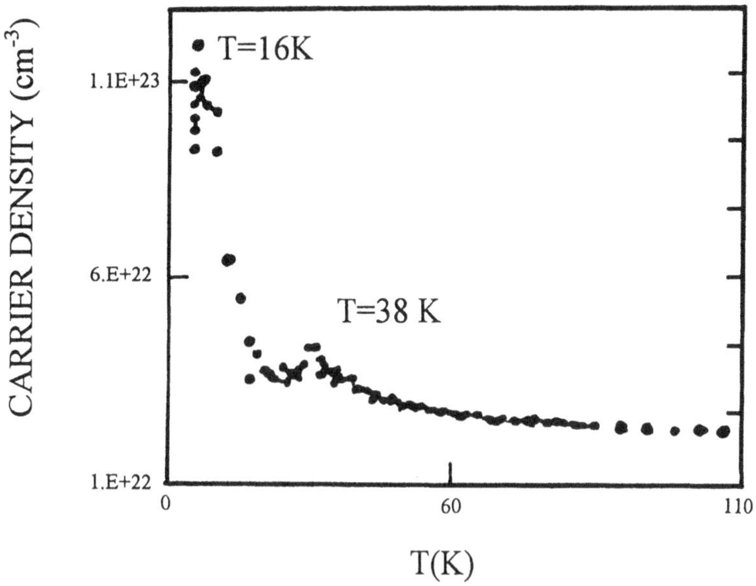


FIGURE 7 Carrier density n(T) of terbium silicide films (after [8]).

## REFERENCES


[1] M.-A. Nicolet, S.S. Lau [ in *VLSI Electronics-Microstructure Science no.6* Eds N.G. Einspruch, G.B. Larrabee (1983) and references therein ]

[2] [ *Physical and Chemical Constants of Rare Earths* (Industry of Metallurgy Press, Beijing, 1978) and references therein ]

[3] J.Y. Duboz, P.A. Badoz, F. Arnaud, J.A. Chroboczek [ *Appl. Phys. Lett. (USA)* vol.55 (1989) p.84-6 ]

[4] J.Y. Duboz et al [ *Appl. Surf. Sci. (Netherlands)* vol.38 (1989) p.171-7 ]

[5] F. Arnaud D'Avitaya et al [ *Thin Solid Films (Switzerland)* vol.184 (1990) p.283-93 ]

[6] I. Sagnes, G. Vincent, P.A. Badoz [ *J. Appl. Phys. (USA)* vol.72 (1992) p.4295-9 ]

[7] P.L. Janega, J. McCaffrey, D. Landheer [ *Appl. Phys. Lett. (USA)* vol.55 (1989) p.1415-7 ]

[8] F.H. Kaatz, W.R. Graham, J. Van der Spiegel, W. Joss, J.A. Chroboczek [ *J. Vac. Sci. Technol. A (USA)* vol.9 (1991) p.426-9 ]

[9] A. Travlos, P. Aloupogiannis, E. Rokofyllou, C. Papastaikoudis, G. Weber, A. Traverse [ *J. Appl. Phys. (USA)* vol.72 (1992) p.948-52 ]

CHAPTER 6

# OPTICAL PROPERTIES

# 6.1 Reflectance, transmittance and ellipsometric spectroscopies of TM silicides

G. Guizetti and F. Marabelli

December 1994

## A INTRODUCTION

Optical techniques are one of the most widely used and varied tools to investigate the physical-chemical properties of materials. The 'classical' optical experiments, e.g. transmittance (absorption) and reflectance measurements, provide the basic data to investigate the fundamental properties. Besides reflectance (R) and transmittance (T), ellipsometric spectroscopy (SE) is now available over a larger spectral range than in the past. As we will see in the following Datareviews, these three techniques have been the most used to determine the optical functions of transition metal silicides.

The optical spectral functions which completely describe the macroscopic and linear response of a medium (homogeneous, isotropic and non-magnetic) to an electromagnetic wave are [1]: the complex refractive index $\tilde{n} = n + ik$, where n is the ordinary refractive index and k the extinction coefficient, respectively; the complex dielectric function $\tilde{\varepsilon} = \varepsilon_1 + i\varepsilon_2$, where $\varepsilon$ denotes permittivity, which is related to $\tilde{n}$ by the simple relation $\tilde{\varepsilon} = \tilde{n}^2$; moreover, sometimes it is useful to work with the absorption coefficient $\alpha = 4\pi k/\lambda$, where $\lambda$ is the wavelength in vacuum.

n and k can be derived from R, T and SE measurements through the Fresnel formulae. More information about ellipsometric measurements can be found in [2,3].

Although their physical background is the same, different methods can be used to extract the optical functions from the experimental R, T or SE measurements; in particular the treatment of data is different for bulk samples from that used for thin films (see below). Moreover, the accurate determination of the intrinsic optical properties of a material require not only accurate instrumentation but also well-characterised samples. The optical response depends on:

(a) the electronic structure of the material, e.g. the density of states and the character of the electronic bands, through the complex dielectric function;

(b) the geometrical configuration of the experiment; that is to say the angle of incidence $\phi$, the polarisation of the light and the orientation of the crystallographic axes of the sample with respect to both the surface and the direction of polarisation;

(c) the quality of surfaces and interfaces; roughness at interfaces scatters the light of wavelength comparable to or smaller than the dimension of the structural defects. Such an effect can affect the absolute value of the measured reflectance significantly, whereas ellipsometry, measuring the ratio of two reflection coefficients on the same surface, is, in general and for the shorter wavelengths, much less sensitive to this problem [3];

(d) oxide layers or transition layers at surfaces and interfaces; their effects depend on the penetration depth of the light (and hence on the wavelength). Ellipsometry, which is sensitive also to very thin films (up to fractions of monolayers) can be affected by such extra layers if they are not readily characterised and accounted for in determining the optical response of the system [3];

(e) the number of impurities (dopants) and/or defects; these can influence the optical response by introducing spurious structures and by changing the free-carrier density, but, within a certain concentration (a few parts per thousand), such an influence is negligible in the metallic compounds. The problem arises in the semiconducting compounds for which the contribution of extrinsic free carriers significantly affects the properties below the absorption edge. The effect is particularly evident for single crystalline or bulk silicides whose growth conditions do not permit very accurate control of purity of the material.

An overview of the optical properties of silicides and their correlation with electronic and vibrational structures has been given in [4,5].

## B DATA ANALYSIS FOR BULK MATERIALS

The data obtained by the different techniques must be correctly analysed in order to extract reliable and quantitative information. The scheme of analysis used depends, obviously, also on the geometrical and structural features of the samples.

In the spectral region of greatest interest optical absorption is generally quite high so that in bulk materials (single crystals, massive polycrystals or sufficiently thick films) a negligibly small fraction of incident light is transmitted. For example, many semiconducting silicides in the visible and UV regions or metallic silicides in the far infrared (FIR) region have a mean absorption depth of the order of a few hundred Angstrom. It is not always feasible to make films so thin that they possess a sufficient quality to obtain meaningful data. Thus, most experiments consist of reflectivity (R) measurements on a plane surface of the sample.

To determine n and k (or $\varepsilon_1$ and $\varepsilon_2$) at least two independent parameters should be measured at each wavelength. For R measurements at oblique incidence, the most useful methods can be divided into two classes [6]:

(1) R measurements at one angle of incidence with light polarised perpendicular ($R_s$) and parallel ($R_p$) to the plane of incidence, or R (natural or polarised) at two angles;

(2) R measurements at any angle and measurement of the Brewster (or pseudo-Brewster) angle, which supplies the additional information.

R measurement at two angles is a relatively simple technique, because it does not use polarised light. However, it requires relatively large samples, the algorithm of the solution is quite complicated and can give multiple solutions, and moreover it becomes insensitive when the difference in the angles is not large enough or the absorption of the material is large (as happens in metallic compounds).

Another way to extract n and k from R measurements is to use normal incidence and the causality relations, or Kramers-Kronig relations [7,8]. These imply knowledge of the whole R spectrum or, at least, of a large part of it. The relations have a quasi-local character, so one can reasonably extrapolate the R spectrum beyond the experimental limits. A reliable extrapolation can be obtained by fixing the solution of the Kramers-Kronig equations in one or more energies by means of results obtained independently [9], e.g. with ellipsometric measurements. This is the case for most of the data reported in the Tables.

Spectroscopic ellipsometry is certainly the more precise and accurate method to directly obtain n and k from near infrared (NIR) to vacuum ultraviolet (VUV) for bulk material [3].

## C DATA ANALYSIS FOR THIN FILMS

The results obtained from optical measurements on thin films can be affected by several contributions originating from the presence of a substrate. In order to extract the correct information on the film one must use characterised substrates or independently characterise the substrate.

For qualitative analysis, and when the only interest is in determining the excitation energies (e.g. for luminescence, Raman and, in some cases, IR transmittance measurements), it is enough to identify the spectral features of the substrate response and normalise the measured spectra with respect to this contribution.

However, one must, in general, take into account the optical response of the substrate and the interference effects due to the geometrical arrangement of the sample [10]. Hence, one needs a reliable multilayer model for the optical response of the film/substrate system and a good characterisation of all the interfaces (air/sample, sample/substrate and substrate/air) as well as knowledge of the film and substrate thicknesses.

As in the case of bulk samples one needs two different measurements. Such a condition is either fulfilled by ellipsometry or it can be obtained by measuring both reflectance and transmittance of a sample or by considering several samples with different film thicknesses [11].

Finally, we stress that, apart from intrinsic differences in films of the same compound depending on the growth technique, the optical constants and the thickness measured on the same film with different methods can differ by more than 10% [11].

## D EFFECTS OF SURFACE CONDITIONS, CRYSTALLINE ORDER, STRESS

The models and the formulae used to obtain the optical functions from photometric and ellipsometric measurements assume homogeneous materials, samples with flat, planar and mirror-like surfaces and films with sharp and planar-parallel interfaces. Actually, this is not always the case and the optical response of real samples is modified by surface roughness, crystalline order or stress effects. This is a handicap in extracting the bulk properties of the material, but it makes optics a powerful tool to investigate the structural properties and to test the sample quality.

The above-mentioned anomalies or processes give rise to different effects:

(a) Roughness of the surfaces scatters the light reducing the amplitude of the detected signal. The diffusion increases with increasing photon energy (and decreasing wavelength): in the Rayleigh regime the scattered light is proportional to the fourth power of the energy. Thus diffusion depends on the scale of roughness as well as on the energy range investigated. The effect on results obtained by reflectance measurements is a reduction of the spectral weight of the structures at the higher energies. The energy and the bandwidth of such structures are, in general, not changed in a subtle way. In the case of roughness at the sample/substrate interface in thin films, the scattering tends to reduce the amplitude of the interference fringes, giving a spurious absorption-like effect.

(b) Microscopic disorder of the material is reflected in a smearing and a broadening of all the spectral features (regardless of the photon energy). A decrease of the structure amplitudes together with an increase of their bandwidths occurs in the disordered samples (amorphous or polycrystalline films) with respect to the more ordered ones (single crystals or epitaxial films). In semiconducting compounds crystallographic defects can act as dopants, changing the carrier concentration and hence the response at the lowest energies. Furthermore, disorder can make optically observable vibrational modes usually forbidden, giving a spectral contribution roughly proportional to the phonon density.

(c) Strain (uniaxial or biaxial) in thin films epitaxially grown, due to the lattice mismatch between epilayer and substrate, modifies the optical response mainly by changing the interband energy gaps as well as the optical phonon energies. Moreover, in some cases, it induces splitting of the structures by removing symmetry degeneracies. Elastic accommodation of the mismatch by strain is the preferred mechanism for thin films up to a critical thickness $t_c$; above $t_c$, misfit dislocations are generated which relax the built-in strain.

## REFERENCES


[1] M. Born, E. Wolf [ *Principles of Optics* (Pergamon Press, Oxford, 1980) ]

[2] R.M.A. Azzam, N.M. Bashara [ *Ellipsometry and Polarized Light* (North-Holland, Amsterdam, 1977) ]

[3] D.E. Aspnes [ *Handbook of Optical Constants of Solids* Ed. E.D. Palik (Academic Press, Orlando, 1985) ch.5 p.89-112 ]

[4] G. Guizzetti, A. Piaggi [ *Proc. of Int. School of Physics E.Fermi* Eds A. Stella, L. Miglio (North-Holland, Amsterdam, 1993) p.123-55 ]

[5] F. Nava et al [ *Mater. Sci. Rep. (Netherlands)* vol.9 (1993) p.141-200 ]

[6] W.R. Hunter [ *Handbook of Optical Constants of Solids* Ed. E.D. Palik (Academic Press, Orlando, 1985) ch.4 p.69-88 ]

[7] F. Stern [ *Solid State Physics* vol.15 Eds F. Seitz, D. Turnbull (Academic Press, New York, 1963) p.299-408 ]

[8] F. Wooten [ *Optical Properties of Solids* (Academic Press, New York, 1972) ]
[9] R. Hulthén [ *J. Opt. Soc. Am. (USA)* vol.72 (1982) p.794 ]
[10] O.S. Heavens [ *Optical Properties of Thin Solid Film* (Dover, New York, 1965) ]
[11] E. Pelletier [ *Handbook of Optical Constants of Solids* vol.2, Ed. E.D. Palik (Academic Press, Orlando, 1991) ch.3 p.57-73 ]

# 6.2 Optical functions: refractory TM silicides

G. Guizetti and F. Marabelli

December 1994

## A INTRODUCTION

Refractory metals form metallic silicides. The usual stoichiometry considered is $MSi_2$, even if other forms also exist. Vanadium silicides are the most studied, together with $TiSi_2$ and $WSi_2$. A great deal of literature exists starting from the 1970s and, due to the strong technological interest, sample preparation has been considerably improved in recent years. We refer for the data in the tables to the most recent and complete measurements, and, where available, to measurements on single crystals. Refractory metal disilicides crystallise in slightly different structures: orthorhombic C54 ($TiSi_2$), hexagonal C40 (V-, Nb-, Ta- and low temperature Mo-disilicide) and body-centred tetragonal structure C11b ($MoSi_2$ at room temperature and $WSi_2$). Measurements on single crystals have been performed with light polarised along the crystallographic axes.

## B $TiSi_2$

Optical data on thin films obtained from reflectivity and ellipsometric measurements are reported in [1] and [2], respectively. Measurements on single crystals are presented in [3] (unpolarised light) and, in a very extended energy range, in [4] (polarised light). TABLE 1 summarises the results.

TABLE 1 Refractive index (n) and extinction coefficient (k) for $TiSi_2$. The data have been obtained from measurements on a single crystal. The light was polarised along the crystallographic axes a and b of the sample (orthorhombic C54 structure). The high energy data have been obtained with synchrotron radiation [4].

| eV | $n_a$ | $k_a$ | $n_b$ | $k_b$ |
|---|---|---|---|---|
| 0.02 | 19.570 | 196.770 | 12.676 | 196.950 |
| 0.03 | 13.379 | 149.825 | 8.913 | 149.125 |
| 0.04 | 7.188 | 102.880 | 5.149 | 101.300 |
| 0.05 | 5.646 | 86.091 | 4.146 | 84.703 |
| 0.06 | 4.105 | 69.302 | 3.143 | 68.107 |
| 0.07 | 3.496 | 60.798 | 2.729 | 59.726 |
| 0.08 | 2.887 | 52.295 | 2.316 | 51.345 |
| 0.09 | 2.529 | 47.137 | 2.053 | 46.269 |
| 0.10 | 1.809 | 35.673 | 1.520 | 34.997 |
| 0.20 | 1.106 | 21.248 | 0.984 | 20.833 |
| 0.30 | 0.948 | 13.629 | 0.878 | 13.355 |
| 0.40 | 0.989 | 9.703 | 0.930 | 9.498 |
| 0.50 | 1.279 | 7.306 | 1.208 | 7.131 |
| 0.60 | 1.647 | 5.791 | 1.559 | 5.537 |
| 0.70 | 1.905 | 4.859 | 2.023 | 4.565 |
| 0.80 | 2.061 | 4.128 | 2.380 | 4.017 |

Table 1 continued

| eV | $n_a$ | $k_a$ | $n_b$ | $k_b$ |
|---|---|---|---|---|
| 0.90 | 2.300 | 3.476 | 2.596 | 3.716 |
| 1.00 | 2.667 | 3.071 | 2.623 | 3.507 |
| 1.10 | 3.044 | 2.949 | 2.659 | 3.267 |
| 1.20 | 3.152 | 3.069 | 2.724 | 3.089 |
| 1.30 | 3.113 | 3.075 | 2.768 | 2.963 |
| 1.40 | 3.052 | 3.079 | 2.758 | 2.926 |
| 1.50 | 2.936 | 3.022 | 2.739 | 2.790 |
| 1.60 | 2.894 | 2.919 | 2.760 | 2.780 |
| 1.70 | 2.891 | 2.881 | 2.700 | 2.738 |
| 1.80 | 2.862 | 2.899 | 2.667 | 2.703 |
| 1.90 | 2.780 | 2.934 | 2.611 | 2.681 |
| 2.00 | 2.648 | 2.943 | 2.541 | 2.647 |
| 2.10 | 2.504 | 2.901 | 2.471 | 2.600 |
| 2.20 | 2.384 | 2.824 | 2.410 | 2.540 |
| 2.30 | 2.294 | 2.734 | 2.369 | 2.475 |
| 2.40 | 2.226 | 2.644 | 2.352 | 2.426 |
| 2.50 | 2.179 | 2.560 | 2.332 | 2.400 |
| 2.60 | 2.148 | 2.486 | 2.302 | 2.384 |
| 2.70 | 2.125 | 2.423 | 2.262 | 2.366 |
| 2.80 | 2.114 | 2.371 | 2.226 | 2.345 |
| 2.90 | 2.105 | 2.338 | 2.189 | 2.329 |
| 3.00 | 2.089 | 2.321 | 2.143 | 2.310 |
| 3.10 | 2.054 | 2.302 | 2.103 | 2.280 |
| 3.20 | 2.024 | 2.274 | 2.069 | 2.251 |
| 3.30 | 1.998 | 2.243 | 2.046 | 2.215 |
| 3.40 | 1.986 | 2.216 | 2.033 | 2.193 |
| 3.50 | 1.978 | 2.202 | 2.030 | 2.178 |
| 3.60 | 1.966 | 2.202 | 2.030 | 2.185 |
| 3.70 | 1.940 | 2.204 | 2.005 | 2.195 |
| 3.80 | 1.907 | 2.203 | 1.987 | 2.203 |
| 3.90 | 1.885 | 2.194 | 1.956 | 2.221 |
| 4.00 | 1.862 | 2.194 | 1.928 | 2.225 |
| 4.10 | 1.834 | 2.198 | 1.908 | 2.240 |
| 4.20 | 1.805 | 2.202 | 1.878 | 2.271 |
| 4.30 | 1.775 | 2.209 | 1.835 | 2.301 |
| 4.40 | 1.738 | 2.218 | 1.779 | 2.329 |
| 4.50 | 1.701 | 2.224 | 1.716 | 2.351 |
| 4.60 | 1.659 | 2.231 | 1.647 | 2.360 |
| 4.70 | 1.614 | 2.236 | 1.579 | 2.367 |
| 4.80 | 1.570 | 2.237 | 1.508 | 2.365 |
| 4.90 | 1.523 | 2.242 | 1.437 | 2.358 |
| 5.00 | 1.471 | 2.242 | 1.369 | 2.341 |
| 5.10 | 1.420 | 2.243 | 1.308 | 2.325 |
| 5.20 | 1.363 | 2.239 | 1.244 | 2.305 |
| 5.30 | 1.306 | 2.229 | 1.183 | 2.281 |
| 5.40 | 1.251 | 2.216 | 1.128 | 2.258 |
| 5.50 | 1.193 | 2.196 | 1.067 | 2.231 |
| 5.60 | 1.143 | 2.174 | 1.011 | 2.200 |
| 5.70 | 1.093 | 2.152 | 0.958 | 2.167 |
| 5.80 | 1.045 | 2.128 | 0.909 | 2.132 |
| 5.90 | 0.998 | 2.108 | 0.861 | 2.102 |
| 6.00 | 0.946 | 2.081 | 0.813 | 2.062 |
| 6.10 | 0.895 | 2.048 | 0.764 | 2.020 |

Table 1 continued.

| eV | $n_a$ | $k_a$ | $n_b$ | $k_b$ |
|---|---|---|---|---|
| 6.20 | 0.851 | 2.013 | 0.721 | 1.977 |
| 6.30 | 0.807 | 1.977 | 0.683 | 1.931 |
| 6.40 | 0.768 | 1.937 | 0.649 | 1.883 |
| 6.50 | 0.730 | 1.894 | 0.621 | 1.836 |
| 6.60 | 0.699 | 1.853 | 0.594 | 1.794 |
| 6.70 | 0.673 | 1.810 | 0.569 | 1.747 |
| 6.80 | 0.650 | 1.771 | 0.549 | 1.703 |
| 6.90 | 0.629 | 1.731 | 0.532 | 1.660 |
| 7.00 | 0.609 | 1.695 | 0.515 | 1.620 |
| 7.10 | 0.592 | 1.662 | 0.500 | 1.580 |
| 7.20 | 0.573 | 1.626 | 0.488 | 1.538 |
| 7.30 | 0.554 | 1.595 | 0.475 | 1.503 |
| 7.40 | 0.537 | 1.560 | 0.464 | 1.463 |
| 7.50 | 0.520 | 1.527 | 0.456 | 1.428 |
| 7.60 | 0.503 | 1.490 | 0.447 | 1.391 |
| 7.70 | 0.491 | 1.456 | 0.442 | 1.355 |
| 7.80 | 0.480 | 1.421 | 0.440 | 1.321 |
| 7.90 | 0.470 | 1.388 | 0.437 | 1.291 |
| 8.00 | 0.462 | 1.362 | 0.434 | 1.267 |
| 8.10 | 0.451 | 1.324 | 0.430 | 1.233 |
| 8.20 | 0.444 | 1.291 | 0.426 | 1.203 |
| 8.30 | 0.437 | 1.260 | 0.426 | 1.173 |
| 8.40 | 0.431 | 1.228 | 0.426 | 1.147 |
| 8.50 | 0.427 | 1.195 | 0.424 | 1.121 |
| 8.60 | 0.426 | 1.164 | 0.425 | 1.094 |
| 8.70 | 0.425 | 1.135 | 0.426 | 1.069 |
| 8.80 | 0.424 | 1.106 | 0.428 | 1.045 |
| 8.90 | 0.424 | 1.076 | 0.430 | 1.022 |
| 9.00 | 0.429 | 1.047 | 0.432 | 1.001 |
| 9.10 | 0.434 | 1.021 | 0.434 | 0.979 |
| 9.20 | 0.440 | 0.995 | 0.437 | 0.958 |
| 9.30 | 0.446 | 0.972 | 0.439 | 0.938 |
| 9.40 | 0.454 | 0.950 | 0.442 | 0.917 |
| 9.50 | 0.462 | 0.929 | 0.445 | 0.897 |
| 9.60 | 0.472 | 0.910 | 0.450 | 0.877 |
| 9.70 | 0.483 | 0.893 | 0.458 | 0.858 |
| 9.80 | 0.493 | 0.879 | 0.466 | 0.841 |
| 9.90 | 0.506 | 0.869 | 0.477 | 0.828 |
| 10.00 | 0.513 | 0.865 | 0.485 | 0.819 |
| 10.10 | 0.516 | 0.859 | 0.488 | 0.813 |
| 10.20 | 0.520 | 0.854 | 0.492 | 0.804 |
| 10.30 | 0.516 | 0.851 | 0.490 | 0.798 |
| 10.40 | 0.512 | 0.843 | 0.489 | 0.789 |
| 10.50 | 0.506 | 0.834 | 0.484 | 0.779 |
| 10.60 | 0.498 | 0.822 | 0.480 | 0.765 |
| 10.70 | 0.494 | 0.805 | 0.477 | 0.749 |
| 10.80 | 0.491 | 0.789 | 0.477 | 0.732 |
| 10.90 | 0.492 | 0.772 | 0.479 | 0.715 |
| 11.00 | 0.492 | 0.756 | 0.484 | 0.698 |
| 11.10 | 0.496 | 0.739 | 0.490 | 0.683 |
| 11.20 | 0.501 | 0.724 | 0.501 | 0.671 |
| 11.30 | 0.510 | 0.712 | 0.506 | 0.664 |
| 11.40 | 0.516 | 0.703 | 0.512 | 0.655 |

Table 1 continued.

| eV | $n_a$ | $k_a$ | $n_b$ | $k_b$ |
|---|---|---|---|---|
| 11.50 | 0.520 | 0.693 | 0.519 | 0.648 |
| 11.60 | 0.527 | 0.684 | 0.525 | 0.643 |
| 11.70 | 0.534 | 0.679 | 0.529 | 0.640 |
| 11.80 | 0.538 | 0.676 | 0.532 | 0.636 |
| 11.90 | 0.539 | 0.673 | 0.532 | 0.634 |
| 12.00 | 0.540 | 0.669 | 0.531 | 0.630 |
| 12.10 | 0.540 | 0.666 | 0.531 | 0.626 |
| 12.20 | 0.539 | 0.663 | 0.529 | 0.622 |
| 12.30 | 0.536 | 0.658 | 0.527 | 0.617 |
| 12.40 | 0.533 | 0.653 | 0.523 | 0.612 |
| 12.50 | 0.529 | 0.647 | 0.521 | 0.606 |
| 12.60 | 0.526 | 0.639 | 0.516 | 0.600 |
| 12.70 | 0.524 | 0.633 | 0.513 | 0.591 |
| 12.80 | 0.520 | 0.626 | 0.511 | 0.584 |
| 12.90 | 0.516 | 0.617 | 0.507 | 0.576 |
| 13.00 | 0.515 | 0.608 | 0.505 | 0.566 |
| 13.10 | 0.513 | 0.600 | 0.504 | 0.558 |
| 13.20 | 0.511 | 0.592 | 0.503 | 0.550 |
| 13.30 | 0.509 | 0.584 | 0.502 | 0.542 |
| 13.40 | 0.509 | 0.576 | 0.500 | 0.534 |
| 13.50 | 0.507 | 0.568 | 0.500 | 0.526 |
| 13.60 | 0.507 | 0.561 | 0.499 | 0.519 |
| 13.70 | 0.504 | 0.555 | 0.497 | 0.513 |
| 13.80 | 0.502 | 0.547 | 0.496 | 0.505 |
| 13.90 | 0.500 | 0.539 | 0.493 | 0.498 |
| 14.00 | 0.498 | 0.532 | 0.493 | 0.490 |
| 14.10 | 0.495 | 0.525 | 0.489 | 0.484 |
| 14.20 | 0.492 | 0.517 | 0.487 | 0.476 |
| 14.30 | 0.490 | 0.509 | 0.483 | 0.468 |
| 14.40 | 0.487 | 0.501 | 0.480 | 0.459 |
| 14.50 | 0.484 | 0.492 | 0.478 | 0.450 |
| 14.60 | 0.482 | 0.482 | 0.477 | 0.441 |
| 14.70 | 0.480 | 0.474 | 0.474 | 0.432 |
| 14.80 | 0.478 | 0.464 | 0.472 | 0.422 |
| 14.90 | 0.476 | 0.455 | 0.471 | 0.412 |
| 15.00 | 0.475 | 0.445 | 0.470 | 0.403 |
| 15.10 | 0.474 | 0.435 | 0.469 | 0.393 |
| 15.20 | 0.473 | 0.426 | 0.468 | 0.382 |
| 15.30 | 0.472 | 0.416 | 0.469 | 0.373 |
| 15.40 | 0.472 | 0.406 | 0.468 | 0.363 |
| 15.50 | 0.472 | 0.397 | 0.469 | 0.354 |
| 15.60 | 0.472 | 0.388 | 0.469 | 0.345 |
| 15.70 | 0.473 | 0.378 | 0.469 | 0.335 |
| 15.80 | 0.473 | 0.370 | 0.470 | 0.325 |
| 15.90 | 0.473 | 0.360 | 0.471 | 0.316 |
| 16.00 | 0.474 | 0.351 | 0.473 | 0.307 |
| 16.10 | 0.473 | 0.342 | 0.473 | 0.298 |
| 16.20 | 0.475 | 0.332 | 0.474 | 0.288 |
| 16.30 | 0.476 | 0.324 | 0.476 | 0.279 |
| 16.40 | 0.477 | 0.315 | 0.477 | 0.271 |
| 16.50 | 0.478 | 0.306 | 0.478 | 0.261 |
| 16.60 | 0.480 | 0.298 | 0.481 | 0.252 |
| 16.70 | 0.481 | 0.290 | 0.482 | 0.244 |

Table 1 continued

| eV | $n_a$ | $k_a$ | $n_b$ | $k_b$ |
|---|---|---|---|---|
| 16.80 | 0.482 | 0.281 | 0.484 | 0.235 |
| 16.90 | 0.483 | 0.272 | 0.486 | 0.226 |
| 17.00 | 0.484 | 0.264 | 0.489 | 0.217 |
| 17.10 | 0.485 | 0.255 | 0.490 | 0.209 |
| 17.20 | 0.486 | 0.246 | 0.492 | 0.200 |
| 17.30 | 0.487 | 0.238 | 0.494 | 0.191 |
| 17.40 | 0.488 | 0.229 | 0.495 | 0.181 |
| 17.50 | 0.489 | 0.219 | 0.498 | 0.172 |
| 17.60 | 0.491 | 0.209 | 0.501 | 0.163 |
| 17.70 | 0.493 | 0.199 | 0.503 | 0.153 |
| 17.80 | 0.495 | 0.189 | 0.505 | 0.142 |
| 17.90 | 0.497 | 0.178 | 0.508 | 0.132 |
| 18.00 | 0.500 | 0.167 | 0.512 | 0.120 |
| 18.10 | 0.504 | 0.157 | 0.518 | 0.110 |
| 18.20 | 0.507 | 0.146 | 0.522 | 0.100 |
| 18.30 | 0.512 | 0.134 | 0.528 | 0.088 |
| 18.40 | 0.517 | 0.122 | 0.534 | 0.077 |
| 18.50 | 0.524 | 0.111 | 0.542 | 0.066 |
| 18.60 | 0.531 | 0.100 | 0.550 | 0.055 |
| 18.70 | 0.538 | 0.088 | 0.560 | 0.044 |
| 18.80 | 0.550 | 0.078 | 0.571 | 0.035 |
| 18.90 | 0.558 | 0.068 | 0.582 | 0.026 |
| 19.00 | 0.571 | 0.058 | 0.595 | 0.018 |
| 19.10 | 0.583 | 0.051 | 0.609 | 0.012 |
| 19.20 | 0.597 | 0.045 | 0.623 | 0.007 |
| 19.30 | 0.611 | 0.041 | 0.638 | 0.005 |
| 19.40 | 0.624 | 0.039 | 0.653 | 0.005 |
| 19.50 | 0.640 | 0.040 | 0.668 | 0.008 |

## C $ZrSi_2$

The only available data have been obtained in a limited spectral range by spectroscopic ellipsometry on polycrystalline films [5] and are reproduced in TABLE 2.

TABLE 2 Refractive index (n) and extinction coefficient (k) for $ZrSi_2$. The data have been obtained from measurements on polycrystalline films [5].

| eV | n | k |
|---|---|---|
| 1.8 | 3.91 | 3.36 |
| 1.9 | 3.70 | 3.27 |
| 2.0 | 3.66 | 3.11 |
| 2.1 | 3.55 | 3.28 |
| 2.2 | 3.44 | 3.16 |
| 2.3 | 3.33 | 3.18 |
| 2.4 | 3.26 | 3.16 |

| eV | n | k |
|---|---|---|
| 2.5 | 3.23 | 3.15 |
| 2.6 | 3.04 | 3.11 |
| 2.7 | 2.92 | 3.11 |
| 2.8 | 2.86 | 3.05 |
| 2.9 | 2.79 | 2.95 |
| 3.0 | 2.71 | 2.96 |

## D $HfSi_2$

TABLE 3 summarises the results for $HfSi_2$ from [6].

TABLE 3 Refractive index (n) and extinction coefficient (k) for $HfSi_2$. The data have been obtained by reflectivity and ellipsometric measurements on polycrystalline films [6].

| eV | n | k |
|---|---|---|
| 0.01 | 46.881 | 75.950 |
| 0.02 | 24.463 | 44.364 |
| 0.03 | 19.068 | 35.287 |
| 0.04 | 15.044 | 29.530 |
| 0.05 | 12.360 | 25.382 |
| 0.06 | 10.397 | 22.198 |
| 0.07 | 9.292 | 19.609 |
| 0.08 | 8.744 | 18.057 |
| 0.09 | 8.093 | 16.457 |
| 0.10 | 6.701 | 12.961 |
| 0.20 | 4.995 | 8.895 |
| 0.30 | 3.830 | 6.117 |
| 0.40 | 3.478 | 4.403 |
| 0.50 | 3.498 | 3.332 |
| 0.60 | 3.636 | 2.752 |
| 0.70 | 3.821 | 2.274 |
| 0.80 | 4.045 | 2.036 |
| 0.90 | 4.245 | 1.980 |
| 1.00 | 4.394 | 2.138 |
| 1.10 | 4.291 | 2.292 |
| 1.20 | 4.209 | 2.369 |
| 1.30 | 4.085 | 2.424 |
| 1.40 | 3.943 | 2.469 |
| 1.50 | 3.773 | 2.483 |
| 1.60 | 3.621 | 2.446 |
| 1.70 | 3.495 | 2.396 |
| 1.80 | 3.387 | 2.322 |
| 1.90 | 3.301 | 2.253 |
| 2.00 | 3.263 | 2.170 |
| 2.10 | 3.249 | 2.116 |
| 2.20 | 3.251 | 2.072 |
| 2.30 | 3.252 | 2.051 |
| 2.40 | 3.236 | 2.060 |
| 2.50 | 3.244 | 2.002 |
| 2.60 | 3.306 | 1.986 |
| 2.70 | 3.367 | 2.056 |
| 2.80 | 3.385 | 2.153 |
| 2.90 | 3.381 | 2.250 |
| 3.00 | 3.376 | 2.350 |
| 3.10 | 3.342 | 2.481 |
| 3.20 | 3.279 | 2.614 |
| 3.30 | 3.142 | 2.760 |
| 3.40 | 2.960 | 2.835 |
| 3.50 | 2.809 | 2.815 |
| 3.60 | 2.683 | 2.816 |
| 3.70 | 2.597 | 2.807 |
| 3.80 | 2.519 | 2.805 |
| 3.90 | 2.439 | 2.815 |
| 4.00 | 2.351 | 2.830 |
| 4.10 | 2.274 | 2.857 |
| 4.20 | 2.149 | 2.882 |
| 4.30 | 1.996 | 2.880 |
| 4.40 | 1.841 | 2.823 |
| 4.50 | 1.733 | 2.752 |
| 4.60 | 1.656 | 2.703 |
| 4.70 | 1.595 | 2.632 |
| 4.80 | 1.536 | 2.557 |
| 4.90 | 1.497 | 2.488 |
| 5.00 | 1.466 | 2.435 |
| 5.10 | 1.439 | 2.416 |
| 5.20 | 1.419 | 2.394 |
| 5.30 | 1.384 | 2.357 |
| 5.40 | 1.335 | 2.328 |
| 5.50 | 1.296 | 2.312 |
| 5.60 | 1.256 | 2.273 |
| 5.70 | 1.225 | 2.243 |
| 5.80 | 1.201 | 2.201 |
| 5.90 | 1.168 | 2.183 |
| 6.00 | 1.131 | 2.149 |
| 6.10 | 1.097 | 2.112 |
| 6.20 | 1.077 | 2.090 |
| 6.30 | 1.049 | 2.065 |
| 6.40 | 1.020 | 2.038 |
| 6.50 | 0.994 | 2.023 |
| 6.60 | 0.967 | 1.995 |
| 6.70 | 0.934 | 1.975 |
| 6.80 | 0.920 | 1.938 |
| 6.90 | 0.899 | 1.920 |
| 7.00 | 0.876 | 1.899 |
| 7.10 | 0.851 | 1.871 |
| 7.20 | 0.820 | 1.846 |
| 7.30 | 0.792 | 1.816 |
| 7.40 | 0.776 | 1.788 |
| 7.50 | 0.748 | 1.761 |
| 7.60 | 0.715 | 1.720 |
| 7.70 | 0.700 | 1.683 |
| 7.80 | 0.683 | 1.655 |
| 7.90 | 0.673 | 1.641 |
| 8.00 | 0.661 | 1.624 |
| 8.10 | 0.644 | 1.594 |
| 8.20 | 0.631 | 1.574 |
| 8.30 | 0.622 | 1.560 |
| 8.40 | 0.596 | 1.526 |
| 8.50 | 0.586 | 1.490 |
| 8.60 | 0.559 | 1.476 |
| 8.70 | 0.535 | 1.464 |
| 8.80 | 0.533 | 1.404 |
| 8.90 | 0.525 | 1.385 |
| 9.00 | 0.516 | 1.364 |
| 9.10 | 0.497 | 1.317 |
| 9.20 | 0.500 | 1.300 |
| 9.30 | 0.505 | 1.275 |
| 9.40 | 0.478 | 1.262 |
| 9.50 | 0.448 | 1.248 |

## E $VSi_2$

Vanadium silicides have been studied for a long time, particularly in the $V_3Si$ superconducting phase. The most stable $VSi_2$ phase currently attracts the majority of interest for technological applications. A set of data obtained from reflectivity measurements on polycrystalline samples with different stoichiometries ($V_3Si$, $V_5Si_3$, $VSi_2$) was given in [7], whereas a comparison between disilicides of column V is made in [8]. Single crystal data listed in TABLE 4 are taken from [9], where reflectivity and ellipsometry have been used together.

TABLE 4 Refractive index (n) and extinction coefficient (k) for $VSi_2$. The data have been obtained from measurements on single crystals with light polarised along the two crystallographic orientations perpendicular (⊥) and parallel (||) to the c-axis (hexagonal C40 structure) [9].

| eV | $n_\perp$ | $k_\perp$ | $n_\parallel$ | $k_\parallel$ |
|---|---|---|---|---|
| 0.01 | 69.357 | 96.302 | 58.339 | 74.676 |
| 0.02 | 42.369 | 65.296 | 37.059 | 51.362 |
| 0.03 | 31.268 | 51.587 | 28.142 | 41.044 |
| 0.04 | 24.926 | 43.404 | 22.961 | 34.875 |
| 0.05 | 20.772 | 37.831 | 19.539 | 30.673 |
| 0.06 | 17.828 | 33.738 | 17.218 | 27.646 |
| 0.07 | 15.645 | 30.589 | 15.052 | 25.715 |
| 0.08 | 13.956 | 28.293 | 12.541 | 23.416 |
| 0.09 | 12.046 | 25.647 | 10.890 | 21.485 |
| 0.10 | 9.136 | 21.774 | 8.787 | 17.203 |
| 0.20 | 5.082 | 14.650 | 6.694 | 12.054 |
| 0.30 | 3.340 | 9.564 | 4.522 | 8.640 |
| 0.40 | 2.642 | 6.707 | 3.670 | 6.104 |
| 0.50 | 2.612 | 4.623 | 3.668 | 4.308 |
| 0.60 | 2.969 | 3.344 | 4.012 | 3.408 |
| 0.70 | 3.595 | 2.455 | 4.238 | 2.918 |
| 0.80 | 4.144 | 2.145 | 4.518 | 2.540 |
| 0.90 | 4.573 | 1.992 | 4.790 | 2.441 |
| 1.00 | 4.810 | 2.029 | 4.810 | 2.362 |
| 1.10 | 5.025 | 2.273 | 5.042 | 2.365 |
| 1.20 | 5.050 | 2.620 | 5.125 | 2.561 |
| 1.30 | 4.797 | 2.772 | 5.101 | 2.777 |
| 1.40 | 4.522 | 2.740 | 4.965 | 2.940 |
| 1.50 | 4.474 | 2.688 | 4.752 | 3.114 |
| 1.60 | 4.433 | 2.686 | 4.567 | 3.201 |
| 1.70 | 4.370 | 2.723 | 4.292 | 3.232 |
| 1.80 | 4.308 | 2.782 | 4.049 | 3.190 |
| 1.90 | 4.187 | 2.863 | 3.859 | 3.108 |
| 2.00 | 3.997 | 2.878 | 3.683 | 2.987 |
| 2.10 | 3.854 | 2.822 | 3.561 | 2.855 |
| 2.20 | 3.774 | 2.763 | 3.496 | 2.727 |
| 2.30 | 3.711 | 2.711 | 3.470 | 2.610 |
| 2.40 | 3.670 | 2.674 | 3.472 | 2.522 |
| 2.50 | 3.632 | 2.654 | 3.485 | 2.463 |
| 2.60 | 3.588 | 2.634 | 3.510 | 2.424 |
| 2.70 | 3.554 | 2.602 | 3.526 | 2.405 |
| 2.80 | 3.546 | 2.581 | 3.550 | 2.397 |
| 2.90 | 3.548 | 2.572 | 3.565 | 2.406 |
| 3.00 | 3.562 | 2.575 | 3.593 | 2.422 |
| 3.10 | 3.593 | 2.613 | 3.609 | 2.468 |
| 3.20 | 3.601 | 2.683 | 3.611 | 2.505 |
| 3.30 | 3.593 | 2.750 | 3.620 | 2.541 |
| 3.40 | 3.550 | 2.845 | 3.629 | 2.607 |
| 3.50 | 3.486 | 2.930 | 3.628 | 2.692 |
| 3.60 | 3.431 | 3.002 | 3.616 | 2.781 |
| 3.70 | 3.380 | 3.052 | 3.599 | 2.858 |
| 3.80 | 3.323 | 3.118 | 3.558 | 2.938 |
| 3.90 | 3.221 | 3.196 | 3.497 | 3.043 |
| 4.00 | 3.141 | 3.247 | 3.434 | 3.120 |
| 4.10 | 3.023 | 3.302 | 3.359 | 3.221 |
| 4.20 | 2.916 | 3.335 | 3.263 | 3.302 |
| 4.30 | 2.854 | 3.328 | 3.185 | 3.342 |
| 4.40 | 2.733 | 3.346 | 3.068 | 3.381 |
| 4.50 | 2.633 | 3.346 | 2.977 | 3.452 |
| 4.60 | 2.580 | 3.312 | 2.862 | 3.505 |
| 4.70 | 2.525 | 3.316 | 2.714 | 3.509 |
| 4.80 | 2.490 | 3.282 | 2.659 | 3.397 |
| 4.90 | 2.502 | 3.265 | 2.688 | 3.368 |

## F NbSi$_2$

Data on polycrystalline films are shown in [8] and [10] (reflectivity and ellipsometry). In [8,9] the data are presented together with those of $VSi_2$.

TABLE 5 Refractive index (n) and extinction coefficient (k) for $NbSi_2$. The data have been obtained from measurements on single crystals with light polarised along the two crystallographic orientations perpendicular (⊥) and parallel (||) to the c-axis (hexagonal C40 structure) [9].

| eV | $n_\perp$ | $k_\perp$ | $n_\parallel$ | $k_\parallel$ |
|---|---|---|---|---|
| 0.01 | 82.761 | 114.429 | 99.572 | 161.870 |
| 0.02 | 39.984 | 72.109 | 35.940 | 97.150 |
| 0.03 | 26.633 | 55.482 | 20.674 | 69.554 |
| 0.04 | 20.014 | 45.121 | 14.090 | 53.985 |
| 0.05 | 15.460 | 37.682 | 10.531 | 43.526 |
| 0.06 | 13.512 | 32.774 | 9.315 | 36.560 |

Table 5 continued

| eV | $n_{\perp}$ | $k_{\perp}$ | $n_{\parallel}$ | $k_{\parallel}$ |
|---|---|---|---|---|
| 0.07 | 11.777 | 28.751 | 8.984 | 31.898 |
| 0.08 | 10.424 | 25.508 | 7.972 | 28.389 |
| 0.09 | 9.567 | 22.899 | 7.278 | 25.480 |
| 0.10 | 8.157 | 18.291 | 6.096 | 20.245 |
| 0.20 | 6.279 | 12.011 | 4.249 | 13.092 |
| 0.30 | 5.476 | 9.035 | 3.305 | 9.154 |
| 0.40 | 4.398 | 7.204 | 2.850 | 6.571 |
| 0.50 | 3.653 | 5.708 | 2.638 | 4.956 |
| 0.60 | 3.466 | 4.501 | 2.824 | 3.841 |
| 0.70 | 3.481 | 3.504 | 2.964 | 3.147 |
| 0.80 | 3.829 | 2.877 | 3.179 | 2.608 |
| 0.90 | 4.094 | 2.650 | 3.422 | 2.254 |
| 1.00 | 4.217 | 2.530 | 3.700 | 2.038 |
| 1.10 | 4.321 | 2.409 | 3.977 | 2.078 |
| 1.20 | 4.429 | 2.352 | 4.087 | 2.138 |
| 1.30 | 4.518 | 2.363 | 4.179 | 2.241 |
| 1.40 | 4.575 | 2.422 | 4.190 | 2.410 |
| 1.50 | 4.581 | 2.499 | 4.091 | 2.560 |
| 1.60 | 4.565 | 2.573 | 3.950 | 2.635 |
| 1.70 | 4.559 | 2.664 | 3.833 | 2.650 |
| 1.80 | 4.508 | 2.771 | 3.764 | 2.668 |
| 1.90 | 4.445 | 2.881 | 3.696 | 2.713 |
| 2.00 | 4.366 | 2.986 | 3.614 | 2.775 |
| 2.10 | 4.249 | 3.124 | 3.484 | 2.830 |
| 2.20 | 4.097 | 3.240 | 3.367 | 2.858 |
| 2.30 | 3.898 | 3.324 | 3.247 | 2.876 |
| 2.40 | 3.689 | 3.349 | 3.131 | 2.881 |
| 2.50 | 3.494 | 3.320 | 3.015 | 2.873 |
| 2.60 | 3.341 | 3.254 | 2.916 | 2.852 |
| 2.70 | 3.239 | 3.185 | 2.827 | 2.832 |
| 2.80 | 3.160 | 3.143 | 2.746 | 2.812 |
| 2.90 | 3.083 | 3.108 | 2.670 | 2.791 |
| 3.00 | 3.027 | 3.080 | 2.612 | 2.775 |
| 3.10 | 2.981 | 3.060 | 2.553 | 2.751 |
| 3.20 | 2.920 | 3.061 | 2.503 | 2.739 |
| 3.30 | 2.849 | 3.073 | 2.443 | 2.739 |
| 3.40 | 2.760 | 3.074 | 2.379 | 2.726 |
| 3.50 | 2.672 | 3.060 | 2.321 | 2.722 |
| 3.60 | 2.586 | 3.047 | 2.268 | 2.707 |
| 3.70 | 2.490 | 3.020 | 2.214 | 2.712 |
| 3.80 | 2.414 | 2.967 | 2.156 | 2.708 |
| 3.90 | 2.357 | 2.933 | 2.103 | 2.705 |
| 4.00 | 2.295 | 2.891 | 2.043 | 2.721 |
| 4.10 | 2.252 | 2.855 | 1.961 | 2.724 |
| 4.20 | 2.198 | 2.820 | 1.881 | 2.711 |
| 4.30 | 2.160 | 2.780 | 1.803 | 2.688 |
| 4.40 | 2.131 | 2.741 | 1.739 | 2.653 |
| 4.50 | 2.110 | 2.711 | 1.684 | 2.620 |
| 4.60 | 2.096 | 2.689 | 1.638 | 2.595 |
| 4.70 | 2.085 | 2.679 | 1.583 | 2.570 |
| 4.80 | 2.069 | 2.680 | 1.543 | 2.533 |
| 4.90 | 2.050 | 2.687 | 1.508 | 2.511 |

Table 5 continued

| eV | $n_{\perp}$ | $k_{\perp}$ | $n_{\parallel}$ | $k_{\parallel}$ |
|---|---|---|---|---|
| 5.00 | 2.024 | 2.698 | 1.471 | 2.489 |
| 5.10 | 1.993 | 2.710 | 1.437 | 2.467 |
| 5.20 | 1.958 | 2.722 | 1.407 | 2.446 |
| 5.30 | 1.923 | 2.735 | 1.382 | 2.428 |
| 5.40 | 1.880 | 2.748 | 1.360 | 2.416 |
| 5.50 | 1.838 | 2.754 | 1.341 | 2.406 |
| 5.60 | 1.806 | 2.755 | 1.337 | 2.403 |
| 5.70 | 1.793 | 2.757 | 1.357 | 2.416 |
| 5.80 | 1.862 | 2.776 | 1.339 | 2.536 |
| 5.90 | 1.655 | 2.996 | 1.158 | 2.614 |
| 6.00 | 1.400 | 3.200 | 0.976 | 2.693 |

## G $TaSi_2$

As for the previous disilicides, reflectivity measurements on polycrystalline films have been performed on $TaSi_2$, and the results are presented in [1] and [11]. Ellipsometric measurements have given the dielectric functions in the mid infrared [12]. The data listed in TABLE 6 have been obtained by reflectivity measurements on a single crystal [13]. The high energy data have been obtained with synchrotron radiation.

TABLE 6 Refractive index (n) and extinction coefficient (k) for $TaSi_2$. The light polarisation has been set perpendicular (⊥) and parallel (||) to the crystallographic c-axis of the sample (hexagonal C40 structure) [13].

| eV | $n_{\perp}$ | $k_{\perp}$ | $n_{\parallel}$ | $k_{\parallel}$ |
|---|---|---|---|---|
| 0.02 | 24.723 | 51.874 | 26.634 | 53.598 |
| 0.03 | 22.160 | 41.143 | 23.698 | 42.789 |
| 0.04 | 19.597 | 30.413 | 20.763 | 31.981 |
| 0.05 | 17.441 | 27.576 | 18.413 | 29.024 |
| 0.06 | 15.285 | 24.739 | 16.064 | 26.067 |
| 0.07 | 14.455 | 22.282 | 15.123 | 23.532 |
| 0.08 | 13.625 | 19.825 | 14.183 | 20.997 |
| 0.09 | 12.930 | 19.080 | 13.399 | 20.193 |
| 0.10 | 10.662 | 16.994 | 10.928 | 17.934 |
| 0.20 | 6.509 | 12.512 | 6.564 | 13.132 |
| 0.30 | 4.383 | 9.162 | 4.357 | 9.587 |
| 0.40 | 3.270 | 7.132 | 3.219 | 7.451 |
| 0.50 | 2.637 | 5.623 | 2.582 | 5.884 |
| 0.60 | 2.320 | 4.396 | 2.261 | 4.647 |
| 0.70 | 2.270 | 3.385 | 2.158 | 3.581 |
| 0.80 | 2.468 | 2.548 | 2.350 | 2.704 |
| 0.90 | 2.872 | 1.982 | 2.754 | 2.703 |
| 1.00 | 3.335 | 1.796 | 3.274 | 1.800 |
| 1.10 | 3.626 | 1.858 | 3.693 | 1.792 |
| 1.20 | 3.745 | 1.945 | 4.062 | 1.993 |
| 1.30 | 3.760 | 2.035 | 4.182 | 2.397 |
| 1.40 | 3.737 | 2.129 | 3.960 | 2.846 |
| 1.50 | 3.678 | 2.062 | 3.491 | 2.933 |

| eV | $n_{\perp}$ | $k_{\perp}$ | $n_{\parallel}$ | $k_{\parallel}$ |
|---|---|---|---|---|
| 1.60 | 3.748 | 2.074 | 3.207 | 2.766 |
| 1.70 | 3.783 | 2.150 | 3.086 | 2.578 |
| 1.80 | 3.797 | 2.251 | 3.082 | 2.438 |
| 1.90 | 3.780 | 2.380 | 3.123 | 2.382 |
| 2.00 | 3.718 | 2.521 | 3.165 | 2.404 |
| 2.10 | 3.614 | 2.652 | 3.164 | 2.487 |
| 2.20 | 3.468 | 2.769 | 3.096 | 2.588 |
| 2.30 | 3.285 | 2.838 | 2.978 | 2.665 |
| 2.40 | 3.113 | 2.849 | 2.829 | 2.694 |
| 2.50 | 2.959 | 2.836 | 2.702 | 2.683 |
| 2.60 | 2.832 | 2.801 | 2.584 | 2.656 |
| 2.70 | 2.729 | 2.749 | 2.484 | 2.621 |
| 2.80 | 2.655 | 2.721 | 2.398 | 2.573 |
| 2.90 | 2.588 | 2.687 | 2.340 | 2.524 |
| 3.00 | 2.532 | 2.667 | 2.287 | 2.495 |
| 3.10 | 2.490 | 2.664 | 2.241 | 2.472 |
| 3.20 | 2.431 | 2.666 | 2.189 | 2.452 |
| 3.30 | 2.368 | 2.674 | 2.139 | 2.435 |
| 3.40 | 2.310 | 2.700 | 2.089 | 2.425 |
| 3.50 | 2.225 | 2.715 | 2.025 | 2.411 |
| 3.60 | 2.121 | 2.724 | 1.969 | 2.373 |
| 3.70 | 2.033 | 2.711 | 1.931 | 2.341 |
| 3.80 | 1.924 | 2.687 | 1.906 | 2.309 |

Table 6 continued

| eV | $n_\perp$ | $k_\perp$ | $n_\parallel$ | $k_\parallel$ |
|---|---|---|---|---|
| 3.90 | 1.734 | 2.534 | 1.866 | 2.333 |
| 4.00 | 1.694 | 2.509 | 1.798 | 2.378 |
| 4.10 | 1.655 | 2.489 | 1.714 | 2.391 |
| 4.20 | 1.604 | 2.470 | 1.622 | 2.373 |
| 4.30 | 1.560 | 2.449 | 1.548 | 2.342 |
| 4.40 | 1.517 | 2.419 | 1.496 | 2.293 |
| 4.50 | 1.491 | 2.406 | 1.467 | 2.258 |
| 4.60 | 1.443 | 2.404 | 1.437 | 2.236 |
| 4.70 | 1.399 | 2.393 | 1.408 | 2.227 |
| 4.80 | 1.347 | 2.384 | 1.366 | 2.222 |
| 4.90 | 1.290 | 2.378 | 1.322 | 2.219 |
| 5.00 | 1.234 | 2.361 | 1.270 | 2.207 |
| 5.10 | 1.176 | 2.339 | 1.213 | 2.188 |
| 5.20 | 1.122 | 2.312 | 1.165 | 2.159 |
| 5.30 | 1.064 | 2.280 | 1.121 | 2.136 |
| 5.40 | 1.020 | 2.253 | 1.080 | 2.112 |
| 5.50 | 0.959 | 2.216 | 1.031 | 2.088 |
| 5.60 | 0.921 | 2.181 | 0.983 | 2.056 |
| 5.70 | 0.877 | 2.142 | 0.941 | 2.021 |
| 5.80 | 0.840 | 2.107 | 0.906 | 1.979 |
| 5.90 | 0.798 | 2.070 | 0.875 | 1.936 |
| 6.00 | 0.766 | 2.032 | 0.857 | 1.899 |
| 6.10 | 0.730 | 1.994 | 0.838 | 1.867 |
| 6.20 | 0.703 | 1.952 | 0.825 | 1.840 |
| 6.30 | 0.675 | 1.916 | 0.809 | 1.818 |
| 6.40 | 0.649 | 1.883 | 0.789 | 1.801 |
| 6.50 | 0.623 | 1.848 | 0.768 | 1.785 |
| 6.60 | 0.594 | 1.810 | 0.743 | 1.767 |
| 6.70 | 0.567 | 1.774 | 0.717 | 1.752 |
| 6.80 | 0.547 | 1.729 | 0.684 | 1.735 |
| 6.90 | 0.532 | 1.690 | 0.654 | 1.708 |
| 7.00 | 0.518 | 1.653 | 0.623 | 1.687 |
| 7.10 | 0.506 | 1.621 | 0.593 | 1.658 |
| 7.20 | 0.491 | 1.590 | 0.567 | 1.627 |
| 7.30 | 0.478 | 1.557 | 0.540 | 1.600 |
| 7.40 | 0.465 | 1.528 | 0.516 | 1.568 |
| 7.50 | 0.448 | 1.493 | 0.491 | 1.537 |
| 7.60 | 0.437 | 1.458 | 0.468 | 1.504 |
| 7.70 | 0.431 | 1.426 | 0.446 | 1.468 |
| 7.80 | 0.424 | 1.401 | 0.431 | 1.438 |
| 7.90 | 0.412 | 1.366 | 0.412 | 1.396 |
| 8.00 | 0.408 | 1.333 | 0.395 | 1.359 |
| 8.10 | 0.403 | 1.305 | 0.385 | 1.320 |
| 8.20 | 0.399 | 1.278 | 0.378 | 1.285 |
| 8.30 | 0.391 | 1.250 | 0.367 | 1.250 |
| 8.40 | 0.392 | 1.219 | 0.363 | 1.212 |
| 8.50 | 0.394 | 1.196 | 0.362 | 1.178 |
| 8.60 | 0.388 | 1.176 | 0.356 | 1.146 |
| 8.70 | 0.384 | 1.149 | 0.355 | 1.108 |
| 8.80 | 0.384 | 1.123 | 0.362 | 1.074 |
| 8.90 | 0.388 | 1.102 | 0.366 | 1.044 |
| 9.00 | 0.380 | 1.083 | 0.373 | 1.013 |

| eV | $n_\perp$ | $k_\perp$ | $n_\parallel$ | $k_\parallel$ |
|---|---|---|---|---|
| 9.10 | 1.834 | 2.640 | 1.897 | 2.295 |
| 9.20 | 1.772 | 2.584 | 1.884 | 2.303 |
| 9.30 | 0.383 | 1.058 | 0.389 | 0.987 |
| 9.40 | 0.383 | 1.038 | 0.396 | 0.968 |
| 9.50 | 0.385 | 1.018 | 0.408 | 0.947 |
| 9.60 | 0.384 | 1.001 | 0.417 | 0.931 |
| 9.70 | 0.378 | 0.980 | 0.423 | 0.915 |
| 9.80 | 0.380 | 0.955 | 0.435 | 0.900 |
| 9.90 | 0.388 | 0.936 | 0.440 | 0.891 |
| 10.00 | 0.383 | 0.920 | 0.441 | 0.880 |
| 10.10 | 0.392 | 0.899 | 0.447 | 0.866 |
| 10.20 | 0.392 | 0.885 | 0.451 | 0.856 |
| 10.30 | 0.397 | 0.868 | 0.453 | 0.845 |
| 10.40 | 0.397 | 0.854 | 0.457 | 0.834 |
| 10.50 | 0.401 | 0.837 | 0.458 | 0.826 |
| 10.60 | 0.399 | 0.821 | 0.455 | 0.815 |
| 10.70 | 0.407 | 0.802 | 0.459 | 0.802 |
| 10.80 | 0.409 | 0.789 | 0.457 | 0.792 |
| 10.90 | 0.416 | 0.771 | 0.459 | 0.777 |
| 11.00 | 0.425 | 0.757 | 0.465 | 0.765 |
| 11.10 | 0.431 | 0.746 | 0.469 | 0.757 |
| 11.20 | 0.439 | 0.733 | 0.470 | 0.748 |
| 11.30 | 0.451 | 0.725 | 0.476 | 0.738 |
| 11.40 | 0.457 | 0.722 | 0.479 | 0.732 |
| 11.50 | 0.459 | 0.717 | 0.482 | 0.727 |
| 11.60 | 0.466 | 0.710 | 0.482 | 0.722 |
| 11.70 | 0.467 | 0.707 | 0.481 | 0.715 |
| 11.80 | 0.468 | 0.701 | 0.482 | 0.708 |
| 11.90 | 0.471 | 0.696 | 0.482 | 0.703 |
| 12.00 | 0.471 | 0.693 | 0.483 | 0.698 |
| 12.10 | 0.473 | 0.688 | 0.480 | 0.694 |
| 12.20 | 0.473 | 0.685 | 0.477 | 0.687 |
| 12.30 | 0.471 | 0.683 | 0.478 | 0.681 |
| 12.40 | 0.468 | 0.678 | 0.474 | 0.677 |
| 12.50 | 0.463 | 0.673 | 0.470 | 0.669 |
| 12.60 | 0.465 | 0.666 | 0.472 | 0.662 |
| 12.70 | 0.459 | 0.663 | 0.466 | 0.659 |
| 12.80 | 0.454 | 0.656 | 0.463 | 0.651 |
| 12.90 | 0.453 | 0.649 | 0.459 | 0.645 |
| 13.00 | 0.451 | 0.644 | 0.455 | 0.637 |
| 13.10 | 0.444 | 0.638 | 0.454 | 0.630 |
| 13.20 | 0.442 | 0.629 | 0.448 | 0.624 |
| 13.30 | 0.437 | 0.623 | 0.446 | 0.616 |
| 13.40 | 0.434 | 0.615 | 0.440 | 0.609 |
| 13.50 | 0.433 | 0.608 | 0.440 | 0.600 |
| 13.60 | 0.428 | 0.603 | 0.434 | 0.594 |
| 13.70 | 0.424 | 0.595 | 0.431 | 0.585 |
| 13.80 | 0.419 | 0.587 | 0.430 | 0.577 |
| 13.90 | 0.417 | 0.579 | 0.422 | 0.570 |
| 14.00 | 0.411 | 0.571 | 0.421 | 0.559 |
| 14.10 | 0.407 | 0.561 | 0.417 | 0.551 |
| 14.20 | 0.405 | 0.552 | 0.416 | 0.541 |

Table 6 continued

| eV | $n_\perp$ | $k_\perp$ | $n_\parallel$ | $k_\parallel$ |
|---|---|---|---|---|
| 14.30 | 0.402 | 0.544 | 0.413 | 0.535 |
| 14.40 | 0.398 | 0.535 | 0.408 | 0.527 |
| 14.50 | 0.396 | 0.525 | 0.406 | 0.516 |
| 14.60 | 0.393 | 0.516 | 0.403 | 0.506 |
| 14.70 | 0.392 | 0.507 | 0.403 | 0.498 |
| 14.80 | 0.389 | 0.499 | 0.400 | 0.491 |
| 14.90 | 0.385 | 0.489 | 0.395 | 0.481 |
| 15.00 | 0.385 | 0.479 | 0.394 | 0.470 |
| 15.10 | 0.383 | 0.471 | 0.394 | 0.462 |
| 15.20 | 0.379 | 0.462 | 0.391 | 0.454 |
| 15.30 | 0.379 | 0.452 | 0.388 | 0.445 |
| 15.40 | 0.376 | 0.443 | 0.387 | 0.434 |
| 15.50 | 0.375 | 0.433 | 0.385 | 0.425 |
| 15.60 | 0.375 | 0.425 | 0.385 | 0.416 |
| 15.70 | 0.372 | 0.416 | 0.383 | 0.407 |
| 15.80 | 0.372 | 0.406 | 0.381 | 0.397 |
| 15.90 | 0.370 | 0.397 | 0.383 | 0.389 |
| 16.00 | 0.370 | 0.388 | 0.379 | 0.382 |
| 16.10 | 0.367 | 0.379 | 0.377 | 0.372 |
| 16.20 | 0.366 | 0.369 | 0.377 | 0.362 |
| 16.30 | 0.365 | 0.359 | 0.376 | 0.353 |
| 16.40 | 0.366 | 0.349 | 0.375 | 0.343 |
| 16.50 | 0.366 | 0.341 | 0.375 | 0.335 |
| 16.60 | 0.365 | 0.332 | 0.373 | 0.326 |
| 16.70 | 0.362 | 0.322 | 0.371 | 0.315 |
| 16.80 | 0.363 | 0.311 | 0.374 | 0.305 |
| 16.90 | 0.363 | 0.302 | 0.371 | 0.296 |

| eV | $n_\perp$ | $k_\perp$ | $n_\parallel$ | $k_\parallel$ |
|---|---|---|---|---|
| 17.00 | 0.364 | 0.293 | 0.373 | 0.285 |
| 17.10 | 0.364 | 0.284 | 0.372 | 0.276 |
| 17.20 | 0.363 | 0.275 | 0.373 | 0.266 |
| 17.30 | 0.363 | 0.264 | 0.371 | 0.257 |
| 17.40 | 0.365 | 0.254 | 0.373 | 0.246 |
| 17.50 | 0.365 | 0.246 | 0.373 | 0.237 |
| 17.60 | 0.365 | 0.236 | 0.373 | 0.227 |
| 17.70 | 0.364 | 0.226 | 0.374 | 0.216 |
| 17.80 | 0.365 | 0.215 | 0.376 | 0.206 |
| 17.90 | 0.367 | 0.206 | 0.377 | 0.197 |
| 18.00 | 0.367 | 0.196 | 0.378 | 0.188 |
| 18.10 | 0.368 | 0.185 | 0.378 | 0.176 |
| 18.20 | 0.371 | 0.175 | 0.381 | 0.165 |
| 18.30 | 0.371 | 0.164 | 0.384 | 0.155 |
| 18.40 | 0.375 | 0.153 | 0.386 | 0.145 |
| 18.50 | 0.377 | 0.143 | 0.388 | 0.135 |
| 18.60 | 0.380 | 0.133 | 0.391 | 0.124 |
| 18.70 | 0.380 | 0.122 | 0.394 | 0.114 |
| 18.80 | 0.386 | 0.110 | 0.396 | 0.103 |
| 18.90 | 0.387 | 0.100 | 0.401 | 0.091 |
| 19.00 | 0.393 | 0.087 | 0.405 | 0.081 |
| 19.10 | 0.399 | 0.078 | 0.412 | 0.071 |
| 19.20 | 0.402 | 0.068 | 0.415 | 0.062 |
| 19.30 | 0.407 | 0.056 | 0.421 | 0.050 |
| 19.40 | 0.416 | 0.047 | 0.429 | 0.042 |
| 19.50 | 0.422 | 0.039 | 0.435 | 0.034 |

## H $MoSi_2$

Refractive index data for $MoSi_2$ have been determined only in the energy range 1-5 eV by ellipsometric measurements on single crystals [14,15]. Reflectivity measurements showing the metallic character of the compound have been performed in the infrared on single crystals [15] and in the range 0.05-6 eV on polycrystalline films (also $Mo_3Si$) [16]. A morphological study of $CrSi_2$, $MoSi_2$ and PtSi with ellipsometry is presented in [17].

TABLE 7 Refractive index (n) and extinction coefficient (k) for $MoSi_2$. The data have been obtained from ellipsometric measurements on a single crystal. The indexes refer to crystallographic directions perpendicular (⊥) and parallel (||) to the c-axis (body centred tetragonal C11b structure) [14].

| eV | $n_\perp$ | $k_\perp$ | $n_\parallel$ | $k_\parallel$ |
|---|---|---|---|---|
| 1.00 | 4.778 | 1.152 | 5.212 | 1.634 |
| 1.10 | 4.798 | 1.269 | 5.192 | 1.755 |
| 1.20 | 4.787 | 1.375 | 5.145 | 1.881 |
| 1.30 | 4.791 | 1.487 | 5.082 | 2.000 |
| 1.40 | 4.785 | 1.577 | 5.016 | 2.104 |
| 1.50 | 4.765 | 1.641 | 4.953 | 2.195 |
| 1.60 | 4.744 | 1.690 | 4.898 | 2.275 |

| eV | $n_\perp$ | $k_\perp$ | $n_\parallel$ | $k_\parallel$ |
|---|---|---|---|---|
| 1.70 | 4.732 | 1.741 | 4.854 | 2.353 |
| 1.80 | 4.732 | 1.807 | 4.822 | 2.437 |
| 1.90 | 4.739 | 1.893 | 4.800 | 2.533 |
| 2.00 | 4.744 | 1.971 | 4.781 | 2.647 |
| 2.10 | 4.734 | 2.100 | 4.755 | 2.781 |
| 2.20 | 4.700 | 2.249 | 4.707 | 2.932 |
| 2.30 | 4.639 | 2.405 | 4.621 | 3.091 |

Table 7 continued

| eV | $n_{\perp}$ | $k_{\perp}$ | $n_{\parallel}$ | $k_{\parallel}$ |
|---|---|---|---|---|
| 2.40 | 4.550 | 2.554 | 4.485 | 3.249 |
| 2.50 | 4.437 | 2.686 | 4.298 | 3.389 |
| 2.60 | 4.308 | 2.795 | 4.071 | 3.499 |
| 2.70 | 4.170 | 2.880 | 3.825 | 3.565 |
| 2.80 | 4.033 | 2.943 | 3.583 | 3.584 |
| 2.90 | 3.903 | 2.990 | 3.358 | 3.567 |
| 3.00 | 3.786 | 3.027 | 3.161 | 3.541 |
| 3.10 | 3.684 | 3.062 | 2.991 | 3.469 |
| 3.20 | 3.597 | 3.099 | 2.848 | 3.402 |
| 3.30 | 3.522 | 3.144 | 2.727 | 3.343 |
| 3.40 | 3.453 | 3.199 | 2.626 | 3.292 |

| eV | $n_{\perp}$ | $k_{\perp}$ | $n_{\parallel}$ | $k_{\parallel}$ |
|---|---|---|---|---|
| 3.50 | 3.384 | 3.261 | 2.540 | 3.247 |
| 3.60 | 3.307 | 3.330 | 2.468 | 3.206 |
| 3.70 | 3.216 | 3.398 | 2.407 | 3.164 |
| 3.80 | 3.107 | 3.459 | 2.355 | 3.122 |
| 3.90 | 2.980 | 3.507 | 2.312 | 3.079 |
| 4.00 | 2.840 | 3.533 | 2.278 | 3.041 |
| 4.10 | 2.699 | 3.533 | 2.254 | 3.016 |
| 4.20 | 2.576 | 3.503 | 2.248 | 3.016 |
| 4.30 | 2.502 | 3.446 | 2.282 | 3.059 |
| 4.40 | 2.520 | 3.369 | 2.557 | 3.169 |

## I $WSi_2$

Tungsten disilicide is one of the most studied silicide compounds. A study of the optical response of thin films in the infrared and with spectroscopic ellipsometry in the range 1.4-5 eV, as a function of the annealing temperature of the samples, is given in [18,19]. Data on single crystals are presented in [15] (ellipsometry) and, in a very broad spectral range and together with band structure calculations, in [20] (reflectivity and ellipsometry).

TABLE 8 Refractive index (n) and extinction coefficient (k) for $WSi_2$. The data have been obtained from measurements on a single crystal. The high energy data have been obtained with synchrotron radiation. The light was polarised along the directions perpendicular (⊥) and parallel (||) to the crystallographic c-axis of the sample (body centred C11b structure) [20].

| eV | $n_{\perp}$ | $k_{\perp}$ | $n_{\parallel}$ | $k_{\parallel}$ |
|---|---|---|---|---|
| 0.02 | 20.220 | 0.490 | 20.220 | 0.802 |
| 0.03 | 20.230 | 0.489 | 20.230 | 0.803 |
| 0.04 | 20.240 | 0.489 | 20.240 | 0.804 |
| 0.05 | 20.250 | 0.488 | 20.250 | 0.804 |
| 0.06 | 20.260 | 0.487 | 20.260 | 0.804 |
| 0.07 | 20.270 | 0.486 | 20.270 | 0.805 |
| 0.08 | 20.280 | 0.486 | 20.280 | 0.806 |
| 0.09 | 20.290 | 0.485 | 20.290 | 0.806 |
| 0.10 | 20.320 | 0.485 | 20.320 | 0.808 |
| 0.20 | 20.400 | 0.489 | 20.400 | 0.812 |
| 0.30 | 20.500 | 0.495 | 20.500 | 0.818 |
| 0.40 | 20.600 | 0.500 | 20.600 | 0.820 |
| 0.50 | 20.700 | 0.503 | 20.700 | 0.822 |
| 0.60 | 20.800 | 0.506 | 20.800 | 0.824 |
| 0.70 | 20.900 | 0.511 | 20.900 | 0.825 |
| 0.80 | 21.000 | 0.517 | 21.000 | 0.823 |
| 0.90 | 21.100 | 0.523 | 21.100 | 0.823 |
| 1.00 | 21.200 | 0.528 | 21.200 | 0.827 |
| 1.10 | 21.300 | 0.533 | 21.300 | 0.839 |
| 1.20 | 21.400 | 0.543 | 21.400 | 0.850 |
| 1.30 | 21.500 | 0.559 | 21.500 | 0.861 |
| 1.40 | 21.600 | 0.576 | 21.600 | 0.873 |

| eV | $n_{\perp}$ | $k_{\perp}$ | $n_{\parallel}$ | $k_{\parallel}$ |
|---|---|---|---|---|
| 1.50 | 21.700 | 0.587 | 21.700 | 0.883 |
| 1.60 | 21.800 | 0.591 | 21.800 | 0.895 |
| 1.70 | 21.900 | 0.600 | 21.900 | 0.908 |
| 1.80 | 22.000 | 0.613 | 22.000 | 0.924 |
| 1.90 | 22.100 | 0.633 | 22.100 | 0.946 |
| 2.00 | 22.200 | 0.652 | 22.200 | 0.970 |
| 2.10 | 22.300 | 0.664 | 22.300 | 0.995 |
| 2.20 | 22.400 | 0.671 | 22.400 | 1.022 |
| 2.30 | 22.500 | 0.681 | 22.500 | 1.049 |
| 2.40 | 22.600 | 0.693 | 22.600 | 1.073 |
| 2.50 | 22.700 | 0.710 | 22.700 | 1.093 |
| 2.60 | 22.800 | 0.724 | 22.800 | 1.101 |
| 2.70 | 22.900 | 0.737 | 22.900 | 1.090 |
| 2.80 | 7.907 | 2.367 | 7.831 | 3.254 |
| 2.90 | 4.021 | 2.794 | 3.812 | 3.796 |
| 3.00 | 3.903 | 2.817 | 3.581 | 3.768 |
| 3.10 | 3.793 | 2.817 | 3.369 | 3.714 |
| 3.20 | 3.710 | 2.820 | 3.190 | 3.642 |
| 3.30 | 3.633 | 2.828 | 3.037 | 3.569 |
| 3.40 | 3.559 | 2.831 | 2.894 | 3.492 |
| 3.50 | 3.508 | 2.837 | 2.770 | 3.409 |
| 3.60 | 3.457 | 2.858 | 2.660 | 3.317 |

Table 8 continued

| eV | $n_\perp$ | $k_\perp$ | $n_\parallel$ | $k_\parallel$ |
|---|---|---|---|---|
| 3.70 | 3.411 | 2.887 | 2.577 | 3.222 |
| 3.80 | 3.359 | 2.925 | 2.514 | 3.133 |
| 3.90 | 3.313 | 2.968 | 2.468 | 3.052 |
| 4.00 | 3.248 | 3.036 | 2.439 | 2.986 |
| 4.10 | 3.161 | 3.094 | 2.418 | 2.934 |
| 4.20 | 3.064 | 3.137 | 2.402 | 2.901 |
| 4.30 | 2.967 | 3.163 | 2.381 | 2.881 |
| 4.40 | 2.882 | 3.174 | 2.358 | 2.871 |
| 4.50 | 2.847 | 3.180 | 2.331 | 2.869 |
| 4.60 | 2.808 | 3.299 | 2.301 | 2.878 |
| 4.70 | 2.614 | 3.439 | 2.256 | 2.903 |
| 4.80 | 2.438 | 3.448 | 2.180 | 2.915 |
| 4.90 | 2.259 | 3.378 | 2.110 | 2.901 |
| 5.00 | 2.156 | 3.382 | 2.044 | 2.835 |
| 5.10 | 1.995 | 3.339 | 2.087 | 2.946 |
| 5.20 | 1.893 | 3.257 | 1.989 | 2.944 |
| 5.30 | 1.794 | 3.220 | 1.905 | 2.954 |
| 5.40 | 1.722 | 3.127 | 1.783 | 3.004 |
| 5.50 | 1.660 | 3.128 | 1.645 | 2.979 |
| 5.60 | 1.537 | 3.057 | 1.549 | 2.923 |
| 5.70 | 1.493 | 2.972 | 1.414 | 2.882 |
| 5.80 | 1.443 | 2.936 | 1.346 | 2.783 |
| 5.90 | 1.393 | 2.873 | 1.286 | 2.693 |
| 6.00 | 1.360 | 2.842 | 1.266 | 2.634 |
| 6.10 | 1.295 | 2.822 | 1.229 | 2.594 |
| 6.20 | 1.234 | 2.758 | 1.186 | 2.547 |
| 6.30 | 1.210 | 2.719 | 1.122 | 2.525 |
| 6.40 | 1.159 | 2.702 | 1.073 | 2.461 |
| 6.50 | 1.097 | 2.667 | 1.036 | 2.408 |
| 6.60 | 1.046 | 2.625 | 0.978 | 2.385 |
| 6.70 | 0.987 | 2.587 | 0.925 | 2.296 |
| 6.80 | 0.952 | 2.533 | 0.902 | 2.238 |
| 6.90 | 0.895 | 2.513 | 0.876 | 2.194 |
| 7.00 | 0.838 | 2.455 | 0.831 | 2.145 |
| 7.10 | 0.792 | 2.406 | 0.788 | 2.083 |
| 7.20 | 0.747 | 2.354 | 0.768 | 2.021 |
| 7.30 | 0.706 | 2.304 | 0.735 | 1.955 |
| 7.40 | 0.656 | 2.247 | 0.722 | 1.891 |
| 7.50 | 0.619 | 2.191 | 0.697 | 1.834 |
| 7.60 | 0.578 | 2.111 | 0.682 | 1.757 |
| 7.70 | 0.564 | 2.050 | 0.692 | 1.685 |
| 7.80 | 0.536 | 1.987 | 0.704 | 1.633 |
| 7.90 | 0.530 | 1.922 | 0.713 | 1.587 |
| 8.00 | 0.517 | 1.863 | 0.716 | 1.547 |
| 8.10 | 0.509 | 1.806 | 0.715 | 1.500 |
| 8.20 | 0.512 | 1.750 | 0.728 | 1.457 |
| 8.30 | 0.504 | 1.703 | 0.730 | 1.421 |
| 8.40 | 0.504 | 1.648 | 0.735 | 1.382 |
| 8.50 | 0.510 | 1.600 | 0.745 | 1.346 |
| 8.60 | 0.508 | 1.560 | 0.746 | 1.308 |
| 8.70 | 0.513 | 1.507 | 0.749 | 1.272 |
| 8.80 | 0.521 | 1.468 | 0.761 | 1.231 |

| eV | $n_\perp$ | $k_\perp$ | $n_\parallel$ | $k_\parallel$ |
|---|---|---|---|---|
| 8.90 | 0.529 | 1.426 | 0.776 | 1.197 |
| 9.00 | 0.536 | 1.388 | 0.786 | 1.166 |
| 9.10 | 0.546 | 1.349 | 0.797 | 1.133 |
| 9.20 | 0.553 | 1.312 | 0.811 | 1.101 |
| 9.30 | 0.568 | 1.272 | 0.827 | 1.070 |
| 9.40 | 0.585 | 1.239 | 0.847 | 1.042 |
| 9.50 | 0.602 | 1.208 | 0.862 | 1.015 |
| 9.60 | 0.622 | 1.179 | 0.888 | 0.991 |
| 9.70 | 0.641 | 1.156 | 0.908 | 0.973 |
| 9.80 | 0.660 | 1.131 | 0.930 | 0.958 |
| 9.90 | 0.684 | 1.117 | 0.950 | 0.944 |
| 10.00 | 0.693 | 1.103 | 0.967 | 0.937 |
| 10.10 | 0.716 | 1.085 | 0.983 | 0.925 |
| 10.20 | 0.730 | 1.082 | 0.999 | 0.920 |
| 10.30 | 0.737 | 1.074 | 1.011 | 0.914 |
| 10.40 | 0.743 | 1.064 | 1.023 | 0.915 |
| 10.50 | 0.747 | 1.052 | 1.023 | 0.909 |
| 10.60 | 0.746 | 1.038 | 1.028 | 0.899 |
| 10.70 | 0.755 | 1.010 | 1.040 | 0.898 |
| 10.80 | 0.776 | 0.989 | 1.020 | 0.870 |
| 10.90 | 0.805 | 0.991 | 1.044 | 0.870 |
| 11.00 | 0.773 | 0.999 | 1.035 | 0.861 |
| 11.10 | 0.777 | 0.935 | 1.031 | 0.817 |
| 11.20 | 0.827 | 0.891 | 1.059 | 0.778 |
| 11.30 | 0.885 | 0.893 | 1.102 | 0.754 |
| 11.40 | 0.932 | 0.892 | 1.145 | 0.751 |
| 11.50 | 0.975 | 0.936 | 1.181 | 0.760 |
| 11.60 | 0.989 | 0.972 | 1.212 | 0.778 |
| 11.70 | 0.992 | 1.022 | 1.227 | 0.807 |
| 11.80 | 0.963 | 1.049 | 1.231 | 0.823 |
| 11.90 | 0.941 | 1.056 | 1.236 | 0.834 |
| 12.00 | 0.941 | 1.057 | 1.240 | 0.848 |
| 12.10 | 0.912 | 1.076 | 1.238 | 0.855 |
| 12.20 | 0.904 | 1.063 | 1.244 | 0.859 |
| 12.30 | 0.898 | 1.064 | 1.249 | 0.870 |
| 12.40 | 0.891 | 1.067 | 1.248 | 0.884 |
| 12.50 | 0.880 | 1.072 | 1.245 | 0.888 |
| 12.60 | 0.864 | 1.071 | 1.247 | 0.897 |
| 12.70 | 0.853 | 1.066 | 1.244 | 0.907 |
| 12.80 | 0.844 | 1.064 | 1.239 | 0.916 |
| 12.90 | 0.831 | 1.062 | 1.231 | 0.924 |
| 13.00 | 0.818 | 1.054 | 1.221 | 0.926 |
| 13.10 | 0.809 | 1.045 | 1.215 | 0.927 |
| 13.20 | 0.802 | 1.036 | 1.209 | 0.926 |
| 13.30 | 0.797 | 1.029 | 1.204 | 0.926 |
| 13.40 | 0.791 | 1.022 | 1.201 | 0.921 |
| 13.50 | 0.788 | 1.018 | 1.207 | 0.919 |
| 13.60 | 0.780 | 1.013 | 1.210 | 0.931 |
| 13.70 | 0.778 | 1.008 | 1.204 | 0.940 |
| 13.80 | 0.770 | 1.009 | 1.196 | 0.946 |
| 13.90 | 0.761 | 1.004 | 1.190 | 0.948 |
| 14.00 | 0.752 | 1.002 | 1.186 | 0.954 |

Table 8 continued

| eV | $n_\perp$ | $k_\perp$ | $n_\parallel$ | $k_\parallel$ |
|---|---|---|---|---|
| 14.10 | 0.741 | 0.994 | 1.174 | 0.963 |
| 14.20 | 0.735 | 0.990 | 1.162 | 0.963 |
| 14.30 | 0.724 | 0.987 | 1.153 | 0.965 |
| 14.40 | 0.711 | 0.980 | 1.139 | 0.968 |
| 14.50 | 0.701 | 0.968 | 1.126 | 0.962 |
| 14.60 | 0.694 | 0.959 | 1.121 | 0.955 |
| 14.70 | 0.686 | 0.952 | 1.117 | 0.952 |
| 14.80 | 0.675 | 0.943 | 1.114 | 0.955 |
| 14.90 | 0.666 | 0.931 | 1.102 | 0.964 |
| 15.00 | 0.660 | 0.918 | 1.083 | 0.961 |
| 15.10 | 0.657 | 0.906 | 1.072 | 0.950 |
| 15.20 | 0.652 | 0.897 | 1.065 | 0.941 |
| 15.30 | 0.647 | 0.889 | 1.066 | 0.933 |
| 15.40 | 0.641 | 0.878 | 1.063 | 0.940 |
| 15.50 | 0.638 | 0.867 | 1.045 | 0.943 |
| 15.60 | 0.635 | 0.859 | 1.032 | 0.927 |
| 15.70 | 0.631 | 0.851 | 1.034 | 0.915 |
| 15.80 | 0.626 | 0.844 | 1.035 | 0.919 |
| 15.90 | 0.620 | 0.835 | 1.024 | 0.921 |
| 16.00 | 0.612 | 0.824 | 1.013 | 0.916 |
| 16.10 | 0.611 | 0.810 | 1.005 | 0.910 |
| 16.20 | 0.611 | 0.801 | 0.998 | 0.903 |
| 16.30 | 0.609 | 0.793 | 0.996 | 0.897 |
| 16.40 | 0.607 | 0.785 | 0.992 | 0.896 |
| 16.50 | 0.606 | 0.778 | 0.983 | 0.899 |
| 16.60 | 0.604 | 0.772 | 0.971 | 0.895 |
| 16.70 | 0.599 | 0.767 | 0.959 | 0.889 |
| 16.80 | 0.593 | 0.759 | 0.948 | 0.879 |
| 16.90 | 0.589 | 0.750 | 0.942 | 0.865 |
| 17.00 | 0.587 | 0.743 | 0.943 | 0.857 |
| 17.10 | 0.582 | 0.737 | 0.942 | 0.854 |
| 17.20 | 0.576 | 0.729 | 0.934 | 0.853 |
| 17.30 | 0.571 | 0.720 | 0.927 | 0.849 |
| 17.40 | 0.566 | 0.711 | 0.919 | 0.843 |
| 17.50 | 0.562 | 0.701 | 0.911 | 0.837 |
| 17.60 | 0.560 | 0.691 | 0.904 | 0.830 |
| 17.70 | 0.558 | 0.683 | 0.897 | 0.822 |
| 17.80 | 0.554 | 0.676 | 0.890 | 0.815 |
| 17.90 | 0.548 | 0.668 | 0.884 | 0.808 |
| 18.00 | 0.541 | 0.659 | 0.877 | 0.799 |
| 18.10 | 0.535 | 0.646 | 0.871 | 0.789 |
| 18.20 | 0.534 | 0.632 | 0.868 | 0.780 |
| 18.30 | 0.534 | 0.622 | 0.864 | 0.773 |
| 18.40 | 0.533 | 0.612 | 0.858 | 0.768 |
| 18.50 | 0.532 | 0.603 | 0.850 | 0.759 |
| 18.60 | 0.529 | 0.595 | 0.843 | 0.748 |
| 18.70 | 0.526 | 0.586 | 0.840 | 0.734 |
| 18.80 | 0.523 | 0.577 | 0.840 | 0.725 |
| 18.90 | 0.519 | 0.567 | 0.838 | 0.718 |
| 19.00 | 0.516 | 0.555 | 0.834 | 0.712 |
| 19.10 | 0.514 | 0.543 | 0.828 | 0.703 |
| 19.20 | 0.514 | 0.531 | 0.824 | 0.693 |
| 19.30 | 0.515 | 0.520 | 0.820 | 0.682 |
| 19.40 | 0.516 | 0.512 | 0.817 | 0.671 |
| 19.50 | 0.513 | 0.504 | 0.814 | 0.659 |
| 19.60 | 0.510 | 0.492 | 0.817 | 0.645 |
| 19.70 | 0.511 | 0.480 | 0.819 | 0.638 |
| 19.80 | 0.512 | 0.472 | 0.819 | 0.633 |
| 19.90 | 0.511 | 0.465 | 0.813 | 0.629 |
| 20.00 | 0.506 | 0.457 | 0.804 | 0.619 |
| 20.10 | 0.501 | 0.447 | 0.799 | 0.601 |
| 20.20 | 0.492 | 0.269 | 0.800 | 0.357 |
| 20.30 | 0.486 | 0.050 | 0.807 | 0.050 |
| 20.40 | 0.489 | 0.154 | 0.812 | 0.154 |
| 20.50 | 0.495 | 0.600 | 0.818 | 0.600 |
| 20.60 | 0.500 | 1.100 | 0.820 | 1.100 |
| 20.70 | 0.503 | 1.600 | 0.822 | 1.600 |
| 20.80 | 0.506 | 2.100 | 0.824 | 2.100 |
| 20.90 | 0.511 | 2.600 | 0.825 | 2.600 |
| 21.00 | 0.517 | 3.100 | 0.823 | 3.100 |
| 21.10 | 0.523 | 3.600 | 0.823 | 3.600 |
| 21.20 | 0.528 | 4.100 | 0.827 | 4.100 |
| 21.30 | 0.533 | 4.600 | 0.839 | 4.600 |
| 21.40 | 0.543 | 5.100 | 0.850 | 5.100 |
| 21.50 | 0.559 | 5.600 | 0.861 | 5.600 |
| 21.60 | 0.576 | 6.100 | 0.873 | 6.100 |
| 21.70 | 0.587 | 6.600 | 0.883 | 6.600 |
| 21.80 | 0.591 | 7.100 | 0.895 | 7.100 |
| 21.90 | 0.600 | 7.600 | 0.908 | 7.600 |
| 22.00 | 0.613 | 8.100 | 0.924 | 8.100 |
| 22.10 | 0.633 | 8.600 | 0.946 | 8.600 |
| 22.20 | 0.652 | 9.100 | 0.970 | 9.100 |
| 22.30 | 0.664 | 9.600 | 0.995 | 9.600 |
| 22.40 | 0.671 | 10.100 | 1.022 | 10.100 |
| 22.50 | 0.681 | 10.600 | 1.049 | 10.600 |
| 22.60 | 0.693 | 11.100 | 1.073 | 11.100 |
| 22.70 | 0.710 | 11.600 | 1.093 | 11.600 |
| 22.80 | 0.724 | 12.100 | 1.101 | 12.100 |
| 22.90 | 0.737 | 12.600 | 1.090 | 12.600 |

## REFERENCES


[1] W. Henrion, H. Lange [ *Phys. Status Solidi B (Germany)* vol.151 (1989) p.375 ]

[2] K. Lee, J.T. Lue [ *Phys. Lett. A (Netherlands)* vol.125 (1987) p.271 ]

[3] A. Borghesi, A. Piaggi, G. Guizzetti, F. Lévy, M. Tanaka, H. Fukutani [ *Phys. Rev. B (USA)* vol.40 (1989) p.1611 ]

[4] M. Tanaka, S. Kurita, M. Fujisawa, F. Lévy [ *Phys. Rev. B (USA)* vol.43 (1991) p.9133-7 ]

[5] Shean-Jyeh Mu, Juh-Tzeng Lue, In-Chin Wu [ *J. Phys. Chem. Solids (UK)* vol.49 (1988) p.1389-95 ]

[6] A. Borghesi, F. Marabelli, G. Guizzetti, M. Michelini, F. Nava [ *J. Appl. Phys. (USA)* vol.69 (1991) p.7645-50 ]

[7] A. Borghesi, A. Piaggi, G. Guizzetti, F. Nava, M. Bacchetta [ *Phys. Rev. B (USA)* vol.40 (1989) p.3249 ]

[8] M. Amiotti, A. Borghesi, F. Marabelli, A. Piaggi, G. Guizzetti, F. Nava [ *Appl. Surf. Sci. (Netherlands)* vol.53 (1991) p.230 ]

[9] G. Guizzetti, F. Marabelli, A. Borghesi, F. Nava, U. Gottlieb, O. Laborde [ *Appl. Surf. Sci. (Netherlands)* vol.73 (1993) p.237-42 ]

[10] M. Amiotti, A. Borghesi, F. Marabelli, G. Guizzetti, F. Nava [ *Phys. Rev. B (USA)* vol.44 (1991) p.3757 ]

[11] A. Borghesi, L. Nosenzo, A. Piaggi, G. Guizzetti, C. Nobili, G. Ottaviani [ *Phys. Rev. B (USA)* vol.38 (1988) p.10937 ]

[12] A. Röseler, F. Marabelli, G. Guizzetti, A. Borghesi, A. Piaggi [ *Nuovo Cimento D (Italy)* vol.13 (1991) p.169 ]

[13] M. Tanaka, S. Kurita, M. Fujisawa, F. Lévy [ *Phys. Rev. B (USA)* vol.46 (1992) p.10442-5 ]

[14] Ph. Ged, R. Madar, J.P. Senateur [ *Phys. Rev. B (USA)* vol.29 (1984) p.6981 ]

[15] F. Ferrieu et al [ *Solid State Commun. (USA)* vol.62 (1987) p.455 ]

[16] W. Henrion, G. Beddies, H. Lange [ *Phys. Status Solidi B (Germany)* vol.131 (1985) p.K55 ]

[17] Juh-Tzeng Lue, Shean-Jyeh Mu, In-Chin Wu [ *Phys. Rev. B (USA)* vol.36 (1987) p.1657 ]

[18] M. Amiotti, A. Borghesi, G. Guizzetti, F. Nava, G. Santoro [ *Europhys. Lett. (Switzerland)* vol.14 (1991) p.587 ]

[19] M. Amiotti et al [ *Appl. Phys. A (Germany)* vol.54 (1992) p.181 ]

[20] V.N. Antonov et al [ *Phys. Rev. B (USA)* vol.44 (1991) p.8437-45 ]

# 6.3 Optical functions: near noble TM silicides

G. Guizetti and F. Marabelli

December 1994

## A INTRODUCTION

Near noble silicides are essentially Co- and Ni-silicides. The $MSi_2$ stoichiometry, which crystallises in the cubic calcium-fluorite structure, is the most stable and the most studied of both compounds. In the case of Ni, the whole series of Ni- and Si-rich compounds has been characterised. A complete set of data from 0.062 up to 22.3 eV is reported in a recent paper [1]. The data have been obtained by ellipsometry and reflectance measurements on epitaxial films (MBE) of different thickness on Si (001). Series of data covering different spectral intervals and obtained by different authors are listed in the tables below.

## B $CoSi_2$

Optical data on $CoSi_2$ are listed in TABLES 1 to 4.

TABLE 1 Refractive index (n) and extinction coefficient (k) for $CoSi_2$ between 0 and 1 eV, according to [2]. The data have been obtained by a fit of the transmittances of several epitaxial (MBE) films with different thickness.

| eV | n | k |
|---|---|---|
| 0.10 | 7.474 | 44.647 |
| 0.20 | 3.928 | 25.836 |
| 0.30 | 2.631 | 17.826 |
| 0.40 | 1.999 | 13.951 |
| 0.50 | 1.609 | 11.975 |
| 0.60 | 1.336 | 10.465 |
| 0.70 | 1.130 | 8.971 |
| 0.80 | 0.968 | 7.561 |
| 0.90 | 0.838 | 6.451 |
| 1.00 | 0.730 | 5.827 |

TABLE 2 Refractive index (n) and extinction coefficient (k) for $CoSi_2$ between 0 and 1 eV, according to [3]. The data have been obtained by inversion of transmission and reflection measurements on thin films grown by MBE.

| eV | n | k |
|---|---|---|
| 0.50 | 2.720 | 11.005 |
| 0.60 | 2.121 | 9.113 |
| 0.70 | 1.604 | 7.795 |
| 0.80 | 1.315 | 6.824 |
| 0.90 | 1.113 | 6.044 |
| 1.00 | 1.000 | 5.395 |

TABLE 3 Refractive index (n) and extinction coefficient (k) for $CoSi_2$ between 1 and 2.4 eV, according to [4]. The data have been obtained by ellipsometric measurements on epitaxial (MBE) films.

| eV | n | k | eV | n | k | eV | n | k |
|---|---|---|---|---|---|---|---|---|
| 1.00 | 1.316 | 5.695 | 1.50 | 1.847 | 2.384 | 2.00 | 2.829 | 1.416 |
| 1.10 | 1.225 | 4.788 | 1.60 | 2.027 | 2.020 | 2.10 | 3.039 | 1.409 |
| 1.20 | 1.329 | 4.057 | 1.70 | 2.214 | 1.754 | 2.20 | 3.239 | 1.437 |
| 1.30 | 1.493 | 3.412 | 1.80 | 2.411 | 1.575 | 2.30 | 3.419 | 1.496 |
| 1.40 | 1.670 | 2.850 | 1.90 | 2.617 | 1.467 | 2.40 | 3.566 | 1.586 |

TABLE 4 Refractive index (n) and extinction coefficient (k) for $CoSi_2$ between 1 and 4.4 eV, according to [5]. The data have been obtained by ellipsometric measurements on monocrystalline bulk samples.

| eV | n | k | eV | n | k | eV | n | k |
|---|---|---|---|---|---|---|---|---|
| 1.00 | 1.227 | 4.346 | 2.30 | 3.115 | 2.028 | 3.60 | 2.386 | 2.242 |
| 1.10 | 1.428 | 3.560 | 2.40 | 3.071 | 2.106 | 3.70 | 2.362 | 2.252 |
| 1.20 | 1.630 | 2.912 | 2.50 | 3.000 | 2.188 | 3.80 | 2.334 | 2.267 |
| 1.30 | 1.834 | 2.404 | 2.60 | 2.911 | 2.258 | 3.90 | 2.305 | 2.285 |
| 1.40 | 2.037 | 2.038 | 2.70 | 2.813 | 2.300 | 4.00 | 2.278 | 2.306 |
| 1.50 | 2.238 | 1.811 | 2.80 | 2.718 | 2.316 | 4.10 | 2.253 | 2.327 |
| 1.60 | 2.432 | 1.701 | 2.90 | 2.631 | 2.313 | 4.20 | 2.232 | 2.348 |
| 1.70 | 2.614 | 1.676 | 3.00 | 2.560 | 2.299 | 4.30 | 2.215 | 2.367 |
| 1.80 | 2.778 | 1.703 | 3.10 | 2.507 | 2.280 | 4.40 | 2.201 | 2.383 |
| 1.90 | 2.917 | 1.755 | 3.20 | 2.469 | 2.262 | 4.50 | 2.192 | 2.396 |
| 2.00 | 3.024 | 1.817 | 3.30 | 2.444 | 2.248 | 4.60 | 2.186 | 2.404 |
| 2.10 | 3.094 | 1.884 | 3.40 | 2.426 | 2.240 | 4.70 | 2.183 | 2.406 |
| 2.20 | 3.124 | 1.954 | 3.50 | 2.407 | 2.238 | 4.80 | 2.184 | 2.402 |

## C $Ni_xSi_y$

Optical data on epitaxial films of $NiSi_2$ obtained by ellipsometric measurements are reported in [4]. We list in TABLE 8 the data of [6], determined from reflectivity measurements on polycrystalline films, because of the large spectral range and for homogeneity with the data for $Ni_3Si$, $Ni_2Si$ and NiSi listed in TABLES 5, 6 and 7. Reflectivity spectra taken on nickel silicide films are shown in [7,8].

TABLE 5 Refractive index (n) and extinction coefficient (k) for $Ni_2Si$. The measurements have been performed on polycrystalline films [6].

| eV | n | k | eV | n | k | eV | n | k |
|---|---|---|---|---|---|---|---|---|
| 0.02 | 36.017 | 43.551 | 0.50 | 7.641 | 4.115 | 1.70 | 2.123 | 3.605 |
| 0.03 | 26.488 | 36.225 | 0.60 | 7.096 | 4.783 | 1.80 | 1.979 | 3.367 |
| 0.04 | 16.959 | 28.898 | 0.70 | 6.479 | 5.121 | 1.90 | 1.871 | 3.139 |
| 0.05 | 14.185 | 24.825 | 0.80 | 5.896 | 5.153 | 2.00 | 1.793 | 2.929 |
| 0.06 | 11.411 | 20.751 | 0.90 | 5.382 | 5.318 | 2.10 | 1.736 | 2.734 |
| 0.07 | 10.286 | 18.324 | 1.00 | 4.699 | 5.471 | 2.20 | 1.699 | 2.553 |
| 0.08 | 9.161 | 15.897 | 1.10 | 3.837 | 5.221 | 2.30 | 1.681 | 2.390 |
| 0.09 | 8.619 | 14.292 | 1.20 | 3.352 | 4.891 | 2.40 | 1.678 | 2.247 |
| 0.10 | 7.599 | 10.640 | 1.30 | 3.072 | 4.611 | 2.50 | 1.680 | 2.126 |
| 0.20 | 7.922 | 6.356 | 1.40 | 2.746 | 4.337 | 2.60 | 1.677 | 2.016 |
| 0.30 | 6.980 | 5.612 | 1.50 | 2.527 | 4.088 | 2.70 | 1.689 | 1.916 |
| 0.40 | 6.788 | 3.790 | 1.60 | 2.298 | 3.849 | 2.80 | 1.696 | 1.831 |

Table 5 continued

| eV | n | k |
|---|---|---|
| 2.90 | 1.702 | 1.755 |
| 3.00 | 1.709 | 1.684 |
| 3.10 | 1.714 | 1.618 |
| 3.20 | 1.722 | 1.555 |
| 3.30 | 1.732 | 1.497 |
| 3.40 | 1.744 | 1.446 |
| 3.50 | 1.756 | 1.399 |
| 3.60 | 1.769 | 1.352 |
| 3.70 | 1.790 | 1.317 |
| 3.80 | 1.806 | 1.291 |

| eV | n | k |
|---|---|---|
| 3.90 | 1.818 | 1.271 |
| 4.00 | 1.823 | 1.254 |
| 4.10 | 1.825 | 1.237 |
| 4.20 | 1.821 | 1.221 |
| 4.30 | 1.817 | 1.199 |
| 4.40 | 1.815 | 1.176 |
| 4.50 | 1.817 | 1.153 |
| 4.60 | 1.822 | 1.135 |
| 4.70 | 1.827 | 1.122 |
| 4.80 | 1.827 | 1.113 |

| eV | n | k |
|---|---|---|
| 4.90 | 1.823 | 1.104 |
| 5.00 | 1.815 | 1.093 |
| 5.10 | 1.809 | 1.078 |
| 5.20 | 1.804 | 1.065 |
| 5.30 | 1.797 | 1.052 |
| 5.40 | 1.792 | 1.037 |
| 5.50 | 1.788 | 1.024 |
| 5.60 | 1.782 | 1.011 |
| 5.70 | 1.776 | 0.995 |
| 5.80 | 1.773 | 0.982 |

TABLE 6 Refractive index (n) and extinction coefficient (k) for $Ni_2Si$. The measurements have been performed on polycrystalline films [6].

| eV | n | k |
|---|---|---|
| 0.02 | 48.410 | 59.426 |
| 0.03 | 35.568 | 49.992 |
| 0.04 | 22.726 | 40.558 |
| 0.05 | 18.661 | 35.365 |
| 0.06 | 14.596 | 30.172 |
| 0.07 | 12.918 | 27.055 |
| 0.08 | 11.240 | 23.937 |
| 0.09 | 10.195 | 22.100 |
| 0.10 | 7.853 | 17.554 |
| 0.20 | 5.650 | 11.142 |
| 0.30 | 4.749 | 7.970 |
| 0.40 | 4.107 | 6.289 |
| 0.50 | 3.784 | 5.097 |
| 0.60 | 3.783 | 4.259 |
| 0.70 | 4.094 | 3.899 |
| 0.80 | 3.757 | 3.872 |
| 0.90 | 3.613 | 3.815 |
| 1.00 | 3.372 | 3.626 |
| 1.10 | 3.222 | 3.476 |
| 1.20 | 3.081 | 3.376 |
| 1.30 | 2.923 | 3.298 |
| 1.40 | 2.728 | 3.208 |

| eV | n | k |
|---|---|---|
| 1.50 | 2.570 | 3.075 |
| 1.60 | 2.430 | 2.949 |
| 1.70 | 2.313 | 2.820 |
| 1.80 | 2.209 | 2.696 |
| 1.90 | 2.129 | 2.564 |
| 2.00 | 2.075 | 2.452 |
| 2.10 | 2.027 | 2.360 |
| 2.20 | 1.979 | 2.274 |
| 2.30 | 1.931 | 2.195 |
| 2.40 | 1.889 | 2.115 |
| 2.50 | 1.851 | 2.045 |
| 2.60 | 1.803 | 1.965 |
| 2.70 | 1.784 | 1.882 |
| 2.80 | 1.772 | 1.811 |
| 2.90 | 1.764 | 1.751 |
| 3.00 | 1.758 | 1.699 |
| 3.10 | 1.749 | 1.652 |
| 3.20 | 1.742 | 1.605 |
| 3.30 | 1.739 | 1.562 |
| 3.40 | 1.739 | 1.525 |
| 3.50 | 1.736 | 1.494 |
| 3.60 | 1.733 | 1.447 |

| eV | n | k |
|---|---|---|
| 3.70 | 1.751 | 1.423 |
| 3.80 | 1.760 | 1.407 |
| 3.90 | 1.767 | 1.395 |
| 4.00 | 1.770 | 1.388 |
| 4.10 | 1.771 | 1.386 |
| 4.20 | 1.765 | 1.387 |
| 4.30 | 1.757 | 1.388 |
| 4.40 | 1.743 | 1.389 |
| 4.50 | 1.728 | 1.388 |
| 4.60 | 1.711 | 1.386 |
| 4.70 | 1.695 | 1.384 |
| 4.80 | 1.673 | 1.384 |
| 4.90 | 1.651 | 1.379 |
| 5.00 | 1.627 | 1.372 |
| 5.10 | 1.606 | 1.363 |
| 5.20 | 1.587 | 1.353 |
| 5.30 | 1.567 | 1.345 |
| 5.40 | 1.550 | 1.336 |
| 5.50 | 1.532 | 1.330 |
| 5.60 | 1.512 | 1.325 |
| 5.70 | 1.489 | 1.320 |
| 5.80 | 1.470 | 1.318 |

TABLE 7 Refractive index (n) and extinction coefficient (k) for NiSi. The measurements have been performed on polycrystalline films [6].

| eV | n | k |
|---|---|---|
| 0.02 | 80.622 | 176.925 |
| 0.03 | 51.848 | 138.241 |
| 0.04 | 23.073 | 99.557 |
| 0.05 | 17.577 | 83.852 |
| 0.06 | 12.082 | 68.147 |
| 0.07 | 10.129 | 59.681 |
| 0.08 | 8.176 | 51.215 |
| 0.09 | 7.645 | 46.474 |
| 0.10 | 5.343 | 35.835 |

| eV | n | k |
|---|---|---|
| 0.20 | 2.931 | 21.965 |
| 0.30 | 1.947 | 14.656 |
| 0.40 | 1.437 | 10.832 |
| 0.50 | 1.255 | 8.392 |
| 0.60 | 1.325 | 6.661 |
| 0.70 | 1.477 | 5.535 |
| 0.80 | 1.541 | 4.730 |
| 0.90 | 1.568 | 4.014 |
| 1.00 | 1.682 | 3.344 |

| eV | n | k |
|---|---|---|
| 1.10 | 1.987 | 2.778 |
| 1.20 | 2.388 | 2.475 |
| 1.30 | 2.702 | 2.420 |
| 1.40 | 2.859 | 2.447 |
| 1.50 | 2.973 | 2.496 |
| 1.60 | 2.990 | 2.588 |
| 1.70 | 2.950 | 2.670 |
| 1.80 | 2.850 | 2.734 |
| 1.90 | 2.731 | 2.750 |

Table 7 continued

| eV | n | k |
|---|---|---|
| 2.00 | 2.599 | 2.763 |
| 2.10 | 2.431 | 2.724 |
| 2.20 | 2.299 | 2.616 |
| 2.30 | 2.221 | 2.510 |
| 2.40 | 2.168 | 2.413 |
| 2.50 | 2.133 | 2.332 |
| 2.60 | 2.093 | 2.262 |
| 2.70 | 2.067 | 2.194 |
| 2.80 | 2.043 | 2.125 |
| 2.90 | 2.037 | 2.061 |
| 3.00 | 2.040 | 2.017 |
| 3.10 | 2.039 | 1.986 |
| 3.20 | 2.033 | 1.963 |

| eV | n | k |
|---|---|---|
| 3.30 | 2.020 | 1.940 |
| 3.40 | 2.012 | 1.914 |
| 3.50 | 2.009 | 1.898 |
| 3.60 | 2.007 | 1.869 |
| 3.70 | 2.029 | 1.876 |
| 3.80 | 2.026 | 1.900 |
| 3.90 | 2.009 | 1.926 |
| 4.00 | 1.976 | 1.950 |
| 4.10 | 1.942 | 1.970 |
| 4.20 | 1.896 | 1.991 |
| 4.30 | 1.843 | 2.005 |
| 4.40 | 1.783 | 2.011 |
| 4.50 | 1.725 | 2.010 |

| eV | n | k |
|---|---|---|
| 4.60 | 1.664 | 2.004 |
| 4.70 | 1.601 | 1.992 |
| 4.80 | 1.536 | 1.971 |
| 4.90 | 1.476 | 1.939 |
| 5.00 | 1.425 | 1.900 |
| 5.10 | 1.383 | 1.861 |
| 5.20 | 1.347 | 1.824 |
| 5.30 | 1.313 | 1.790 |
| 5.40 | 1.282 | 1.756 |
| 5.50 | 1.251 | 1.724 |
| 5.60 | 1.221 | 1.692 |
| 5.70 | 1.191 | 1.659 |
| 5.80 | 1.167 | 1.632 |

TABLE 8 Refractive index (n) and extinction coefficient (k) for $NiSi_2$.
The measurements have been performed on polycrystalline films [6].

| eV | n | k |
|---|---|---|
| 0.02 | 60.418 | 45.314 |
| 0.03 | 48.947 | 41.523 |
| 0.04 | 37.476 | 37.731 |
| 0.05 | 32.536 | 34.912 |
| 0.06 | 27.596 | 32.093 |
| 0.07 | 24.792 | 30.254 |
| 0.08 | 21.988 | 28.416 |
| 0.09 | 20.071 | 27.160 |
| 0.10 | 15.067 | 23.936 |
| 0.20 | 8.303 | 17.164 |
| 0.30 | 5.270 | 12.323 |
| 0.40 | 4.042 | 9.502 |
| 0.50 | 3.483 | 7.727 |
| 0.60 | 3.227 | 6.629 |
| 0.70 | 2.788 | 5.873 |
| 0.80 | 2.507 | 5.121 |
| 0.90 | 2.324 | 4.465 |
| 1.00 | 2.229 | 3.897 |
| 1.10 | 2.220 | 3.403 |
| 1.20 | 2.272 | 2.993 |
| 1.30 | 2.376 | 2.672 |
| 1.40 | 2.465 | 2.429 |

| eV | n | k |
|---|---|---|
| 1.50 | 2.618 | 2.221 |
| 1.60 | 2.765 | 2.115 |
| 1.70 | 2.886 | 2.072 |
| 1.80 | 2.968 | 2.063 |
| 1.90 | 3.044 | 2.060 |
| 2.00 | 3.115 | 2.123 |
| 2.10 | 3.117 | 2.217 |
| 2.20 | 3.068 | 2.285 |
| 2.30 | 3.010 | 2.331 |
| 2.40 | 2.956 | 2.368 |
| 2.50 | 2.903 | 2.421 |
| 2.60 | 2.803 | 2.478 |
| 2.70 | 2.702 | 2.509 |
| 2.80 | 2.592 | 2.538 |
| 2.90 | 2.459 | 2.554 |
| 3.00 | 2.308 | 2.547 |
| 3.10 | 2.153 | 2.496 |
| 3.20 | 2.019 | 2.408 |
| 3.30 | 1.915 | 2.304 |
| 3.40 | 1.837 | 2.190 |
| 3.50 | 1.779 | 2.081 |
| 3.60 | 1.752 | 1.935 |

| eV | n | k |
|---|---|---|
| 3.70 | 1.775 | 1.849 |
| 3.80 | 1.786 | 1.777 |
| 3.90 | 1.803 | 1.718 |
| 4.00 | 1.818 | 1.667 |
| 4.10 | 1.840 | 1.623 |
| 4.20 | 1.861 | 1.590 |
| 4.30 | 1.889 | 1.565 |
| 4.40 | 1.916 | 1.556 |
| 4.50 | 1.939 | 1.560 |
| 4.60 | 1.954 | 1.577 |
| 4.70 | 1.957 | 1.603 |
| 4.80 | 1.940 | 1.634 |
| 4.90 | 1.910 | 1.658 |
| 5.00 | 1.867 | 1.674 |
| 5.10 | 1.820 | 1.677 |
| 5.20 | 1.774 | 1.670 |
| 5.30 | 1.730 | 1.656 |
| 5.40 | 1.691 | 1.636 |
| 5.50 | 1.656 | 1.614 |
| 5.60 | 1.625 | 1.588 |
| 5.70 | 1.598 | 1.563 |
| 5.80 | 1.581 | 1.544 |

## REFERENCES


[1] Z.C. Wu, E.T. Arakawa, J.R. Jimenez, L.J. Schowalter [ *Phys. Rev. B (USA)* vol.47 (1993) p.4356 ]

[2] M. Wölfel, M. Schulz, J. Ionally, P.J. Grunthaner [ *Appl. Phys. A (Germany)* vol.50 (1990) p.177-81 ]

[3] J.Y. Duboz, P.A. Badoz, J. Henz, H. von Känel [ *J. Appl. Phys. (USA)* vol.68 (1990) p.2346-50 ]

[4] J.R. Jimenez et al [ *J. Appl. Phys. (USA)* vol.66 (1989) p.2738-41 ]

[5] C. Viguier et al [ *Solid State Commun. (USA)* vol.60 (1986) p.923-6 ]
[6] M. Amiotti, A. Borghesi, G. Guizzetti, F. Nava [ *Phys. Rev. B (USA)* vol.42 (1990) p.8939-45 ]
[7] W. Henrion, H. Lange [ *Phys. Status Solidi B (Germany)* vol.112 (1982) p.K57 ]
[8] A. Humbert, A. Cros [ *J. Phys. Lett. (France)* vol.44 (1983) p.L929 ]

# 6.4 Optical functions: noble TM silicides

G. Guizetti and F. Marabelli

December 1994

## A INTRODUCTION

The only noble silicides investigated with optical techniques are Pt- and Pd-compounds. Several different stoichiometries can be obtained for both metals. The most studied one for Pd is $Pd_2Si$, which crystallises in a hexagonal cell.

## B $Pd_2Si$

Ellipsometric measurements are reported on polycrystalline films in the energy range 1.75-3 eV [1]. The optical response of single crystalline samples has been investigated with reflectivity and ellipsometric measurements in connection with band structure calculations [2] and the results are summarised in TABLE 1.

TABLE 1 Refractive index (n) and extinction coefficient (k) for $Pd_2Si$ obtained from measurements on single crystals with light polarised perpendicular (⊥) and parallel (||) to the crystallographic c-axis [2].

| eV | $n_\perp$ | $k_\perp$ | $n_\parallel$ | $k_\parallel$ |
|---|---|---|---|---|
| 0.05 | 15.818 | 47.415 | 14.803 | 60.811 |
| 0.06 | 14.230 | 42.242 | 13.256 | 53.670 |
| 0.07 | 12.643 | 37.069 | 11.708 | 46.530 |
| 0.08 | 11.849 | 34.482 | 10.934 | 42.960 |
| 0.09 | 10.531 | 30.165 | 9.917 | 37.328 |
| 0.10 | 8.954 | 24.987 | 8.475 | 30.757 |
| 0.20 | 6.557 | 16.442 | 5.916 | 20.052 |
| 0.30 | 5.172 | 11.382 | 4.450 | 13.720 |
| 0.40 | 4.530 | 8.817 | 3.742 | 10.549 |
| 0.50 | 4.122 | 7.014 | 3.245 | 8.386 |
| 0.60 | 4.005 | 6.072 | 2.890 | 7.207 |
| 0.70 | 3.944 | 5.535 | 2.612 | 6.208 |
| 0.80 | 3.748 | 5.141 | 2.422 | 5.402 |
| 0.90 | 3.520 | 4.852 | 2.376 | 4.685 |
| 1.00 | 3.331 | 4.458 | 2.427 | 3.955 |
| 1.10 | 3.217 | 4.099 | 2.700 | 3.313 |
| 1.20 | 3.130 | 3.883 | 3.053 | 3.055 |
| 1.30 | 3.091 | 3.705 | 3.272 | 3.107 |
| 1.40 | 3.044 | 3.591 | 3.343 | 3.223 |
| 1.50 | 2.935 | 3.465 | 3.278 | 3.284 |
| 1.60 | 2.877 | 3.363 | 3.199 | 3.313 |
| 1.70 | 2.823 | 3.264 | 3.064 | 3.347 |
| 1.80 | 2.765 | 3.182 | 2.884 | 3.338 |
| 1.90 | 2.725 | 3.124 | 2.728 | 3.276 |
| 2.00 | 2.674 | 3.072 | 2.564 | 3.173 |
| 2.10 | 2.633 | 3.050 | 2.451 | 3.051 |

Table 1 continued

| eV | $n_{\perp}$ | $k_{\perp}$ | $n_{\parallel}$ | $k_{\parallel}$ |
|---|---|---|---|---|
| 2.20 | 2.567 | 3.049 | 2.384 | 2.944 |
| 2.30 | 2.463 | 3.061 | 2.338 | 2.854 |
| 2.40 | 2.336 | 3.042 | 2.296 | 2.786 |
| 2.50 | 2.210 | 3.000 | 2.254 | 2.742 |
| 2.60 | 2.088 | 2.930 | 2.193 | 2.708 |
| 2.70 | 1.978 | 2.851 | 2.115 | 2.673 |
| 2.80 | 1.900 | 2.771 | 2.039 | 2.626 |
| 2.90 | 1.826 | 2.695 | 1.956 | 2.569 |
| 3.00 | 1.749 | 2.603 | 1.867 | 2.479 |
| 3.10 | 1.687 | 2.509 | 1.815 | 2.364 |
| 3.20 | 1.642 | 2.425 | 1.797 | 2.277 |
| 3.30 | 1.606 | 2.361 | 1.792 | 2.223 |
| 3.40 | 1.570 | 2.274 | 1.778 | 2.171 |
| 3.50 | 1.537 | 2.195 | 1.755 | 2.126 |
| 3.60 | 1.519 | 2.125 | 1.728 | 2.092 |
| 3.70 | 1.505 | 2.080 | 1.703 | 2.062 |
| 3.80 | 1.495 | 2.034 | 1.679 | 2.028 |
| 3.90 | 1.473 | 1.980 | 1.643 | 1.993 |
| 4.00 | 1.461 | 1.929 | 1.607 | 1.952 |
| 4.10 | 1.461 | 1.887 | 1.574 | 1.911 |
| 4.20 | 1.443 | 1.866 | 1.543 | 1.855 |
| 4.30 | 1.437 | 1.827 | 1.533 | 1.818 |
| 4.40 | 1.417 | 1.796 | 1.505 | 1.768 |
| 4.50 | 1.394 | 1.767 | 1.518 | 1.688 |
| 4.60 | 1.388 | 1.721 | 1.507 | 1.705 |
| 4.70 | 1.369 | 1.709 | 1.496 | 1.663 |
| 4.80 | 1.364 | 1.661 | 1.455 | 1.664 |
| 4.90 | 1.360 | 1.644 | 1.449 | 1.578 |
| 5.00 | 1.344 | 1.612 | 1.453 | 1.559 |
| 5.10 | 1.332 | 1.588 | 1.448 | 1.515 |
| 5.20 | 1.324 | 1.573 | 1.445 | 1.494 |
| 5.30 | 1.304 | 1.544 | 1.441 | 1.464 |
| 5.40 | 1.291 | 1.525 | 1.438 | 1.443 |
| 5.50 | 1.278 | 1.497 | 1.431 | 1.426 |
| 5.60 | 1.262 | 1.463 | 1.430 | 1.402 |
| 5.70 | 1.255 | 1.450 | 1.421 | 1.387 |
| 5.80 | 1.245 | 1.430 | 1.398 | 1.377 |
| 5.90 | 1.234 | 1.409 | 1.385 | 1.330 |
| 6.00 | 1.219 | 1.394 | 1.387 | 1.310 |
| 6.10 | 1.203 | 1.378 | 1.387 | 1.286 |
| 6.20 | 1.188 | 1.362 | 1.385 | 1.279 |
| 6.30 | 1.172 | 1.347 | 1.371 | 1.272 |
| 6.40 | 1.145 | 1.323 | 1.368 | 1.235 |
| 6.50 | 1.113 | 1.288 | 1.358 | 1.229 |
| 6.60 | 1.095 | 1.260 | 1.351 | 1.210 |
| 6.70 | 1.072 | 1.228 | 1.342 | 1.185 |
| 6.80 | 1.064 | 1.178 | 1.333 | 1.160 |
| 6.90 | 1.072 | 1.136 | 1.325 | 1.138 |
| 7.00 | 1.075 | 1.115 | 1.336 | 1.095 |
| 7.10 | 1.070 | 1.090 | 1.341 | 1.100 |
| 7.20 | 1.053 | 1.057 | 1.342 | 1.072 |
| 7.30 | 1.076 | 0.989 | 1.340 | 1.072 |

Table 1 continued

| eV | $n_{\perp}$ | $k_{\perp}$ | $n_{\parallel}$ | $k_{\parallel}$ |
|---|---|---|---|---|
| 7.40 | 1.100 | 0.978 | 1.318 | 1.055 |
| 7.50 | 1.106 | 0.963 | 1.320 | 0.996 |
| 7.60 | 1.113 | 0.938 | 1.341 | 0.977 |
| 7.70 | 1.119 | 0.916 | 1.356 | 0.959 |
| 7.80 | 1.124 | 0.903 | 1.363 | 0.953 |
| 7.90 | 1.133 | 0.884 | 1.373 | 0.945 |
| 8.00 | 1.146 | 0.870 | 1.379 | 0.948 |
| 8.10 | 1.161 | 0.847 | 1.379 | 0.930 |
| 8.20 | 1.168 | 0.841 | 1.380 | 0.922 |
| 8.30 | 1.175 | 0.835 | 1.382 | 0.915 |
| 8.40 | 1.186 | 0.821 | 1.393 | 0.887 |
| 8.50 | 1.192 | 0.812 | 1.407 | 0.878 |
| 8.60 | 1.196 | 0.807 | 1.415 | 0.874 |
| 8.70 | 1.206 | 0.792 | 1.432 | 0.872 |
| 8.80 | 1.212 | 0.783 | 1.441 | 0.865 |
| 8.90 | 1.216 | 0.777 | 1.446 | 0.862 |
| 9.00 | 1.233 | 0.754 | 1.465 | 0.859 |
| 9.10 | 1.245 | 0.749 | 1.477 | 0.857 |
| 9.20 | 1.258 | 0.745 | 1.489 | 0.855 |
| 9.30 | 1.270 | 0.745 | 1.502 | 0.875 |
| 9.40 | 1.276 | 0.745 | 1.508 | 0.885 |
| 9.50 | 1.296 | 0.735 | 1.510 | 0.893 |
| 9.60 | 1.305 | 0.737 | 1.512 | 0.898 |
| 9.70 | 1.322 | 0.742 | 1.517 | 0.909 |
| 9.80 | 1.330 | 0.748 | 1.517 | 0.919 |
| 9.90 | 1.342 | 0.757 | 1.516 | 0.934 |
| 10.00 | 1.347 | 0.768 | 1.513 | 0.941 |
| 10.10 | 1.354 | 0.783 | 1.507 | 0.952 |
| 10.20 | 1.351 | 0.796 | 1.500 | 0.962 |
| 10.30 | 1.345 | 0.819 | 1.487 | 0.978 |
| 10.40 | 1.338 | 0.822 | 1.481 | 0.976 |
| 10.50 | 1.327 | 0.826 | 1.472 | 0.973 |
| 10.60 | 1.321 | 0.829 | 1.468 | 0.972 |
| 10.70 | 1.309 | 0.829 | 1.456 | 0.988 |
| 10.80 | 1.302 | 0.829 | 1.450 | 0.996 |
| 10.90 | 1.298 | 0.823 | 1.441 | 0.991 |
| 11.00 | 1.295 | 0.816 | 1.431 | 0.985 |
| 11.10 | 1.295 | 0.820 | 1.428 | 0.991 |
| 11.20 | 1.295 | 0.825 | 1.424 | 0.999 |
| 11.30 | 1.295 | 0.829 | 1.421 | 1.004 |
| 11.40 | 1.289 | 0.839 | 1.408 | 1.017 |
| 11.50 | 1.282 | 0.849 | 1.393 | 1.030 |
| 11.60 | 1.274 | 0.854 | 1.382 | 1.034 |
| 11.70 | 1.261 | 0.862 | 1.364 | 1.041 |
| 11.80 | 1.247 | 0.871 | 1.345 | 1.048 |
| 11.90 | 1.235 | 0.868 | 1.330 | 1.045 |
| 12.00 | 1.222 | 0.865 | 1.314 | 1.042 |
| 12.10 | 1.209 | 0.862 | 1.298 | 1.038 |
| 12.20 | 1.194 | 0.861 | 1.276 | 1.040 |
| 12.30 | 1.179 | 0.860 | 1.255 | 1.042 |

## C PtSi

Reflectivity and thermoreflectance measurements on thin films of PtSi and $Pt_2Si$ are reported in [3,4]. Optical constants of PtSi have been derived only in the infrared from Kramers-Kronig analysis of reflectivity [5] and are summarised in TABLE 2.

TABLE 2 Refractive index (n) and extinction coefficient (k) for PtSi obtained from reflectivity measurements on polycrystalline films [5].

| eV | n | k |
|---|---|---|
| 0.1 | 13.0 | - |
| 0.2 | 6.0 | 18.6 |
| 0.3 | 3.8 | 13.0 |
| 0.4 | 3.0 | 10.0 |
| 0.5 | 2.4 | 8.3 |
| 0.6 | 2.0 | 6.8 |
| 0.7 | 1.8 | 5.6 |
| 0.8 | 1.8 | 4.7 |
| 0.9 | 1.7 | 4.0 |
| 1.0 | 1.8 | 3.5 |

## REFERENCES

[1] Juh-Tzeng Lue, Hong-Wen Chen, Shwa-Ing Lew [ *Phys. Rev. B (USA)* vol.34 (1986) p.5438 ]

[2] M. Amiotti et al [ *Phys. Rev. B (USA)* vol.45 (1992) p.13285-92 ]

[3] A. Borghesi, G. Guizzetti, L. Nosenzo, A. Stella, G. Majni [ *Thin Solid Films (Switzerland)* vol.140 (1986) p.95-8 ]

[4] W. Henrion, H. Lange [ *Phys. Status Solidi B (Germany)* vol.112 (1982) p.K57 ]

[5] J.M. Pimbley, W. Katz [ *Appl. Phys. Lett. (USA)* vol.42 (1983) p.984 ]

# 6.5 Optical functions: semiconducting TM silicides

G. Guizetti and F. Marabelli

December 1994

## A INTRODUCTION

Among the silicides formed with transition metals of the VI, VII and VIII columns a semiconducting phase can be obtained with a particular stoichiometry (in most cases the $MSi_2$ form). The semiconducting one is the most stable phase because the opening of the gap itself occurs in order to minimise the total energy of the crystal and usually can be associated with a slight rearrangement of the crystal structure with respect to the hypothetical one which can be assumed by analogy with similar materials. The optical functions of semiconductors at the lowest energies are affected by the presence of impurities, so, even when single crystals were studied ($CrSi_2$), the best results were obtained on well-characterised polycrystalline or epitaxial films.

The compounds for which well-defined optical data exist are Cr-, Re-, Fe- and Ir-silicides; in particular chromium and iron compounds have been intensively studied either in the semiconducting phase, or on samples with different stoichiometry. Some studies exist for $MnSi_{1.75}$.

## B $CrSi_2$

The optical response of $CrSi_2$ has been thoroughly investigated, mostly near to the absorption edge at 0.35 eV [1-3]. The far-infrared vibrational spectrum has also been studied [2-5]. Reflectance and transmittance were usually measured on polycrystalline films, but in [5] (TABLE 1) a comparison is given between thin films and single crystal data obtained by reflectivity and ellipsometric measurements. High energy ellipsometric data measured with synchrotron radiation and listed in TABLES 2 and 3 are taken from [3] and [6].

TABLE 1 Refractive index (n) and extinction coefficient (k) for $CrSi_2$ obtained from measurements on polycrystalline films [5].

| eV | n | k | eV | n | k | eV | n | k |
|---|---|---|---|---|---|---|---|---|
| 0.01 | 5.688 | 0.000 | 0.20 | 5.818 | 0.000 | 1.20 | 4.537 | 3.618 |
| 0.02 | 5.688 | 0.000 | 0.30 | 6.247 | 0.015 | 1.30 | 3.859 | 3.520 |
| 0.03 | 5.689 | 0.000 | 0.40 | 6.795 | 0.367 | 1.40 | 3.459 | 3.239 |
| 0.04 | 5.689 | 0.000 | 0.50 | 6.939 | 1.208 | 1.50 | 3.203 | 2.954 |
| 0.05 | 5.690 | 0.000 | 0.60 | 6.566 | 2.097 | 1.60 | 3.068 | 2.747 |
| 0.06 | 5.691 | 0.000 | 0.70 | 5.496 | 2.288 | 1.70 | 2.981 | 2.639 |
| 0.07 | 5.692 | 0.000 | 0.80 | 5.462 | 1.766 | 1.80 | 2.878 | 2.560 |
| 0.08 | 5.694 | 0.000 | 0.90 | 5.985 | 1.967 | 1.90 | 2.785 | 2.452 |
| 0.09 | 5.697 | 0.000 | 1.00 | 6.028 | 2.920 | 2.00 | 2.732 | 2.355 |
| 0.10 | 5.681 | 0.000 | 1.10 | 5.252 | 3.475 | 2.10 | 2.702 | 2.262 |

Table 1 continued

| eV | n | k |
|---|---|---|
| 2.20 | 2.708 | 2.209 |
| 2.30 | 2.708 | 2.199 |
| 2.40 | 2.689 | 2.205 |
| 2.50 | 2.663 | 2.201 |
| 2.60 | 2.638 | 2.209 |
| 2.70 | 2.607 | 2.227 |
| 2.80 | 2.569 | 2.241 |
| 2.90 | 2.537 | 2.258 |
| 3.00 | 2.498 | 2.280 |
| 3.10 | 2.458 | 2.305 |

| eV | n | k |
|---|---|---|
| 3.20 | 2.410 | 2.328 |
| 3.30 | 2.359 | 2.364 |
| 3.40 | 2.303 | 2.394 |
| 3.50 | 2.248 | 2.418 |
| 3.60 | 2.186 | 2.432 |
| 3.70 | 2.121 | 2.445 |
| 3.80 | 2.055 | 2.456 |
| 3.90 | 1.990 | 2.456 |
| 4.00 | 1.931 | 2.453 |

| eV | n | k |
|---|---|---|
| 4.10 | 1.856 | 2.440 |
| 4.20 | 1.795 | 2.439 |
| 4.30 | 1.731 | 2.418 |
| 4.40 | 1.682 | 2.421 |
| 4.50 | 1.585 | 2.410 |
| 4.60 | 1.535 | 2.381 |
| 4.70 | 1.504 | 2.410 |
| 4.80 | 1.494 | 2.405 |
| 4.90 | 1.490 | 2.479 |

TABLE 2 Refractive index (n) and extinction coefficient (k) for $CrSi_2$ obtained from ellipsometric measurements with synchrotron radiation on polycrystalline films [3,6]. A $MgF_2$ Rochon rotating analyser was used.

| eV | n | k |
|---|---|---|
| 4.50 | 2.336 | 1.964 |
| 4.60 | 2.273 | 2.045 |
| 4.70 | 2.181 | 2.111 |
| 4.80 | 2.073 | 2.151 |
| 4.90 | 1.962 | 2.164 |
| 5.00 | 1.859 | 2.150 |
| 5.10 | 1.768 | 2.121 |
| 5.20 | 1.685 | 2.085 |
| 5.30 | 1.613 | 2.045 |
| 5.40 | 1.552 | 2.003 |
| 5.50 | 1.502 | 1.959 |
| 5.60 | 1.460 | 1.915 |
| 5.70 | 1.423 | 1.874 |
| 5.80 | 1.388 | 1.838 |
| 5.90 | 1.356 | 1.804 |
| 6.00 | 1.326 | 1.774 |

| eV | n | k |
|---|---|---|
| 6.10 | 1.297 | 1.746 |
| 6.20 | 1.269 | 1.720 |
| 6.30 | 1.240 | 1.695 |
| 6.40 | 1.214 | 1.669 |
| 6.50 | 1.187 | 1.642 |
| 6.60 | 1.159 | 1.614 |
| 6.70 | 1.131 | 1.586 |
| 6.80 | 1.106 | 1.557 |
| 6.90 | 1.081 | 1.528 |
| 7.00 | 1.058 | 1.499 |
| 7.10 | 1.034 | 1.469 |
| 7.20 | 1.012 | 1.438 |
| 7.30 | 0.992 | 1.406 |
| 7.40 | 0.974 | 1.372 |
| 7.50 | 0.958 | 1.336 |

| eV | n | k |
|---|---|---|
| 7.60 | 0.946 | 1.300 |
| 7.70 | 0.937 | 1.263 |
| 7.80 | 0.933 | 1.225 |
| 7.90 | 0.932 | 1.191 |
| 8.00 | 0.935 | 1.159 |
| 8.10 | 0.940 | 1.132 |
| 8.20 | 0.946 | 1.109 |
| 8.30 | 0.951 | 1.089 |
| 8.40 | 0.957 | 1.073 |
| 8.50 | 0.962 | 1.058 |
| 8.60 | 0.968 | 1.044 |
| 8.70 | 0.974 | 1.031 |
| 8.80 | 0.981 | 1.022 |
| 8.90 | 0.987 | 1.016 |
| 9.00 | 0.993 | 1.013 |

TABLE 3 Refractive index (n) and extinction coefficient (k) for $CrSi_2$ obtained from ellipsometric measurements with synchrotron radiation on polycrystalline films [3,6]. A rotating triple reflection gold analyser was used.

| eV | n | k |
|---|---|---|
| 11.00 | 0.314 | 0.294 |
| 11.10 | 0.326 | 0.276 |
| 11.20 | 0.336 | 0.263 |
| 11.30 | 0.346 | 0.254 |
| 11.40 | 0.355 | 0.248 |
| 11.50 | 0.361 | 0.247 |
| 11.60 | 0.365 | 0.248 |
| 11.70 | 0.367 | 0.254 |
| 11.80 | 0.367 | 0.265 |
| 11.90 | 0.365 | 0.279 |
| 12.00 | 0.364 | 0.299 |
| 12.10 | 0.364 | 0.322 |
| 12.20 | 0.366 | 0.349 |

| eV | n | k |
|---|---|---|
| 12.30 | 0.371 | 0.380 |
| 12.40 | 0.377 | 0.414 |
| 12.50 | 0.384 | 0.450 |
| 12.60 | 0.393 | 0.487 |
| 12.70 | 0.404 | 0.526 |
| 12.80 | 0.417 | 0.564 |
| 12.90 | 0.433 | 0.602 |
| 13.00 | 0.451 | 0.638 |
| 13.10 | 0.470 | 0.670 |
| 13.20 | 0.488 | 0.698 |
| 13.30 | 0.506 | 0.722 |
| 13.40 | 0.521 | 0.744 |
| 13.50 | 0.535 | 0.764 |

| eV | n | k |
|---|---|---|
| 13.60 | 0.545 | 0.782 |
| 13.70 | 0.553 | 0.801 |
| 13.80 | 0.558 | 0.818 |
| 13.90 | 0.561 | 0.831 |
| 14.00 | 0.564 | 0.842 |
| 14.10 | 0.566 | 0.848 |
| 14.20 | 0.568 | 0.849 |
| 14.30 | 0.569 | 0.848 |
| 14.40 | 0.570 | 0.845 |
| 14.50 | 0.569 | 0.840 |
| 14.60 | 0.569 | 0.835 |
| 14.70 | 0.567 | 0.830 |
| 14.80 | 0.565 | 0.824 |

Table 3 continued

| eV | n | k |
|---|---|---|
| 14.90 | 0.564 | 0.817 |
| 15.00 | 0.562 | 0.810 |
| 15.10 | 0.559 | 0.802 |
| 15.20 | 0.557 | 0.794 |
| 15.30 | 0.554 | 0.785 |
| 15.40 | 0.551 | 0.777 |
| 15.50 | 0.547 | 0.768 |
| 15.60 | 0.544 | 0.760 |
| 15.70 | 0.540 | 0.752 |
| 15.80 | 0.535 | 0.744 |
| 15.90 | 0.530 | 0.737 |
| 16.00 | 0.526 | 0.729 |
| 16.10 | 0.522 | 0.719 |
| 16.20 | 0.520 | 0.709 |
| 16.30 | 0.519 | 0.698 |
| 16.40 | 0.519 | 0.687 |
| 16.50 | 0.519 | 0.675 |
| 16.60 | 0.519 | 0.664 |
| 16.70 | 0.517 | 0.654 |
| 16.80 | 0.516 | 0.643 |
| 16.90 | 0.513 | 0.633 |
| 17.00 | 0.511 | 0.624 |
| 17.10 | 0.507 | 0.615 |
| 17.20 | 0.504 | 0.606 |
| 17.30 | 0.499 | 0.597 |
| 17.40 | 0.495 | 0.589 |
| 17.50 | 0.490 | 0.581 |
| 17.60 | 0.485 | 0.573 |

| eV | n | k |
|---|---|---|
| 17.70 | 0.479 | 0.564 |
| 17.80 | 0.474 | 0.555 |
| 17.90 | 0.469 | 0.546 |
| 18.00 | 0.464 | 0.537 |
| 18.10 | 0.460 | 0.527 |
| 18.20 | 0.456 | 0.517 |
| 18.30 | 0.453 | 0.507 |
| 18.40 | 0.451 | 0.496 |
| 18.50 | 0.448 | 0.485 |
| 18.60 | 0.446 | 0.474 |
| 18.70 | 0.445 | 0.463 |
| 18.80 | 0.443 | 0.451 |
| 18.90 | 0.441 | 0.440 |
| 19.00 | 0.440 | 0.428 |
| 19.10 | 0.439 | 0.417 |
| 19.20 | 0.438 | 0.406 |
| 19.30 | 0.439 | 0.395 |
| 19.40 | 0.440 | 0.384 |
| 19.50 | 0.442 | 0.373 |
| 19.60 | 0.445 | 0.363 |
| 19.70 | 0.448 | 0.352 |
| 19.80 | 0.452 | 0.341 |
| 19.90 | 0.456 | 0.330 |
| 20.00 | 0.460 | 0.320 |
| 20.10 | 0.464 | 0.311 |
| 20.20 | 0.470 | 0.302 |
| 20.30 | 0.477 | 0.294 |
| 20.40 | 0.484 | 0.286 |

| eV | n | k |
|---|---|---|
| 20.50 | 0.491 | 0.279 |
| 20.60 | 0.498 | 0.272 |
| 20.70 | 0.505 | 0.266 |
| 20.80 | 0.513 | 0.260 |
| 20.90 | 0.520 | 0.255 |
| 21.00 | 0.527 | 0.250 |
| 21.10 | 0.534 | 0.245 |
| 21.20 | 0.542 | 0.241 |
| 21.30 | 0.549 | 0.238 |
| 21.40 | 0.557 | 0.235 |
| 21.50 | 0.565 | 0.233 |
| 21.60 | 0.572 | 0.232 |
| 21.70 | 0.579 | 0.230 |
| 21.80 | 0.586 | 0.229 |
| 21.90 | 0.593 | 0.228 |
| 22.00 | 0.599 | 0.227 |
| 22.10 | 0.606 | 0.227 |
| 22.20 | 0.613 | 0.226 |
| 22.30 | 0.620 | 0.226 |
| 22.40 | 0.627 | 0.227 |
| 22.50 | 0.635 | 0.227 |
| 22.60 | 0.642 | 0.229 |
| 22.70 | 0.649 | 0.230 |
| 22.80 | 0.656 | 0.231 |
| 22.90 | 0.662 | 0.233 |
| 23.00 | 0.668 | 0.235 |
| 23.10 | 0.674 | 0.236 |

## C $Cr_5Si_3$ AND $Cr_3Si$

TABLES 4 and 5 summarise the data on $Cr_5Si_3$ and $Cr_3Si$, respectively.

TABLE 4 Refractive index (n) and extinction coefficient (k) for $Cr_5Si_3$ obtained from reflectivity and ellipsometric measurements on polycrystalline films [7].

| eV | n | k |
|---|---|---|
| 0.10 | 8.613 | 16.090 |
| 0.20 | 5.164 | 11.230 |
| 0.30 | 3.466 | 7.904 |
| 0.40 | 2.816 | 5.849 |
| 0.50 | 2.592 | 4.456 |
| 0.60 | 2.470 | 3.825 |
| 0.70 | 2.467 | 3.140 |
| 0.80 | 2.679 | 2.669 |
| 0.90 | 2.918 | 2.531 |
| 1.00 | 2.995 | 2.576 |
| 1.10 | 2.936 | 2.640 |
| 1.20 | 2.805 | 2.676 |
| 1.30 | 2.666 | 2.685 |

| eV | n | k |
|---|---|---|
| 1.40 | 2.472 | 2.667 |
| 1.50 | 2.323 | 2.595 |
| 1.60 | 2.171 | 2.512 |
| 1.70 | 2.048 | 2.411 |
| 1.80 | 1.954 | 2.316 |
| 1.90 | 1.877 | 2.229 |
| 2.00 | 1.807 | 2.142 |
| 2.10 | 1.745 | 2.069 |
| 2.20 | 1.689 | 2.004 |
| 2.30 | 1.634 | 1.929 |
| 2.40 | 1.593 | 1.860 |
| 2.50 | 1.557 | 1.807 |
| 2.60 | 1.507 | 1.743 |

| eV | n | k |
|---|---|---|
| 2.70 | 1.476 | 1.686 |
| 2.80 | 1.443 | 1.632 |
| 2.90 | 1.415 | 1.573 |
| 3.00 | 1.392 | 1.513 |
| 3.10 | 1.373 | 1.451 |
| 3.20 | 1.361 | 1.399 |
| 3.30 | 1.355 | 1.359 |
| 3.40 | 1.348 | 1.311 |
| 3.50 | 1.343 | 1.262 |
| 3.60 | 1.344 | 1.222 |
| 3.70 | 1.347 | 1.199 |
| 3.80 | 1.348 | 1.174 |
| 3.90 | 1.351 | 1.146 |

Table 4 continued

| eV | n | k |
|---|---|---|
| 4.00 | 1.353 | 1.126 |
| 4.10 | 1.353 | 1.110 |
| 4.20 | 1.349 | 1.095 |
| 4.30 | 1.346 | 1.082 |
| 4.40 | 1.341 | 1.064 |
| 4.50 | 1.336 | 1.043 |
| 4.60 | 1.335 | 1.030 |

| eV | n | k |
|---|---|---|
| 4.70 | 1.333 | 1.017 |
| 4.80 | 1.332 | 1.005 |
| 4.90 | 1.331 | 0.994 |
| 5.00 | 1.330 | 0.984 |
| 5.10 | 1.327 | 0.974 |
| 5.20 | 1.326 | 0.964 |
| 5.30 | 1.323 | 0.956 |

| eV | n | k |
|---|---|---|
| 5.40 | 1.321 | 0.946 |
| 5.50 | 1.320 | 0.939 |
| 5.60 | 1.318 | 0.933 |
| 5.70 | 1.317 | 0.931 |
| 5.80 | 1.315 | 0.928 |
| 5.90 | 1.311 | 0.928 |

TABLE 5 Refractive index (n) and extinction coefficient (k) for $Cr_3Si$ obtained from reflectivity and ellipsometric measurements on polycrystalline films [7].

| eV | n | k |
|---|---|---|
| 0.09 | 7.692 | 22.231 |
| 0.10 | 5.015 | 16.561 |
| 0.20 | 2.992 | 10.625 |
| 0.30 | 2.146 | 6.850 |
| 0.40 | 1.973 | 5.003 |
| 0.50 | 1.996 | 3.845 |
| 0.60 | 1.976 | 3.311 |
| 0.70 | 2.078 | 2.768 |
| 0.80 | 2.273 | 2.501 |
| 0.90 | 2.349 | 2.457 |
| 1.00 | 2.294 | 2.451 |
| 1.10 | 2.178 | 2.422 |
| 1.20 | 2.040 | 2.368 |
| 1.30 | 1.910 | 2.298 |
| 1.40 | 1.761 | 2.212 |
| 1.50 | 1.655 | 2.090 |
| 1.60 | 1.566 | 1.973 |
| 1.70 | 1.498 | 1.856 |
| 1.80 | 1.448 | 1.754 |
| 1.90 | 1.410 | 1.664 |

| eV | n | k |
|---|---|---|
| 2.00 | 1.375 | 1.574 |
| 2.10 | 1.347 | 1.495 |
| 2.20 | 1.324 | 1.428 |
| 2.30 | 1.303 | 1.355 |
| 2.40 | 1.290 | 1.288 |
| 2.50 | 1.283 | 1.233 |
| 2.60 | 1.272 | 1.175 |
| 2.70 | 1.267 | 1.126 |
| 2.80 | 1.262 | 1.080 |
| 2.90 | 1.261 | 1.034 |
| 3.00 | 1.262 | 0.988 |
| 3.10 | 1.268 | 0.943 |
| 3.20 | 1.276 | 0.907 |
| 3.30 | 1.283 | 0.879 |
| 3.40 | 1.294 | 0.848 |
| 3.50 | 1.305 | 0.818 |
| 3.60 | 1.315 | 0.794 |
| 3.70 | 1.326 | 0.778 |
| 3.80 | 1.335 | 0.766 |
| 3.90 | 1.347 | 0.750 |

| eV | n | k |
|---|---|---|
| 4.00 | 1.357 | 0.738 |
| 4.10 | 1.367 | 0.730 |
| 4.20 | 1.375 | 0.724 |
| 4.30 | 1.381 | 0.718 |
| 4.40 | 1.389 | 0.713 |
| 4.50 | 1.397 | 0.708 |
| 4.60 | 1.403 | 0.705 |
| 4.70 | 1.410 | 0.703 |
| 4.80 | 1.417 | 0.705 |
| 4.90 | 1.423 | 0.706 |
| 5.00 | 1.430 | 0.712 |
| 5.10 | 1.434 | 0.718 |
| 5.20 | 1.438 | 0.725 |
| 5.30 | 1.441 | 0.735 |
| 5.40 | 1.443 | 0.748 |
| 5.50 | 1.442 | 0.764 |
| 5.60 | 1.437 | 0.785 |
| 5.70 | 1.432 | 0.796 |
| 5.80 | 1.426 | 0.808 |
| 5.90 | 1.369 | 0.841 |

## D $MnSi_{1.75}$

Few data have been reported on the semiconducting manganese silicide. Reflectivity data are reported in [8,9] and high energy spectra are given in [6].

## E $ReSi_{1.75}$

Optical investigations on $ReSi_{1.75}$, often referred to as $ReSi_2$ (the correct stoichiometry and crystal structure have been the subject of debate), have been performed with reflection and transmission measurements on thin films in order to analyse the absorption edge [10,11]. A complete study on single crystals ($ReSi_{1.96}$) up to 12 eV has been reported in [12].

TABLE 6 Refractive index (n) and extinction coefficient (k) for $ReSi_{1.75}$ obtained from measurements on single crystalline bulk samples [12].

| eV | n | k | eV | n | k | eV | n | k |
|---|---|---|---|---|---|---|---|---|
| 0.01 | 9.544 | 5.917 | 3.50 | 3.558 | 2.600 | 7.80 | 1.309 | 2.101 |
| 0.02 | 8.340 | 4.651 | 3.60 | 3.456 | 2.626 | 7.90 | 1.282 | 2.098 |
| 0.03 | 7.781 | 4.305 | 3.70 | 3.353 | 2.652 | 8.00 | 1.264 | 2.097 |
| 0.04 | 7.076 | 6.750 | 3.80 | 3.302 | 2.665 | 8.10 | 1.248 | 2.058 |
| 0.05 | 1.743 | 4.459 | 3.90 | 3.145 | 2.635 | 8.20 | 1.233 | 2.019 |
| 0.06 | 2.983 | 2.560 | 4.00 | 3.040 | 2.615 | 8.30 | 1.225 | 2.000 |
| 0.07 | 3.190 | 2.039 | 4.10 | 2.957 | 2.614 | 8.40 | 1.205 | 1.958 |
| 0.08 | 3.303 | 1.440 | 4.20 | 2.874 | 2.613 | 8.50 | 1.192 | 1.929 |
| 0.09 | 3.395 | 1.007 | 4.30 | 2.833 | 2.612 | 8.60 | 1.171 | 1.895 |
| 0.10 | 4.004 | 0.398 | 4.40 | 2.738 | 2.581 | 8.70 | 1.150 | 1.860 |
| 0.20 | 4.552 | 0.100 | 4.50 | 2.675 | 2.561 | 8.80 | 1.139 | 1.843 |
| 0.30 | 4.918 | 0.127 | 4.60 | 2.629 | 2.544 | 8.90 | 1.102 | 1.796 |
| 0.40 | 5.096 | 0.260 | 4.70 | 2.583 | 2.528 | 9.00 | 1.077 | 1.764 |
| 0.50 | 5.192 | 0.445 | 4.80 | 2.559 | 2.520 | 9.10 | 1.068 | 1.741 |
| 0.60 | 5.090 | 0.637 | 4.90 | 2.503 | 2.493 | 9.20 | 1.059 | 1.718 |
| 0.70 | 4.965 | 0.728 | 5.00 | 2.465 | 2.475 | 9.30 | 1.055 | 1.707 |
| 0.80 | 4.881 | 0.791 | 5.10 | 2.430 | 2.461 | 9.40 | 1.032 | 1.656 |
| 0.90 | 4.835 | 0.853 | 5.20 | 2.395 | 2.447 | 9.50 | 1.017 | 1.623 |
| 1.00 | 4.808 | 0.901 | 5.30 | 2.377 | 2.440 | 9.60 | 1.004 | 1.603 |
| 1.10 | 4.766 | 0.957 | 5.40 | 2.321 | 2.447 | 9.70 | 0.992 | 1.582 |
| 1.20 | 4.732 | 0.998 | 5.50 | 2.283 | 2.452 | 9.80 | 0.986 | 1.572 |
| 1.30 | 4.716 | 1.018 | 5.60 | 2.223 | 2.447 | 9.90 | 0.976 | 1.527 |
| 1.40 | 4.690 | 1.101 | 5.70 | 2.163 | 2.441 | 10.00 | 0.969 | 1.496 |
| 1.50 | 4.673 | 1.156 | 5.80 | 2.132 | 2.439 | 10.10 | 0.956 | 1.474 |
| 1.60 | 4.659 | 1.232 | 5.90 | 2.075 | 2.433 | 10.20 | 0.943 | 1.452 |
| 1.70 | 4.645 | 1.309 | 6.00 | 2.038 | 2.429 | 10.30 | 0.937 | 1.441 |
| 1.80 | 4.638 | 1.347 | 6.10 | 1.997 | 2.408 | 10.40 | 0.917 | 1.407 |
| 1.90 | 4.585 | 1.449 | 6.20 | 1.956 | 2.387 | 10.50 | 0.903 | 1.384 |
| 2.00 | 4.550 | 1.517 | 6.30 | 1.936 | 2.376 | 10.60 | 0.903 | 1.362 |
| 2.10 | 4.506 | 1.594 | 6.40 | 1.871 | 2.363 | 10.70 | 0.903 | 1.340 |
| 2.20 | 4.463 | 1.672 | 6.50 | 1.827 | 2.353 | 10.80 | 0.903 | 1.329 |
| 2.30 | 4.442 | 1.711 | 6.60 | 1.775 | 2.331 | 10.90 | 0.893 | 1.310 |
| 2.40 | 4.375 | 1.848 | 6.70 | 1.723 | 2.309 | 11.00 | 0.886 | 1.298 |
| 2.50 | 4.331 | 1.939 | 6.80 | 1.697 | 2.298 | 11.10 | 0.879 | 1.285 |
| 2.60 | 4.279 | 2.006 | 6.90 | 1.669 | 2.264 | 11.20 | 0.872 | 1.273 |
| 2.70 | 4.226 | 2.074 | 7.00 | 1.651 | 2.242 | 11.30 | 0.868 | 1.267 |
| 2.80 | 4.200 | 2.107 | 7.10 | 1.586 | 2.204 | 11.40 | 0.858 | 1.247 |
| 2.90 | 4.077 | 2.232 | 7.20 | 1.521 | 2.167 | 11.50 | 0.851 | 1.234 |
| 3.00 | 3.995 | 2.315 | 7.30 | 1.489 | 2.149 | 11.60 | 0.843 | 1.221 |
| 3.10 | 3.918 | 2.373 | 7.40 | 1.398 | 2.138 | 11.70 | 0.836 | 1.208 |
| 3.20 | 3.840 | 2.431 | 7.50 | 1.338 | 2.131 | 11.80 | 0.832 | 1.201 |
| 3.30 | 3.801 | 2.460 | 7.60 | 1.326 | 2.119 | 11.90 | 0.832 | 1.201 |
| 3.40 | 3.655 | 2.544 | 7.70 | 1.315 | 2.107 | 12.00 | 0.832 | 1.201 |

## F $FeSi_2$

Optical functions of $FeSi_2$ in the orthorhombic β-phase (semiconducting) have been measured on polycrystalline thin films, epitaxial films and bulk polycrystalline samples. Some discrepancies among the results have been ascribed to different growth conditions, to the method of measurement and to surface contamination. In particular, the character (direct or indirect) of the gap is still controversial. Values of n and k reported in TABLE 7 have been obtained from reflectance, transmittance and spectroscopic ellipsometry on epitaxial films grown by MBE on both (111) and (100) silicon substrates. No noticeable dependence on the orientation has been detected [13]. The results agree quite well with those obtained in a narrower spectral range on polycrystalline bulk samples and films [14,15]. Ellipsometric measurements with synchrotron radiation have been used to obtain the data listed in TABLES 8 and 9 [3,6]. Far infrared vibrational properties are reported in [2,3,16].

TABLE 7 Refractive index (n) and extinction coefficient (k) for $FeSi_2$ obtained from measurements on epitaxial films [13]. A $MgF_2$ Rochon rotating analyser was used.

| eV | n | k | eV | n | k | eV | n | k |
|---|---|---|---|---|---|---|---|---|
| 0.20 | 5.706 | 0.000 | 1.90 | 4.457 | 2.834 | 3.50 | 3.235 | 2.506 |
| 0.30 | 5.726 | 0.000 | 2.00 | 4.191 | 2.854 | 3.60 | 3.207 | 2.549 |
| 0.40 | 5.758 | 0.000 | 2.10 | 3.998 | 2.775 | 3.70 | 3.176 | 2.600 |
| 0.50 | 5.850 | 0.000 | 2.20 | 3.863 | 2.726 | 3.80 | 3.120 | 2.665 |
| 0.60 | 5.983 | 0.000 | 2.30 | 3.744 | 2.669 | 3.90 | 3.052 | 2.718 |
| 0.70 | 6.164 | 0.000 | 2.40 | 3.646 | 2.606 | 4.00 | 2.973 | 2.767 |
| 0.80 | 6.420 | 0.015 | 2.50 | 3.591 | 2.563 | 4.10 | 2.895 | 2.802 |
| 0.90 | 6.792 | 0.258 | 2.60 | 3.536 | 2.539 | 4.20 | 2.811 | 2.841 |
| 1.00 | 6.273 | 1.210 | 2.70 | 3.468 | 2.519 | 4.30 | 2.729 | 2.874 |
| 1.10 | 5.979 | 1.569 | 2.80 | 3.407 | 2.481 | 4.40 | 2.627 | 2.918 |
| 1.20 | 5.792 | 1.821 | 2.90 | 3.369 | 2.450 | 4.50 | 2.513 | 2.911 |
| 1.30 | 5.583 | 2.021 | 3.00 | 3.352 | 2.436 | 4.60 | 2.426 | 2.928 |
| 1.40 | 5.378 | 2.178 | 3.10 | 3.319 | 2.439 | 4.70 | 2.317 | 2.934 |
| 1.50 | 5.217 | 2.324 | 3.20 | 3.295 | 2.439 | 4.80 | 2.217 | 2.924 |
| 1.60 | 5.033 | 2.486 | 3.30 | 3.277 | 2.450 | 4.90 | 2.103 | 2.912 |
| 1.70 | 4.874 | 2.618 | 3.40 | 3.258 | 2.475 | 5.00 | 2.019 | 2.888 |
| 1.80 | 4.692 | 2.740 | | | | | | |

TABLE 8 Refractive index (n) and extinction coefficient (k) for $FeSi_2$ obtained from ellipsometric measurements with synchrotron radiation on polycrystalline films [3,6]. A rotating triple reflection gold analyser was used.

| eV | n | k | eV | n | k | eV | n | k |
|---|---|---|---|---|---|---|---|---|
| 4.50 | 2.567 | 2.760 | 6.10 | 1.153 | 2.573 | 7.60 | 0.704 | 1.900 |
| 4.60 | 2.542 | 2.827 | 6.20 | 1.102 | 2.524 | 7.70 | 0.667 | 1.858 |
| 4.70 | 2.478 | 2.892 | 6.30 | 1.059 | 2.477 | 7.80 | 0.633 | 1.817 |
| 4.80 | 2.386 | 2.945 | 6.40 | 1.023 | 2.429 | 7.90 | 0.614 | 1.775 |
| 4.90 | 2.276 | 2.977 | 6.50 | 0.987 | 2.381 | 8.00 | 0.604 | 1.731 |
| 5.00 | 2.157 | 2.984 | 6.60 | 0.949 | 2.334 | 8.10 | 0.594 | 1.685 |
| 5.10 | 2.036 | 2.975 | 6.70 | 0.911 | 2.287 | 8.20 | 0.584 | 1.640 |
| 5.20 | 1.917 | 2.959 | 6.80 | 0.876 | 2.242 | 8.30 | 0.573 | 1.596 |
| 5.30 | 1.801 | 2.938 | 6.90 | 0.843 | 2.199 | 8.40 | 0.562 | 1.556 |
| 5.40 | 1.690 | 2.910 | 7.00 | 0.813 | 2.157 | 8.50 | 0.560 | 1.522 |
| 5.50 | 1.589 | 2.874 | 7.10 | 0.786 | 2.115 | 8.60 | 0.560 | 1.492 |
| 5.60 | 1.499 | 2.831 | 7.20 | 0.761 | 2.072 | 8.70 | 0.558 | 1.460 |
| 5.70 | 1.420 | 2.783 | 7.30 | 0.744 | 2.028 | 8.80 | 0.569 | 1.432 |
| 5.80 | 1.347 | 2.730 | 7.40 | 0.735 | 1.984 | 8.90 | 0.600 | 1.410 |
| 5.90 | 1.278 | 2.676 | 7.50 | 0.727 | 1.942 | 9.00 | 0.628 | 1.390 |
| 6.00 | 1.212 | 2.623 | | | | | | |

TABLE 9 Refractive index (n) and extinction coefficient (k) for $FeSi_2$ obtained from ellipsometric measurements with synchrotron radiation on polycrystalline films [3,6].

| eV | n | k |
|---|---|---|
| 11.00 | 0.152 | 0.268 |
| 11.10 | 0.190 | 0.227 |
| 11.20 | 0.232 | 0.203 |
| 11.30 | 0.268 | 0.192 |
| 11.40 | 0.293 | 0.189 |
| 11.50 | 0.308 | 0.193 |
| 11.60 | 0.314 | 0.201 |
| 11.70 | 0.314 | 0.213 |
| 11.80 | 0.309 | 0.233 |
| 11.90 | 0.303 | 0.258 |
| 12.00 | 0.296 | 0.290 |
| 12.10 | 0.292 | 0.329 |
| 12.20 | 0.291 | 0.373 |
| 12.30 | 0.292 | 0.421 |
| 12.40 | 0.296 | 0.471 |
| 12.50 | 0.301 | 0.522 |
| 12.60 | 0.307 | 0.574 |
| 12.70 | 0.315 | 0.625 |
| 12.80 | 0.325 | 0.675 |
| 12.90 | 0.336 | 0.722 |
| 13.00 | 0.350 | 0.766 |
| 13.10 | 0.366 | 0.805 |
| 13.20 | 0.383 | 0.840 |
| 13.30 | 0.399 | 0.872 |
| 13.40 | 0.415 | 0.899 |
| 13.50 | 0.429 | 0.924 |
| 13.60 | 0.440 | 0.948 |
| 13.70 | 0.448 | 0.970 |
| 13.80 | 0.452 | 0.993 |
| 13.90 | 0.454 | 1.014 |
| 14.00 | 0.454 | 1.034 |
| 14.10 | 0.451 | 1.049 |
| 14.20 | 0.448 | 1.060 |
| 14.30 | 0.442 | 1.066 |
| 14.40 | 0.436 | 1.067 |
| 14.50 | 0.428 | 1.065 |
| 14.60 | 0.421 | 1.058 |
| 14.70 | 0.413 | 1.048 |
| 14.80 | 0.407 | 1.034 |
| 14.90 | 0.402 | 1.018 |
| 15.00 | 0.398 | 1.001 |
| 15.10 | 0.395 | 0.983 |
| 15.20 | 0.392 | 0.964 |
| 15.30 | 0.389 | 0.946 |
| 15.40 | 0.385 | 0.929 |
| 15.50 | 0.382 | 0.911 |
| 15.60 | 0.379 | 0.893 |
| 15.70 | 0.376 | 0.874 |
| 15.80 | 0.374 | 0.854 |
| 15.90 | 0.374 | 0.833 |
| 16.00 | 0.375 | 0.812 |
| 16.10 | 0.379 | 0.790 |
| 16.20 | 0.383 | 0.769 |
| 16.30 | 0.389 | 0.748 |
| 16.40 | 0.395 | 0.729 |
| 16.50 | 0.402 | 0.711 |
| 16.60 | 0.407 | 0.695 |
| 16.70 | 0.413 | 0.681 |
| 16.80 | 0.417 | 0.667 |
| 16.90 | 0.421 | 0.655 |
| 17.00 | 0.423 | 0.643 |
| 17.10 | 0.424 | 0.632 |
| 17.20 | 0.424 | 0.623 |
| 17.30 | 0.423 | 0.614 |
| 17.40 | 0.420 | 0.605 |
| 17.50 | 0.415 | 0.597 |
| 17.60 | 0.411 | 0.588 |
| 17.70 | 0.406 | 0.580 |
| 17.80 | 0.401 | 0.572 |
| 17.90 | 0.396 | 0.563 |
| 18.00 | 0.392 | 0.553 |
| 18.10 | 0.388 | 0.543 |
| 18.20 | 0.384 | 0.532 |
| 18.30 | 0.382 | 0.521 |
| 18.40 | 0.380 | 0.509 |
| 18.50 | 0.379 | 0.496 |
| 18.60 | 0.380 | 0.483 |
| 18.70 | 0.381 | 0.470 |
| 18.80 | 0.382 | 0.456 |
| 18.90 | 0.384 | 0.443 |
| 19.00 | 0.385 | 0.430 |
| 19.10 | 0.387 | 0.418 |
| 19.20 | 0.390 | 0.406 |
| 19.30 | 0.393 | 0.394 |
| 19.40 | 0.396 | 0.382 |
| 19.50 | 0.400 | 0.371 |
| 19.60 | 0.405 | 0.360 |
| 19.70 | 0.410 | 0.349 |
| 19.80 | 0.416 | 0.338 |
| 19.90 | 0.423 | 0.328 |
| 20.00 | 0.431 | 0.318 |
| 20.10 | 0.439 | 0.310 |
| 20.20 | 0.448 | 0.303 |
| 20.30 | 0.457 | 0.295 |
| 20.40 | 0.466 | 0.289 |
| 20.50 | 0.475 | 0.283 |
| 20.60 | 0.483 | 0.277 |
| 20.70 | 0.492 | 0.272 |
| 20.80 | 0.500 | 0.267 |
| 20.90 | 0.509 | 0.262 |
| 21.00 | 0.517 | 0.258 |
| 21.10 | 0.526 | 0.254 |
| 21.20 | 0.535 | 0.252 |
| 21.30 | 0.544 | 0.250 |
| 21.40 | 0.553 | 0.249 |
| 21.50 | 0.561 | 0.247 |
| 21.60 | 0.568 | 0.245 |
| 21.70 | 0.575 | 0.244 |
| 21.80 | 0.582 | 0.242 |
| 21.90 | 0.590 | 0.241 |
| 22.00 | 0.598 | 0.240 |
| 22.10 | 0.606 | 0.239 |
| 22.20 | 0.614 | 0.239 |
| 22.30 | 0.623 | 0.240 |
| 22.40 | 0.632 | 0.240 |
| 22.50 | 0.640 | 0.241 |
| 22.60 | 0.649 | 0.243 |
| 22.70 | 0.657 | 0.244 |
| 22.80 | 0.665 | 0.245 |
| 22.90 | 0.672 | 0.246 |
| 23.00 | 0.680 | 0.246 |
| 23.10 | 0.687 | 0.247 |

## G FeSi$_2$ (CsCl DEFECT STRUCTURE)

TABLE 10 summarises the data for the metallic phase $FeSi_2$.

TABLE 10 Refractive index (n) and extinction coefficient (k) for the metallic phase $FeSi_2$ ordered in a defect CsCl structure. The data have been obtained from measurements on epitaxial layers on Si substrates [17].

| eV | n | k | eV | n | k | eV | n | k |
|---|---|---|---|---|---|---|---|---|
| 0.01 | 29.678 | 28.084 | 1.20 | 4.227 | 2.893 | 3.10 | 3.131 | 2.272 |
| 0.02 | 22.389 | 20.159 | 1.30 | 4.088 | 2.834 | 3.20 | 3.113 | 2.310 |
| 0.03 | 18.845 | 17.202 | 1.40 | 3.967 | 2.795 | 3.30 | 3.090 | 2.354 |
| 0.04 | 15.870 | 15.490 | 1.50 | 3.829 | 2.754 | 3.40 | 3.036 | 2.424 |
| 0.05 | 13.803 | 13.957 | 1.60 | 3.697 | 2.687 | 3.50 | 2.943 | 2.469 |
| 0.06 | 12.191 | 12.554 | 1.70 | 3.596 | 2.619 | 3.60 | 2.877 | 2.491 |
| 0.07 | 11.325 | 11.348 | 1.80 | 3.509 | 2.555 | 3.70 | 2.826 | 2.512 |
| 0.08 | 10.675 | 10.216 | 1.90 | 3.437 | 2.494 | 3.80 | 2.770 | 2.529 |
| 0.09 | 10.197 | 9.658 | 2.00 | 3.373 | 2.429 | 3.90 | 2.693 | 2.551 |
| 0.10 | 8.975 | 8.156 | 2.10 | 3.321 | 2.381 | 4.00 | 2.627 | 2.566 |
| 0.20 | 7.393 | 6.212 | 2.20 | 3.279 | 2.330 | 4.10 | 2.562 | 2.583 |
| 0.30 | 6.299 | 4.843 | 2.30 | 3.251 | 2.289 | 4.20 | 2.489 | 2.594 |
| 0.40 | 5.786 | 4.098 | 2.40 | 3.225 | 2.255 | 4.30 | 2.436 | 2.610 |
| 0.50 | 5.522 | 3.697 | 2.50 | 3.208 | 2.232 | 4.40 | 2.345 | 2.614 |
| 0.60 | 5.244 | 3.409 | 2.60 | 3.194 | 2.216 | 4.50 | 2.251 | 2.614 |
| 0.70 | 5.033 | 3.243 | 2.70 | 3.182 | 2.208 | 4.60 | 2.199 | 2.598 |
| 0.80 | 4.863 | 3.116 | 2.80 | 3.176 | 2.217 | 4.70 | 2.118 | 2.618 |
| 0.90 | 4.721 | 3.072 | 2.90 | 3.158 | 2.223 | 4.80 | 2.062 | 2.570 |
| 1.00 | 4.541 | 3.006 | 3.00 | 3.148 | 2.241 | 4.90 | 1.989 | 2.583 |
| 1.10 | 4.379 | 2.946 | | | | | | |

## H $Ir_3Si_5$ ($IrSi_{1.75}$)

The first study on the optical absorption near the gap was presented in [18]. An investigation performed with spectroscopic ellipsometry has given the data listed in TABLE 11 [19]. Synchrotron radiation has been used to obtain the data in TABLES 12 and 13 [3,6].

TABLE 11 Refractive index (n) and extinction coefficient (k) for $Ir_3Si_5$ obtained from measurements on polycrystalline films [19].

| eV | n | k | eV | n | k | eV | n | k |
|---|---|---|---|---|---|---|---|---|
| 0.60 | 3.939 | 2.675 | 2.30 | 3.554 | 1.938 | 4.00 | 2.382 | 2.150 |
| 0.70 | 3.838 | 2.267 | 2.40 | 3.529 | 2.002 | 4.10 | 2.336 | 2.134 |
| 0.80 | 3.912 | 2.263 | 2.50 | 3.476 | 2.063 | 4.20 | 2.303 | 2.117 |
| 0.90 | 3.902 | 2.172 | 2.60 | 3.424 | 2.125 | 4.30 | 2.259 | 2.099 |
| 1.00 | 3.843 | 2.142 | 2.70 | 3.377 | 2.193 | 4.40 | 2.227 | 2.081 |
| 1.10 | 3.810 | 2.109 | 2.80 | 3.236 | 2.234 | 4.50 | 2.196 | 2.071 |
| 1.20 | 3.764 | 2.092 | 2.90 | 3.151 | 2.282 | 4.60 | 2.169 | 2.059 |
| 1.30 | 3.717 | 2.076 | 3.00 | 3.064 | 2.304 | 4.70 | 2.139 | 2.056 |
| 1.40 | 3.641 | 2.006 | 3.10 | 2.964 | 2.303 | 4.80 | 2.108 | 2.042 |
| 1.50 | 3.564 | 1.892 | 3.20 | 2.855 | 2.286 | 4.90 | 2.085 | 2.041 |
| 1.60 | 3.551 | 1.824 | 3.30 | 2.792 | 2.278 | 5.00 | 2.056 | 2.043 |
| 1.70 | 3.564 | 1.782 | 3.40 | 2.752 | 2.272 | 5.10 | 2.024 | 2.044 |
| 1.80 | 3.558 | 1.775 | 3.50 | 2.664 | 2.269 | 5.20 | 1.996 | 2.048 |
| 1.90 | 3.574 | 1.780 | 3.60 | 2.581 | 2.226 | 5.30 | 1.965 | 2.050 |
| 2.00 | 3.581 | 1.813 | 3.70 | 2.523 | 2.205 | 5.40 | 1.948 | 2.047 |
| 2.10 | 3.580 | 1.846 | 3.80 | 2.469 | 2.184 | 5.50 | 1.922 | 1.974 |
| 2.20 | 3.573 | 1.880 | 3.90 | 2.420 | 2.166 | | | |

TABLE 12 Refractive index (n) and extinction coefficient (k) for $Ir_3Si_5$ obtained from ellipsometric measurements with synchrotron radiation on polycrystalline films [6].

| eV | n | k | eV | n | k | eV | n | k |
|---|---|---|---|---|---|---|---|---|
| 4.10 | 2.402 | 1.919 | 5.50 | 2.035 | 2.038 | 6.90 | 1.495 | 2.037 |
| 4.20 | 2.410 | 1.917 | 5.60 | 1.987 | 2.050 | 7.00 | 1.459 | 2.027 |
| 4.30 | 2.413 | 1.918 | 5.70 | 1.940 | 2.057 | 7.10 | 1.424 | 2.013 |
| 4.40 | 2.410 | 1.921 | 5.80 | 1.896 | 2.060 | 7.20 | 1.392 | 1.995 |
| 4.50 | 2.401 | 1.925 | 5.90 | 1.855 | 2.063 | 7.30 | 1.361 | 1.975 |
| 4.60 | 2.386 | 1.932 | 6.00 | 1.817 | 2.067 | 7.40 | 1.333 | 1.955 |
| 4.70 | 2.365 | 1.939 | 6.10 | 1.780 | 2.068 | 7.50 | 1.307 | 1.933 |
| 4.80 | 2.342 | 1.947 | 6.20 | 1.745 | 2.066 | 7.60 | 1.282 | 1.908 |
| 4.90 | 2.317 | 1.955 | 6.30 | 1.710 | 2.062 | 7.70 | 1.260 | 1.881 |
| 5.00 | 2.285 | 1.964 | 6.40 | 1.676 | 2.059 | 7.80 | 1.239 | 1.853 |
| 5.10 | 2.240 | 1.975 | 6.50 | 1.641 | 2.058 | 7.90 | 1.221 | 1.825 |
| 5.20 | 2.189 | 1.990 | 6.60 | 1.604 | 2.057 | 8.00 | 1.207 | 1.796 |
| 5.30 | 2.136 | 2.006 | 6.70 | 1.568 | 2.053 | 8.10 | 1.197 | 1.768 |
| 5.40 | 2.084 | 2.022 | 6.80 | 1.531 | 2.046 | 8.20 | 1.191 | 1.744 |

TABLE 13 Refractive index (n) and extinction coefficient (k) for $Ir_3Si_5$ obtained from ellipsometric measurements with synchrotron radiation on polycrystalline films [6].

| eV | n | k | eV | n | k | eV | n | k |
|---|---|---|---|---|---|---|---|---|
| 11.10 | 0.534 | 0.661 | 14.30 | 0.725 | 0.956 | 17.50 | 0.620 | 0.729 |
| 11.20 | 0.529 | 0.656 | 14.40 | 0.725 | 0.959 | 17.60 | 0.614 | 0.723 |
| 11.30 | 0.523 | 0.652 | 14.50 | 0.725 | 0.961 | 17.70 | 0.609 | 0.717 |
| 11.40 | 0.519 | 0.649 | 14.60 | 0.724 | 0.961 | 17.80 | 0.604 | 0.711 |
| 11.50 | 0.516 | 0.648 | 14.70 | 0.723 | 0.960 | 17.90 | 0.598 | 0.705 |
| 11.60 | 0.513 | 0.648 | 14.80 | 0.721 | 0.957 | 18.00 | 0.593 | 0.699 |
| 11.70 | 0.511 | 0.651 | 14.90 | 0.718 | 0.953 | 18.10 | 0.588 | 0.692 |
| 11.80 | 0.510 | 0.656 | 15.00 | 0.716 | 0.949 | 18.20 | 0.583 | 0.685 |
| 11.90 | 0.511 | 0.661 | 15.10 | 0.712 | 0.944 | 18.30 | 0.579 | 0.678 |
| 12.00 | 0.514 | 0.668 | 15.20 | 0.708 | 0.940 | 18.40 | 0.575 | 0.670 |
| 12.10 | 0.519 | 0.675 | 15.30 | 0.704 | 0.936 | 18.50 | 0.571 | 0.662 |
| 12.20 | 0.527 | 0.684 | 15.40 | 0.699 | 0.931 | 18.60 | 0.568 | 0.654 |
| 12.30 | 0.536 | 0.693 | 15.50 | 0.695 | 0.925 | 18.70 | 0.564 | 0.646 |
| 12.40 | 0.546 | 0.705 | 15.60 | 0.691 | 0.919 | 18.80 | 0.561 | 0.637 |
| 12.50 | 0.558 | 0.717 | 15.70 | 0.686 | 0.913 | 18.90 | 0.558 | 0.629 |
| 12.60 | 0.571 | 0.732 | 15.80 | 0.680 | 0.907 | 19.00 | 0.556 | 0.620 |
| 12.70 | 0.584 | 0.748 | 15.90 | 0.674 | 0.902 | 19.10 | 0.554 | 0.611 |
| 12.80 | 0.597 | 0.764 | 16.00 | 0.668 | 0.895 | 19.20 | 0.551 | 0.602 |
| 12.90 | 0.612 | 0.779 | 16.10 | 0.664 | 0.885 | 19.30 | 0.550 | 0.593 |
| 13.00 | 0.626 | 0.794 | 16.20 | 0.661 | 0.873 | 19.40 | 0.548 | 0.584 |
| 13.10 | 0.640 | 0.809 | 16.30 | 0.658 | 0.859 | 19.50 | 0.547 | 0.575 |
| 13.20 | 0.653 | 0.822 | 16.40 | 0.656 | 0.845 | 19.60 | 0.547 | 0.567 |
| 13.30 | 0.667 | 0.833 | 16.50 | 0.654 | 0.830 | 19.70 | 0.546 | 0.558 |
| 13.40 | 0.680 | 0.843 | 16.60 | 0.652 | 0.816 | 19.80 | 0.546 | 0.549 |
| 13.50 | 0.692 | 0.853 | 16.70 | 0.649 | 0.803 | 19.90 | 0.547 | 0.541 |
| 13.60 | 0.702 | 0.865 | 16.80 | 0.647 | 0.790 | 20.00 | 0.548 | 0.531 |
| 13.70 | 0.709 | 0.879 | 16.90 | 0.646 | 0.777 | 20.10 | 0.550 | 0.522 |
| 13.80 | 0.713 | 0.896 | 17.00 | 0.644 | 0.765 | 20.20 | 0.553 | 0.512 |
| 13.90 | 0.715 | 0.914 | 17.10 | 0.641 | 0.754 | 20.30 | 0.557 | 0.502 |
| 14.00 | 0.717 | 0.930 | 17.20 | 0.637 | 0.746 | 20.40 | 0.562 | 0.491 |
| 14.10 | 0.720 | 0.942 | 17.30 | 0.632 | 0.739 | 20.50 | 0.566 | 0.481 |
| 14.20 | 0.722 | 0.950 | 17.40 | 0.626 | 0.734 | 20.60 | 0.570 | 0.472 |

Table 13 continued

| eV | n | k | eV | n | k | eV | n | k |
|---|---|---|---|---|---|---|---|---|
| 20.70 | 0.574 | 0.463 | 22.00 | 0.647 | 0.372 | 23.30 | 0.733 | 0.316 |
| 20.80 | 0.579 | 0.454 | 22.10 | 0.653 | 0.367 | 23.40 | 0.738 | 0.313 |
| 20.90 | 0.583 | 0.446 | 22.20 | 0.661 | 0.361 | 23.50 | 0.743 | 0.311 |
| 21.00 | 0.588 | 0.438 | 22.30 | 0.668 | 0.356 | 23.60 | 0.748 | 0.308 |
| 21.10 | 0.594 | 0.430 | 22.40 | 0.676 | 0.351 | 23.70 | 0.753 | 0.306 |
| 21.20 | 0.600 | 0.422 | 22.50 | 0.684 | 0.346 | 23.80 | 0.758 | 0.303 |
| 21.30 | 0.607 | 0.415 | 22.60 | 0.691 | 0.342 | 23.90 | 0.763 | 0.301 |
| 21.40 | 0.613 | 0.408 | 22.70 | 0.698 | 0.338 | 24.00 | 0.768 | 0.299 |
| 21.50 | 0.620 | 0.401 | 22.80 | 0.704 | 0.334 | 24.10 | 0.773 | 0.297 |
| 21.60 | 0.626 | 0.395 | 22.90 | 0.710 | 0.330 | 24.20 | 0.778 | 0.295 |
| 21.70 | 0.631 | 0.389 | 23.00 | 0.716 | 0.326 | 24.30 | 0.782 | 0.293 |
| 21.80 | 0.636 | 0.383 | 23.10 | 0.722 | 0.322 | 24.40 | 0.787 | 0.291 |
| 21.90 | 0.641 | 0.378 | 23.20 | 0.728 | 0.319 | 24.50 | 0.791 | 0.290 |

## REFERENCES


[1] M.C. Bost, J.E. Mahan [ *J. Appl. Phys. (USA)* vol.63 (1991) p.63 ]

[2] H. Lange, M. Giehler, W. Henrion, F. Feske, I. Sieber, G. Oertel [ *Phys. Status Solidi B (Germany)* vol.171 (1992) p.63 ]

[3] W. Henrion, H. Lange, E. Jahne, M. Giehler [ *Appl. Surf. Sci. (Netherlands)* vol.70/71 (1993) p.569-72 ]

[4] A. Borghesi, A. Piaggi, A. Fraschini, G. Guizzetti, F. Nava, G. Santoro [ *Europhys. Lett. (Switzerland)* vol.11 (1989) p.61-5 ]

[5] V. Bellani et al [ *Phys. Rev. B (USA)* vol.46 (1992) p.9380-9 ]

[6] H. Lange, W. Henrion, E. Jahne, M. Giehler, O. Guenther, J. Schumann [ *Mater. Res. Soc. Symp. Proc. (USA)* vol.324 (1994) ]; O. Guenther, E. Janhe, H. Lange [ private communication - to be published ]

[7] G. Guizzetti, F. Marabelli [ private communication ]

[8] V.K. Zaitsev, S.V. Ordin, V.I. Tarasov, M.I. Fedorov [ *Sov. Phys.-Solid State (USA)* vol.21 (1979) p.1454 ]

[9] V.K. Zaitsev, M.I. Fedorov, K.A. Rakhimov, A.E. Engalychev, V.V. Popov [ *Sov. Phys.-Solid. State (USA)* vol.26 (1984) p.494 ]

[10] R.G. Long, M.C. Bost, J.E. Mahan [ *Thin Solid Films (Switzerland)* vol.162 (1988) p.29-40 ]

[11] C. Krontiras et al [ *Thin Solid Films (Switzerland)* vol.161 (1988) p.197-206 ]

[12] T. Siegrist, F. Hullinger, G. Travaglini [ *J. Less-Common Met. (Switzerland)* vol.92 (1983) p.119 ]

[13] V. Bellani, G. Guizzetti, F. Marabelli, S. Lagomarsino, H. von Känel [ *Solid State Commun. (USA)* (1994) submitted ]

[14] M.C. Bost, J.E. Mahan [ *J. Appl. Phys. (USA)* vol.58 (1985) p.2696; and vol.64 (1988) p.2034 ]

[15] C.A. Dimitriadis, J.H. Werner, S. Logothetidis, M. Stutzmann, J. Weber, R. Nesper [ *J. Appl. Phys. (USA)* vol.68 (1990) p.1726 ]

[16] G. Guizzetti, F. Marabelli, M. Patrini, Y. Mo, N. Onda, H. von Känel [ *Mater. Res. Soc. Symp. Proc. (USA)* vol.324 (1994) p.127-32 ]

[17] G. Guizzetti, F. Marabelli, N. Onda, H. von Känel [ *Solid State Commun. (USA)* vol.86 (1993) p.217-9 ]

[18] S. Petersson et al [ *J. Appl. Phys. (USA)* vol.53 (1982) p.3342-3 ]

[19] Th. Zettler, J. Schumann, H. Lange [ private communication - to be published ]

# 6.6 Raman scattering in TM silicides

G. Guizetti and F. Marabelli

December 1994

Although Raman scattering (RS) has made a notable contribution to understanding structural and vibrational properties of bulk, thin film, crystalline and amorphous materials, it has been applied only in a few cases to silicides. In particular RS has been used to investigate metal/silicon interfaces for some time [1-3] and, due to the capability of probing more than a few hundred Angstrom below the surface, to characterise silicide formation.

In particular, RS has been used [2,3] to identify PtSi thin film, at the interface with Si (100), estimate its thickness (ranging from 1 to 200 nm), and describe its crystallographic order (the symmetry is $D_{4h}$). The spectra between 50 and 350 $cm^{-1}$ show two sharp lines at 82 and 140 $cm^{-1}$ with weaker structures at 150 and 260 $cm^{-1}$.

Analogous characterisation has been performed on Ti silicide thin films on Si (100), as-grown and after rapid isothermal annealing, or in-situ in a UHV system [4]. After annealing from 550 to 750°C, a strong line is observed at 305 $cm^{-1}$, corresponding to TiSi formation, while annealing to temperatures higher than 800°C produces two sharp peaks at ~280 $cm^{-1}$, attributed to $TiSi_2$.

The Raman microprobe spectrum of tetragonal $WSi_2$, obtained from an annealed tungsten film on a crystalline silicon substrate [5], exhibits two characteristic peaks, due to the longitudinal optical phonons: the first at 331 $cm^{-1}$ is fairly sharp, whereas the second, at 451 $cm^{-1}$, is broad. The position, intensity and full width at half maximum (FWHM) of these peaks have been used to characterise $WSi_2$ thin films, formed by either rapid thermal annealing or laser annealing, and to estimate the microcrystallite size [6].

Three kinds of semiconducting β-$FeSi_2$ sample have been analysed by RS [7]:

(i) films on Si (111) substrate;
(ii) polycrystalline bulk samples;
(iii) films on FeSi (110) substrate.

In the last sample, 14 Raman lines (generally grouped two by two or three by three) have been clearly observed in the range from 150 to 550 $cm^{-1}$, which systematically shift toward lower frequencies on moving from sample (iii) to sample (i).

These lines have been compared with theoretical predictions on the basis of factor group analysis. β-$FeSi_2$ crystallises in the orthorhombic space group $D_{2h}^{18}$ with 48 atoms per cell. The primitive cell contains 24 atoms, i.e. 8 molecules. In total there are 36 Raman active modes which may be differentiated into 12 internal modes (corresponding to atoms moving differently inside the molecule) and 24 external modes (the molecules move as a unit). The internal modes probably give rise to most of the 14 lines observed experimentally. For polarisation of the exciting field parallel to that of the scattered light, the three strongest lines occur at 179, 197 and 253 $cm^{-1}$, respectively; for crossed polarisation, three very strong lines ($A_g$ modes) occur at 197, 252 and 346 $cm^{-1}$.

RS spectra on semiconducting $CrSi_2$ polycrystalline films on Si [8,9], with thicknesses from 100 to 400 nm, display a strong band at 300-310 $cm^{-1}$, with a low-frequency shoulder, two lower bands at about 400 and 410 $cm^{-1}$, and a structured peak of medium intensity at about 350 $cm^{-1}$. On the basis of the symmetry analysis of $CrSi_2$ (space group $P_2622$ and point group isomorphic to $D_2$), and by comparison with FIR absorption spectra, the peaks at ~300 and ~350 $cm^{-1}$ have been associated with phonons of $E_1$ symmetry [8].

A variety of RS studies have been reported on $V_3Si$ with the A15 structure [10-13], which has a superconducting-transition temperature of 17 K. Single crystal samples have been grown by Czochralski [10,12] and floating zone [11,13] techniques. The samples with exact stoichiometry (25 at. % Si) exhibit a cubic to tetragonal transformation at ~22 K, while no transformation has been observed in samples with ~24 at. % Si.

In the A15 compounds there are eight optical phonons at Γ, but only two are Raman active: the $E_g$ and the $F_{2g}$ modes. The $E_g$ line is very intense and for transforming $V_3Si$ can be well fitted by an asymmetric Lorentzian shape [10,12]. The phonon frequency is ~281.4 $cm^{-1}$ at 340 K, decreases upon cooling and reaches a minimum of ~275.4 $cm^{-1}$ at ~120 K, and then increases upon further cooling [12]. The line width has a more striking temperature dependence with an FWHM of 47.5 $cm^{-1}$ at T = 340 K which doubles as T decreases to 50 K. The $F_{2g}$ mode has a scattering intensity much lower than that of $E_g$ and has been shown only in [13]. Its peak is located at ~175 $cm^{-1}$ for T = 310 K, and shifts to a lower frequency of ~169 $cm^{-1}$ when the temperature is reduced to ~8 K.

Disorder induced Raman scattering has been observed in epitaxial films of $NiSi_2$ on Si (111) [14]. Raman spectra show the major features of the whole phonon density of states and can be compared with inelastic neutron scattering results. In the same paper [14] the only available data for the Raman line of $CoSi_2$ at 267 $cm^{-1}$ are cited.

## REFERENCES


[1] R.J. Nemanich, C.C. Tsai, M.J. Thomson, T.W. Sigmon [ *J. Vac. Sci. Technol. (USA)* vol.19 (1981) p.685 ]

[2] R.J. Nemanich, T.W. Sigmon, N.M. Johnson, M.D. Moyer, S.S. Lau [ in *Laser and Electron-Beam Solid Interactions and Materials Processing* Eds J.F. Gibbons, L.D. Hess, T.W. Sigmon (North-Holland, New York, 1981) p.541 ]

[3] J.C. Tsang, Y. Yokota, R. Matz, G.W. Rubloff [ *Appl. Phys. Lett. (USA)* vol.44 (1984) p.430-2 ]

[4] R.J. Nemanich, R.T. Fulk, B.L. Stafford, H.A. Vander Plas [ *Appl. Phys. Lett. (USA)* vol.46 (1985) p.670-2 ]

[5] P.J. Codella, F. Adar, Y.S. Liu [ *Appl. Phys. Lett. (USA)* vol.46 (1985) p.1076-8 ]

[6] R. Vuppuladhadium, H.E. Jackson, J.T. Boyd [ *J. Appl. Phys. (USA)* vol.73 (1993) p.7887-93 ]

[7] K. Lefki et all [ *Solid State Commun. (USA)* vol.80 (1991) p.791-5 ]

[8] A. Borghesi, A. Piaggi, A. Franchini, G. Guizzetti, F. Nava, G. Santoro [ *Europhys. Lett. (Switzerland)* vol.11 (1990) p.61-5 ]

[9] H. Lange, M. Giehler, W. Henrion, F. Fenske, I. Sieber, G. Oertel [ *Phys. Status Solidi B (Germany)* vol.171 (1992) p.63-76 ]

[10] H. Wipf, M.W. Klein, B.S. Chandrasekhar, T.H. Geballe, J.H. Wernick [ *Phys. Rev. Lett. (USA)* vol.41 (1978) p.1752-4 ]

[11] S. Schicktanz, R. Kaiser, E. Schneider, W. Glaser [ *Phys. Rev. B (USA)* vol.22 (1980) p.2386-91 ]

[12] S.B. Dierker, R. Merlin, M.V. Klein, G.W. Webb, Z. Fisk [ *Phys. Rev. B (USA)* vol.27 (1983) p.3577-91 ]

[13] R. Hackl, R. Kaiser, A. Muller, W. Weber [ *J. Phys. F, Met. Phys. (UK)* vol.16 (1986) p.1299-1306 ]

[14] F. Li, N. Lustig, P. Klosowski, J.S. Lannin [ *Phys. Rev. B (USA)* vol.41 (1990) p.10210 ]

# 6.7 Plasma frequency and scattering time of free carriers in metallic TM silicides

G. Guizetti and F. Marabelli

December 1994

The electronic transitions are more directly related to the complex dielectric function $\tilde{\varepsilon}(\omega) = \varepsilon_1(\omega) + i\varepsilon_2(\omega)$ instead of the complex refractive index $\tilde{n} = n + ik = (\tilde{\varepsilon})^{1/2}$ (see Datareview 6.1). The free-electron region of metal-like silicides can be classically described by the Drude model [1], yielding the following dielectric functions:

$$\varepsilon_1(\omega) = \varepsilon_\infty - \omega_p^2/(\omega^2 + \Gamma^2) \qquad (1)$$

$$\varepsilon_2(\omega) = \omega_p^2\Gamma/\omega(\omega^2 + \Gamma^2) \qquad (2)$$

Here $\omega_p$ is the plasma frequency $\omega_p = (4\pi Ne^2/m^*)^{1/2}$ with N the free-charge density and m* the effective optical mass. The parameter $\Gamma$ describes the damping of the electron motion due to impurities, phonons, etc. It is related to the scattering time, $\tau$, by $\Gamma = 1/\tau$. $\varepsilon_\infty$ is the high-frequency interband contribution to $\varepsilon_1$.

At low frequencies ($\omega << \Gamma$) $\varepsilon_1$ is constant, large and negative, and $\varepsilon_2$ is very large.

The optical conductivity $\sigma(\omega)$, defined as $\sigma(\omega) = \omega\varepsilon_2/4\pi$, coincides, for $\omega = 0$, with the DC conductivity $\sigma_{DC} = Ne^2\tau/m^*$, as obtained from the classical transport equation [2,3].

The applicability of the Drude model can be tested by plotting $\varepsilon_1$ versus $\omega\varepsilon_2$, which should give a straight line with slope $1/\Gamma = \tau$.

Using $\varepsilon_1 = \varepsilon_\infty - \omega_p^2/\omega^2$ in the region where $\omega_p >> \omega >> \Gamma$ (for metals, at near-infrared frequencies), $\omega_p^2$ and the optical mass can be determined from the slope of the plot of $1/|\varepsilon_1|$ versus $\omega^2$. Using the expression $\varepsilon_2 = \omega_p^2\Gamma/\omega^3$ and plotting $1/\omega\varepsilon_2$ versus $\omega^2$, $\Gamma$ may also be determined from the slope of the curve.

Deviations from the Drude theory arise from low-energy interband transitions, from scattering mechanisms which are inadequately described in terms of an energy-independent scattering time, and when photon energies become large compared with free carrier energies [2,3]. A discussion of the free carrier optical response of silicides and a comparison between optical and transport results was given in [4].

TABLE 1 Plasma frequencies and scattering times for several metallic silicides.

| Compound | $\omega_p$ (eV) | $\tau$ ($10^{-14}$ s) | Ref |
|---|---|---|---|
| $TiSi_2$ | 4.2 | 4.5 | [5] |
| $HfSi_2$ | 1.5 - 2 | 2.4 | [6] |
| $VSi_2$ | (⊥) 1.6 | | [7] |
| | (‖) 2.94 | | [7] |
| | 2.75 | 0.94 | [8] |
| | 2.7 | | [9] |
| $NbSi_2$ | (⊥) 2.0 | | [7] |
| | (‖) 2.3 | | [7] |
| | 2.3 | 1.3 | [8] |
| | 2.3 | 1.2 | [10] |
| $TaSi_2$ | 3.45 | 0.7 | [11] |
| | 2.7 | 1.1 | [12] |
| | 2.6 | 1.1 | [8] |
| $WSi_2$ | (⊥) 1.7 | 3.29 | [13] |
| | (‖) 1.8 | 3.66 | [13] |
| $CoSi_2$ | 5.4 - 7.6 | 0.16 - 0.66 | [14] |
| | 5.8 | 0.26 | [15] |
| | 7 - 8 | 0.51 - 0.66 | [16] |
| $Ni_2Si$ | 2.4 | 1.4 | [17] |
| $Pd_2Si$ | (⊥) 2.8 - 3.36 | 2.19 | [18] |
| | (‖) 3.6 - 3.69 | 2.19 | [18] |

## REFERENCES


[1] F. Wooten [ *Optical Properties of Solids* (Academic Press, New York, 1972) ]

[2] W.P. Dumkee [ *Phys. Rev. (USA)* vol.124 (1961) p.1813-7 ]

[3] B. Jensen [ in *Handbook of Optical Constants of Solids* Ed. E.D. Palik (Academic Press, Orlando, 1985 ) ch.9 p.169-88 ]

[4] F. Nava et al [ *Mater. Sci. Rep. (Netherlands)* vol.9 (1993) p.141-200 ]

[5] A. Borghesi, A. Piaggi, G. Guizzetti, F. Lévy, M. Tanaka, H. Fukutani [ *Phys. Rev. B (USA)* vol.40 (1989) p.1611 ]

[6] A. Borghesi, F. Marabelli, G. Guizzetti, M. Michelini, F. Nava [ *J. Appl. Phys. (USA)* vol.69 (1991) p.7645-50 ]

[7] G. Guizzetti, F. Marabelli, A. Borghesi, F. Nava, U. Gottlieb, O. Laborde [ *Appl. Surf. Sci. (Netherlands)* vol.73 (1993) p.237-42 ]

[8] A. Borghesi, A. Piaggi, G. Guizzetti, F. Nava, M. Bacchetta [ *Phys. Rev. B (USA)* vol.40 (1989) p.3249 ]

[9] M. Amiotti, A. Borghesi, F. Marabelli, A. Piaggi, G. Guizzetti, F. Nava [ *Appl. Surf. Sci. (Netherlands)* vol.53 (1991) p.230 ]

[10] M. Amiotti, A. Borghesi, F. Marabelli, G. Guizzetti, F. Nava [ *Phys. Rev. B (USA)* vol.44 (1991) p.3757 ]

[11] A. Roseler, F. Marabelli, G. Guizzetti, A. Borghesi, A. Piaggi [ *Nuovo Cimento D (Italy)* vol.13 (1991) p.169 ]

[12] A. Borghesi, L. Nosenzo, A. Piaggi, G. Guizzetti, C. Nobili, G. Ottaviani [ *Phys. Rev. B (USA)* vol.38 (1988) p.10937 ]

[13] V.N. Antonov et al [ *Phys. Rev. B (USA)* vol.44 (1991) p.8437-45 ]

[14] M. Wölfel, M. Schulz, J. Ionally, P.J. Grunthaner [ *Appl. Phys. A (Germany)* vol.50 (1990) p.177-81 ]

[15] J.Y. Duboz, P.A. Badoz, J. Henz, H. von Känel [ *J. Appl. Phys. (USA)* vol.68 (1990) p.2346-50 ]

[16] J.R. Jimenez et al [ *J. Appl. Phys. (USA)* vol.66 (1989) p.2738-41 ]

[17] M. Amiotti, A. Borghesi, G. Guizzetti, F. Nava [ *Phys. Rev. B (USA)* vol.42 (1990) p.8939-45 ]

[18] M. Amiotti et al [ *Phys. Rev. B (USA)* vol.45 (1992) p.13285-92 ]

# 6.8 Energy gap and static dielectric constant in semiconducting TM silicides

G. Guizetti and F. Marabelli

December 1994

The absorption coefficient $\alpha$ is calculated from $\alpha = 4\pi k/\lambda$, where k is the extinction coefficient and $\lambda = 2\pi c/\omega$ is the wavelength. Above the absorption edge, $\alpha$ increases in a manner revealing the fundamental interband transition. An established method of determining whether a transition is indirect is to plot the square root of absorption coefficient versus energy.

For an indirect transition the data should obey the following relationships:

$$\alpha = 0 \qquad h\nu < E_g - E_p \qquad (1a)$$
$$\alpha = A\,(h\nu - E_g + E_p)^2 \qquad E_g + E_p > h\nu > E_g - E_p \qquad (1b)$$
$$\alpha = A\,(h\nu - E_g + E_p)^2 + B\,(h\nu - E_g - E_p)^2 \qquad h\nu > E_g + E_p \qquad (1c)$$

where $E_g$ is the forbidden energy gap, $E_p$ is the value of the phonon associated with the indirect transition, h is Planck's constant, A and B are constants determined by the effective mass of the band, and $\nu$ is the frequency of absorbed light.

For a direct transition the relationships

$$\alpha = 0 \qquad h\nu < E_d \qquad (2a)$$
$$\alpha = C\,(h\nu - E_d)^{1/2} \qquad h\nu > E_d \qquad (2b)$$

should be satisifed, where $E_d$ is the energy of the direct transition and C is a constant.

The value of the optical gap ($E_d$ or $E_g$) should correspond to twice the activation energy obtained by transport measurements. Such a relation is usually fulfilled in semiconducting silicides.

Finally, semiconductors are often characterised by an undispersed value of the index of refraction, $n_0$, or by $\varepsilon_0 = (n_0)^2$, which is known as the high frequency dielectric constant. This value is most easily determined by plotting $1/(n^2 - 1)$ versus the reciprocal of the square of the wavelength The intercept at $1/\lambda^2 = 0$ gives the value of $n_0$. The relationship modelled by this plot is meaningful only if there is no significant absorption at the wavelengths being considered [1].

TABLE 1 Values of the optical gap ($E_g$) and of the dielectric constant ($\varepsilon_0$) for several semiconducting silicides.

| Compound | $E_g$ (eV) | Character | $\varepsilon_0$ | Ref |
|---|---|---|---|---|
| $CrSi_2$ | 0.35<br>0.67 | Indirect<br>Direct | 28.1 | [1] |
| | 0.5<br>0.9 | Indirect<br>Direct | - | [2,3] |
| | - | - | 30 | [4] |
| $MnSi_{1.75}$ | 0.66 | Direct | 34 | [5] |
| $ReSi_2$ | 0.12<br>0.36 | Indirect<br>Direct | - | [6] |
| | 0.195 | - | 15 | [7] |
| $FeSi_2$ | 0.89 | Direct | 31.4 | [8] |
| | 0.85 | Direct | | [9] |
| | 0.78<br>0.83 | Indirect<br>Direct | - | [10] |
| $IrSi_{1.75}$ | 1.2 | Direct | - | [11] |

## REFERENCES


[1] M.C. Bost, J.E. Mahan [ *J.Appl.Phys. (USA)* vol.63 (1991) p.63 ]

[2] H. Lange, M. Giehler, W. Henrion, F. Feske, I. Sieber, G. Oertel [ *Phys. Status Solidi B (Germany)* vol.171 (1992) p.63 ]

[3] W. Henrion, H. Lange, E. Jahne, M. Giehler [ *Appl. Surf. Sci. (Netherlands)* vol.70/71 (1993) p.569-72 ]

[4] V. Bellani et al [ *Phys. Rev. B (USA)* vol.46 (1992) p.9380-9 ]

[5] V.K. Zaitsev, S.V. Ordin, V.I. Tarasov, M.I. Fedorov [ *Sov. Phys.-Solid State (USA)* vol.21 (1979) p.1454 ]

[6] R.G. Long, M.C. Bost, J.E. Mahan [ *Thin Solid Films (Switzerland)* vol.162 (1988) p.29-40 ]

[7] T. Siegrist, F. Hullinger, G. Travaglini [ *J. Less-Common Met. (Switzerland)* vol.92 (1983) p.119 ]

[8] M.C. Bost, J.E. Mahan [ *J. Appl. Phys. (USA)* vol.58 (1985) p.2696; and vol.64 (1988) p.2034 ]

[9] C.A. Dimitriadis, J.H. Werner, S. Logothetidis, M. Stutzmann, J. Weber, R. Nesper [ *J. Appl. Phys. (USA)* vol.68 (1990) p.1726 ]

[10] K. Radermacher, R. Carius, S. Mantl [ *Nucl. Instrum. Methods Phys. Res. B (Netherlands)* vol.84 (1944) p.163-7 ]

[11] S. Petersson et al [ *J. Appl. Phys. (USA)* vol.53 (1982) p.3342-3 ]

## 6.9 Optical properties of RE silicides

Xu Zhenjia (C.C. Hsu)

June 1994

Recently, Sagnes et al [1] reported the first study of the complex index of $ErSi_2$ at energies below the silicon bandgap (0.5 - 1 eV) at room temperature. The real part is almost constant with a mean value of 4. The imaginary part decreases with increasing energy. At 1 eV, the corresponding absorption length is about 300 Å. The near-infrared optical properties of [1] are reproduced in FIGURES 1 to 4.

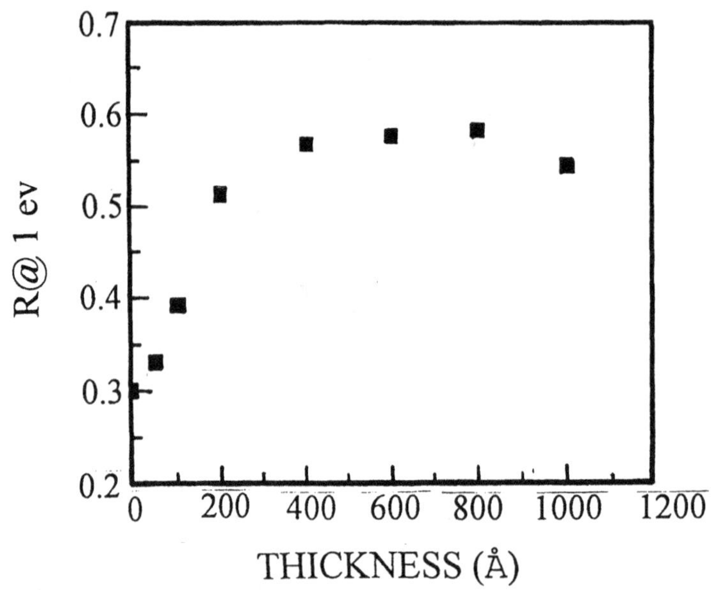


FIGURE 1 Reflection coefficient (R) for a film of $ErSi_2$ at 300 K vs. film thickness at 1 eV (after [1]).

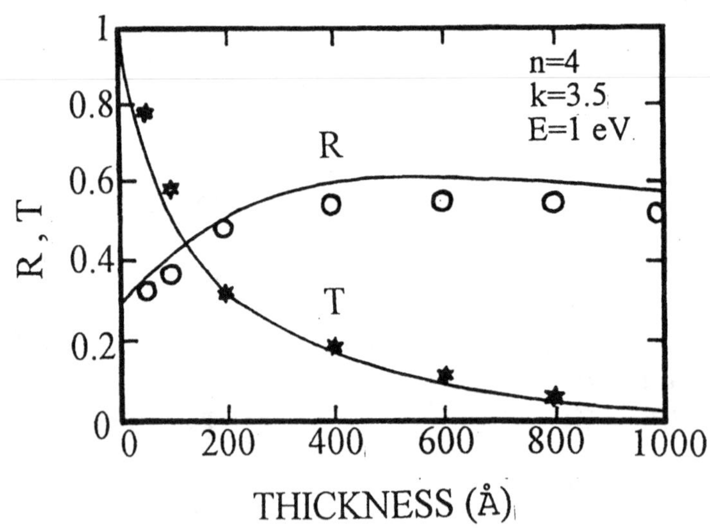


FIGURE 2 Reflection (R) and transmission (T) vs. film thickness of $ErSi_2$ at 300 K. The solid lines are fitted curves based on Abele's formalism (after [1]).

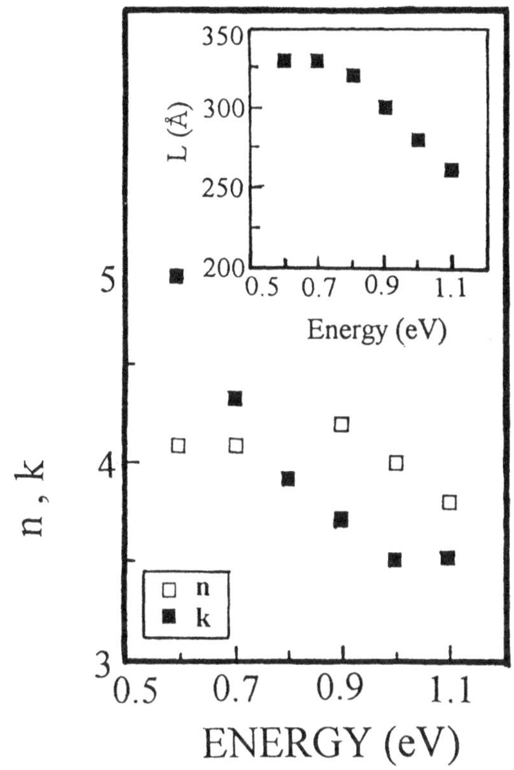


FIGURE 3 Measured real (n) and imaginary (k) parts of the complex index vs. photon energy for $ErSi_2$ at 300 K. The insert shows the absorption length associated with k (after [1]).

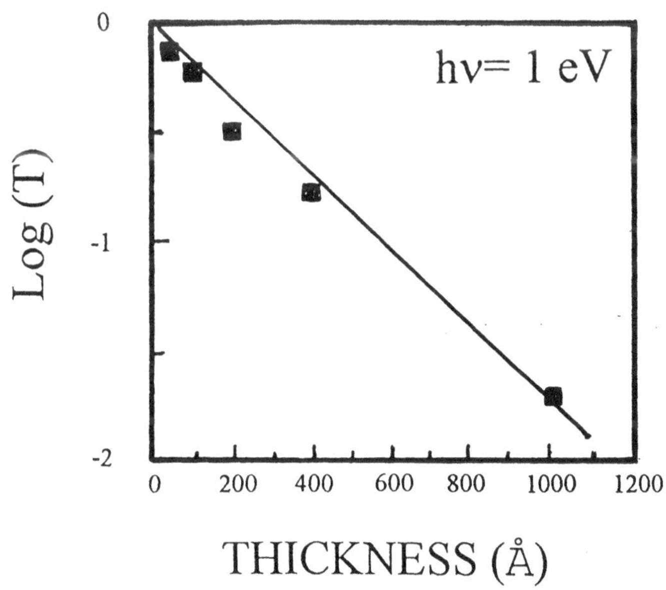


FIGURE 4 Logarithm of transmission [log(T)] for an $ErSi_2$ film vs. thickness at 300 K with a photon energy of 1 eV. The plotted transmission (squares) is defined as the ratio between the flux crossing the (film + silicon) combination, and the flux crossing the silicon alone. Following Lambert's law, the fit by a straight line gives an absorption length of 260 Å (after [1]).

## REFERENCE

[1] I. Sagnes, G. Vincent, P.A. Badoz [ *J. Appl. Phys. (USA)* vol.72 (1992) p.4295-9 ]

CHAPTER 7

# DIFFUSION

# 7.1 Kinetics of formation of TM silicide thin films: self-diffusion

P. Gas and F.M. d'Heurle

January 1995

## A INTRODUCTION

The simplest method to form a silicide thin film is to deposit a film of metal on a silicon substrate and to induce the formation by annealing. It is thus of prime importance to understand and control the kinetics of reaction between metal and silicon. Hence, numerous studies (starting in the early 70s) have been devoted to this subject. The important reservoir of data published allows a systematic analysis of this phenomenon (reactive-diffusion) and, since diffusion is in most cases the controlling step, provides information about self-diffusion in silicides.

In this Datareview we shall describe the different kinetics of formation observed (Section B), focus on diffusion controlled kinetics (Section C), and discuss the content of Sections B and C in terms of diffusion mechanisms (Section D).

## B KINETICS OF FORMATION

The basic experiment is the analysis by RBS (and XRD) of the evolution of a thin metal film (some 100 nm) deposited on a Si single crystal ((100) or (111)) after different annealing schedules. The evolution at a given temperature as a function of time indicates the nature of the phase formed (silicide) and its formation kinetics (linear, parabolic, etc.) while evolution as a function of temperature provides the activation energy of the process.

The Ni/Si system is a perfect example of the behaviour observed:

- At about 250°C the Ni film transforms into $Ni_2Si$. The growth is parallel to the initial Ni/Si interface and the square of the thickness of the silicide formed varies linearly with time. This process continues until total consumption of the Ni film.

- An increase in temperature (300°C) or in time leads to the formation of NiSi by reaction between $Ni_2Si$ and the Si substrate. Again, the square of the thickness varies linearly with time and the reaction proceeds until total consumption of the $Ni_2Si$ film.

- At much higher temperature (about 800°C) NiSi transforms into $NiSi_2$. At this temperature the reaction is very rapid and its kinetics cannot be easily monitored.

The main characteristics of the metal/silicon thin film reactions are included in this example:

- Layer by layer growth.

- Sequential appearance of phases, instead of the apparently simultaneous formation generally observed in bulk diffusion couples.

- Absence of certain phases (stable referring to the phase diagram; e.g. $Ni_5Si_2$, $Ni_3Si_2$).

- Kinetics of reactions which can be classified in two groups [1]:

  (i) Diffusion-controlled kinetics. The formation of the new phase is limited by the diffusion of the most mobile species (metal or silicon) in the growing phase and a progressive and parabolic growth is observed. They are the most frequent and the most reproducible formation kinetics.

  (ii) Nucleation-controlled kinetics. The silicide forms only at relatively high temperatures. Nucleation-controlled reactions are observed when the gain in free energy produced by the formation of the new phase is small and cannot compensate the additional interfacial energy associated with phase formation [2]. Nucleation difficulties are encountered for intermediate or silicon-rich phases. Silicides formed by direct metal/silicon reaction (first phase formed) are excluded since the associated free energy change is high.

The silicides for which kinetics have been reported to be diffusion- or nucleation-controlled are respectively listed in TABLES 1 and 2 together with the temperature at which they appear.

TABLE 1 Silicides for which formation is 'diffusion-controlled' and temperature (°C) at which this formation is observed. From [17].

| $TiSi_2$-425* | $VSi_2$-600 | $CrSi_2$-520 | MnSi-425 | FeSi-450 | $Co_2Si$-350<br>CoSi-350<br>$CoSi_2$-400* | $Ni_2Si$-200<br>NiSi-275<br>$NiSi_2$-350* | $Cu_3Si$-150 |
|---|---|---|---|---|---|---|---|
| $Zr_5Si_4$-670 | $NbSi_2$-645 | $MoSi_2$-850 | | $Ru_2Si_3$-350 | RhSi-350 | $Pd_2Si$-175 | |
| HfSi-525 | | $WSi_2$-800 | | $Os_2Si_3$-450 | IrSi-400 | $Pt_2Si$-250<br>PtSi-300 | |

* Reaction with amorphous (evaporated) silicon.

TABLE 2 Silicides for which formation is 'nucleation-controlled' and temperature (°C) at which this formation is observed. From [17].

| $ScSi_{1.7}$-900 | $TiSi_2$ | $Mn_{11}Si_{19}$-555 | $FeSi_2$ | $CoSi_2$-500 | $NiSi_2$-800 |
|---|---|---|---|---|---|
| $YSi_{1.7}$ | $ZrSi_2$-625 | | | $Rh_4Si_5$-825<br>$Rh_3Si_4$-925 | PdSi-735 |
| | $HfSi_2$-685 | | $OsSi_2$-740 | $IrSi_3$-940 | |

+ some rare earth silicides ~300 - 500°C [2]

Linear kinetics have also been reported. In such cases, the reaction is theoretically limited by the interfacial reaction, namely the capacity of the interface to form the new phase [3]; a good example of this limitation is the initial stage of $SiO_2$ formation. One should mention that linear 'kinetics' have been observed for 'ill-behaved' reactions such as Cr/Si or Ti/Si. They are not easily reproducible and may be the consequence of experimental factors such as a diffusion barrier ($SiO_2$) at the metal/silicon interface [4]. Nucleation problems should also be suspected; a complex kinetic behaviour may be the consequence of a combination of diffusion and nucleation-controlled kinetics [5].

## C DIFFUSION-CONTROLLED KINETICS

Diffusion-controlled kinetics should obviously provide information about self-diffusion in the growing layer. The two quantities experimentally determined are:

- The growth rate, equal to the slope of the line plotting the square of the thickness vs. time.
- The activation energy (variation with temperature of the rates of formation).

These data may be used to predict the conditions in which a given silicide will appear and grow. Their meaning in terms of diffusion mechanisms is somewhat more complex. A discussion will be proposed in Section D. It is important to bear in mind that:

- Generally, diffusion of both species contributes to the growth. However, for most of the silicides, marker experiments [6,7] have shown that only one species is mobile during growth so that rates of formation and activation energies are characteristic of the dominant diffusing species (DDS).
- The driving force for diffusion in a reactive system is the gradient of chemical potential. The appropriate diffusion equation to analyse such a process is the Nernst-Einstein equation [8] which, contrary to Fick's law, explicitly includes this thermodynamic term. Use of this equation shows that the growth rate is in fact the product of an effective diffusion coefficient (taking into account all the diffusion paths) and a chemical term ($\Delta G/kT$) proportional to the free energy of formation of the silicide formed.
- Growth rates depend on all the possible diffusion processes. This includes grain boundaries but also extra point defects eventually created at the reactive interface [9].

In the following tables the principal experimental quantities deduced from the diffusion-controlled growth of silicide thin films are given. The temperature of formation ($T_f$) is listed instead of the rates of formation for the following reason:

- The silicide thicknesses are always included between ~10 and ~100 nm, the lower limit being given by the RBS limit of resolution and the upper limit fixed by the difficulty in growing thicker films without de-cohesion. As most of the studies have been carried out with conventional furnaces, diffusion times are between one hour and one day. The measured rates of formation are thus always equal to $10^{-14}$ to $10^{-13}$

$cm^2/s$. Experimentally, silicides differ by the temperature at which these rates are obtained.

For the same reason (limitation in analysed thicknesses), the activation energies are measured over a limited temperature range (~100°C) with two practical consequences: (1) the uncertainty in these values is relatively high, (2) extrapolations out of the analysed temperature range are doubtful. To overcome this limitation other techniques have been used: rapid thermal annealing [10], lateral silicide formation [11], and in-situ characterisation of silicide formation [12,13]. The results generally confirm the 'conventional' measurements.

## C1 Metal Rich Silicides

Available data on the diffusion-controlled growth of metal rich silicides are given in TABLE 3. It includes: silicide formed (by direct metal/silicon reaction); melting temperature; range of temperature of formation (with a growth rate ~$10^{-14}$ - $10^{-13}$ $cm^2/s$), activation energy, and dominant diffusing species (DDS) during growth. Notes are added to describe the specificity of some experiments (thicker layers, rapid thermal annealing, reaction with amorphous silicon, etc.). Since the sequence of silicide formation usually proceeds from metal-rich silicides to silicon-rich silicides, these data refer to the first phase formed. They are certainly the most reproducible.

TABLE 3 Diffusion-controlled growth of silicide thin films: metal-rich silicides. Successively: silicide formed, melting temperature, range of temperature of formation (with a growth rate of $10^{-14}$ to $10^{-13}$ $cm^2/s$), activation energy, dominant diffusing species (DDS).

| Silicide | $T_m$ (K) | $T_f$ (°C) | Q (eV) | Ref | DDS | Ref | Note |
|---|---|---|---|---|---|---|---|
| $V_3Si$ | 2208 | 730-820 | 2 | [31] | V | [32] | |
| $Mo_3Si$ | 2560 | 1500-1700 | 3.7 | [33] | | | (a) |
| $Cu_3Si$ | 2560 | 150-300 | | [34] | Cu | | (a) |
| $Co_2Si$ | 1605 | 350-500 | 1.5 | [35] | Co | [36] | |
| | 385-490 | | 1.7-1.85 | [37] | | | (b) |
| $Ni_2Si$ | 1580 | 200-325 | 1.5-1.6 | [38,39] | Ni | [40-43] | (c) |
| | 225-325 | | 1.3 | [38] | | | (b) |
| | 420-700 | | 1.4 | [11,44] | | | (d) |
| | 210-335 | | 1.5 | [45] | | | (e) |
| | 300-430 | | 1.3 | [46] | | | |
| $Ni_2Ge$ | | 250 | 1.6 | [47] | Ni | | |
| $Pd_2Si$ | 1670 | 175-450 | 1.4 | [48] | Pd-Si | [49-58] | |
| | | | 0.9-1.4 | [59] | | | (c) |
| | | 310-460 | 1.4 | [60] | | | (f) |
| | | 125-400 | 1.27 | [61] | | | |
| | | 200-275 | 1.5 | [62] | | | |
| | | 160-220 | 1.1 | [63] | | | |
| $Pd_2Ge$ | 230 | 1.1 | [47] | | | | |
| | 175-240 | 1.5 | [64] | | | | |

(a) on metal bulk samples, rates are in the $10^{-10}$ $cm^2/s$ region; (b) on evaporated or poly Si; (c) rate depends on substrate orientation; (d) lateral formation, rates are in the $10^{-10}$ $cm^2/s$ region; (e) on <100> and evaporated Si; (f) rapid thermal annealing; (g) with oxygen impurity; (h) on $SiO_2$.

Table 3 continued

| Silicide | $T_m$ (K) | $T_f$ (°C) | Q (eV) | Ref | DDS | Ref | Note |
|---|---|---|---|---|---|---|---|
| $Pt_2Si$ | 1372 | 250-325 | 1.5 | [65] | Pt | [66] | |
| | | 280-350 | 1.6 | [67] | | | |
| | | 200-500 | 1.04 | [68] | | | |
| | | | 1.3 | [69] | | | (b) |
| | | 400-700 | 1.3 | [70] | | | (d) |
| | | 200-300 | 1.3 | [71] | | | |
| | | 200-300 | 1.4 | [45] | | | (e) |
| $Pt_2Si$/Ge | | 200-400 | 1.2 | [72] | | | |
| $Pt_2Ge$ | 1073 | 250 | 1.3 | [47] | Pt | | |
| $Ti_5Si_3$ | 2400 | 650 | | [73] | | | (g) |
| $Nb_5Si_3$ | 2678 | 1140-1465 | 2.5 | [74] | | | (a) |
| $Ta_5Si_3$ | 2773 | 1040-1465 | 2.5 | [74] | | | (a) |
| $Mo_5Si_3$ | 2400 | 1192-1705 | 3.7 | [33] | | | (a) |
| $W_5Si_3$ | 2600 | 1350-1870 | 3.7 | [33] | | | (a) |
| $Zr_5Si_4$ | 2583 | 670-730 | 2.8 | [75] | | | (h) |

(a) on metal bulk samples, rates are in the $10^{-10}$ $cm^2/s$ region; (b) on evaporated or poly Si; (c) rate depends on substrate orientation; (d) lateral formation, rates are in the $10^{-10}$ $cm^2/s$ region; (e) on <100> and evaporated Si; (f) rapid thermal annealing; (g) with oxygen impurity; (h) on $SiO_2$.

## C2 Monosilicides

Available data on the diffusion-controlled growth of monosilicides are given in TABLE 4.

TABLE 4 Diffusion-controlled growth of silicide thin films: monosilicides. Successively: silicide formed, melting temperature, range of temperature of formation (with a growth rate of $10^{-14}$ to $10^{-13}$ $cm^2/s$), activation energy, dominant diffusing species (DDS).

| Silicide | $T_m$ (K) | $T_f$ (°C) | Q (eV) | Ref | DDS | Ref | Note |
|---|---|---|---|---|---|---|---|
| MnSi | 1548 | 425-500 | 1.9 | [76] | | | |
| FeSi | 1683 | 450-525 | 1.7 | [77] | Si | | |
| CoSi | 1688 | 400-550 | 1.75 | [78] | Si | [36] | |
| | | 375-490 | 1.9 | [35] | | | |
| | | 385-490 | 1.8-1.9 | [37] | | | (b) |
| NiSi | 1273 | 275-400 | 1.7 | [79] | Ni | [80] | |
| | | 290-370 | 1.2-1.8 | [81] | | | (c) |
| | | 300-430 | 1.4 | [82] | | | |
| | | 250-400 | 1.55 | [45] | | | (e) |
| RhSi | 1723 | 300-425 | 1.9 | [46,83] | Si | [84,85] | |
| | | 375-450 | 1.7, 1.9 | [7] | | | |
| IrSi | | 400-550 | 1.9 | [86] | Si | [87] | |

(b) on evaporated or poly Si; (c) rate depends on substrate orientation; (e) on <100> and evaporated Si.

Table 4 continued

| Silicide | $T_m$ (K) | $T_f$ (°C) | Q (eV) | Ref | DDS | Ref | Note |
|---|---|---|---|---|---|---|---|
| PtSi | 1502 | 300-360<br>350-410<br>300-540<br>300-450<br>200-300<br>250-370 | 1.5<br>1.6<br>1.5<br>1.5<br>1.5<br>1.5 | [65]<br>[67]<br>[68]<br>[69]<br>[71]<br>[45] | Pt-Si? | [88,89] | (e) |
| (Ru,Rh)Si | 2073 | 400-475 | 2.4 | [90] | | | |
| HfSi | 2373 | 525-650 | 2.5 | [91] | Si | [92] | |

(b) on evaporated or poly Si; (c) rate depends on substrate orientation;
(e) on <100> and evaporated Si.

## C3 Disilicides

The formation kinetics of these silicon-rich silicides (listed in TABLE 5) are certainly of poorer quality that those of metal-rich or monosilicides. The main reason is that they are usually the last phase to be observed and are thus more subject to nucleation problems (e.g. $NiSi_2$, $CoSi_2$). Special techniques such as reaction with amorphous (evaporated) silicon have been used in order to obtain good parabolic kinetics [14]. These nucleation difficulties may also affect the kinetics of formation of disilicides which are 'apparently' the first phase to appear (e.g. $TiSi_2$), as mentioned in Section B.

TABLE 5 Diffusion-controlled growth of silicide thin films: silicon-rich silicides. Successively: silicide formed, melting temperature, range of temperature of formation (with a growth rate of $10^{-14}$ to $10^{-13}$ $cm^2/s$), activation energy, dominant diffusing species (DDS).

| Silicide | $T_m$ (K) | $T_f$ (°C) | Q (eV) | Ref | DDS | Ref | Note |
|---|---|---|---|---|---|---|---|
| $Ru_2Si_3$ | 1973 | 375-450 | 1.8 | [90] | Si | [93] | |
| $Os_2Si_3$ | | 450-525 | 1.8 | [90] | Si | [93] | |
| $Ir_3Si_5$ | | | | [86] | Si | [93] | |
| $TbSi_{1.7}$ | ~1873 | 250-300 | 1.25 | [94] | Si | [95] | (i,j) |
| $ErSi_{1.7}$ | ~1873 | 250-300 | 1.70 | [94] | Si | [5] | (i,j) |
| $TiSi_2$ | 1813 | 425-550<br>550-650<br><br>575-750 | 1.8<br>2.5<br>1.9<br>2.5 | [96]<br>[97]<br>[98]<br>[99] | Si | [40] | (b) |
| $VSi_2$ | ~1950 | 600-700 | 2.9 | [100] | Si | [88,101] | |
| $CrSi_2$ | 1773 | 520-650 | 1.4 | [102] | Si | [42,43,103,104] | |
| $NbSi_2$ | 2223 | 643 | | [105] | Si | [88,106] | |
| $MoSi_2$ | ~2300 | 474-550<br>850-1100 | 2.4<br>2.2 | [107]<br>[108] | <br>Si | <br>[108-111] | (h)<br>(k) |

(b) on evaporated or poly Si; (f) rapid thermal annealing; (h) on $SiO_2$; (i) after irradiation of the interface; (j) by analogy with $GdSi_{2-x}$ and $SmSi_{2-x}$ which have melting points of 1873 K and 1873-2073 K respectively; (k) laser heated; (l) not a congruent melt, temperature of decomposition.

Table 5 continued

| Silicide | $T_m$ (K) | $T_f$ (°C) | Q (eV) | Ref | DDS | Ref | Note |
|---|---|---|---|---|---|---|---|
| $WSi_2$ | 2453 | 850-1100 | 2.2 | [108] | Si | [108,109] | |
| | | 600-800 | 3.4 | [112] | | | |
| | | 1200-1400 | 2.2 | [112] | | | (f) |
| | | | 2.4 | [110] | | | (l) |
| | | 1200-1350 | 2.1 | [113] | | | |
| | | 1040-1130 | 1.9 | [114] | | | |
| | | 690-740 | 3 | [115] | | | (i) |
| $CoSi_2$ | ~1600 | 405-500 | 2.3 | [116] | Co | [117,118] | (b) |
| | | | 2.3 | [118] | | | (k) |
| $NiSi_2$ | 1298 | 350-425 | 1.65 | [119] | Ni | [118-120] | (b,l) |

(b) on evaporated or poly Si; (f) rapid thermal annealing; (h) on $SiO_2$; (i) after irradiation of the interface; (j) by analogy with $GdSi_{2-x}$ and $SmSi_{2-x}$ which have melting points of 1873 K and 1873-2073 K respectively; (k) laser heated; (l) not a congruent melt, temperature of decomposition.

## D DATA ANALYSIS

A critical look at TABLES 3 to 5 shows the following points:

(a) The dominant diffusing species is generally metal for metal-rich silicides and silicon for silicon-rich silicides. This is a consequence of the so-called $Cu_3Au$ ordered rule which states that in an intermetallic ordered structure the species with the highest diffusion coefficient will be the majority element [1,15,16].

(b) The temperatures of formation of the different silicides are relatively low and scale with their melting temperature: 300°C for silicides with a low melting temperature ($Ni_2Si$, $Co_2Si$, etc.), 800 to 900°C for refractory silicides. In FIGURE 1 the temperature of formation of the silicide is plotted as a function of its melting temperature. A reasonable agreement is obtained between these two values. A linear extrapolation gives: $T_f = 0.4\ T_m$ or r (rate of formation) ~$10^{-14}$ - $10^{-13}$ $cm^2/s$ at $0.4\ T_m$.

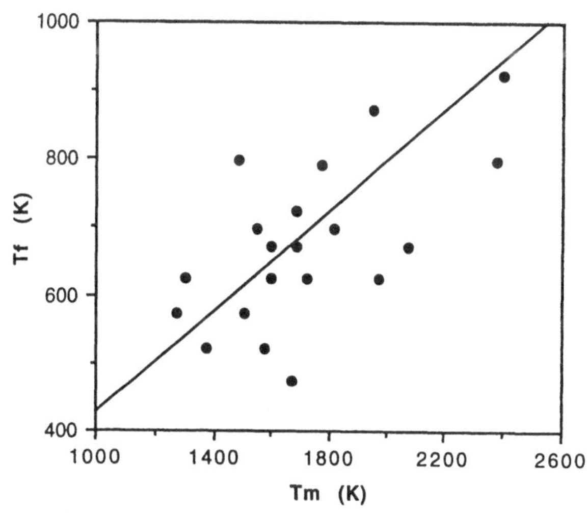


FIGURE 1 Temperature of formation of silicide thin films as a function of their melting temperature. These temperatures correspond to a rate of $10^{-13}$ to $10^{-14}$ $cm^2/s$. From [17].

The same type of correlation can be established for the activation energies [17]. This is done in FIGURE 2 and again a reasonable agreement is obtained

$$Q\ (eV) = T_m\ (K)/1000 \iff Q/RT_m \sim 11.5$$

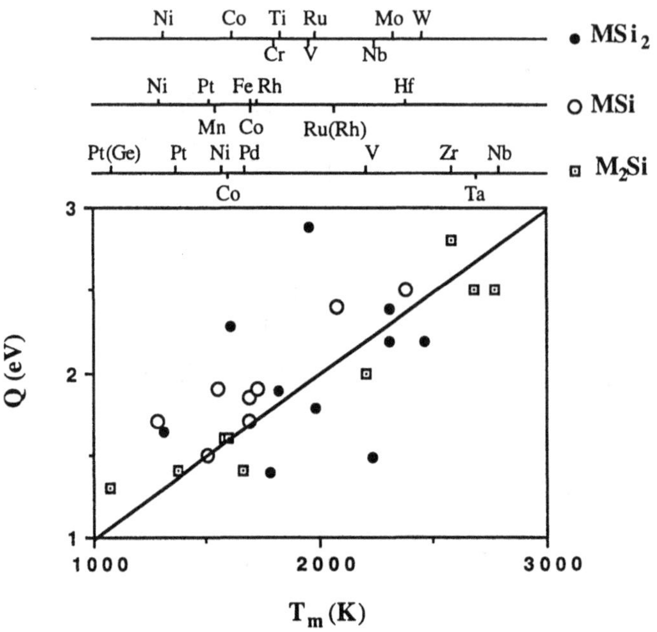


FIGURE 2 Activation energy for 'diffusion-controlled' silicide formation as a function of the silicide melting temperature. Squares correspond to metal-rich silicides, open circles to monosilicides and filled circles to disilicides. From [17].

A wide scatter is observed for disilicides (see Section C3). FIGURE 2 also shows that the monosilicides are all located in the upper part of the curve and thus have activation energies which are slightly larger than metal-rich or silicon-rich silicides, in agreement with the $Cu_3Au$ rule.

An interesting point about these correlations is that the pertinent data (rates, activation energies) for diffusion-controlled kinetics do not depend strongly on the crystallographic structure of the silicide. This is rather surprising since volume self-diffusion in intermetallics has been shown to be very dependent on structure [18] with a large spread in activation energies [19]: 9 ($Cu_3Sn$) < $Q/RT_m$ < 22 ($NbNi_3$), the higher limit being closer to self-diffusion in metals $Q/RT_m \sim 17$ [20]. The possibility of very rapid diffusion kinetics in intermetallics is linked to the existence of structural defects (vacancies or anti-sites). Hence there is a paradox: DDS indicates a possibility of very rapid self-diffusion in silicides but this rapid diffusion is generally insensitive to structural changes. Many studies have been devoted to understanding the diffusion mechanisms which can explain the rapid growth of silicide thin films.

## D1 Mobility of the Less Mobile Species

The basic idea behind these experiments is to analyse the kinetics of mixing two layers of different silicides which have similar crystallographic structures [21]. For example $CrSi_2$ and $MoSi_2$ share the same crystallographic structure (H-$CrSi_2$) and they both form by silicon motion at about 600°C. When two thin layers are put into contact, annealed and analysed by RBS, one observes a significant metal mixture only at temperatures of the order of 900°C which means that the mobilities of the two species at the formation temperature are different by 3 to 4 orders of magnitude. One should, however, not forget that the diffusion conditions during reaction and after reaction are quite different and that the driving forces are not the same: reaction in the first case, mixing in the second. This may explain this large difference in mobility. TABLE 6 gives the different silicides studied by this technique and the relative mobility of the species.

TABLE 6 Mobility of the less mobile species. Successively: bilayer used, silicides formed, temperature of formation of the silicides, temperature at which the mixing of the elements of the bilayer is observed.

| Bilayer | Silicides | T (formation) (°C) | T (mixing) (°C) | Ref |
|---|---|---|---|---|
| Dominant Diffusing Species = Si | | | | |
| V-Cr | $VSi_2$-$CrSi_2$ | 500-600 | 900 | [101] |
| Cr-Mo | $CrSi_2$-$MoSi_2$ | 500-600 | 900 | [103] |
| Cr-W | $CrSi_2$-$WSi_2$ | 500-700 | 900 | [103] |
| V-Nb | $VSi_2$-$NbSi_2$ | 600-650 | 1000 | [88] |
| Nb-Ta | $NbSi_2$-$TaSi_2$ | 650 | <1000 | [88] |
| Mo-W | $MoSi_2$-$WSi_2$ | 700 | >1100 | [109] |
| W-Ta | $WSi_2$-$TaSi_2$ | >750 | >850 | [121] |
| Mo-Ti | $MoSi_2$-$TiSi_2$ | 650 | 900 | [122] |
| Ru-Os | $Ru_2Si_3$-$Os_2Si_3$ | 450 | >700 | [90] |
| Rh-Ir | RhSi-IrSi | 500 | 1000 | [86] |
| Y-Tb | $YSi_{1.7}$-$TbSi_{1.7}$ | 400 | >900 | [95] |
| Y-Er | $YSi_{1.7}$-$ErSi_{1.7}$ | 425 | >900 | [95] |
| Si-Ge-Tb | $TbSi_{1.7}$-$TbGe_{1.7}$ | 425 | 425 | [95] |
| Si-Ge-Ti | $TiSi_2$-$TiGe_2$ | 580 | 580 | [5] |
| Dominant Diffusing Species = M (Si) | | | | |
| Ni-Pt | $Ni_2Si$-$Pt_2Si$ | 300 | 300 | [88] |
| Pd-Pt | $Pd_2Si$-$Pt_2Si$ | 300-400 | 400 | [88] |

## D2 Lattice vs. Grain Boundary Mobility

Different kinds of experiments were conducted in thin films in order to determine the respective contribution of grain boundaries and volume to the overall mobility during growth. Details of such experiments, which are beyond the scope of this Datareview, can be found in [22]. They can be classified as follows:

(a) Use of a metal bilayer (e.g. Pt/Ni) deposited on silicon. A heat treatment leads successively to the formation of $Ni_2Si$ and to the formation of $Pt_2Si$ at the $Ni_2Si$/Si interface [23]. One can thus conclude that Pt diffuses into $Ni_2Si$ without exchange which is strong evidence of a grain boundary diffusion mechanism. The same experiment was recently conducted for NiSi formation: when a Cu layer is deposited on a NiSi film, the formation of $Cu_3Si$ is observed at the NiSi/Si interface [24].

(b) Use of a radioactive tracer (generally $^{31}Si$). The basic idea is to use a radioactive tracer and to analyse its redistribution after growth. An idealised representation is shown in FIGURE 3 where a thin layer of radioactive A* is deposited on top of inactive A (a). If after formation the position of the two layers is inverted (b), then the compound (silicide) has grown by pure grain boundary diffusion (no exchange between A* and A).

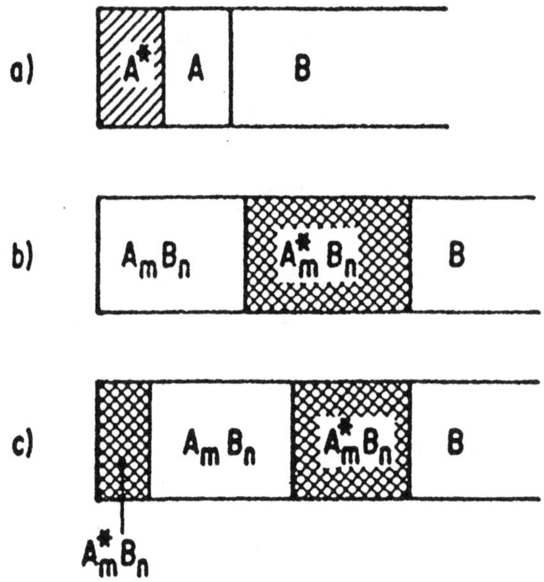


FIGURE 3 Sketch illustrating the position of an isotopic marker A* as a result of phase ($A_mB_n$) growth via grain boundary diffusion: (a) initial condition, (b) after phase growth with A and A* alone being mobile (C type) and (c) with both A (and A*) and B being mobile (again type C diffusion). According to [22].

Even where this method has been applied to silicides (see list in TABLE 7), no simple profiles have been obtained and their interpretation is extremely difficult. Recent theoretical analysis of this problem [25,26] shows that the complexity is linked to the fact that the driving forces for growth and for mixing are different. The only way to obtain a profile such as the one presented in FIGURE 3(b) is to have pure type C diffusion (no diffusion from the grain boundary to the lattice) and to limit the experiment to very short periods of time. If one adds to these theoretical difficulties the fact that some experiments with radioactive silicon were conducted under poorly controlled or doubtful conditions, one can understand that in spite of much effort this approach remains unsatisfactory.

TABLE 7 Experiments conducted with radioactive tracers ($^{31}$Si or *M) in order to establish the diffusion mechanism during silicide growth.

| | $Co_2Si$ | CoSi | $Ni_2Si$ | NiSi | $NiSi_2$ | $Pd_2Si$ | $Pt_2Si$ | PtSi | $CrSi_2$ | $TiSi_2$ | $ZrSi_2$ |
|---|---|---|---|---|---|---|---|---|---|---|---|
| $^{31}$Si | [123] | [124] | [53] [58] [125] | [124] | | [53] [58] [125] [126] | [53] [66] | [53] [58] [89] [124] | [127] | [128] | [128] |
| *M | | | [129] | [129] | [129] | [57] | | | | | |

(c) Separation of natural isotopes: again a very elegant experiment. The basic idea is that the distribution after growth of two natural isotopes should depend on the diffusion path (lattice or grain boundary) [27]. In practice this approach, which has been applied to the growth of Ni, Pt, Mg and V silicides [28-30], is severely limited by experimental as well as theoretical difficulties [29].

## E CONCLUSION

The analysis of silicide formation kinetics shows that the growth of thin films by metal/silicon reaction is, in most cases, controlled by a diffusion process (parabolic kinetics). The pertinent parameters of this diffusion-controlled growth are the nature of the mobile species, the rates of formation and the activation energies. These data, quite useful for practical applications, are known for most of the silicides. Their interpretation in terms of diffusion mechanism is, however, difficult since reactive diffusion is a complex process involving interactive and poorly known elements: chemical driving force, diffusion paths, constitutional point defects, etc. Because of the high density of grain boundaries in the growing films, enhancement by grain boundary diffusion should be suspected. However, except for some rare exceptions, experiments conducted on silicide thin films do not provide an unambiguous proof of the role of grain boundaries.

## REFERENCES

[1] F.M. d'Heurle, P. Gas [ *J. Mater. Res. (USA)* vol.1 (1986) p.205 ]
[2] F.M. d'Heurle [ *J. Mater. Res. (USA)* vol.3 (1988) p.167 ]
[3] M. Millares, B. Pierragi, E. Lelievre [ *Scr. Metall. (USA)* vol.27 (1992) p.1777 ]
[4] L.S. Hung, J. Gyulai, J.W. Mayer, S.S. Lau, M.-A. Nicolet [ *J. Appl. Phys. (USA)* vol.54 (1983) p.5076 ]
[5] O. Thomas, F.M. d'Heurle, S. Delage [ *J. Mater. Res. (USA)* vol.5 (1990) p.1453 ]
[6] F. Brown, W.D. Macintosh [ *J. Electrochem. Soc. (USA)* vol.120 (1973) p.1096 ]
[7] M.-A. Nicolet, S.S. Lau [ *Formation and Characterisation of Transition Metal Silicides* in VLSI Electronics, Microstructure Science (Academic Press, New York, 1983) ]
[8] F.M. d'Heurle, C.S. Petersson, J.E. Baglin, S.J. LaPlaca, C.Y. Wong [ *J. Appl. Phys. (USA)* vol.55 (1984) p.4208 ]
[9] D.S. Wen, P.L. Smith, C.M. Osburn, G.A. Rozgonyi [ *Appl. Phys. Lett. (USA)* vol.51 (1987) p.1182 ]
[10] F.M. d'Heurle, R.T. Hodgson, C.Y. Ting [ *Mater. Res. Soc. Symp. Proc. (USA)* vol.52 (1986) p.261 ]
[11] L.R. Zheng, E. Zingu, J.W. Mayer [ *Mater. Res. Soc. Symp. Proc. (USA)* vol.25 (1984) p.75 ]
[12] C. Nobili, M. Bosi, G. Ottaviani, G. Queirolo, L. Bacci [ *Appl. Surf. Sci. (Netherlands)* vol.53 (1991) p.219 ]
[13] P. Knauth, A. Charaï, C. Bergman, P. Gas [ *Appl. Surf. Sci. (Netherlands)* vol.53 (1991) p.120 ]
[14] C.D. Lien, M.-A. Nicolet, S.S. Lau [ *Appl. Phys. A (Germany)* vol.34 (1984) p.249 ]
[15] F.M. d'Heurle, R. Ghez [ *Thin Solid Films (Switzerland)* vol.215 (1992) p.19 ]
[16] F.M. d'Heurle, R. Ghez [ *Thin Solid Films (Switzerland)* vol.215 (1992) p.26 ]
[17] P. Gas, F.M. d'Heurle [ *Appl. Surf. Sci. (Netherlands)* vol.73 (1993) p.153 ]
[18] H. Bakker [ in *Landolt-Bornstein, New Series III/26 - Self diffusion in homogeneous binary alloys and intermediate phases* (Springer Verlag, Berlin, 1990) ]
[19] T. Barge [ Thesis, University Aix-Marseille (1993) ]
[20] J. Philibert [ *Diffusion et transport dans les solides* (Editions de Physique, France, 1985) ]
[21] J.E.E. Baglin, F.M. d'Heurle, C.S. Petersson [ *Appl. Phys. Lett. (USA)* vol.33 (1978) p.289 ]
[22] F.M. d'Heurle, P.Gas, J. Philibert [ *Mater. Res. Soc. Symp. Proc. (USA)* vol.343 (1994) p.181 ]
[23] T. Finstad [ *Thin Solid Films (Switzerland)* vol.51 (1978) p.411 ]
[24] J. Li, Y. Shacham-Diamand, J.W. Mayer [ *Mater. Sci. Rep. (Netherlands)* vol.9 (1992) p.1 ]

[25] S.L. Zhang, F.M. d'Heurle, P. Gas [ *Appl. Surf. Sci. (Netherlands)* vol.53 (1991) p.103 ]
[26] Y.M. Mishin, G. Borchardt [ *J. Phys. III (France)* vol.3 (1993) p.863 and p.945 ]
[27] J.P. Stark [ *Acta Metall. (USA)* vol.31 (1983) p.2083 ]
[28] E.-S.M. Aly, J.P. Stark [ *Acta Metall. (USA)* vol.32 (1984) p.907 ]
[29] P. Gas, C. Zaring, B.G. Svensson, M. Östling, C.S. Petersson, F.M. d'Heurle [ *J. Appl. Phys. (USA)* vol.67 (1990) p.2390 ]
[30] B.S. Lim, J.P. Stark [ *Acta Metall. (USA)* vol.32 (1984) p.915 ]
[31] H. Kräutle, M.-A. Nicolet, J.W. Mayer [ *Thin Solid Films (Switzerland)* vol. 25 (1975) p.403 ]
[32] R.J. Schultz, L.R. Testardi [ *J. Appl. Phys. (USA)* vol.50 (1979) p.5573 ]
[33] R.W. Bartlett, P.R. Gage, P. Larssen [ *Trans. Metall. Soc. AIME (USA)* vol.230 (1964) p.1528 ]
[34] L. Stolt, F.M. d'Heurle [ *Thin Solid Films (Switzerland)* vol.189 (1990) p.269-74 ]
[35] S.S. Lau, J.W. Mayer, K.N. Tu [ *J. Appl. Phys. (USA)* vol.49 (1978) p.4005 ]
[36] G.J. van Gurp, W.F. van der Weg, D. Sigurd [ *J. Appl. Phys. (USA)* vol.49 (1978) p.4011 ]
[37] C.-D. Lien, M.-A. Nicolet, C.S. Pai, S.S. Lau [ *Appl. Phys. A (Germany)* vol.36 (1985) p.153 ]
[38] J.O. Olowolafe, M.-A. Nicolet, J.W. Mayer [ *Thin Solid Films (Switzerland)* vol.38 (1976) p.143 ]
[39] K.N. Tu, G. Ottaviani, U. Gösele, H. Föll [ *J. Appl. Phys. (USA)* vol.54 (1983) p.758 ]
[40] W.K. Chu, H. Kraütle, J.W. Mayer, H. Müller, M.-A. Nicolet, K.N. Tu [ *Appl. Phys. Lett. (USA)* vol.25 (1974) p.454 ]
[41] W.K. Chu, S.S. Lau, J.W. Mayer, H. Müller, K.N. Tu [ *Thin Solid Films (Switzerland)* vol.28 (1975) p.393 ]
[42] L.S. Hung, J.W. Mayer, C.S. Pai, S.S. Lau [ *J. Appl. Phys. (USA)* vol.58 (1985) p.1527 ]
[43] K. Affolter, X.-A. Zhao, M.-A. Nicolet [ *J. Appl. Phys. (USA)* vol.58 (1985) p.3087 ]
[44] L.R. Zheng, L.S. Hung, J.W. Mayer, G. Majini, G. Ottaviani [ *Appl. Phys. Lett. (USA)* vol.41 (1982) p.646 ]
[45] C.-D. Lien, M.-A. Nicolet, S.S. Lau [ *Thin Solid Films (Switzerland)* vol.143 (1986) p.63 ]
[46] D.J. Coe, E.H. Rhoderick [ *J. Phys D, Appl. Phys. (UK)* vol.9 (1976) p.965 ]
[47] E.D. Marshall, C.S. Pai, D.M. Scott, S.S. Lau [ *Mater. Res. Soc. Symp. Proc. (USA)* vol.47 (1985) p.161 ]
[48] D.J. Fertig, G.Y. Robinson [ *Solid-State Electron. (UK)* vol.19 (1976) p.407 ]
[49] C.D. Lien, M.-A. Nicolet, C.S. Pai [ *J. Appl. Phys. (USA)* vol.57 (1985) p.224 ]
[50] K.T. Ho, C.D. Lien, U. Shreter, M.-A. Nicolet [ *J. Appl. Phys. (USA)* vol.57 (1985) p.227 ]
[51] C.M. Comrie, J.M. Egan [ *J. Appl. Phys. (USA)* vol.64 (1988) p.1173 ]
[52] C.M. Comrie, J.M. Egan [ *J. Vac. Sci. Technol. A (USA)* vol.7 (1989) p.1492 ]
[53] R. Pretorius, C.L. Ramiller, M.-A. Nicolet [ *Nucl. Instrum. Methods (Netherlands)* vol.149 (1978) p.629 ]
[54] E.C. Zingu, J.W. Mayer, C. Comrie, R. Pretorius [ *Phys. Rev. B (USA)* vol.30 (1984) p.5916 ]
[55] K.N. Tu [ *Appl. Phys. Lett. (USA)* vol.27 (1975) p.221 ]
[56] M. Bartur, M.-A. Nicolet [ *J. Appl. Phys. (USA)* vol.54 (1983) p.5404 ]
[57] J. Farmer, M.A.E. Wandt, R. Pretorius [ *Appl. Phys. Lett. (USA)* vol.56 (1990) p.1643 ]
[58] R. Pretorius [ *J. Electrochem. Soc. (USA)* vol.128 (1981) p.107 ]
[59] N.W. Cheung, M.-A. Nicolet, M. Wittmer, C.A. Evans Jr., T.T. Sheng [ *Thin Solid Films (Switzerland)* vol.79 (1981) p.51 ]
[60] C.S. Wei, J. Van der Spiegel, J.J. Santiago [ *J. Electrochem. Soc. (USA)* vol.135 (1988) p.446 ]
[61] G.A. Hutchins, A. Shepela [ *Thin Solid Films (Switzerland)* vol.18 (1973) p.343 ]
[62] R.W. Bower, D. Sigurd, R.E. Scott [ *Solid-State Electron. (UK)* vol.16 (1973) p.1461 ]
[63] B. Coulman, H. Chen [ *J. Appl. Phys. (USA)* vol.59 (1986) p.3467 ]

[64] G. Ottaviani, C. Canali, G. Ferrari, G. Majini, M. Prudenziati, S.S. Lau [ *Thin Solid Films (Switzerland)* vol.47 (1977) p.187 ]

[65] C. Canali, C. Catellani, M. Prudenziati, W.H. Wadlin, C. Evans Jr. [ *Appl. Phys. Lett. (USA)* vol.31 (1977) p.43 ]

[66] M.A.E. Wandt, C.M. Comrie, J.E. McLeod, R. Pretorius [ *J. Appl. Phys. (USA)* vol.67 (1990) p.230 ]

[67] J.M. Poate, T.C. Tisone [ *Appl. Phys. Lett. (USA)* vol.24 (1974) p.391 ]

[68] H. Muta, D. Shinoda [ *J. Appl. Phys. (USA)* vol.43 (1972) p.2913 ]

[69] H. Takai, P.A. Psaras, K.N. Tu [ *J. Appl. Phys. (USA)* vol.58 (1985) p.4165 ]

[70] L.R. Zheng, L.S. Hung, J.W. Mayer [ *Mater. Res. Soc. Symp. Proc. (USA)* vol.18 (1982) p.207 ]

[71] C.A. Crider, J.M. Poate [ *Appl. Phys. Lett. (USA)* vol.36 (1980) p.417 ]

[72] Q.Z. Hong, J.W. Mayer [ *J. Appl. Phys. (USA)* vol.66 (1989) p.611 ]

[73] M. Berti et al [ *J. Appl. Phys. (USA)* vol.55 (1984) p.3558 ]

[74] W. Bartlett [ *Trans. Metall. Soc. AIME (USA)* vol.236 (1966) p.1230 ]

[75] S.Q. Wang, J.W. Mayer [ *J. Appl. Phys. (USA)* vol.64 (1988) p.4711 ]

[76] M. Eizenberg, K.N. Tu [ *J. Appl. Phys. (USA)* vol.53 (1982) p.6885 ]

[77] S.S. Lau, J.S.Y. Feng, J.O. Olowolafe, M.-A. Nicolet [ *Thin Solid Films (Switzerland)* vol.25 (1975) p.415 ]

[78] G.J. van Gurp, C. Langereis [ *J. Appl. Phys. (USA)* vol.46 (1975) p.4301 ]

[79] F.M. d'Heurle, C.S. Petersson, J.E.E. Baglin, S.J. LaPlaca, C.Y. Wong [ *J. Appl. Phys. (USA)* vol.55 (1984) p.4208 ]

[80] T. Finstad [ *Phys. Status Solidi A (Germany)* vol.63 (1981) p.223 ]

[81] D.M. Scottland, M.-A. Nicolet [ *Phys. Status Solidi A (Germany)* vol.66 (1981) p.773 ]

[82] G. Majini, F. Della Valle, C. Nobili [ *J. Phys. D, Appl. Phys. (UK)* vol.17 (1984) p.L77 ]

[83] L. Schellenberg, J. Jorda, J. Mueller [ *J. Less-Common Met. (Switzerland)* vol.109 (1985) p.261 ]

[84] C.S. Petersson et al [ *J. Appl. Phys. (USA)* vol.51 (1980) p.373 ]

[85] P.A. Psaras, R.D. Thomson, S.R. Herd, K.N. Tu [ *J. Appl. Phys. (USA)* vol.55 (1984) p.3526 ]

[86] C.S. Petersson et al [ *J. Appl. Phys. (USA)* vol.50 (1979) p.3357 ]

[87] F.M. d'Heurle [ *Thin Solid Films (Switzerland)* vol.151 (1987) p.41 ]

[88] J.E.E. Baglin, F.M. d'Heurle, W.M. Hammer, C.S. Petersson [ *Nucl. Instrum. Methods (Netherlands)* vol.168 (1980) p.491 ]

[89] R. Pretorius, A.P. Botha, J.C. Lombaard [ *Thin Solid Films (Switzerland)* vol.79 (1981) p.61 ]

[90] C.S. Petersson, J.E.E. Baglin, J.J. Dempsey, F.M. d'Heurle, S.J. La Placa [ *J. Appl. Phys. (USA)* vol.53 (1982) p.4866 ]

[91] J.F. Ziegler, J.W. Mayer, C.J. Kircher, T.N. Tu [ *J. Appl. Phys. (USA)* vol.44 (1973) p.3851 ]

[92] C.J. Kircher, J.W. Mayer, T.N. Tu, J.F. Ziegler [ *Appl. Phys. Lett. (USA)* vol.22 (1973) p.81 ]

[93] F.M. d'Heurle [ *Thin Solid Films (Switzerland)* vol.41 (1987) p.151 ]

[94] B.Y. Tsaur, L.S. Hung [ *Appl. Phys. Lett. (USA)* vol.37 (1980) p.922 ]

[95] J.E.E. Baglin, F.M. d'Heurle, C.S. Petersson [ *J. Appl. Phys. (USA)* vol.52 (1981) p.2841 ]

[96] L.S. Hung, J. Gyulai, J.W. Mayer, S.S. Lau, M.-A. Nicolet [ *J. Appl. Phys. (USA)* vol.54 (1983) p.5076 ]

[97] C.A. Pico, M.G. Lagally [ *J. Appl. Phys. (USA)* vol.64 (1988) p.4957 ]

[98] J.P. Ponpon, A. Saulnier [ *Appl. Surf. Sci. (Netherlands)* vol.40 (1990) p.315 ]

[99] I.J.M.M. Raaijmakers [ PhD Thesis, Technical University of Eindhoven, Netherlands, 1988 ]

[100] K.N. Tu, J.F. Ziegler, C.J. Kircher [ *Appl. Phys. Lett. (USA)* vol.23 (1973) p.493 ]

[101] T.J. Finstad, O. Thomas, F.M. d'Heurle [ *Appl. Surf. Sci. (Netherlands)* vol.38 (1989) p.106 ]

[102] L.R. Zheng, E. Zingu, J.W. Mayer [ *Mater. Res. Soc. Symp. Proc. (USA)* vol.25 (1984) p.75 ]

[103] O. Thomas, T.J. Finstad, F.M. d'Heurle [ *J. Appl. Phys. (USA)* vol.67 (1990) p.2410 ]

[104] A. Martinez, D. Esteve, A. Guivarc'h, P. Auvray, P. Henoc, G. Pelous [ *Solid-State Electron. (UK)* vol.23 (1980) p.55 ]

[105] R.J. Wagner, S.S. Lau, J.W. Mayer, J.A. Roth [ *Proc. Electrochem. Soc. (USA)* vol.78 no.2 (1978) p.59 ]

[106] T.W. Sigmon, S.S. Wakita, J.F. Gibbons [ *Layered Structures and Interface Kinetics* Ed. S. Furukawa (KTH Scientific Publishers, Tokyo, 1985) p.237 ]

[107] A. Guivarc'h et al [ *J. Appl. Phys. (USA)* vol.49 (1978) p.233 ]

[108] P.R. Gage, R.W. Bartlett [ *Trans. Metall. Soc. AIME (USA)* vol.233 (1965) p.832 ]

[109] J.E.E. Baglin, F.M. d'Heurle, C.S. Petersson [ *Appl. Phys. Lett. (USA)* vol.33 (1978) p.289 ]

[110] H.S. Lee, G.J. Wolga [ in *Selected Topics in Electronic Materials* Eds B.R. Appleton, D.K. Biegelsen, W.L. Brown, J.A. Knapp (Materials Research Society, 1988) p.27 ]

[111] G. Bomchil et al [ *Appl. Phys. Lett. (USA)* vol.41 (1982) p.46 ]

[112] J. Lajzerowicz Jr., J. Torres, G. Goltz, R. Pantel [ *Thin Solid Films (Switzerland)* vol.140 no.1 (1986) p.23 ]

[113] V.I. Zmii, A.S. Seryugina [ *Protect. Coat. Met. (USA)* vol.2 (1970) p.158 ]

[114] N. Hashimoto [ *Trans. Metall. Soc. AIME (USA)* vol.239 (1967) p.1109 ]

[115] E. Ma, B.S. Lim, M.-A. Nicolet, N.S. Alvi, A.H. Hamdi [ *J. Electron. Mater. (USA)* vol.17 (1967) p.207 ]

[116] C.-D. Lien, M.-A. Nicolet, S.S. Lau [ *Appl. Phys. A (Germany)* vol.34 (1984) p.249 ]

[117] F.M. d'Heurle, C.S. Petersson [ *Thin Solid Films (Switzerland)* vol.128 (1985) p.283 ]

[118] C.-D. Lien, M. Bartur, M.-A. Nicolet [ *Mater. Res. Soc. Symp. Proc. (USA)* vol.25 (1984) p.51 ]

[119] C.-D. Lien, M.-A. Nicolet, S.S. Lau [ *Phys. Status Solidi A (Germany)* vol.81 (1984) p.123 ]

[120] F.M. d'Heurle, C.S. Petersson, L. Stolt, B. Stritzker [ *J. Appl. Phys. (USA)* vol.53 (1982) p.5378 ]

[121] A. Applebaum, M. Eizenberg, R. Brenner [ *Vacuum (UK)* vol.33 (1983) p.227 ]

[122] T. Misawa, K. Asayama, H. Homma [ *J. Electrochem. Soc. (USA)* vol.137 (1990) p.713 ]

[123] A.P. Botha, R. Pretorius [ *Proc. Conf. Materials Research* Eds P.S. Ho, K.N. Tu (North Holland Publishing, Amsterdam, 1982) p.129 ]

[124] R. Pretorius, M.A.E. Wandt, J.E. McLeod, A.P. Botha, C.M. Comrie [ *J. Electrochem. Soc. (USA)* vol.136 (1989) p.839 ]

[125] R. Pretorius, C.L. Ramiller, S.S. Lau, M.-A. Nicolet [ *Appl. Phys. Lett. (USA)* vol.30 (1977) p.501 ]

[126] J.M. Egan, C.M. Comrie [ *Phys. Rev. B (USA)* vol.40 (1989) p.11670 ]

[127] A.P. Botha, R. Pretorius, S. Kritzinger [ *Appl. Phys. Lett. (USA)* vol.40 (1982) p.412 ]

[128] A.P. Botha, R. Pretorius [ *Thin Solid Films (Switzerland)* vol.93 (1982) p.127 ]

[129] J.E.E. Baglin, H.A. Atwater, D. Gupta, F.M. d'Heurle [ *Thin Solid Films (Switzerland)* vol.93 (1982) p.255 ]

# 7.2 Self-diffusion in TM silicides: tracer experiments

P. Gas, T. Barge and F.M. d'Heurle

January 1995

## A INTRODUCTION

Most of the information concerning self-diffusion in silicides is derived from the kinetics of formation of thin films (Datareview 7.1). As discussed in Datareview 7.1 Section D the relation between these data and the self-diffusion properties of silicides is far from being obvious.

On the other hand, volume diffusion in intermetallic compounds has been the subject of much interest and despite the difficulty in preparing and analysing these materials many studies have been performed. In his recent and complete survey of the literature on self-diffusion in homogeneous binary alloys and intermediate phases, Bakker [1,2] describes about 25 different intermediate phases. It is concluded [3] that diffusion in intermetallics is an unsolved problem and that "the diffusion behaviour differs in essential aspects from the one to the other compound". Moreover: (i) no silicide appears in this review and (ii) studies have been limited to volume diffusion and there are no data available on grain boundary (gb) diffusion.

In this Datareview we will present recent experiments carried out on bulk silicides in order to obtain the missing information.

## B EXPERIMENTAL PROCESSES

Volume and grain boundary self-diffusion coefficients in silicides have been measured using the sectioning method based on the use of radioactive tracers. In this method a thin layer of a radioactive tracer is deposited on the surface of a bulk polycrystalline silicide. After a diffusion treatment the sample is cut in a direction parallel to the initial surface and the activity is measured using (depending on the characteristics of the radiotracer) a $\beta$ or a $\gamma$ detector. Since the diffusion conditions are extremely simple (tracer diffusion in a semi-infinite medium), the diffusion profiles obtained obey Fick's law [4]. By changing the experimental conditions it is possible to determine the volume diffusion coefficient and/or the grain boundary diffusion coefficient. If volume diffusion prevails, the logarithm of the activity varies as the square of the distance from the initial deposit while in the case where grain boundary diffusion is predominant the logarithm of the activity varies as the 6/5 power of the depth [5]. Usually, lattice diffusion is measured at high temperatures while grain boundary diffusion is measured at lower temperatures ($T \leq 0.5T_m$). However, with well chosen diffusion conditions it is possible to determine both diffusion coefficients on the same penetration plot as shown in FIGURE 1. It is important to mention that except in very rare cases where grain boundary diffusion has been studied independently of volume diffusion [6], the determination of the grain boundary diffusion coefficient is always linked to a model of the grain boundary and of grain boundary diffusion. The simplest is to suppose that the grain boundary is a uniform slab of thickness δ (usually 0.5 nm) in which the diffusion coefficient is constant and equal to $D_{gb}$ [7].

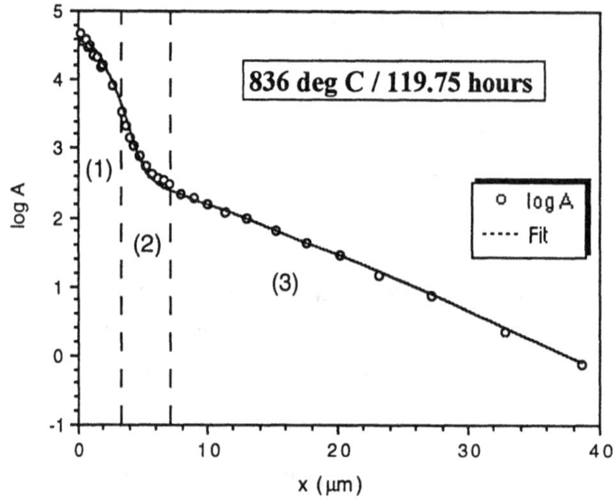


FIGURE 1 Penetration plot of $^{60}Co$ in $CoSi_2$ after a heat treatment at 836°C for 120 hours. Circles are experimental measurements; the solid line is a numerical simulation using the following diffusion coefficients: $D_v = 2.9 \times 10^{-14}$ cm²/s; $D_{gb} = 4.2 \times 10^{-9}$ cm²/s. From [11].

TABLES 1 and 2 present the results obtained in three silicides: $Cr_3Si$ [8], $Ni_2Si$ [9], and $CoSi_2$ [10-12]. In the first case, only Cr volume diffusion has been measured while in the other two silicides volume and grain boundary diffusion of both elements have been analysed. Because of the very short half-life of the Si radioactive isotopes, a Ge radiotracer ($^{68}Ge$) is substituted for the Si tracer. This method which has been used for silicon self-diffusion 'in silicon' is somehow more questionable for silicides since the solubility of germanium in silicides is generally limited. However, it should provide a correct order of magnitude of volume diffusion coefficients (which is the main aim of such studies). The situation is slightly more complex for grain boundary diffusion since, in such a case, one measures the product of the diffusion coefficient and the tracer exchange coefficient between grain boundary and volume diffusion [13,14]. The latter is equal to 1 only for 'true' self-diffusion (A in A), and in any other case it takes greater values. Thus the grain boundary diffusion coefficients given for Ge represent upper limits of the Si grain boundary diffusion coefficients.

TABLE 1 Volume self-diffusion coefficients in bulk silicides. Columns give successively: the silicide and its main characteristics (structure, melting temperature), the element for which diffusion is studied, the temperature field analysed, the volume diffusion coefficients at the limits of this temperature field, the activation energy for diffusion and the corresponding reference.

| Silicide | Elem. | $T_{inf}$ < (°C) < $T_{sup}$ | | $D_{inf}$ < (cm²/s) < $D_{sup}$ | | Q (eV) | Ref |
|---|---|---|---|---|---|---|---|
| $Cr_3Si$ - A15 $T_m$ = 1730°C | Cr | 1400 | 1550 | $1.75 \times 10^{-11}$ | $1.35 \times 10^{-10}$ | 3.58 | [8] |
| $Ni_2Si$ - O $T_m$ = 1200°C | Ni<br>Si* | 650<br>710 | 910<br>850 | $1.05 \times 10^{-13}$<br>$5.0 \times 10^{-15}$ | $1.0 \times 10^{-10}$<br>$7.5 \times 10^{-13}$ | 2.48 | [9] |
| $CoSi_2$ - $CaF_2$ $T_m$ = 1327°C | Ci<br>Si* | 700<br>700 | 1100<br>1100 | $6.2 \times 10^{-16}$<br>$1.6 \times 10^{-16}$ | $1.1 \times 10^{-10}$<br>$1.2 \times 10^{-11}$ | 2.78<br>3.23 | [10,11]<br>[11,12**] |

*a germanium tracer ($^{68}Ge$) is used in lieu of a Si tracer,
**in [12], diffusion coefficients are deduced from SIMS measurements of Ge implantation profiles.

TABLE 2 Grain boundary self-diffusion coefficients in bulk silicides. Columns give successively: the silicide and its main characteristics (structure, melting temperature), the element for which diffusion is studied, the temperature field analysed, the grain boundary diffusion coefficients at the limits of this temperature field, the activation energy for diffusion and the corresponding reference.

| Silicide | Elem. | $T_{inf}$ < (°C) < $T_{sup}$ | | $D_{inf}$ < ($cm^2/s$) < $D_{sup}$** | | Q (eV) | Ref |
|---|---|---|---|---|---|---|---|
| $Ni_2Si$ - O | Ni | 530 | 710 | $3.45 \times 10^{-10}$ | $3.2 \times 10^{-8}$ | 1.71 | [9] |
| $T_m$ = 1200°C | Si* | | 710 | | $3.4 \times 10^{-9}$ | | [9] |
| $CoSi_2$ - $CaF_2$ | Co | 700 | 1100 | $1.1 \times 10^{-10}$ | $6.6 \times 10^{-7}$ | 2.50 | [11] |
| $T_m$ = 1327°C | Si* | 700 | 1100 | $2.0 \times 10^{-9}$ | $2.5 \times 10^{-5}$ | 2.71 | [11] |

*a germanium tracer ($^{68}$Ge) is used in lieu of a Si tracer,
**the grain boundary width δ is taken as 0.5 nm.

## C VOLUME AND GRAIN BOUNDARY SELF-DIFFUSION

TABLE 1 shows that volume diffusion in these silicides is slow. The ratio $Q/RT_m$ is of the order of 20 (20 for Ni in $Ni_2Si$ and for Co in $CoSi_2$, 21 for Cr in $Cr_3Si$, 23 for Ge in $CoSi_2$). If one compares these values to those previously obtained in intermediate phases ($9 < Q/RT_m < 22$), these silicides appear among the phases where mass transport is slow and comparable to metals ($Q/RT_m = 17$) [15]. This means that diffusion is not affected by constitutional point defects (antisite atoms or structural vacancies) which may increase diffusion markedly. Is this the case for all silicides formed by metal/silicon reactions? Given the few silicides listed in TABLE 1 and even if one adds the recent result obtained for Ni self-diffusion in $Ni_3Ge$ ($Q/RT_m = 21$) [16], a positive answer to this question is doubtful. However, the following facts would support it:

- $Cr_3Si$, $Ni_2Si$ and $CoSi_2$ belong to very different classes of silicides, with different structures (A15, O-anti $PbCl_2$, $CaF_2$) and M/Si ratio,

- rapid diffusion in intermetallics is always associated with important departures from stoichiometry. This is not the case for silicides formed by metal/silicon thin film reactions which generally have extremely narrow composition limits. The largest departures are observed for $Cu_3Si$ <1 at. %, $Co_2Si$ ~2%, $Pd_2Si$ ~2%, CoSi ~3%, and $VSi_2$ ~8%.

TABLE 2 shows that grain boundary diffusion coefficients are much higher than lattice diffusion coefficients. Their ratio (about $10^5$ to $10^6$ at $T/T_m = 0.5$) is comparable with the same ratio measured in pure metals [17], which indicates that this diffusion mechanism should be taken into account to understand mass transport. The ratio of the activation energies for both diffusion mechanisms is also interesting: it ranges from 0.7 (Ni in $Ni_2Si$) to 0.9 (Co in $CoSi_2$), values slightly higher than those measured in metals (from 0.5 (Ag) to 0.7 (Ni) [18]). The high values obtained for silicides, even if they raise a fundamental question about this diffusion mechanism, indicate that grain boundary mass transport remains effective at relatively high temperatures.

## D COMPARISON BETWEEN TRACER MEASUREMENTS (this Datareview) AND GROWTH KINETICS (Datareview 7.1)

A direct comparison between the diffusion coefficients measured by tracer and the rate of growth deduced from the diffusion-controlled growth of thin films [19,20] is seen in FIGURE 2 for $Ni_2Si$. The volume diffusion coefficients are really too low to explain the very rapid growth of thin films. This is strong evidence that growth is controlled by grain boundary diffusion. The agreement between the activation energies is also quite good (1.5 - 1.6 for thin films and 1.71 for grain boundary diffusion).

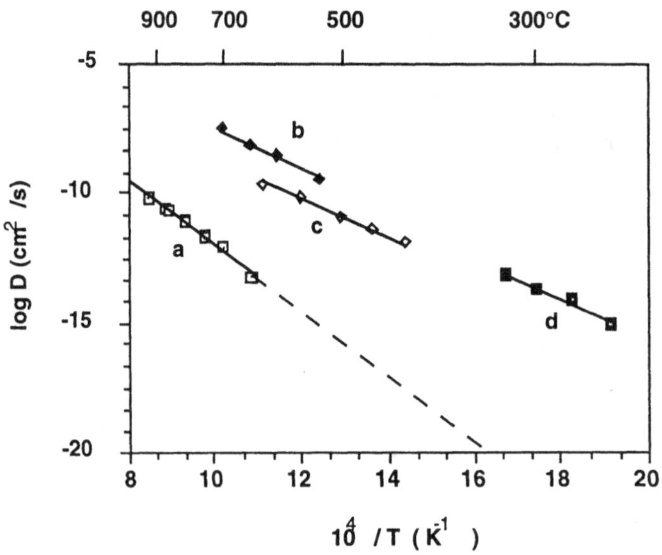


FIGURE 2 Diffusion in $Ni_2Si$. (a) The diffusion coefficient of Ni in the lattice of $Ni_2Si$ and (b) in grain boundaries. The growth rates of $Ni_2Si$ (c) in 'lateral formation' [19] and (d) in 'normal formation' [20]. From [9].

The same is also true for $CoSi_2$. On the basis of the diffusion coefficients (for Co and Ge) given in TABLES 1 and 2 and knowing the free energies of formation of $CoSi_2$ from CoSi and Si, it is possible to estimate the rate of growth of $CoSi_2$ as a function of its grain size [11,21]. The data are presented in FIGURE 3, where one sees that the experimental rates of growth [21-23] agree reasonably well with calculations for respective grain sizes of 10 μm (bulk diffusion couples) and 0.01 μm (thin films). This last comparison takes into account, through the free energy of formation, the basic difference between tracer and reactive diffusion, namely the extra chemical driving force.

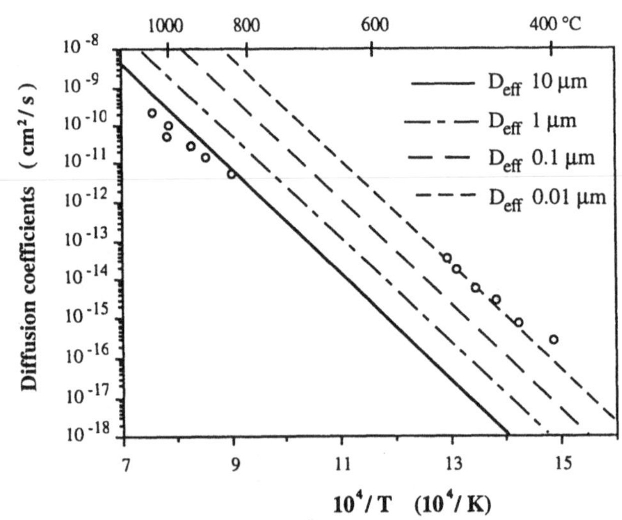


FIGURE 3 The effective diffusion coefficients, including Co and Si, lattice and grain boundaries, as a function of the grain size. It is seen that the data points for thin film growth (400°C to 600°C [22]) and bulk (800°C to 1000°C [11,23]) agree quite well with the coefficients for 0.01 μm and 10 μm respectively. From [21].

## E CONCLUSION

Tracer experiments provide the 'real' diffusion coefficients and are thus the key element in understanding any phenomena in which long range atomic migration is involved. The results published for intermetallic compounds show that diffusion in ordered compounds depends on various factors such as crystal structure, degree of atomic order, composition range, etc. Diffusion behaviour thus differs in essential aspects from one compound to another. Only very few data exist on silicides. They show, however, a diffusion behaviour similar to that of pure metals: volume diffusion of both elements is slow, grain boundaries are rapid diffusion paths and their diffusion coefficient is $10^6$ times higher than the volume diffusion coefficient. Comparison between 'tracer' diffusion coefficients and the rate of growth of thin films indicates that the very rapid formation of thin films is essentially due to diffusion along grain boundaries.

## REFERENCES

[1] H. Bakker [ *Landolt-Bornstein New Series III/26* Self diffusion in homogeneous binary alloys and intermediate phases (Springer Verlag, Berlin, 1990) ]

[2] H. Bakker [ *Tracer Diffusion in Concentrated Alloys* Eds A.S. Nowick, G.E. Murch (Academic Press, New York, 1984) p.189 ]

[3] H. Bakker, D.M.R. Lo Cascio [ *Defect and Diffusion Forum* (Trans. Tech. Publications, USA, 1993) vol.95-98 p.803 ]

[4] Y. Adda, J. Philibert [ *La diffusion dans les solides* (Presses Universitaires de France, Paris, 1966) ]

[5] I. Kaur, W. Gust [ *Fundamentals of grain and interphase boundary diffusion* (Ziegler Press, Stuttgart, 1988) ]

[6] P. Gas, S. Poize, J. Bernardini [ *Colloq. Phys. (France)* vol.C1 (1990) p.477 ]

[7] J.C. Fisher [ *J. Appl. Phys. (USA)* vol.22 (1951) p.74 ]

[8] M. Jurisch, D. Bergner [ *Diffusion and Defect Monograph Series* vol.7 (Trans. Tech. Publications, USA, 1983) p.465 ]

[9] J.-C. Ciccariello, S. Poize, P. Gas [ *J. Appl. Phys. (USA)* vol.67 (1990) p.3315 ]

[10] T. Barge, S. Poize, J. Bernardini, P. Gas [ *Appl. Surf. Sci. (Netherlands)* vol.53 (1991) p.180 ]

[11] T. Barge [ Thesis, Marseille (1993) ]

[12] O. Thomas, P. Gas, A. Charaï, F.K. Le Goues, A. Michel, G. Scillia [ *J. Appl. Phys. (USA)* vol.64 (1988) p.2973 ]

[13] J. Bernardini, G. Martin [ *Scr. Metall. (USA)* vol.10 (1976) p.833 ]

[14] P. Gas, S. Poize, J. Bernardini, F. Cabane [ *Acta Metall. (USA)* vol.37 (1989) p.17 ]

[15] J. Philibert [ *Diffusion et transport dans les solides* (Editions de Physique, France, 1985) ]

[16] H. Yasuda, H. Nakajima, M. Koiwa [ *Defect and Diffusion Forum* (Trans. Tech. Publications, USA, 1993) vol.95-98 p.822 ]

[17] G. Martin, B.Perraillon [ *J. Phys. (France)* vol.36-C4 (1975) p.165 ]

[18] I. Kaur, W. Gust, L. Kozma [ *Handbook of Grain and Interphase Boundary Diffusion Data* (Ziegler Press, Stuttgart, 1989) ]

[19] C.R.Zheng, E. Zingo, J.W. Mayer [ *Mater. Res. Soc. Symp. Proc. (USA)* vol.25 (1984) p.75 ]

[20] C.D. Lien, M.-A. Nicolet, S.S. Lau [ *Phys. Status Solidi A (Germany)* vol.81 (1984) p.123 ]

[21] T. Barge, P. Gas, F.M. d'Heurle [ *J. Mater. Res. (USA)* vol.10 (1995) p.1134 ]

[22] C.D. Lien, M.-A. Nicolet, S.S. Lau [ *Appl. Phys. A (Germany)* vol.34 (1984) p.249 ]

[23] C.H. Jan, C.-P. Chen, Y.A. Chen [ *J. Appl. Phys. (USA)* vol.73 (1993) p.1168 ]

# 7.3 Impurity diffusion in TM silicide thin films

P. Gas, O. Thomas and F.M. d'Heurle

January 1995

## A INTRODUCTION

In microelectronic applications silicides are often used in close contact with doped silicon. Any heat treatment will thus produce a redistribution of the dopant between the silicon and the silicide. This redistribution may induce an alteration of the dopant concentration at the silicide/silicon interface and a modification of the electrical characteristics of the contact. In cases where silicides are used in interconnections, dopants may be transported from p-type contacts to n-type ones, again with undesirable consequences. One may also use this redistribution for beneficial purposes: a dopant initially implanted in the silicide may diffuse into the silicon giving a very shallow junction [1-5].

Knowledge of the diffusion coefficients of impurities (dopants) in silicides is of prime importance to understand and possibly control these redistributions. However, the problem is extremely complex since the redistribution of a dopant, which obviously depends on kinetic parameters (diffusion in the silicon and in the silicide), is also influenced by equilibrium parameters (solubility limits [6], compound formation [7-9], segregation at grain boundaries and interfaces [10]) which act as driving forces for diffusion.

Because of this complexity the interests of the scientist and of the technologist are slightly different: the first wants to know precise values of diffusion coefficients (and/or dopant solubilities) while the other may be satisfied to know which redistribution will result from a certain heat treatment. This explains the diversity of the publications from which information on diffusion of dopants in silicides may be extracted. In the following we will try to mention most of the available data on this topic and to analyse their meaning in terms of diffusion.

## B MOBILITY OF DOPANTS (AND GERMANIUM) IN SILICIDES

### B1 Experimental Approaches

Because of practical considerations attention has been focused on diffusion of Si dopants (B, P, As, Sb, Ga) in disilicides. Diffusion of Ge has also been analysed for practical reasons (Ge is frequently used in Si technology) and for fundamental reasons (information on self-diffusion).

In most cases, the impurity is introduced by implantation in the silicide film (300 to 800 nm thick) and its redistribution after different heat treatments is analysed by Secondary Ion Mass Spectrometry (SIMS) (e.g. [6]) or Rutherford Backscattering (RBS). Redistribution of dopants initially located in the Si during silicide formation may also provide information on the diffusion in the silicide layer (e.g. [11]). Such experiments (reviewed in [12]) should be considered with caution regarding diffusion since the redistribution process is very complex and involves dopant solubility and diffusion in a growing silicide. One should also mention an early study based on the use of radioactive tracers ($^{32}P$ and $^{33}P$) and determination of activity profiles [13,14]. All these approaches allow the determination of dopant mobility in a direction perpendicular to the Si substrate. They have the main advantage of providing a detailed description of the dopant profile in the silicide layer but the range of penetration depth which may be analysed is limited by the thickness of the silicide film.

Recently, an original method to measure lateral dopant diffusion has been proposed [15]. The idea is to monitor the change in the electrical characteristics of a silicide/polysilicon interface due to dopant diffusion along a silicide line. This ingenious method has the following advantages: (i) diffusion is measured in a material which is initially free of dopant, (ii) diffusion may be followed over larger distances and thus with different experimental conditions (thermal budget). It has, however, the disadvantage of being an indirect method and is only sensitive to the overall flux of dopant without giving information on diffusion mechanisms. Lateral dopant diffusion has also been studied using three-dimensional SIMS [16].

## B2 Diffusion Profiles

Experimental conditions to analyse impurity diffusion must be extremely rigorous in order to avoid artefacts associated with solubility limits, precipitation or segregation to interfaces. Such conditions are practically impossible to achieve in thin films and the experimental redistributions obtained have to be analysed with great caution in order to prevent misinterpretations. FIGURES 1 to 6 represent series of diffusion profiles obtained in conditions typical of the various studies developed. These profiles illustrate the influence of the different phenomena acting on dopant mobility and the potentialities of the different studies in terms of diffusion. In all cases, the dopant is introduced by implantation and analysed by SIMS.

FIGURE 1 represents the modification of a B profile implanted in a bulk sample of $CoSi_2$ after different heat treatments [17]. Such samples have large grain sizes (> mm) and at the scale of the analysis can be considered as single crystals. The modifications observed are extremely simple: (i) the initial Gaussian-like profile spreads and globally keeps its shape, (ii) the mode of the profile is displaced to greater depth as the heat treatment duration and temperature increase, indicating a surface effect. These changes are typical of a pure diffusion process by a single diffusion mechanism and allow the determination of the volume diffusion coefficients of B in $CoSi_2$ providing an adequate solution of the diffusion equation, which takes into

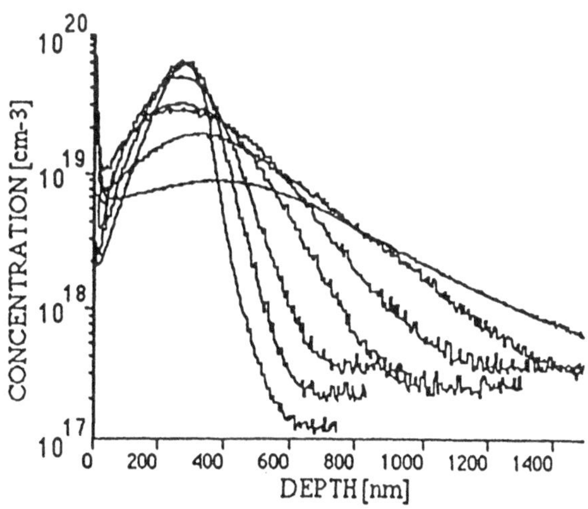


FIGURE 1 Secondary ion mass spectrometry profiles of boron implanted in bulk $CoSi_2$, respectively as implanted, and after heat treatments at 450°C for 48 hr, 500°C for 18.3 hr, 550°C for 5 hr, 600°C for 2 hr, 634°C for 1 hr and 700°C for 30 min. From [17].

account surface effects, is used [18,19]. The diffusion coefficients obtained range from $6 \times 10^{-17}$ $cm^2/s$ at 450°C to $4 \times 10^{-13}$ $cm^2/s$ at 700°C, values several orders of magnitude larger than the corresponding values for B volume diffusion in Si. The validity of these coefficients and the fact that implantation defects do not influence the redistribution were confirmed by analysis of diffusion from a deposited B source [17].

FIGURE 2 represents the modifications of a B profile after different heat treatments but in this case the sample is a $CoSi_2$ thin film (850 nm) obtained by solid state reaction at 900°C between a Co film and a Si substrate [20]. One can observe the following differences: (i) broadening of the profile around the implantation peak, (ii) accumulation of B at the silicide/silicon interface, (iii) uniform increase of B concentration in the centre of the layer. Such a redistribution can only be explained if one takes into account the different diffusion paths existing in the silicide layer. A schematic view of the diffusion process is given in FIGURE 3: the primary diffusion source is the implantation peak, its broadening is similar to the one observed in bulk samples and is mainly influenced by volume diffusion, $CoSi_2$ grain boundaries connect this primary diffusion source to the $CoSi_2$/Si interface, and rapid grain boundary diffusion supplies B to this interface which then acts as a secondary diffusion source. Complete homogenisation of the dopant in the silicide layer can be obtained, as shown, after the 800°C annealing.

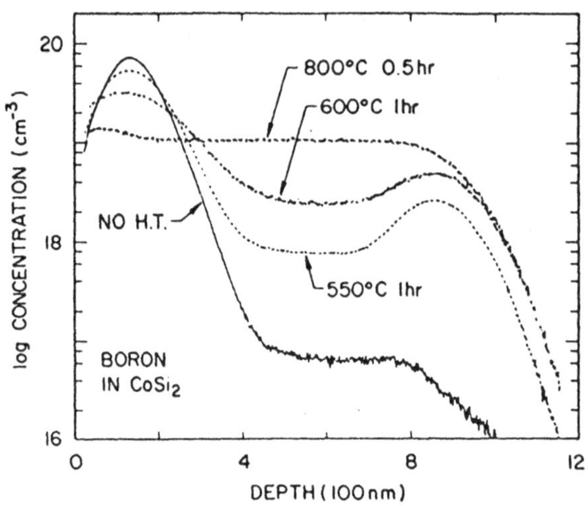


FIGURE 2 Secondary ion mass spectrometry profiles of boron implanted in $CoSi_2$ layers ~840 nm thick, respectively as implanted, and after heat treatments at 550°C for 1 hr, 600°C for 1 hr and 800°C for 0.5 hr. From [20].

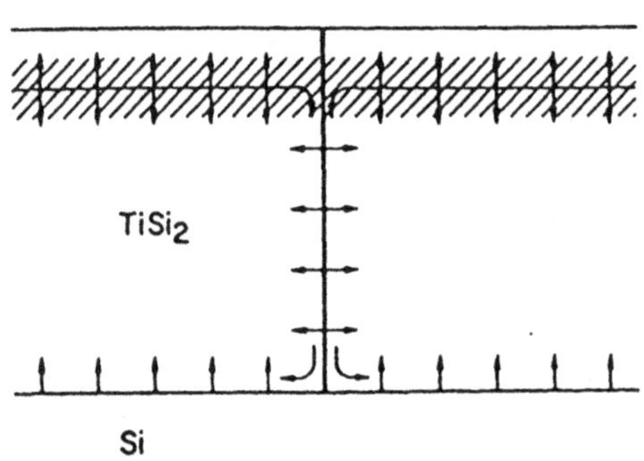


FIGURE 3 Schematic representation of the diffusion processes in silicide layers. The main source of dopant is the implanted zone (hatched area); however, with respect to the time scale for volume diffusion, the grain boundaries and the silicide-silicon interface are almost instantly saturated with the dopant. They act as secondary diffusion sources. From [23].

From such a redistribution two quantities may be obtained: the volume diffusion coefficient (from the modification of the implanted peak) and an overall mobility taking into account all the possible diffusion paths (from the temperature at which complete homogenisation is obtained).

FIGURES 4 and 5 represent similar experiments conducted for As [20,5] in $CoSi_2$. In this case the diffusion profile is quite different. None of the applied heat treatments lead to a homogenisation inside the silicide layer. However, one can observe a pile-up at the $CoSi_2$/Si interface and a penetration of As into the Si substrate indicating As mobility inside the silicide layer. The interpretation of such profiles is essentially similar to that proposed previously except that no lattice diffusion of As is possible because of a limited solubility causing precipitation of As. The only efficient diffusion paths are thus $CoSi_2$ grain boundaries which feed the $CoSi_2$/Si interface. In this case the pile-up will depend on the silicide grain size which is experimentally observed as shown by the comparison between FIGURES 4 and 5 [20,5]. Such a distribution only provides qualitative information on diffusion (efficiency of grain boundary). However, it gives an order of magnitude of the solubility limit of As which must be below a value of $10^{19}$ at/cm$^3$ and indicates that despite this low solubility $CoSi_2$ can be used as a diffusion source for As [5].

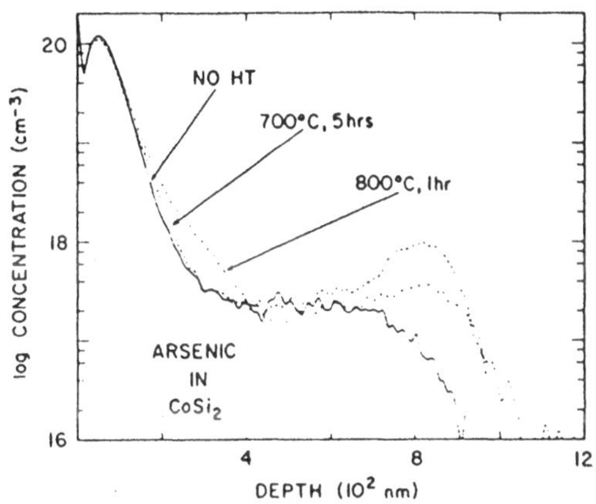


FIGURE 4 Secondary ion mass spectrometry profiles of arsenic implanted in $CoSi_2$ layers ~840 nm thick, respectively as implanted, and after heat treatments at 700°C for 5 hr and 800°C for 1 hr. From [35].

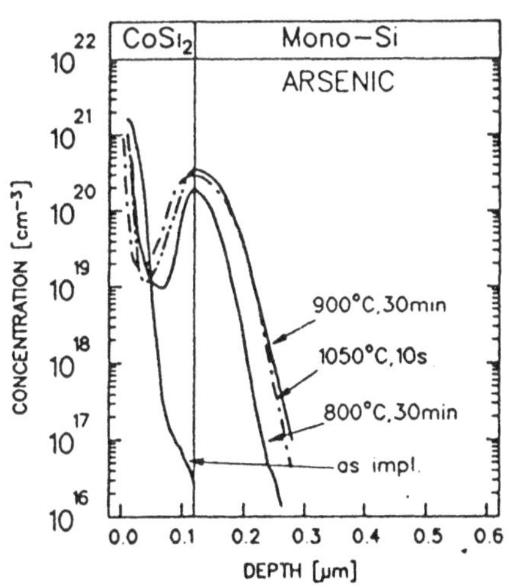


FIGURE 5 Secondary ion mass spectrometry profiles of the arsenic redistribution in $CoSi_2$-Si structures, respectively as implanted, and after heat treatments at 800°C for 30 min, 900°C for 30 min and 1050°C for 10 s. From [5].

FIGURE 6 gives the profiles of As implanted in an amorphous $WSi_2$ layer obtained by chemical vapour deposition (CVD) [21]. The behaviour is essentially similar to that observed in FIGURE 1, indicating a single diffusion mechanism. Transmission electron microscopy (TEM) reveals that heat treatments also induce crystallisation. The diffusion observed is thus more a matter of grain boundary diffusion in a material with extremely fine grains than of diffusion in an amorphous matrix. This is confirmed by the almost total absence of lattice diffusion in well-annealed $WSi_2$ layers at temperatures as high as 900°C. Analysis of the diffusion profile leads, in that case, to the estimation of grain boundary diffusion coefficients and of their activation energy. One should, however, be careful with these values since the transport of impurities along grain boundaries is controlled by both grain boundary diffusion and grain boundary segregation [22]. Thus, the coefficients and activation energies are unlikely to be valid for diffusion considered alone.

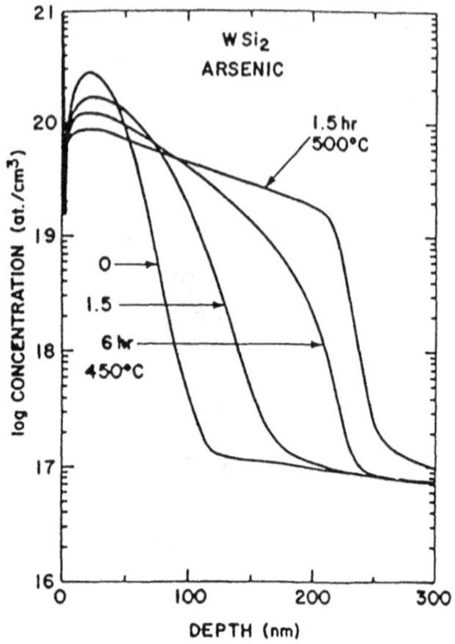


FIGURE 6 Secondary ion mass spectrometry profiles of arsenic implanted in $WSi_2$ 'amorphous' layers, respectively as implanted, and after heat treatments at 450°C for 6 hr and at 500°C for 1.5 hr. From [21].

These different examples show the possibilities and the difficulties presented by the analysis of dopant redistribution in thin films in terms of diffusion. Since, in most cases, implanted profiles are used, no information can be obtained for impurities with very low solubility limits ($<10^{20}$ at/cm$^3$). In other cases, close examination of the profile coupled with a determination of the microstructure of the sample (by TEM) may indicate the respective roles played by grain boundary and volume diffusion. Depending on the experimental conditions, diffusion-related quantities may be obtained: overall mobilities of the dopant, volume or grain boundary diffusion coefficients and corresponding activation energies. Indirect methods (e.g. [15]) only provide information on diffusivities.

## C DATA ANALYSIS

The results presented in TABLE 1 constitute a compilation of reports from various sources, some of limited reliability. One finds successively in that table: the silicide identity, the diffusing impurity, the temperature range analysed, the overall mobility (taking into account all possible diffusion paths), the volume diffusion coefficient at a specific temperature (when measured), and the activation energy. Because of the very different experimental approaches (temperature range, microstructure of the film), direct comparisons between data are difficult. However, the following points appear.

TABLE 1 Summary of available data on the diffusion of dopant elements (and Ge) in various silicides. In order: silicide identity, diffusing impurity (Imp.), temperature range (Temp.), mobility at a specific temperature (Mob. - T), volume diffusion coefficient at a specific temperature ($D_v$ - T), reported activation energy for volume or grain boundary diffusion = 'mobility' (Q), references (Ref) and footnotes (Note).

| Silicide | Imp. | Temp. (°C) | Mob. - T $10^{-13}$ cm²/s | $D_v$ - T $10^{-13}$ cm²/s | Q (eV) | Ref | Note (a) |
|---|---|---|---|---|---|---|---|
| $TiSi_2$ | B | 600-800 | No mob. | | | [6] | (b) |
| | | 900-1100 | ~No mob. | | | [25] | (c,d) |
| | | 900-1100 | ~100-1100 | | | [26] | |
| | | 400-650 | ~2-500 | | | [27] | (e,k) |
| | P | 550-800 | >10-800 | 0.2-800 | 2.0 | [6,28] | (f) |
| | | 800-900 | $2 \times 10^4$-900 | | 2.7 | [15] | (g) |
| | | 400-600 | 2-500 | | | [27] | (e,k) |
| | Ga | 600-800 | <5-800 | | | [23] | |
| | Ge | 600-800 | >5-800 | 0.5-800 | 3.0 | [23] | (f) |
| | | 400-600 | 2-300 | | | [27] | (e,k) |
| | As | 600-800 | >5-800 | 0.17-800 | 1.8 | [23] | (f) |
| | | 850 | ~5-850 | | | [29] | |
| | | 675-900 | Not evaluated | | 0.95 | [30,11] | (h) |
| | | 600-800 | ~5-800 | | | [31] | |
| | | 750-950 | $4 \times 10^4$-900 | | 2.1 | [15] | (g) |
| | | 400-600 | 2-500 | | | [27] | (e,k) |
| | Sb | 600-800 | No mob. | | | [23] | (b) |
| | | 600-850 | Not evaluated | | 1.9 | [32] | (h) |
| | | 400-600 | 2-600 | | | [27] | (e,k) |
| $CrSi_2$ | B | 500/27hr | No mob. | | | [33] | |
| | P | 500/27hr | No mob. | | | [33] | |
| | As | 500/27hr | No mob. | | | [33] | |
| | | 500/0.5hr | Low mob. | | | [34] | |
| | | 720/0.5hr | <1.4-720 | | | [34] | |
| $CoSi_2$ | B | 400-800 | >50-800 | 0.15-600 | 2 | [2] | (f) |
| | | 450-700 | | 4-700 | 2.1 | [17] | (i,f) |
| | | 800-950 | $5 \times 10^4$-900 | | 2.1 | [15] | (g) |
| | P | 600-800 | <1-800 | 0.06-800 | 2.6 | [20] | (f) |
| | | 800-950 | $6 \times 10^3$-900 | | 2.9 | [15] | (g) |
| | Ga | 600-800 | ~1-800 | | | [20] | |
| | Ge | 600-800 | ~1-800 | 0.1-800 | 3.1 | [20] | (f) |
| | As | 600-800 | Very low mob. | | | [35] | (b,j) |
| | | 1000/30s | Not evaluated | | | [36] | (c) |
| | | 800/2hr | Low mob. | | | [37] | (j) |
| | | 800-950 | $5 \times 10^3$-960 | | 2.9 | [15] | (g) |
| | | 800-1050 | Mob. along gb | | | [5] | (b,j) |
| | Sb | 600-800 | Very low mob. | | | [20] | (b,j) |
| $Ni_2Si$ | As | 275/0.5hr | No mob. | | | [34] | |
| $MoSi_2$ | B | 1060/6s | >700-1060 | 8-1060 | | [38] | |
| | | 950/1hr | Not evaluated | 0.1-950 | | [39] | |
| | | 450-800 | Almost no mob. | | | [40] | (b) |
| | As | 550-700 | 0.6-650 | 0.6-650 | 1.4 | [40] | |

Table 1 continued

| Silicide | Imp. | Temp. (°C) | Mob. - T $10^{-13}$ $cm^2/s$ | $D_v$ - T $10^{-13}$ $cm^2/s$ | Q (eV) | Ref | Note (a) |
|---|---|---|---|---|---|---|---|
| $TaSi_2$ | B | 900/0.5hr | No mob. | | | [39] | (b) |
| | | 700-900 | Not evaluated | | | [41] | (k) |
| | | 900/0.67hr | >1-900 | $10^{-3}$-900 | | [42] | (k) |
| | P | 644-913 | 8.8-814 | | 0.61 | [13] | (k,l) |
| | | | | | 0.75 | [14] | (f) |
| | As | 650/0.5hr | Not evaluated | | | [33] | |
| | | 500-900 | 0.01-600 | 3 x $10^{-4}$-700 | | [42] | (n) |
| | | 900/1hr | Not evaluated | | | [43] | |
| $WSi_2$ | B | 600-1000 | Not evaluated | | | [44] | |
| | | 350-400 | 0.1-400 | | | [21,45] | (e,k) |
| | | 850-1000 | 1 x $10^5$-900 | | 1.2 | [15] | |
| | | 800-900 | 1 x $10^4$-900 | | | [16] | (o) |
| | P | 850-1000 | Not evaluated | | | [46] | (p) |
| | | 350-400 | 0.01-400 | | 2.4 | [21] | (e,k,m) |
| | | 800-1000 | 33-800 | | | [47] | |
| | | 800-950 | 3 x $10^4$-900 | | 2.1 | [15] | (g) |
| | As | 450-550 | 0.1-500 | | 2.4 | [21] | (e,k) |
| | | 850-1000 | 2 x $10^4$-900 | | 2.1 | [15] | (g) |
| | Sb | 450-600 | 0.01-500 | | 2.4 | [21] | (e,k) |

(a) The mobility (Mob) takes into account the contribution of all the diffusion mechanisms possible in the silicide layer.
(b) In the experimental conditions used, the non-mobility is a consequence of a low solubility ($<10^{18}$ at/$cm^3$).
(c) RTA.
(d) Evidence for Ti-B compound formation.
(e) Dopant implanted in an amorphous silicide
(f) Activation energy for lattice diffusion.
(g) Lateral dopant diffusion.
(h) Dopant initially implanted in Si; the diffusivity is deduced from the out-diffusion of Sb through the silicide. Activation energy for the overall mobility.
(i) In bulk samples.
(j) The mobility is due to grain boundary diffusion; no lattice diffusion could be observed.
(k) Crystallisation of the silicide layer and dopant diffusion occur simultaneously.
(l) Two activation energies have been found (one for Recrys + Diff, the other for Diff). The value mentioned here is supposed to represent P diffusivity in stabilised $TaSi_2$.
(m) Activation energy for grain boundary diffusion.
(n) A comparison between amorphous and polycrystalline $TaSi_2$ shows that at the same temperature As is more mobile in poly $TaSi_2$.
(o) 3D SIMS.
(p) If P is implanted in $WSi_2$, there is precipitation of a $WP_2$(Si) phase and the mobility of elemental P cannot be studied. However, when phosphorus is introduced in the underlying Si, the data show a high mobility of P in $WSi_2$ at temperatures as low as 800°C.

## C1 Mobilities

(1) The behaviour of the various dopants in a given silicide is relatively similar. This is well illustrated by the mobilities measured for P, As, Ga and Ge in $TiSi_2$ or B, P, Ga and Ge in $CoSi_2$. In these silicides the observation of a limited mobility is the

consequence of dopant precipitation due to a low solubility (e.g. B in $TiSi_2$ [6], As in $CoSi_2$ [5,20]) and thus has no real meaning concerning diffusion.

(2) Mobilities (and volume diffusion coefficients) scale relatively well with the silicide melting temperature. In well-annealed silicide layers where the grain size is commensurate with the layer thickness, dopants have a significant mobility in $CoSi_2$ and $TiSi_2$ (with respective melting temperatures of 1600 and 1800 K) while no movement is observable in $WSi_2$ ($T_m$ = 2450 K).

(3) Grain boundary diffusion plays a dominant role in the observed mobilities and explains the very rapid homogenisation of dopant profiles inside silicide films. The grain size of the silicide layer is the important parameter. If the grain size is small enough, measurable mobilities may be observed in refractory silicides even in the absence of lattice diffusion. This grain boundary diffusion also explains the very high mobilities observed in lateral dopant diffusion experiments [5].

## C2 Lattice and Grain Boundary Diffusion Coefficients

(1) The values of the dopant volume diffusion coefficients (measured in $CoSi_2$ and $TiSi_2$) are at least three orders of magnitude higher than volume diffusion of the same dopant in Si. They are comparable to the volume diffusion coefficients measured for self-diffusion (Ge in $TiSi_2$ [23], Co and Ge in $CoSi_2$ [20,24]) which implies a substitutional mechanism for dopant diffusion.

(2) Diffusivities measured in initially amorphous layers (e.g. [21]) or in lateral diffusion experiments [15] provide estimations of grain boundary diffusion coefficients. Apparently, there is a negative correlation between these grain boundary diffusion coefficients and the atomic size of the moving species, small atoms such as B diffusing faster than large ones such as Sb [21].

(3) The grain boundary diffusion coefficients thus obtained are about 5 orders of magnitude larger than the corresponding volume diffusion coefficients. This ratio is similar to that observed for self-diffusion in silicides (see Datareview 7.2).

## C3 Activation Energies

(1) Activation energies for volume diffusion are, depending on the silicide and on the dopant included, between 2 and 3 eV. Such values, which are comparable to those obtained for silicide self-diffusion, confirm the assumption of a substitutional diffusion mechanism of dopants.

(2) Activation energies obtained from the diffusivities are probably more characteristic of grain boundary diffusion. They also range between 2 and 3 eV. Small differences between grain boundary and volume activation energies are also observed for self-diffusion (see Datareview 7.2). However, in the present case one should be very careful and one should keep in mind the fact that these activation energies are measured over a limited temperature range and in experimental conditions that are far from rigorous. The small values obtained in certain studies [11,13,14] are very suspicious.

## D CONCLUSION

Dopant diffusion in silicides is several orders of magnitude faster than in Si. The analyses of the dopant redistributions observed in silicide/silicon structures have to take into account both the solubility of the dopant and the microstructure of the silicide film. Only three silicides ($TiSi_2$, $CoSi_2$ and $WSi_2$) may be said to have been investigated in a systematic fashion with respect to the diffusion of dopants. Mobilities in these silicides scale with the melting temperature. The dopant diffusion coefficients are commensurate with the self-diffusion coefficients, indicating a probable substitutional diffusion mechanism. In all cases, grain boundary diffusion is very rapid and explains the high mobility of dopants in silicide thin films.

## REFERENCES


[1] K. Maex [ *Appl. Surf. Sci. (Netherlands)* vol.53 (1991) p.328 ]

[2] C. Shone, K.C. Saraswat, J.D. Plummer [ *IEDM Techn. Dig. (USA)* (1985) p.407 ]

[3] R. Liu, D.S. Williams, W.T. Lynch [ *J. Appl. Phys. (USA)* vol.63 (1988) p.1990 ]

[4] R. Angelucci et al [ *J. Appl. Phys. (USA)* vol.69 (1991) p.3962 ]

[5] V. Probst, H. Schaber, A. Mitwalsky, H. Kabza [ *J. Appl. Phys. (USA)* vol.70 (1991) p.708 ]

[6] P. Gas, V. Deline, F.M. d'Heurle, A. Michel, G. Scilla [ *J. Appl. Phys. (USA)* vol.60 (1986) p.1634 ]

[7] K. Maex, R.F. De Keersmaecker, G. Ghosh, L. Delaey, V. Probst [ *J. Appl. Phys. (USA)* vol.66 (1989) p.5327 ]

[8] M. Setton, J. Van der Spiegel [ *J. Appl. Phys. (USA)* vol.69 (1991) p.994 ]

[9] V. Probst, H. Schaber, A. Mitwalsky, H. Kabza, B. Hoffmann [ *J. Appl. Phys. (USA)* vol.70 (1991) p.693 ]

[10] A. Malchere, P. Gas, C. Haut, A. Larere, T.T. Nguyen, S. Poize [ *Appl. Surf. Sci. (Netherlands)* vol.53 (1991) p.132 ]

[11] J. Amano, P. Merchant, T. Koch [ *Appl. Phys. Lett. (USA)* vol.44 (1984) p.744 ]

[12] M. Wittmer, K.N. Tu [ *Phys. Rev. B (USA)* vol.29 (1984) p.2110 ]

[13] J. Pelleg, S.P. Murarka [ *J. Appl. Phys. (USA)* vol.54 (1983) p.1337 ]

[14] J. Pelleg [ *Thin Solid Films (Switzerland)* vol.110 (1983) p.115 ]

[15] C.L. Chu, K.C. Saraswat, S.S. Wong [ *IEEE Trans. Electron Devices (USA)* vol.39 (1993) p.2333 ]

[16] K. Suzuki, H. Horie, Y. Yamashita, Y. Kataoka [ *Appl. Phys. Lett. (USA)* vol.57 (1990) p.1019 ]

[17] P. Gas, C. Zaring, B.G. Svensson, M. Östling, H.J. Whitlow, T. Barge [ *Mater. Res. Soc. Symp. Proc. (USA)* vol.187 (1990) p.131 ]

[18] R. Ghez [ *A Primer of Diffusion Problems* (John Wiley & Sons Inc., New York, 1988) ]

[19] R. Ghez, G.S. Oeherlein, T.O. Sedgewick, F.F. Morehead, Y.H. Lee [ *Appl. Phys. Lett. (USA)* vol.45 (1984) p.881 ]

[20] O. Thomas et al [ *J. Appl. Phys. (USA)* vol.64 (1988) p.2973 ]

[21] C. Stanis et al [ *J. Vac. Sci. Technol. A (USA)* vol.10 (1992) p.907 ]

[22] J. Bernardini, P. Gas [ *Defect and Diffusion Forum* (Trans. Tech. Publications, USA, 1993) vol.95-98 p.393 ]

[23] P. Gas, G. Scilla, A. Michel, F.K. LeGoues, O. Thomas, F.M. d'Heurle [ *J. Appl. Phys. (USA)* vol.63 (1988) p.5335 ]

[24] T. Barge [ Thesis, University Aix-Marseille 3 (1993) ]

[25] V. Probst, H. Schaber, R. Lippens, L. Van den Hove, R. De Keersmaecker [ *Appl. Phys. Lett. (USA)* vol.52 (1988) p.1803 ]

[26] Y.H. Ku, S.K. Lee, D.L. Kwong, P. Chu [ *Appl. Phys. Lett. (USA)* vol.54 (1989) p.1684 ]

[27] F.M. d'Heurle, J. Cotte, P. Gas, G. Göltz, C. Stanis, O. Thomas [ *Appl. Surf. Sci. (Netherlands)* vol.73 (1993) p.167 ]

[28] P. Gas, F.M. d'Heurle, F.K. LeGoues, A. Michel, G. Scilla [ *Le Vide, les Couches Minces (France)* vol.42-236 (1987) p.83 ]

[29] M. Östling et al [ *Thin Solid Films (Switzerland)* vol.110 (1983) p.281 ]

[30] J. Amano, P. Merchant, T.R. Cass, J.N. Miller [ *J. Appl. Phys. (USA)* vol.59 (1986) p.2689 ]

[31] A.H. van Ommen, H.J.W. van Houtum, A.M.L. Theunissen [ *J. Appl. Phys. (USA)* vol.60 (1986) p.627 ]

[32] P. Revesz, J. Gyimesi, E. Zsoldos [ *J. Appl. Phys. (USA)* vol.54 (1983) p.1860 ]

[33] L.R. Zheng, L.S. Hung, J.W. Mayer [ *J. Appl. Phys. (USA)* vol.58 (1985) p.1505 ]

[34] A. Rockett, J.E. Greene, H. Jiang, M. Östling, C.S. Petersson [ *J. Appl. Phys. (USA)* vol.64 (1988) p.4187 ]

[35] O. Thomas, P. Gas, F.M. d'Heurle, F.K. LeGoues, A. Michel, G. Scilla [ *J. Vac. Sci. Technol. A (USA)* vol.6 (1988) p.1736 ]

[36] P.B. Moynagh, A.A. Brown, P.J. Rosser [ *Symp. on Rapid Thermal Annealing/Chemical Vapor Deposition and Integrated Processing* San Diego, California, USA, 25-28 April 1989, Eds D. Hodul, J.C. Gelpey, M.L. Green, T.E. Seidal (Mater. Res. Soc., Pittsburgh, USA, 1989) p.261-6 ]

[37] R. Liu, D.S. Williams, W.T. Lynch [ *J. Appl. Phys. (USA)* vol.63 (1988) p.1990 ]

[38] J. Kato, A. Fujisawa, M. Asahina, H. Shimura, Y. Yamamoto [ *J. Appl. Phys. (USA)* vol.59 (1986) p.4186 ]

[39] V. Probst, H. Kabza, A. Mitwalsky, H. Goebel, B. Hoffmann [ *Selected Topics in Electronic Materials* Eds B.R. Appleton, B.K. Biegelsen, W.L. Brown, J.A. Knapp (Materials Research Society, USA, 1988) p.79 ]

[40] S. Solmi, R. Angelucci, G. Cicognani, R. Canteri [ *Appl. Surf. Sci. (Netherlands)* vol.53 (1991) p.186 ]

[41] P. Eichinger, E. Frenzel, F. Neppl [ *Mater. Res. Soc. Symp. Proc. (USA)* vol.25 (1984) p.165 ]

[42] U. Schwalke, C. Mazuré, F. Neppl [ *J. Vac. Sci. Technol. B (USA)* vol.7 (1989) p.120 ]

[43] S. Valdya, T.F. Retajczyk, R.V. Knoell [ *J. Vac. Sci. Technol. B (USA)* vol.3 (1985) p.846 ]

[44] F. Jahnel et al [ *J. Appl. Phys. (USA)* vol.53 (1982) p.7372 ]

[45] O. Thomas et al [ *Appl. Surf. Sci. (Netherlands)* vol.53 (1991) p.165 ]

[46] J. Torres, O. Thomas, D. Jourdain, R. Madar, A. Perio, J.P. Senateur [ *J. Appl. Phys. (USA)* vol.63 (1988) p.732 ]

[47] P. Pan, N. Hsieh, H.J. Geipel Jr., G.L. Slusser [ *J. Appl. Phys. (USA)* vol.53 (1982) p.3059 ]

CHAPTER 8

# HYPERFINE INTERACTIONS

# 8.1 General remarks on the characterisation of TM silicides with hyperfine interactions

G. Langouche

January 1995

## A INTRODUCTION

The hyperfine interaction is the electromagnetic interaction between the atomic nucleus and its surrounding charge distribution. It manifests itself in the hyperfine structure of atomic transitions - electron paramagnetic resonance (EPR) resolves this very well - and also in the hyperfine splitting of nuclear levels. When nuclear techniques are used to study this hyperfine structure, they can either focus on stable isotopes in their ground states, as is done in nuclear magnetic resonance (NMR) spectroscopy, or on radioactive isotopes, as is done in various nuclear solid state physics techniques such as Mössbauer Spectroscopy (MS), Perturbed Angular Correlations (PAC) and extremely low temperature Nuclear Orientation (NO) [1]. Although the manipulation of radioactive materials requires special infrastructure, the use of radioactive isotopes offers distinct advantages. One basic one is the fact that not only the ground state but also excited nuclear states of isotopes are used in hyperfine interaction studies. Many ground states have zero nuclear quadrupole moment, and therefore show a zero nuclear quadrupole hyperfine interaction. This is not the case for most excited states. Also, the fact that these methods are based on the detection of radiation emitted from these isotopes allows ultra-sensitivity in terms of the number of probe atoms needed. A typical number of (radioactive) atoms needed for such a study is of the order of $10^{10}$ only, so that extremely low concentrations of probe atoms can be studied. Another typical property of nuclear hyperfine interaction studies is their short range nature. In contrast to the long-range magnetic interaction, the electric quadrupole interaction decreases as $r^{-3}$, so that this interaction is made up by the neighbouring atomic shells only. This is a very important advantage in the study of local defects in materials, a field in which nuclear hyperfine interaction techniques are particularly strong, but which is only occasionally mentioned in this Chapter.

## B MÖSSBAUER SPECTROSCOPY

Mössbauer spectroscopy is the only hyperfine interaction technique which has focused intensively on metal silicides, and in particular on iron- and cobalt-silicides. The reason for this is simple. All nuclear hyperfine interaction methods have their preferred isotopes. These are isotopes with nuclear parameters (lifetime, spin, multipole moments etc.) which are best suited for one particular method. $^{111}$In is the favourite isotope for PAC studies. $^{57}$Fe, and automatically its parent $^{57}$Co, are the favourite isotopes for Mössbauer studies. Every iron-containing compound has been, or will run the risk of being, closely studied by the Mössbauer technique. The same holds true to a smaller degree for Co-containing compounds, but here the level of effort is much smaller, as the Mössbauer study of Co-containing compounds involves radio-chemistry, $^{57}$Co being the radioactive parent of $^{57}$Fe. Although about one hundred nuclear transitions are suitable for Mössbauer

Spectroscopy, involving about 40 different elements, the number of intensively used Mössbauer isotopes is limited to favourites such as Fe, the 5sp-elements (Sb, Sn, Te, I, Xe) and the rare-earth elements.

Observable Mössbauer parameters are the position of the resonance, its splitting and its intensity. The position of the Mössbauer resonance is called the isomer shift, $\delta$. The splitting is either magnetic in origin, and proportional to the hyperfine field, H, or electric in origin, and proportional to the electric field gradient (EFG), the major component of this tensor quantity being $V_{zz}$. The intensity of the resonance is proportional to the recoilless fraction, f, and governed by the vibrational properties of the Mössbauer atom imbedded in the lattice. All three parameters can be used to characterise the material under investigation. $\delta$ and $V_{zz}$ are especially sensitive to the electronic structure of the compound, as the isomer shift is proportional to the s-electronic density at the nucleus, and the electric field gradient is very sensitive to the symmetry of the electron distribution. If, for example, the point symmetry around the Mössbauer atom is lower than cubic, a non-zero EFG results, giving rise to a quadrupole splitting of the Mössbauer resonance.

We follow the standard procedure and express the isomer shifts in this Chapter as absorber isomer shifts with respect to $\alpha$-Fe. For $^{57}$Fe the proportionality constant between the isomer shift and the s-electron density at the nucleus is negative. For $^{57}$Fe, and for other Mössbauer isotopes where the nuclear transition involves a transition between a spin $I = 3/2$ and an $I = 1/2$ state, a Mössbauer spectrum which is split due to an electric field gradient consists of a simple doublet. Also here, we follow the tradition not to cite the EFG value, but the quadrupole splitting $\Delta$, which is given by

$$\Delta = \left| \frac{e\,Q\,V_{zz}}{2} \sqrt{\frac{1+\eta^2}{3}} \right|$$

Q being the nuclear quadrupole moment of the $I = 3/2$ state and $\eta$ being the asymmetry parameter of the EFG, given by

$$\eta = \frac{V_{xx} - V_{yy}}{V_{zz}}$$

## REFERENCE

[1] A recent review text on MS and PAC, focusing on semiconductors, can be found in G. Langouche [ *Hyperfine Interaction of Defects in Semiconductors* (Elsevier, 1993) ]; for an introduction to NO we refer the reader to H. Postma, N. Stone [ *Low Temperature Nuclear Orientation* (Elsevier, 1986) ]

# 8.2 Hyperfine interaction parameters of stable Co-silicides

G. Langouche

January 1995

## A METALLIC Co

Pure cobalt shows a Mössbauer spectrum which is split due to a magnetic interaction and an additional quadrupole interaction [1]. This quadrupole interaction, which is due to the hexagonal structure of metallic Co, is almost negligibly small [2], as can be seen in TABLE 1. From the relative intensity of the different lines in the Mössbauer spectrum one can determine the direction of the hyperfine field in the sample. This is especially interesting for thin layer studies, as Mössbauer spectroscopy is able to measure the hyperfine fields in materials as a function of depth with submonolayer sensitivity. This is illustrated in a recent study [3] where the magnitude and the direction of the magnetic hyperfine field could be determined in bulk Fe, monolayer by monolayer from the surface. As none of the known stable CoSi phases is magnetic, the presence of a magnetically split component in the Mössbauer spectrum always indicates metastable phases, or, more likely, metallic Co particles.

TABLE 1 Mössbauer parameters of stable cobalt silicide phases and Co-defect structures in Si. The spectra are obtained from $^{57}$Co-doped sources using a standard single line sodiumferrocyanide absorber. Isomer shifts, δ, are quoted as absorber isomer shifts with respect to α-Fe. (H is the hyperfine field and Δ the quadrupole splitting.)

| Compound | H (T) | Δ (mm/s) | δ (mm/s) | Ref |
|---|---|---|---|---|
| Co | -32.18(4) | 0.064(5) | +0.01(1) | [1,2] |
| $Co_2Si$ | 0 | 0.46(2) | +0.17(1) | [4-6] |
| CoSi | 0 | 0.17(2) | +0.26(1) | [4-6] |
| $CoSi_2$ | 0 | 0 | -0.10(1) | [4-6] |
| Interstitial Co in Si | 0 | 0 | +0.84(1) | [9,10] |
| Substitutional Co in Si | 0 | 0 | -0.07(3) | [9,10] |
| $Co_2$ pairs in Si | 0 | 2.65(5) | +0.17(3) | [11] |
| Co in amorphised Si | 0 | 0.95(5) | +0.20(3) | [6] |

## B $Co_2Si$

There are two distinct lattice positions for Co atoms in the orthorhombic $Co_2Si$ structure, both with non-cubic point symmetry. The Mössbauer spectrum therefore consists of two superimposed quadrupole doublets, with a different isomer shift [4]. The differences between these two spectrum components are so small, however, that the spectra are normally analysed with a somewhat broadened and asymmetric doublet only, the parameters of which are given in TABLE 1.

## C CoSi

The Mössbauer spectrum obtained from a $^{57}Co$ in CoSi source shows a slightly broadened single line, due to a small unresolved quadrupole splitting. It has the largest isomer shift of all the stable cobalt silicides, and therefore the smallest s-electron density at the nucleus. The Mössbauer parameters [4] are listed in TABLE 1.

## D $CoSi_2$

$CoSi_2$ has the $CaF_2$ structure with a cubic point symmetry around the Co atoms. One therefore expects a single line Mössbauer spectrum. This is indeed observed and it has a negative isomer shift with the largest s-electron density at the nucleus of all cobalt silicides. In many experiments [5,6] an extra component appears in the Mössbauer spectrum, as shown in FIGURE 1, illustrating the difficulty in obtaining single-phase material. Mössbauer spectroscopy seems to be one of the few techniques which is sensitive to such impurity phases. The exact microscopic assignment [4] of this fairly well-defined spectrum component ($\delta = +0.28(2)$ mm/s; $\Delta = 0.28(2)$) remains a matter of discussion. It becomes more prominent after high temperature treatment of samples. Its hyperfine parameters are very close to those of CoSi, although X-ray diffraction does not show any trace of cobalt monosilicide in the samples. One group [7] claims a correlation between the presence of this spectrum component and the area of the surface of $CoSi_2$ platelets in Si, and therefore ascribes this spectrum component to Co atoms at the $CoSi_2$/Si interface.

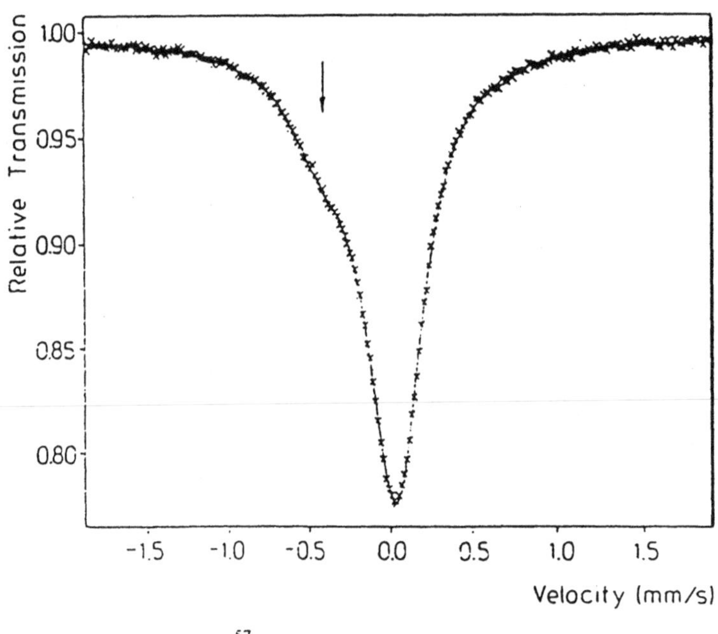


FIGURE 1 Mössbauer spectrum of a $^{57}CoSi_2$ source using a standard single line sodiumferrocyanide absorber (from [4]). The arrow indicates the anomalous component in the spectrum.

Another interesting phenomenon in $CoSi_2$, which so far has only been observed by Mössbauer spectroscopy, is an apparent low temperature phase transition between 150 K and 220 K [8].

## E OFF-STOICHIOMETRIC $Co_xSi_y$

Off-stoichiometric $Co_xSi_y$ samples were also synthesised and studied [5] by Mössbauer spectroscopy and X-ray diffraction. Only the above-mentioned stable phases were discovered after annealing. The existence of a $CoSi_3$ phase could therefore not be confirmed.

## F EPITAXIAL COBALT SILICIDES

In annealing studies of Co-covered Si, the subsequent formation of the above phases could be followed through their Mössbauer parameters [4]. Simultaneous Rutherford Backscattering Spectroscopy (RBS) measurements indicated that they were formed as parallel layers on the Si substrate. The major difference between Mössbauer spectroscopy and X-ray diffraction observations is that the subsequent phases already form at somewhat (~50°C) lower temperatures. This can be understood in view of the very local sensitivity of Mössbauer spectroscopy, which can detect the formation of new phases before these are crystallised over ranges long enough to give rise to good X-ray diffraction data.

## G INTERSTITIAL AND SUBSTITUTIONAL Co IN Si

An interesting Mössbauer spectroscopy result deals with the diffusion of small concentrations of $^{57}Co$ atoms in silicon. All Mössbauer spectra, even after the fastest quench, are dominated by the presence of silicides. Apparently, interstitial Co in Si cannot be formed or at least kept in solution upon diffusion and quench. So far, only low-temperature ion implantation [9] seems to be able to produce interstitial Co in Si, which transforms into substitutional Co after annealing to temperatures at which implantation-produced vacancies migrate [10]. Annealing of room temperature implanted Si leads again to silicide formation, with the formation of $Co_2$ dimers as a first precipitation phase [11].

## H CONCLUSION

We have shown in this Datareview that metal silicides can be recognised by their Mössbauer signature. Some results that have been obtained so far from Mössbauer studies, such as the subsequent formation as a function of temperature of different silicides in epitaxial structures, are very similar to the results obtained with other techniques, such as X-ray diffraction. One of the unique capabilities of Mössbauer spectroscopy should be stressed, however, which is the small number of probe atoms needed. When studying Co-silicides, using radioactive Co, $10^{-5}$ of a monolayer is enough to obtain a Mössbauer spectrum.

## REFERENCES


[1] I.N. Nikolaev, I.Y. Berzstosnyi [ *Sov. Phys.-Solid State (USA)* vol.21 (1979) p.1679 ]

[2] J.G. Stevens, V.E. Stevens [ *Mössbauer Effect Data Index 1970* (Adam Hilger, London, 1972) ]

[3] M. Przybylski, J. Korecki, U. Gradmann [ *Appl. Phys. A (Germany)* vol.52 (1991) p.33 ]

[4] A. Vantomme, I. Dézsi, G. Langouche [ *Hyperfine Interact. (Switzerland)* vol.41 (1988) p.725 ]

[5] I. Dézsi, H. Engelmann, U. Gonser, G. Langouche [ *Hyperfine Interact. (Switzerland)* vol.33 (1987) p.161 ]

[6] G. Langouche [ in *Mössbauer Spectroscopy Applied to Inorganic Chemistry* Eds G.J. Long, F. Grandjean, vol.3 (Plenum, New York, 1989) ch.10 p.445-512 ]

[7] M. Pasemann, W. Bergholz, W. Schröter [ *Phys. Status Solidi A (Germany)* vol.81 (1984) p.273 ]

[8] W.G. Spanninga, P.J.M. Smulders, L. Niesen [ *Nucl. Instrum. Methods B (Netherlands)* vol.63 (1992) p.138 ]

[9] G. Langouche, M. de Potter [ *Nucl. Instrum. Methods B (Netherlands)* vol.19/20 (1987) p.322 ]

[10] G. Langouche [ *Hyperfine Interact. (Switzerland)* vol.72 (1992) p.217 ]

[11] G. Langouche, M. de Potter, D. Schroyen [ *Phys. Rev. Lett. (USA)* vol.53 (1984) p.1364 ]

# 8.3 Hyperfine interaction parameters of stable Fe-silicides

G. Langouche

January 1995

## A METALLIC Fe

The Mössbauer spectrum of $^{57}Fe$ in Fe is probably the best known Mössbauer spectrum in the literature, as it is the standard calibration spectrum that every Mössbauer spectroscopist records regularly. Its hyperfine interaction parameters [1] are listed in TABLE 1.

TABLE 1 Mössbauer parameters of stable and metastable iron silicide phases and Fe defect structures in Si. The data for the iron silicide phases were obtained from bulk absorbers and from thin epitaxial layers using a standard single line source. The Fe-defect data were obtained from in-beam experiments. Isomer shifts, δ, are quoted as absorber isomer shifts with respect to α-Fe. (H is the hyperfine field and Δ the quadrupole splitting.)

| Compound | H (T) | Δ (mm/s) | δ (mm/s) | Ref |
|---|---|---|---|---|
| Fe | -33.2(1) | 0 | 0 | [1] |
| $Fe_3Si$ | 31.0 | 0 | 0.1 | [2,7] |
| | 19.9 | 0 | 0.35 | |
| ε-FeSi | 0 | 0.51(2) | +0.26(1) | [3] |
| β-$FeSi_2$ | 0 | 0.40(2) | +0.14(2) | [4] |
| | 0 | 0.44(2) | +0.03(2) | |
| α-$FeSi_2$ | 0 | 0.47(2) | +0.23(2) | [3] |
| | | 0.73(2) | +0.26(2) | |
| FeSi (CsCl structure) | 0 | 0 | +0.26(2) | [4] |
| Interstitial Fe in Si | 0 | 0 | +0.84(1) | [13] |
| Substitutional Fe in Si | 0 | 0 | -0.27(2) | [13] |
| Fe in amorphised Si | 0 | 0.7 to 1.0 | +0.15(2) | [11] |

## B $Fe_3Si$

$Fe_3Si$ has a cubic $DO_3$ structure, containing two different Fe-sites in its unit cell. They are characterised by quite different hyperfine fields, as measured by different groups. The result of a recent study [2] is quoted in TABLE 1.

## C ε-FeSi

Although ε-FeSi has a cubic structure, the point symmetry around the Fe-atom is lower than cubic so that an EFG results. The Mössbauer spectrum indeed shows a quadrupole doublet. Its hyperfine interaction parameters, as measured in a recent study [3] on bulk material, are given in TABLE 1.

## D $\beta$-FeSi$_2$

This is the low-temperature phase of iron disilicide. Because of its semiconducting character with a direct bandgap of 0.87 eV and its close lattice-match with Si, it is intensively investigated for optoelectronic purposes. Two iron sites are equally populated, and their hyperfine interaction parameters [4], given in TABLE 1, are only slightly different.

## E $\alpha$-FeSi$_2$

This is the metallic high-temperature phase of iron disilicide. It is observed in Mössbauer spectra after quenching the high-temperature phase to room temperature. Several Fe-sites can be specified in this structure. A good fit to the Mössbauer spectrum can be obtained with two superimposed quadrupole doublets with different parameters [3] as given in TABLE 1.

## F EPITAXIAL IRON SILICIDES

In search of the synthesis of epitaxial semiconducting $\beta$-FeSi$_2$ on Si, the reaction of deposited and implanted Fe layers with Si is being intensively investigated. FIGURES 1 and 2 show the results of such a Mössbauer study [5], where 30 Å $^{57}$Fe was deposited by MBE on (111)Si. FIGURE 1 shows the Mössbauer spectrum of the as-deposited metallic Fe layer and the subsequent formation of a metastable Fe-Si-phase upon heating the sample, which will be discussed below. After heating to 400°C $\varepsilon$-FeSi is formed, as shown in the Mössbauer spectrum of FIGURE 2(a). Further heating to 600°C transforms this layer completely into $\beta$-FeSi$_2$ as can be seen in FIGURE 2(c). This transformation occurs gradually in thin layers of this thickness, so that after heating at 500°C a mixture of $\varepsilon$-FeSi and nucleation centres of the $\beta$-FeSi$_2$ phase are observed.

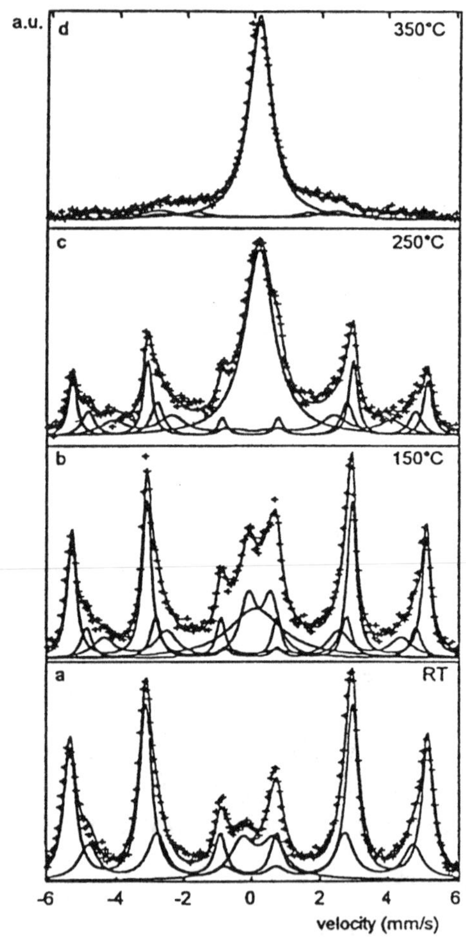


FIGURE 1 Mössbauer spectra of a 30 Å MBE-deposited $^{57}$Fe layer on (111)Si, as-deposited and annealed to 150°C, 250°C and 350°C (from [5]).

## G Fe-RICH Fe$_{1-x}$Si$_x$ ALLOYS

This system has been thoroughly discussed in the literature, especially in the context of steel manufacturing. Systematic investigations of $^{57}$Fe hyperfine fields in such Fe$_{1-x}$Si$_x$ alloys indicate that the Fe hyperfine field diminishes by ~2.6 T [6,7] for each Si atom in the first neighbour shell around the $^{57}$Fe probe.

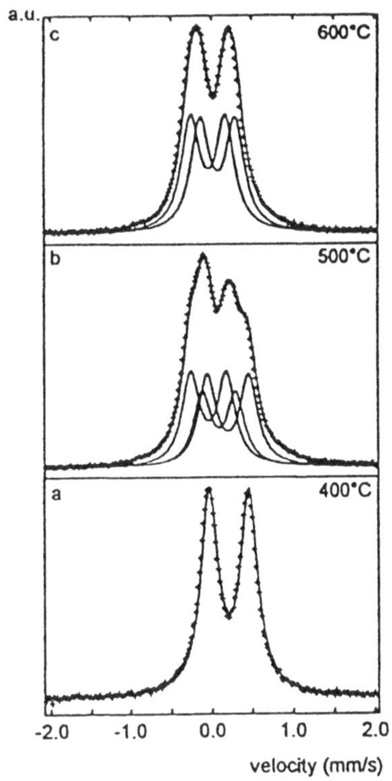


FIGURE 2 Mössbauer spectra of a 30 Å MBE-deposited $^{57}Fe$ layer on (111)Si, annealed to 400°C, 500°C and 600°C (from [5]).

## H Si-RICH $Si_{1-x}Fe_x$ ALLOYS

At the opposite end of the SiFe phase diagram, amorphous Si-rich $Si_{1-x}Fe_x$ alloys have been studied by different groups, as discussed in [8]. In contrast to the Fe-rich alloys, these samples do not show a magnetic splitting, but are characterised by superimposed quadrupole doublets. The parameters of one of the dominating spectrum components in amorphous Si-rich alloys are [9] $\delta = 0.12(2)$ mm/s and $\Delta = 0.62(3)$ mm/s. It was observed by Sawicka et al [10,11] that high dose ion implantation of $^{57}Fe$ into Si leads to a Mössbauer spectrum with very similar parameters, the quadrupole splitting being somewhat Fe concentration-dependent. This led to the conclusion that the atomic configuration around the Fe atoms is identical in both systems. Although the hyperfine interaction parameters of this Fe-configuration are well-defined, no microscopic model has been put forward for this Fe defect structure in Si, which probably involves dangling bonds and possibly vacancies.

## I INTERSTITIAL AND SUBSTITUTIONAL Fe IN Si

Low dose (<$10^{14}$ atoms/cm$^2$) $^{57}Fe$ ion implanted Si samples do not contain enough $^{57}Fe$ atoms to be studied by Mössbauer spectroscopy in standard Mössbauer transmission experiments. This contrasts strongly with the (radioactive) $^{57}Co$ implanted samples, discussed in Datareview 8.2, which can be studied for implantation doses down to $10^{10}$ atoms/cm$^2$. A sophisticated in-beam experiment can be set up, however, at a heavy ion accelerator, in a so-called Coulomb Excitation Recoil Implantation geometry [12], which produces the $^{57}Fe$ Mössbauer spectrum for implantation doses in the $10^{10}$ atoms/cm$^2$ range. In such a recent experiment [13], the Mössbauer resonance for interstitial $^{57}$Fe could be very convincingly determined. Its Mössbauer parameters, as well as the ones for substitutional $^{57}$Fe in Si, are given in TABLE 1. These values agree very well with the isomer shift values for $^{57}$Fe in the decay of $^{57}$Co, given in TABLE 1 of Datareview 8.2, but measured in a completely different type of experiment.

## J CONCLUSION

We have shown in this Datareview that metal silicides can be recognised by their Mössbauer signature. Some results that have been obtained so far from Mössbauer studies, such as the subsequent formation as a function of temperature of different silicides in epitaxial structures, are very similar to the results obtained with other techniques, such as X-ray diffraction. One of the unique capabilities of Mössbauer spectroscopy should be stressed, however, which is the small number of probe atoms needed. For Fe-silicides, one monolayer is enough to record a Mössbauer spectrum.

## REFERENCES


[1] J.G. Stevens, V.E. Stevens [ *Mössbauer Effect Data Index 1970* (Adam Hilger, London, 1972) ]

[2] G. Principi et al [ *Hyperfine Interact. (Switzerland)* vol.69 (1991) p.627 ]

[3] Ö. Helgason, T.I. Sigfússon [ *Hyperfine Interact. (Switzerland)* vol.45 (1989) p.415 ]

[4] T.I. Sigfússon, Ö. Helgason [ *Hyperfine Interact. (Switzerland)* vol.54 (1990) p.861 ]

[5] S. Degroote et al [ *Mater. Res. Soc. Symp. Proc. (USA)* vol.320 (1994) p.133 ]

[6] M.B. Stearns [ *Phys. Rev. (USA)* vol.129 (1963) p.1136 ]

[7] M. Miyazaki, M. Ichikawa, T. Komatsu, K. Matusita [ *J. Appl. Phys. (USA)* vol.71 (1992) p.2368 ]

[8] G. Langouche [ in *Mössbauer Spectroscopy Applied to Inorganic Chemistry* Eds G.J. Long, F. Grandjean, vol.3 (Plenum, New York, 1989) ch.10 p.445-512 ]

[9] F.S. Nasredinov, M.M. Mezdrogina, V.P. Podkhalyuzin, P.P. Seregin [ *Sov. Phys.-Solid State (USA)* vol.28 (1986) p.1425 ]

[10] J.A. Sawicki, B.D. Sawicka, S. Lazarski, E. Maydell-Ondrusz [ *Phys. Status Solidi B (Germany)* vol.57 (1973) p.K143 ]

[11] B.D. Sawicka, J.A. Sawicki, J. Stanek [ *J. Phys. Colloq. (France)* vol.37 no.C6 (1976) p.879 ]

[12] A recent review text on MS and PAC, focusing on semiconductors, can be found in G. Langouche [ *Hyperfine Interaction of Defects in Semiconductors* (Elsevier, 1993) ]; for an introduction to NO we refer the reader to H. Postma, N. Stone [ *Low Temperature Nuclear Orientation* (Elsevier, 1986) ]

[13] P. Schwalbach et al [ *Phys. Rev. Lett. (USA)* vol.64 (1990) p.1274 ]

# 8.4 Hyperfine interaction parameters of metastable silicides

G. Langouche

January 1995

The epitaxial growth of layers onto substrates with slightly different lattice parameters or different atomic geometries can give rise to the formation of metastable phases. A very illustrative example is the formation of bcc Co, a phase which does not exist otherwise in nature, when grown on a suitable substrate like Fe-covered GaAs [1]. Also, the attempts to grow Fe-silicides on Si give rise to the formation of metastable silicides. This can easily be recognised in FIGURE 1 of Datareview 8.3. All stable Fe-silicides are characterised by quadrupole splittings or magnetic splittings, as discussed previously and shown in TABLE 1 of Datareview 8.3. Therefore, the single line, present in the Mössbauer spectra of Fe-covered Si after low temperature annealing, must be connected with a metastable phase, with a cubic point symmetry for the Fe atoms. This particular phase is most likely FeSi in the CsCl structure [2,3]. This follows from a comparison with extensive literature data, including X-ray diffraction data solidus RD-data of samples synthesised under similar conditions. Its isomer shift (TABLE 1 of Datareview 8.3) is quite similar to that of ε-FeSi, indicating a similar s-electron density at the Fe nuclei, but the quadrupole splitting is totally lacking. Another intermediate metastable phase can equally be observed in the spectra shown in FIGURE 1 of Datareview 8.3. It has a large quadrupole splitting, $\Delta = 0.80(10)$ mm/s, and a large isomer shift, $\delta = 0.35(5)$ mm/s [2]. These parameters again do not match any of the known iron silicide phases, but an identification of this phase is not yet possible.

It is clear that this type of on-going study allows the identification of hitherto unknown metastable phases. Investigations of the Mössbauer spectra of these spectrum components at different temperatures will reveal more information about these new systems, e.g. phase transitions and possible magnetic ordering at low temperatures. The isomer shift and quadrupole splitting, on the other hand, shed light on their electronic structure.

Hyperfine interaction studies of metal silicides have focused so far on Co- and Fe-silicides only. There is scope for Mössbauer spectroscopists to study rare earth silicides or nickel silicides by this technique. We have shown in Datareviews 8.2 and 8.3 that metal silicides can be recognised by their Mössbauer signature. This Datareview illustrates how metastable metal silicides can be recognised and how information on their electronic and magnetic structure can be obtained, for monolayer or submonolayer systems.

**REFERENCES**

[1] J. Dekoster, E. Jedryka, C. Mény, G. Langouche [ *Europhys. Lett. (Switzerland)* vol.22 (1993) p.433 ]

[2] S. Degroote et al [ *Mater. Res. Soc. Symp. Proc. (USA)* vol.320 (1994) p.133 ]

[3] T. Kobayashi, J. Dekoster, S. Degroote, M.H. Langelaar, L. Niesen, G. Langouche [ *Proc. 4th Int. Conf. on Formation of Semiconductor Interfaces*, Jülich, 1993, Eds B. Lengeler, H. Lüth, W. Mönch, J. Pollmann (World Scientific, Singapore, 1994) p.455 ]

www.ingramcontent.com/pod-product-compliance
Lightning Source LLC
LaVergne TN
LVHW080020210926
848481LV00011B/3072
* 9 7 8 0 8 6 3 4 1 7 7 6 4 *